PATTERNS OF EVOLUTION

as illustrated by the fossil record

Developments in Palaeontology and Stratigraphy, 5

PATTERNS OF EVOLUTION

AS ILLUSTRATED BY THE FOSSIL RECORD

Edited by

A. Hallam

Department of Geological Sciences, University of Birmingham, Birmingham, Great Britain

ELSEVIER SCIENTIFIC PUBLISHING COMPANY

Amsterdam, Oxford, New York, 1977

ELSEVIER SCIENTIFIC PUBLISHING COMPANY
335 Jan van Galenstraat
P.O. Box 211, 1000 AE Amsterdam, The Netherlands

Distributors for the United States and Canada:

ELSEVIER/NORTH-HOLLAND INC.
52, Vanderbilt Avenue
New York, N.Y. 10017

First edition 1977
Second impression 1978

Library of Congress Cataloging in Publication Data
Main entry under title:

Patterns of evolution as illustrated by the fossil
   record.

   (Developments in palaeontology and stratigraphy ;
5)
   Includes bibliographies and indexes.
   1. Paleontology. 2. Evolution. I. Hallam,
Anthony. II. Series.
QE721.P34   575   77-2819
ISBN 0-444-41495-9

ISBN: 0-444-41495-9 (Vol.5)
ISBN: 0-444-41142-9 (Series)

Printed in The Netherlands

PREFACE

Fossils are of great scientific value for three quite distinct reasons. They are the best means we have of determining the relative ages of sedimentary rocks and hence have allowed correlation between different regions and the erection of a stratigraphic framework into which the succession of geological events can be fitted. They are in many instances excellent indicators of past environments. Thirdly, they are the only direct evidence we have of the history of life on our planet, and in particular of the course of evolution. Yet this three-fold interest is not proportionally reflected in the work of palaeontologists, the great majority of whom have been mainly preoccupied with the first two, geological, aspects. Thus, if one looks at the contents of such a well-known journal as *Evolution* one finds very few papers by palaeontologists. The situation is hardly different in their specialist journals, which over the decades have contained relatively little of interest to the student of evolution.

The shining exception to this generalization is, of course, George Gaylord Simpson. *Tempo and Mode in Evolution* (1944) and *The Major Features of Evolution* (1953) rank as the outstanding palaeontological contribution to evolutionary thought in this century. Nearly a quarter of a century has passed, however, since Simpson's second book appeared, and the time is ripe for a modern evaluation of what the fossil record has to reveal about evolution.

The last few years have in fact seen the beginnings of a major change in research orientation of students of fossils, a significant number of whom would now rather be called palaeobiologists. There has been a gratifyingly positive response to the explosion of knowledge and ideas in a number of fields pertinent to the history of life. Whether on the grand scale of plate tectonics or the smaller scale of detailed sedimentological facies analysis, we now know much more than our predecessors of the nature of environmental changes through time. Moreover, the taxonomic documentation provided by such publications as the continuing multivolume *Treatise on Invertebrate Paleontology* (Geological Society of America) and *The Fossil Record* (Geological Society of London, 1967) are invaluable data bases for studies of diversity, origination and extinction. Sophisticated studies of functional morphology allied with evolutionary change, and widespread application of stimulating new ecological concepts, are also helping to revitalize an old discipline.

It has been customary to be somewhat apologetic about the limitations of the fossil record, but the time has surely come for us to emphasize the more positive aspects. Certainly, elucidation of evolutionary *mechanisms* must remain the province of geneticists and ecologists, but these scientists are denied the invaluable time dimension which allows us to investigate evolutionary *patterns* in a comprehensive and meaningful way. Such topics as crucial to a full understanding of evolution as the nature of diversity change through time, rates of origination and extinction, progressive colonization of and adaptation to ecological niches, convergence and parallelism, paedomorphosis, size change, the temporal aspect of speciation and origination of new higher taxa, and radiation and extinction in relation to changing macro-environments, are decidedly the realm of the palaeontologist.

In recent years some of the most spectacular advances in our knowledge of evolution have come from molecular biology (e.g. Ayala, 1976). It would be wrong to assume from this, however, that the focus of interest has shifted away from fossils, because organic evolution has always been and must remain a multidisciplinary subject in which no single research field can be expected to yield more than a partial picture. Whereas palaeontology cannot throw much light on such important current controversies as the selectionist versus neutralist hypotheses to account for the large amount of polymorphism and heterozygosity now known to occur in natural populations, or the relative importance of structural and regulatory genes, it can provide independent corroboration of phylogenetic models based on protein sequencing and immunological distance measurements. Furthermore, palaeontology can indicate important problems for the molecular evolutionist to tackle. Evolutionary theory must account for the immense diversity of phenotypes and the multiplicity of adaptations in the organic world, for which fossils provide critical evidence.

This volume represents an assessment by a number of distinguished palaeontologists of the evolutionary patterns that can be discerned from the fossil record, interpreted in the light of evolutionary theory and our present knowledge of the changing environments of our planet. After an opening historical essay, a general account of metazoan evolution and a discussion of stochastic models, many of the most important invertebrate and vertebrate groups are dealt with successively by specialists. There is no attempt at comprehensiveness. Not only would this have given rise to a book of inordinate length, but specialists with the appropriate interests and experience would have been hard to find for every fossil group. In particular, only one chapter is devoted to plants, essentially because plant fossils have never in the past contributed much to our knowledge of evolution. An exception has been made, however, in the case of the angiosperms, because modern palynological work has begun to throw significant new light on the origin and early evolution of this important and highly interesting group.

Authors were given some general directives but otherwise encouraged to present their material in a way that reflected their own interests and the nature of their subject matter. The wide diversity in style, ranging from the comprehensive review to the essay on one or more particular themes, and from the cautious to the boldly imaginative, clearly points to the variety of fruitful approaches that can be adopted by the evolution-minded palaeontologist.

Finally, I am indebted to Tom Schopf for volunteering to undertake the daunting task of writing a concluding summary, and to my wife for doing most of the copy editing.

References

Ayala, F.J. (Editor), 1976. Molecular Evolution. Sinauer Assoc. Inc., Sunderland, Mass., 277 pp.
Simpson, G.G., 1944. Tempo and Mode in Evolution. Columbia University Press, New York, N.Y., 237 pp.
Simpson, G.G., 1953. The Major Features of Evolution. Columbia University Press, New York, N.Y.,
    434 pp.

A. HALLAM
Oxford
June 1976

LIST OF CONTRIBUTORS

R.T. BAKKER Department of Earth and Planetary Sciences, The Johns Hopkins University, Baltimore, Md. 21218 (U.S.A.)

R.L. CARROLL Redpath Museum, McGill University, Montreal 101, Que. (Canada)

J.A. DOYLE Museum of Paleontology, University of Michigan, Ann Arbor, Mich. 48109 (U.S.A.)

N. ELDREDGE Department of Invertebrates, American Museum of Natural History, Central Park West, New York, N.Y. 10024 (U.S.A.)

P.D. GINGERICH Museum of Paleontology, University of Michigan, Ann Arbor, Mich. 48109 (U.S.A.)

S.J. GOULD Museum of Comparative Zoology, Harvard University, Cambridge, Mass. 02138 (U.S.A.)

J. HURST Greenland Geological Survey, Copenhagen (Denmark)

W.J. KENNEDY Department of Geology and Mineralogy, University of Oxford, Oxford (Great Britain)

C.R.C. PAUL Jane Herdman Laboratory of Geology, University of Liverpool, Liverpool (Great Britain)

D.M. RAUP Department of Geological Sciences, University of Rochester, Rochester, N.Y. 14627 (U.S.A.)

R.B. RICKARDS Department of Geology, Sedgwick Museum, University of Cambridge, Cambridge (Great Britain)

T.J.M. SCHOPF Department of Geophysical Sciences, University of Chicago, Chicago, Ill. 60637 (U.S.A.)

A. SEILACHER Institut für Geologie und Paläontologie, University of Tübingen, 7400 Tübingen (German Federal Republic)

S.M. STANLEY Department of Earth and Planetary Sciences, The Johns Hopkins University, Baltimore, Md. 21218 (U.S.A.)

K.S. THOMSON Peabody Museum of Natural History, Yale University, New Haven, Conn. 06520 (U.S.A.)

J.W. VALENTINE Department of Geology, University of California, Davis, Calif. 95616 (U.S.A.)

A. WILLIAMS Principal, University of Glasgow, Glasgow (Great Britain)

CONTENTS

CHAPTER 1

ETERNAL METAPHORS OF PALAEONTOLOGY

STEPHEN JAY GOULD

The Content of Palaeontology

Of eternity

Alexander wept at the height of his triumphs because he had no new worlds to conquer. Whitehead declared that all of philosophy had been a footnote to Plato. The Preacher exclaimed (Ecclesiastes 1:10): "Is there *any* thing whereof it may be said, See, this is new? It hath been already of old time, which was before us."

The essential questions of a discipline are usually specified by the first competent thinkers to enter it. The intense professional activity of later centuries can often be identified as so many variations on a set of themes. The arrow of history specifies a sequence of changing contexts within which the same old questions are endlessly debated — "The thing that hath been, it is that which shall be; and that which is done is that which shall be done" (Ecclesiastes 1:9).

In this paper, I wish to propose that (1) the basic questions palaeontologists have asked about the history of life are three in number; (2) the formulation of these questions preceded evolutionary thought and found no resolution within the Darwinian paradigm; (3) the major contemporary issues in palaeobiology represent the latest reclothing of these ancient questions. Although fencesitting, waffling, and middle-of-the-roadism have a pedigree as long as the three questions themselves, palaeontologists have generally come down clearly on one side or the other of each issue. The issues are so pervasive and general that chosen opinions reflect social and cultural climates as well as a supposedly "objective" reading of the fossil record. There are strong correlations among the issues, but all eight possible combinations (for two opinions on each of three questions) can and have been maintained by palaeontologists. Adherence to one or the other side of each issue has usually been expressed in metaphor — hence the somewhat cryptic title of this chapter.

Three antitheses as the subject of palaeontology

I do not contend that these three questions inform all the work that palaeontologists have done. They are issues for the major features of life's history. One can spend a lifetime taxonomizing local faunas and correlating sections within a basin and never encounter them in any serious way. I identify the three questions and their polar opinions as the following:

(1) Does the history of life have definite directions; does time have an arrow

specified by some vectorial property of the organic world (increasing complexity of structure, or numbers of species, e.g.). No question received more discussion during the establishment of modern palaeontology in the early to mid-nineteenth century. Despite the misunderstanding of most modern geological textbooks, it was the central issue in the debate between "uniformitarianism" and "catastrophism". The catastrophists were, for the most part, progressionists who viewed each new episode of life as a distinct improvement leading inexorably towards the modern creation dominated by *Homo sapiens*. Lyell, on the other hand, had a vision metaphorically linked to the Newtonian timelessness of endlessly revolving planets. Species came and went; land rose and fell; seas went in and out — but the world was ever the same. During the late nineteenth century, the battle raged again in an evolutionary context, with vitalists and finalists speaking of inevitable direction (e.g., Osborn's "aristogenesis"), and some strict Darwinians maintaining that evolution means nothing more than successive adaptation to changing local environments. In contemporary palaeobiology, we find the neo-Lyellian metaphor in the stochastic, equilibrium models of Raup et al. (1973), while Huxley's notion of anagenesis and successive grades (1958) upholds the idea of direction. I will refer to the vectorial view as *directionalist* (rather than progressionist, because the direction might be down), and the Lyellian vision as *steady-statist*.

(2) What is the motor of organic change? More specifically, how are life and the earth related? Does the external environment and its alterations set the course of change, or does change arise from some independent and internal dynamic within organisms themselves? In pre-evolutionary times, debate raged on the importance to life of geological change. The opposing sides included some strange bedfellows — or so we judge it, wearing the anachronistic spectacles of later categorizations. (In fact, our reluctance to see Buckland and Lyell as joint protagonists, for example, leads directly to our total blindness towards this major debate during the early nineteenth century.) Buckland and Lyell believed that alterations in the physical environment provided the primary impetus for organic change (for Buckland, the earth changed in a directional fashion; for Lyell it did not — hence their differences on the question of directionalism—steady-statism. They were united on the issue of physical control). Agassiz, on the other hand (though conventionally a "catastrophist" like Buckland), argued vigorously that life changed by a dynamic of its own, quite unrelated to modification of the physical environment. Direct control by the physical world, Agassiz claimed, would be a crassly vulgar way for the creator to pursue his evident design for the history of life. In a later evolutionary context, various orthogenetic trend theories upheld Agassiz's position, while the most Darwinian of palaeontologists — Dollo and Matthew in particular — argued that evolution proceeded when changes in the physical environment established selective pressures for new adaptation. Today, the Buckland—Lyell thesis reemerges in Valentine's vision (1973) of the external environment as a motor for all major evolutionary change, while claims for an internal dynamic can be seen in Stanley's invocation of the cropping principle to explain the Cambrian explosion (1973) and Flessa and Levinton's

belief (1975) that the potential for finer niche partitioning may be virtually limitless. I will refer to the belief in external control as *environmentalist* and to claims for an inherent cause of change as *internalist*.

(3) What is the tempo of organic change? Does it proceed gradually in a continuous and stately fashion, or is it episodic? In pre-evolutionary times, Lyell was a dogmatic gradualist in arguing for continual exercise of the creative power at about the same rate through time (with consequently constant rates of extinction). The more colourful catastrophists, with their claims of nearly complete destruction followed by wholesale creation of new faunas, occupied the other pole. With the guidance of evolution, Darwin followed the Lyellian metaphor of gradualism, while D'Arcy Thompson invoked the spirit of Pythagoras to support macromutational steps between idealized forms (Gould, 1971). The current debate between Eldredge and Gould's notion of "punctuated equilibria" (1972) and Gingerich's defense of phyletic gradualism (1976) merely replays an old tale. I will label stately, Lyellian change as *gradualist*, and episodic views as *punctuational*. I recognize — and this will be important for the concluding section — that scaling has its usual effect and that one can be a gradualist at one level (the origin of species, for example) and a punctuationist at another (by asserting the importance of mass extinctions as a general control of diversity).

Apologia

I was assigned to write a general history of palaeontological attitudes towards evolution. Instead, I am denying that evolution had any influence in shaping the major questions of palaeontology (though it changed completely and permanently the context in which they are debated). I am also writing a most peculiarly anhistorical history — no chronologies, no sequential accounts of the worthy bricklayers who built our modern edifice. It might even seem to verge on a structuralist denial of history, by asserting the timelessness of central themes (to which the true history of changing contexts is merely so much superficial recasting). So be it. I would only say that the changing contexts *are* the subject matter of palaeontology, and that I do not view their historical sequence as superficial in any respect. I also regard this theme of "eternal questions" as a legitimate way to present a coherent, conceptual account of palaeontological history in the allotted space.

Some will be offended at what might seem to be a claim of patrimony. A modern directionalist may well reject an "ancestor" like Buckland, claiming with invincible logic either that he never heard of the man, or that he chooses not to be ranked with the author of "Geology and mineralogy considered with reference to natural theology" (Buckland, 1836). But I am not making a claim of homology via chains of physical descent like pedigrees of kinship. Pedigrees of ideas can be "homologous" — indoctrination of child by parent, or student by teacher, for example. But the chains of which I speak are forged of analogous links only. Basic ideas, like idealized geometric figures, are few in number. They are eternally available for consumption, and the sequential list of their consumers is no pedigree, but the convergence of independent minds to one of a very limited set of basic attitudes. They are, of course, consumed in very different contexts: Buckland's benevolent God and Valentine's drifting continents are phenomena of such a different order that it becomes difficult to see how they function for both men in a similar way — as the "motor" for changes in the physical world that control patterns of organic form and diversity. I am even tempted to claim that the task of history is to explain the contexts so clearly that they can be separated and subtracted, thus permitting us to see the unchanging themes.

A Taxonomic Interlude

Before I proceed in more conventional, chronological fashion, I would like to defend my categories by demonstrating that all eight combinations of the three antitheses have been upheld by famous students of the history of life. I will defend my assignments in later sections and confine myself here to thumbnail sketches drawn from different periods of time. (The accompanying diagram, Table I, is set up as a contingency table, not a flow chart. I do not believe that the structure of the three basic questions is hierarchical.)

(1) SEP (steady-statist—environmentalist—punctuational). D'Arcy Thompson (1917, 1942) believed that physical forces shaped organisms directly — a radical environmentalism (see Gould, 1971). No viable intermediate forms exist between many basic designs (*Baupläne*), just as many geometrical figures can-

TABLE I

Classification of early palaeontologists' evolutionary beliefs

Direction of change	Mode of change	Tempo of change	Code	Name and school	Frequency
steady state (S)	environmentalist (E)	punctuational (P)	SEP	D'Arcy Thompson	rare
		gradualist (G)	SEG	early Lyell, part of Darwin, "strict uniformitarianism"	fairly common
	internalist (I)	punctuational (P)	SIP	late Agassiz	very rare
		gradualist (G)	SIG	Lamarck	very rare
directional (D)	environmentalist (E)	punctuational (P)	DEP	Buckland "catastrophism"	common
		gradualist (G)	DEG	late Lyell, part of Darwin	common
	internalist (I)	punctuational (P)	DIP	early Agassiz, Oken, most of "Natur-philosophie"	moderately common
		gradualist (G)	DIG	Osborn "orthogenesis"	common

not be smoothly transformed one into the other; transitions must be by macro-mutation — punctuationism. Since physical forces shape organisms and have not varied through time, life has no direction — steady-statism.

(2) SEG. Lyell's school of strict uniformitarianism. The mean complexity and diversity of life does not vary through time (S); geological changes regulate the extinction and origination of new taxa (E); rates of origination and extinction are slow and fairly constant through time (G). Lyell never abandoned E and G, but accumulating evidence for direction in the history of vertebrates led him to surrender S when he converted to evolutionism (Wilson, 1970; Gould, 1970). Darwin also never wavered on E and G, but had an ambiguous attitude toward directionalism—steady-statism. He argued vigorously that nothing in his theory of natural selection itself permitted any belief in inherent progress or direction, for natural selection refers only to adaptation in local environments. But natural selection did not forbid progress as an empirical result if local adaptation led occasionally to general improvement in structural design.

(3) SIP. Agassiz held firm to his belief in punctuations (glaciers as "God's great plough," for example), and in the independence of newly created forms from physical control (P and I). For most of his career, Agassiz was a firm progressionist as well (D). But after Darwin's *Origin of Species* (1859) he began to see progressionism as an argument for the dreaded notion of evolution, and he turned against it to favour the idea that mean complexity of life had not changed since the Cambrian explosion (S).

(4) SIG. Few will dispute a characterization of Lamarck's evolutionary theory as gradualist and internalist (progress by the "sentiment intérieur" — the "force which tends incessantly to complicate organization"). But how, given all his talk of progress up the ladder of life, can he be called a steady-statist. In a brilliant essay, Simpson (1961) has shown that Lamarck's more global view is non-directional. An individual lineage exhibits undisputed progress from amoeba towards man. But the highest forms of organization are inevitably degraded to their basic constituents, and must start the upward trek anew by spontaneous generation at the bottom. Upward motion is just one phase of an endlessly repeating cosmic cycle of advance and degradation.

(5) DEP. This is a characteristic combination of early nineteenth century catastrophism, as found, for example, in William Buckland. Each new creation, following a mass extinction (P), is an improvement on the previous inhabitants (D). The improvements place the new inhabitants into harmony with an altered earth (E).

(6) DEG. Lyell reluctantly submitted to the empirical evidence for progressionism (D), but held firm to his uniformitarian beliefs in gradual change (G) for the physical environments that set the course of organic evolution (E).

(7) DIP. Another group of catastrophists shared Buckland's convictions about progressive and discontinuous change (P and D). But they saw no relationship between this advance and any change in the earth's physical appearance. The creator, according to Agassiz, had his own plans for displaying the order and progress of his thoughts; he would not resort to so vulgar a device as simply fitting life to its external situations (I). L. Oken, a prominent German

Naturphilosoph, spoke of drastic reformulation of each higher stage (P) from the primal zero (D) as spirit strove ever upward in its inherent attempt to become man (I).

(8) DIG. Most "orthogenetic" beliefs of the late nineteenth century viewed straight-line evolution as an upwardly mobile affair, proceeding gradually through time without reference to any environmental control exerted by such unimportant factors as natural selection.

Finally, I emphasize that the filling of all eight categories does not deny that attitudes are strongly correlated from one question to the next. The guestimated column on frequencies reflects these correlations. Directionalism and internalism form the strongest link (DI). Very few internalists lack a belief in progress — why postulate an inherent mechanism for change unless it is leading somewhere?

Pre-evolutionary Establishment of the Three Questions

Introduction

The "eternity" of my title is prospective only. Philosophy needed its Plato to write the documents for later annotation. Palaeontology could not pose its three basic questions until it had resolved two prior issues in favour of the beliefs that established its modern character. The likes of Steno and Hooke, Smith and Cuvier deserve all the usual accolades for the resolution. But their story is not the tale I have chosen for this work; need I say more in their honour than that eternity begins with them?

(1) The organic nature of fossils had to be established. This debate, largely over by the mid-eighteenth century, had consumed much attention during the early years of the Royal Society (late seventeenth century). We praise Steno for the meticulous arguments of his *Prodromus* and *Canis carchariae dissectum caput* (1667, reprinted 1958, with its appendix on the organic nature of fossil shark's teeth), especially his proofs that sedimentary rocks are the historical products of deposition. But we should recognize that the opposing views of Lister, Lhwyd and Plot were reasonable in the context of their times (Rudwick, 1972), and that the cause of poor Beringer's undoing was vanity rather than stupidity (Jahn and Woolf, 1963; Beringer considered the possibility that his figured stones were hoaxes, but finally concluded that the Lord in his wisdom had salted the mountain with these wondrous objects for him, i.e., Beringer, to find).

(2) Fossils had to be recognized as the sequential products of an extended history. Organic status was not enough; fossils had to be imbued with a history, i.e., their potential use for establishing a time scale had to be recognized. Many palaeontologists are not aware that the question of extinction was not settled until Cuvier's time, and that Thomas Jefferson himself had argued in the opening years of the nineteenth century that no animals could disappear from the earth. If fossils only record a single creation (one still with us), then none of

the three questions can be posed at all. Even a two-stage classification of pre and post Noah will not suffice. Fossils must record a long and varied history of organic change before any rocks can be correlated and any of the three questions asked.

The modern character of palaeontology begins not with the acceptance of evolutionary theory in the mid nineteenth century, but with the establishment of a methodology for temporal ordering by Smith and Cuvier during the earliest years of the nineteenth century. By Lyell's time, the three questions had been posed and debated extensively. Since then, we have not stopped — and I doubt that we ever will.

Progression vs. steady state as the essence of the uniformitarian catastrophist debate

Contrary to popular belief, no serious nineteenth century scientist — not even the most theological catastrophist — argued for the direct intervention of God in the earth's affairs. All accepted the constancy of natural law. God had ordained unchanging principles when he wound up the universal clock; he did not need to meddle by miracle with the subsequent history of the earth. None-the-less, God was scarcely banished from the thoughts of leading palaeontologists; for opinion in the debate of steady-statists and directionalists reflected one of two basic positions on the nature of divine benevolence . Either God ordained his laws so that the earth's history would be a steady march from simple beginnings towards divine glory; or, like a timeless presence, he had set the original world as he wanted it for all time — earth "history" would be as anhistorical as the endless cycling of Newton's planets.

Empirically, the fossil record as known to Cuvier and his contemporaries — and it is not much different now — displayed a sequence of faunas, each separated from the one below by a major break both in organic form and geological environment. Moreover, as more and more fossil vertebrates were described, their sequence seemed to mirror the conventional ladder of advance among the classes — fish first of all, followed by amphibians and reptiles, then by mammals and finally by man (whose fossil remains had yet to be found at all).

Cuvier, again contrary to his popular image as an overt, theological apologist, was a rigid empiricist who proffered no opinion on the largely metaphysical question of organic progress. But most of his catastrophist colleagues did not feel so constrained. Rudwick (1972) has shown convincingly that the progressionist world view had become a coherent, recognized and well articulated geological paradigm during the years preceding Lyell's *Principles*.

William Buckland, Dean of Oxford Cathedral and first Reader of Geology at the University, stated "that the more perfect forms of animals become gradually more abundant, as we advance from the older into the newer series of depositions" (1836, p. 115). (He hastened to add that "imperfect" did not mean ill-adapted, but only simpler in structure. All God's creatures are ideally fit for their appointed stations — 1836, pp. 107—108.)

Louis Agassiz articulated a well-developed theory of progress with his "three-fold parallelism" between the stages of ontogeny, the sequences of comparative anatomy and the successive introduction of higher types into the fossil record (Gould, 1977). Of vertebrate history, he writes (1857, in 1962, p. 108): "Through all these intricate relations there runs an evident tendency towards the production of higher and higher types, until at last Man crowns the whole series".

James Hutton had rather little to say about life, but his "world machine" of endlessly cycling erosion, consolidation and uplift provided a timeless, non-directional stage for Lyell's steady state of geological and biological history. In asserting that the mean diversity and complexity of life had not changed through time, Lyell encountered the empirical dilemma of an apparently progressive vertebrate history. Lyell's response was typical of his methodology — he argued that prior theory had to be imposed upon an imperfect record to establish the plausibility of his view in the face of opposing documents. Some may brand this as dogmatism; I regard it as sound and imaginative science, for it transfers the argument to the prior theory itself (which must be falsifiable to be scientific). (And Lyell was surely no uncompromising dogmatist, for he abandoned his steady state by the 1850s when predicted evidence against progression did not develop.) Lyell presents three main arguments for not accepting the empirical evidence of progression.

(1) The fossil record is notoriously imperfect. Fossil mammals will eventually be found in all Palaeozoic periods; the rocks investigated so far would not preserve their remains:

"We must not, however, too hastily infer from the absence of fossil bones of mammalia in the older rocks, that the highest class of vertebrated animals did not exist in remoter ages. There are regions at present, in the Indian and Pacific oceans, co-extensive in area with the continents of Europe and North America [domain of virtually all Palaeozoic rocks studied by Lyell's time], where we might dredge the bottom and draw up thousands of shells and corals, without obtaining one bone of a land quadruped" (1842, p. 231 — these quotations, from the widely read 6th edition of the *Principles*, differ in no substantive way from the 1st edition of 1830).

Lyell (1842, pp. 238–239) took great comfort from the recent discovery of a single mammalian fauna in Jurassic beds of the Stonesfield Slate. If the Mesozoic had fallen, the Palaeozoic could not be far behind. Moreover, he saw no evidence for improvement in the Tertiary, where the record was at least adequate, if not good:

"In this succession of quadrupeds, we cannot detect any signs of a progressive development of organization, — any clear indication that the Eocene fauna was less perfect than the Miocene, or the Miocene than that of the Older or Newer Pliocene periods" (1842, p. 249).

(2) Lyell offered another potential answer to the directionalist claim that the earth's temperature had been decreasing as a result of cooling from the primal nebula (and that this cooling had inspired such progressive adaptations as the creation of warm-blooded mammals). Lyell admitted the potential decrease

through time, but disputed the inherent and irreversible directional mechanism. He argued instead that cooling had been a contingent and reversible result of the distribution of land. The Northern Hemisphere had been largely oceanic during the warm Carboniferous, and had become progressively continental since then. Climates followed continents, and continental disposition varied cyclically or stochastically through time. The cooling had occurred, but it had been part of a grand cycle, not the first stages of an ineluctable trend (Ospovat, 1975). We are now, he stated, in "the winter of the 'great year,' or geological cycle". Warmer times will come again, and when they do, Lyell states in what must be the most striking passage of the entire *Principles*:

"Then might those genera of animals return, of which the memorials are preserved in the ancient rocks of our continents. The huge iguanodon might reappear in the woods, and the ichthyosaur in the sea, while the pterodactyle [*sic*] might flit again through the umbrageous groves of tree-ferns" (1842, p. 193).

I shall have more to say later of the ichthyosaur's return.

(3) But what of Man? No human fossils had been found, and Lyell was not about to deny our superiority. In a resolution, uncomfortable even in his own context, Lyell argued that the introduction of Man marked a discontinuity to which his system did not apply — an imposition of a new moral order upon the steady state of the material world. As an event within the moral sphere, it did not come under his purview. The Lord had ordained consciousness so that the steady state external to it might be appreciated as a sign of divine order.

"To pretend that such a step, or rather leap, can be part of a regular series of changes in the animal world, is to strain analogy beyond all reasonable bounds" (1842, p. 252).

Our own arrival has not interrupted the steady state of geological and biological change:

"And so the earth might be conceived to have become, at a certain period, a place of moral discipline, and intellectual improvement to man, without the slightest derangement of a previously existing order of change in its animate and inanimate productions" (1842, p. 258).

The motor of change and the nature of creative power: internal and environmental theories

For a brief period during the early nineteenth century, geology was the queen of the sciences. The greatest minds of Europe flocked to it, and even became palaeontologists. The relationship between geological and biological change became a major subject and source of contention. The world reflected a divine order, to be sure, but how did it do so? A school of "environmental determinists" (Ospovat, 1975) saw greatest evidence of design in the perfect fitting of each organism to its appointed station in the environment. The geological alteration of environments became the primary cause of organic change

— an unavoidable conclusion if newly created forms are constrained to fit new environments perfectly. But other palaeontologists rejected this link between a material earth and the history of life. Their vision of organic order required an independent dynamic — a separation of life from the earth and an affirmation of self-directed, organic uniqueness. The extreme versions are visions, not empirical readings of the fossil record. Some men are attracted to the idea of harmonious interaction and oneness, others to a separate and irreducible status for life.

William Buckland took the environmentalist position; life progressed because it had to fit an improving physical world:

> "We may collect an infinity of arguments, to show that the creatures from which all these fossils are derived were constructed with a view to the varying conditions of the surface of the earth, and to its gradually increasing capabilities of sustaining more complex forms of organic life, advancing through successive stages of perfection" (1836, p. 107).

Lyell agreed with Buckland's tie of life to the earth. But since his earth did not progress, neither did his history of life. And so we return to the importance of what must strike most modern readers as Lyell's utterly absurd statement about returning ichthyosaurs. (Indeed, it also struck some of his contemporaries as absurd — witness Henry de la Beche's cartoon of the future Professor *Ichthyosaurus* lecturing to his students about a fossil human skull, Fig. 1.) What could this founder of geology, this great and sober uniformitarian, have meant by such nonsense. In an 1830 letter to Gideon Mantell (quoted in Ospovat, 1975), he was even more explicit and definite:

> "I will not tell you how, till the book is out — but without help from a comet, or any astronomical change, or any cooling down of the original red-hot nucleus, or any change of inclination of axis or central heat, or volcanic hot vapours and waters and other nostrums, but all easily and naturally. I will give you a recipe for growing tree ferns at the pole, or if it suits me, pines at the equator; walruses under the line, and crocodiles in the arctic circle. And now, as I shall say no more, I am sure you will keep the secret. All these changes are to happen in future again, and iguanodons and their congeners must as assuredly live again in the latitude of Cuckfield as they have done so."

Pines at the equator, crocodiles in the arctic and iguanodons again in England. Each environment has its characteristic forms of life. Organisms will follow the migrations of their environments over the earth; they will even reappear — at least at the generic level of basic design — when their environment, now present nowhere on earth, forms again during the returning phase of the great cycle.

"Catastrophism" was not a monolithic dogma. We find Buckland united with the uniformitarian Lyell in support of environmentalism. Louis Agassiz, America's greatest catastrophist, argued vehemently for an intrinsic and independent control of life's direction. The history of life displays the thoughts of the Creator in their most sublime embodiment; life does not merely map a sequence of environments. Physical history is catastrophic, but it has no direction; organic history, on the other hand, is inherently progressive:

Fig. 1. Professor *Ichthyosaurus*, at some future time after his triumphant return, lecturing to his students on the fossil skull of a peculiar Quaternary mammal. Satirical figure by de la Beche, commenting on Lyell's strict environmentalism and non-progressionism (from Buckland, 1890).

"While the material world is ever the same through all ages in all its combinations, as far back as direct investigations can trace its existence, organized beings, on the contrary, transform these same materials into ever new forms and new combinations. . . This identity of the products of physical agents in all ages totally disproves any influence on their part in the production of these ever changing beings which constitute the organic world, and which exhibits, as a whole, such striking evidence of connected thoughts" (1857, 1962 ed., pp. 97—99).

The tempo of organic change

Cuvier hestitated to write about what he could not see. He carefully avoided the issue of directionalism, but hesitated not at all in expressing his belief in the episodic nature of both geological and organic change. The literal record of the rocks required it. "The surface of the earth", he wrote (1817, p. 7), "has been much convulsed by successive revolutions and various catastrophes." The fossil record of life is equally episodic: "If the species have changed by degrees. . . we ought to find traces of this gradual modification. . . We should be able to discover some intermediate forms; and yet no such discovery has ever been made" (1817, p. 115). Its episodes, unsurprisingly, match the physical revolutions, as waves of extinction prepare an earth to receive new forms. (Cuvier did not believe that physical revolutions wiped out all of life; some new forms of a subsequent fauna may have migrated into an area from a pool of survivors elsewhere):

"Numberless living beings have been the victims of these catastrophes; some have been destroyed by sudden inundations, others have been laid dry in consequence of the bottom of the seas being instantaneously elevated. Their races even have become extinct, and have left no memorial of them except some fragment which the naturalist can scarcely recognize" (1817, pp. 16—17).

How could Lyell assert his gradualist notions of slow and constant tempos for extinction and creation against the weight of Cuvier's evidence? Lyell defended gradualism by invoking his distinctive principle of scientific methodology (as he did in arguing for a steady state of life against the evidence of vertebrate progression). When evidence is imperfect, a scientist must probe behind appearance. He must establish and defend a general theory, predicting from it the expectations of evidence. When empirical evidence does not meet these expectations, a scientist must decide whether the evidence is illusory or the theory false. Disconfirming, but inconclusive, evidence should be reinterpreted in the light of theory; as long as the reinterpretation works, the theory should be maintained. In this case, Lyell stuck with his uniformitarianism by invoking the incontestable incompleteness of the geological record — the preservation of a few words per page of the earth's book, our prejudices as inhabitants of dry land where sedimentation is so sporadic, etc. In a remarkable passage, Lyell defends his method. At first reading, it seems riddled with the folly of dogmatism; but the message grows on you. And when we recognize that no observations are independent of theory — that, in fact, theory is a proper guide to observation — it begins to make sense. The antidote to dogmatism is a willingness (that Lyell had) to abandon theory, not an assertion of the primacy of unbiased observation:

"It is only by becoming sensible of our natural disadvantages that we shall be roused to exertion, and prompted to seek out opportunities of discovering such of the operations now in progress, as do not present themselves readily to view. We are called upon, in our researches into the state of the earth, as in our endeavors to comprehend the mechanism of the heavens, to invent means for overcoming the limited range of our vision. We are perpetually required to bring, as far as possible, within the sphere of observation, things to which the eye, unassisted by art, could never obtain access" (1842, p. 121).

The Altered Context of Evolutionary Theory

For palaeontology, evolutionary theory was not the watershed that later reconstructions usually proclaim. It did enlighten some issues — gradualism at the lowest level of the species became possible for the first time (even Lyell's extreme gradualism at higher levels had required the immediate creation of each new species). But it provoked virtually no change in taxonomic or strati-graphic practice — the backbone of our day-to-day activity. The fossil record, with its literal evidence of catastrophe and sudden faunal transition, proved more of a problem for Darwin than a benefit. His chapter bore no such trium-phant title as "proof of evolution from the recorded history of life", but rather the apologetic: "on the imperfection of the geological record."

Evolutionary theory swept through palaeontology as rapidly as it triumphed in other disciplines. A few, like Agassiz, took their creationism to the grave, but the battle was over within a decade after 1859. However, by accepting the fact of evolution, none of the three great issues was resolved. I cannot even say that the relative frequency of belief changed markedly for any of the three sets of alternative positions. Directionalism or steady-statism, internalism or environ-mentalism, gradualism or punctuationism — all could be easily encompassed within evolutionary theory.

Darwin convinced his colleagues about the fact of evolution. If his theory of natural selection had found equal favour, palaeontology might have reached a consensus on at least two of its three great issues. In a famous statement, Dar-win said that he felt as if his books had come half out of Lyell's brain. The extensive accuracy of this statement has rarely been appreciated, and many commentators have read it as just one more exercise in sham, Victorian mod-esty. Darwin's position on the three central issues was that of Lyell, updated for an evolutionary context. He was a convinced gradualist at all levels, and this is the only opinion on the three issues that most nineteenth century evolu-tionary palaeontologists shared with him. He, like Lyell, was an environmen-talist. Internal forces do little more than provide a spectrum of random varia-bility for the creative agent of natural selection. Local environments determine the direction of selection.

On the third issue of steady state vs. direction, Darwin experienced Lyell's ambiguity; but whereas Lyell espoused the two positions sequentially, Darwin managed to hold them simultaneously. By the mid 1850s, Lyell was ready to bow to the evidence for progression in the phyletic history of vertebrates. It is widely believed that he converted to evolution about the same time because Darwin had convinced him. I maintain that he accepted evolution as a "mini-mum retreat" from his former set of beliefs, once he had been forced to accept the fact of progression (Gould, 1970). Progressionistic theories of life had been the property of catastrophists who, to a man, were strongly committed to the punctuationist theory of paroxysms — catastrophes removed much of life, and new, higher creations repopulated the void. For Lyell, evolution provided a way to accept progressionism without abandoning either the gradualism or environmentalism so vital to his "uniformitarianism."

Ironically, progression was not an important argument for Darwin's personal belief in evolution. In fact, I maintain that an explicit denial of innate progres-sion is the most characteristic feature separating Darwin's theory of natural selection from other nineteenth century evolutionary theories. Natural selec-tion speaks only of adaptation to local environments, not of directed trends or inherent improvement. Darwin wrestled with the issue of progression and finally concluded that his theory provided no rationale for a belief in evolutionary directions — adaptation to local environments meant just that and nothing more. He wrote to the American directionalist Alpheus Hyatt: "After long reflection I cannot avoid the conviction that no innate tendency to progressive development exists" (quoted in F. Darwin, 1903, p. 344).

But we cannot label Darwin as a pure steady-statist; for, while he denied progress as an intrinsic ingredient of evolutionary mechanisms, he did not deny the possibility of direction as an outcome of the operation of natural selection. Evolution is adaptation to local environments, but one pathway to adaptation lies in structural "improvement" conferring a more general success upon its bearer:

> "The inhabitants of each successive period in the world's history have beaten their predecessors in the race for life, and are, in so far, higher in the scale of nature; and this may account for that vague yet ill-defined sentiment, felt by many paleontologists, that organization on the whole has progressed" (1859, p. 345).

A few palaeontologists accepted Darwin's theory of natural selection — some of the German ammonite workers and a few functional morphologists, Kovalevsky and Dollo, in particular. But it was as much a minority view among students of fossils as it was in the community of evolutionists at large. The standard reasons given for its disfavour cite strictly scientific difficulties — particularly, Darwin's inability to explain the genetic nature of variation. I do not deny the importance of these arguments; nonetheless, I believe that natural selection was rejected during the nineteenth century primarily because it was philosophically too radical (Gould, 1974a) and that this radicalism is primarily expressed in Darwin's position on the three issues. Comfortable Victorians had no quarrel with gradualism; if things had to change, they might as well do it in a slow and orderly manner. But with his environmentalism and his partial steady-statism, most of Darwin's colleagues could muster no agreement. They deplored the coarsely materialistic nature of natural selection, a theory that granted no special uniqueness to living material and that identified the agent of change not within organisms themselves, but in a fluctuating external environment. They were not willing to view organisms as being directed nowhere and buffeted continually by external forces not under their control. If Darwin had removed a benevolent God from the domain of life, then purpose and uniqueness would have to be sought in the very nature of life itself.

Thus, the common thread of anti-Darwinian evolutionary theories during the late nineteenth and early twentieth centuries is their "DI bias"; i.e., they proclaim an intrinsic direction for evolution and they locate the directional force within organisms themselves. Evolution can be rescued from Darwin's materialism; life can remain separate and inviolate.

It has often been said — and truly — that Darwin's main competition within palaeontology came from orthogenetic, vitalistic and finalistic theories of life (Simpson, 1949, for example). It has rarely been noted that the common denominator of these theories is their DI bias. We may specify three "levels" of anti-Darwinian DI theories.

(1) Claims for universal direction. Henry Fairfield Osborn spent a lifetime trying to induce the laws of evolution from the history of fossil vertebrates. Most of his conclusions were negative. He flirted with Lamarckism as a young man, but rejected it when no mechanism could be found for the inheritance of acquired characters. He saw a role for natural selection in eliminating the unfit, but did not understand how it could create the fit. The fossil record did teach him that change was slow and continuous, and that it seemed to lead in definite directions — larger titanotheres, with bigger horns, for example.

"It has taken me thirty-three years of uninterrupted observation in many groups of mammals and reptiles to reach the conclusion that the origin of new characters is invariably orthogenetic" (Osborn, 1922, p. 135).

Osborn developed a finalistic theory of "aristogenesis" to explain the direction of life. Since evolution is slow and gradual, structures useful only at large size must go through a long prior history of gradual and non-adaptive increase. (Titanothere horns must begin as tiny, useless nubbins and increase directionally through a long non-adaptive phase towards the rather large size of their incipient utility.) "Aristogenes" — little more than a name for ignorance — control the origin and direction of traits with only distantly prospective value. Moreover, Osborn drew his phylogenies to support his DI theory. Osbornian phylogeny is always highly polyphyletic — a bush with numerous independent branches joining only at the base (Fig. 2). Traits marking the predetermined paths of evolution can arise independently and progress directionally in any number of lineages. Though environments can enhance or eliminate, they do not direct the trend.

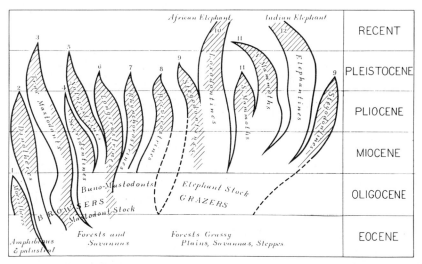

Fig. 2. Osborn's phylogeny of elephants (from Osborn, 1921). Independent, largely pre-programmed trends characterize the phylogeny of each lineage. Similar characters do not, therefore, reflect common descent and nearly every lineage springs independently from the basal stock.

Finalism — the control of evolution by preset goals — is not entirely dead. The omega-point of the Jesuit palaeontologist Teilhard de Chardin reflects its most recent popularity, while the finalistic gospel of human evolution according to Clarke—Kubrick formed a major theme of the story and film *2001*.

(2) The theory of racial life cycles. Alpheus Hyatt believed that lineages, as individuals, had appointed cycles of birth, youthful vigour, confident maturity, eventual decline and final death. His "old age theory", as he fondly called it, located the preset and inevitable history of lineages in the inherent properties of what later generations would call "germ plasm". Hyatt argued explicitly against any environmental control of these programmed trends. In his monograph on the Tertiary species of *Planorbis* at Steinheim (1880), for example, he thought he could discern five independent lineages living sympatrically through an identical sequence of environments. But three of these lineages displayed "progressive" tendencies: they increase in size, strength of ornament and degree of regular, helical coiling. The two other lineages regress: they become smaller, thinner and begin to coil so irregularly that they seem to reflect the agonies of impending phyletic death. How could the same environments permit such different trends in related lineages? Hyatt answers that the environment exerts no control over an internally directed evolution; it may eliminate the inadaptive by selection, but it plays no role in initiating any phyletic change. Very simple, the three progressive lineages were in vigorous phyletic youth and could meet and overcome any environmental challenge; the regressive lineages were in phyletic senility.

Of the internal mechanism of phyletic change, Hyatt claimed to understand very little. But the inductive study of phylogeny provided one important hint: acceleration in development is a universal feature of life. Adult traits of ancestors appear earlier and earlier on the ontogeny of descendants, leaving "room" at the end of ontogeny for the addition of new features. Early in the history of a lineage, these added features exhibit the vigour of youth; later on, the senile traits of impending extinction join the treadmill. Phylogeny is a programmed ontogeny (see more detailed analysis of Hyatt's schema in Gould, 1977). Darwin stated his belief in "no innate tendency to progressive development" during a long exchange with Hyatt about the principle of acceleration.

(3) Orthogenesis within individual lineages. A lineage evolving in a definite and inadaptive direction furnishes the best inductive argument for DI theories against the SE aspect of Darwinian natural selection. (Adaptive orthogenesis can be controlled by selection; but directed evolution towards doom cannot be Darwinian.) Thus, the "overcoiling" of *Gryphaea*, and the hypertely of antlers in Irish Elks, canines in saber-toothed cats, and tusks in mammoths became the standard palaeontological examples of internally directed, non-Darwinian evolution. I have restudied the two most famous cases and, unsurprisingly, have found nothing to support the DI claim. *Gryphaea* did not evolve towards greater coiling at all, and the antlers of the Irish Elk were adaptive in intraspecific competition among males (Hallam, 1968; Gould, 1972, 1974b; Hallam and Gould, 1975). The British orthogeneticist W.D. Lang argued that overcoiling in *Gryphaea* clearly precluded any control by external environments.

"These trends. . . are soon out of the environment's control; they are lapses which may overtake *Ostrea* [*Gryphaea*'s ancestor in Lang's view] at any moment of its evolution — trends which having once started continue inevitably to the point when their exaggeration puts the organism so much out of harmony with its environment as to cause its extinction" (Lang, 1923, p. 11).

As his major achievement in drawing palaeontology under the umbrella of the modern synthesis, Simpson (1944, 1953) discredited the idea that macro-evolution required laws of its own, separate from those regulating change in gene frequencies within local populations. Thus, he synthesized the vastness of Phanerozoic history with selection experiments in *Drosophila* bottles. All could be encompassed within Darwin's basic theory, and with Darwin's own positions on the three issues — ambiguity on directionalism vs. steady-statism (see Simpson, 1974, on evolutionary "progress"), environmental control by natural selection, and gradualism (but see 1944, pp. 206–207 on "quantum evolution" for relatively rapid transitions across adaptive zones). I would epitomize Simpson's achievement by stating that he invalidated the "classical" DI arguments for the independence of macro-evolution. I say "classical" because I believe that macro-evolution has a legitimate independence from the laws of gene substitution in natural populations, but that its independence is not anti-Darwinian, and not tied to DI arguments.

Contemporary Palaeobiology Recycles the Ancient Issues

The hottest modern debates in palaeobiology are, as they almost have to be, extended discussions of the three great issues in new contexts. Since my assignment specified history, I will not analyse my contemporaries in detail; still, two recent examples should exemplify my contention:
(1) Valentine on diversity and drifting continents: a radical environmentalism.
Valentine (1973, and nearly a score of recent papers) believes that world-wide continental configurations hold the key to the history of diversity, primarily through their influence upon the relative stability of trophic resource regimes. When continents are fragmented and widely separated, world-wide diversity increases by pronounced endemism, and increased predictability of trophic resources resulting from climatic stabilization. When continents coalesce, climatic instability leads to unpredictable trophic resource regimes; previously endemic faunas merge and decimate each other in competitive interaction, and the area of shallow seas is reduced by suturing and eustatic regression.

"The biotic revolutions in past oceans are related to environmental restructuring and ecological repatterning. The most likely source for such changes lies in the changing geographic patterns of continental dispersion, fragmentation, assembly, submergence and emergence, and oceanic fragmentation and interconnection" (Valentine, 1973, p. 462).

Valentine views the empirical curve of taxonomic diversity through time

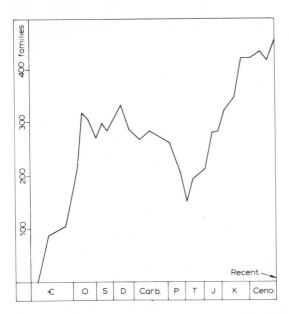

Fig. 3. Valentine's empirical curve of familial diversity among Phanerozoic marine invertebrates. This curve is a "map" of continental positions through time.

(Fig. 3) as, almost literally, a "map" of continental positions (see Raup, 1972,
for an alternative interpretation based on biases of preservation). Following the
early Phanerozoic rise (for which Valentine also proposes an environmentalist
rationale — see below), modest mid Palaeozoic diversity reflects moderately
separated continents; late Palaeozoic decline mirrors the suturing of continents
with the latest Permian debacle as a sign of the final coalescence into Pangaea;
steadily rising diversity since the Triassic records the break-up of Pangaea and
the ever widening dispersal of its fragments.   ,
    But Valentine's boldest argument by far concerns the origin of the coelom
and the basic ground plans of invertebrate design — I'll bet ten to one that he's
wrong, but I admire both the daring, and the consistency it imparts to his
entire system. If any subject had seemed to lie safely in the domain of classical
morphology, and free from an environmentalist interpretation, this was surely
it. But Valentine argues that the coelom is an adaptation for burrowing (fair
enough so far — see Clark, 1964), and that its origin in the late Proterozoic
must record a destabilization in trophic resource regimes — forcing soft, epibenthic creatures into the more stable substrate to lead new lives as burrowing,
detritus feeders. Likewise, the Cambrian explosion must represent a skeletonization following the re-invasion of epibenthic sites when trophic resources
restabilized. Since Valentine has identified continental assembly and fragmentation as a primary cause of fluctuation in trophic resource regimes, he is led to
postulate (with no direct evidence) an assembly of a late Proterozoic supercontinent and its subsequent break-up near the Cambrian boundary (1973, p. 454).
    (2) The Lyellian steady state rides again.

Raup et al. (1973) have provided the latest explicit defense of Lyell's steady state, at least as applied to the history of diversity (see also May, 1973 for ecological support). Raup et al. construct a model to simulate evolutionary trees by treating all times and taxa alike, and by setting a constant value for equilibrial diversity within time units. They study the waxing and waning of diversity in major groups (clades) and find remarkable similarities between these stochastic patterns and the fluctuations in real clades that have inspired so many causal hypotheses. Of course, Raup et al. do not claim that total diversity has undergone no real fluctuation through time. They do maintain — and this is the central element of Lyell's steady state — that there are no intrinsic trends towards increasing (or decreasing) diversity. Ecological roles are, in a sense, "preset" by the nature of environments and the topological limits to species packing; they are filled soon after the Cambrian explosion. Thereafter, inhabitants change continually, but the roles remain. Fluctuations are not intrinsic, but a result of externally imposed configurations. If shallow seas disappear, a large number of roles are temporarily eliminated and a mass extinction ensues.

The critics of Raup et al. have reasserted a DI hypothesis to counter the notion of equilibrium. Flessa and Levinton (1975) propose a "continuous diversification" model in which potentials for "speciation and niche division are limitless" (1975, p. 245). This is an intriguing type of *ceteris paribus* (all other things being equal) argument, in which intrinsic direction asserts itself only in constant conditions; i.e. Flessa and Levinton do not maintain that diversity has really increased steadily, but only that it would do so in constant environments. In the real world, major and periodic environmental "zaps" — like the coalescence of Pangaea — disrupt the trend and keep total diversity in check. (Obviously, as one of the *et alii* in Raup's paper, I favour the equilibrium.)

Even Valentine, for all his talk of environmental control, accepts a DI argument for both increasing complexity and diversity through time. He believes that the primary control by continental fragmentation and assembly is so strong that we don't see the "vector" of increasing niche partitioning in Phanerozoic diversity curves (except, perhaps, as a co-contributor with increasing endemism for the post-Permian rise). Valentine considers the ancient subject of direction as sufficiently important to set the final lines of his book (in which he argues that a "standardized" environment through time would display increased diversity and morphological complexity of its species):

"A sort of moving picture of the biological world with its selective processes that favor increasing fitness and that lead to "biological improvement" is projected upon an environmental background that itself fluctuates. . . The resulting ecological images expand and contract, but, when measured at some standardized configuration, have a gradually rising average complexity and exhibit a gradually expanding ecospace."

I hope that readers will now appreciate why I chose to designate positions on the three issues as eternal "metaphors".

An Editorial Comment in Conclusion

I hope that no one reading this account has been inspired to ask: "Well, which set of positions is right after all?" — because the cryptic answer can only be: "either none of them, or all of them". They all capture a part of complex reality; if they did not, they would not have been vying for our attention and support for so long.

But I am urging neither a council of despair, nor a cynical claim that science cannot solve any deep problems. I only state that the world is too complex to permit the total triumph in all situations of any extreme position on such fundamental questions. Thus we have the ancient wisdom of Aristotle's *aurea mediocritas* — the golden mean. None the less, I believe (with all scientists) that any specific claim for one position or another, stated in the context of verifiable hypotheses, has a potentially definite resolution. Thus, it either is or is not true that the evolution of the coelom is correlated with a destabilization of trophic resource regimes produced by continental coalescence.

The psychological issue of why and how scientists choose their basic attitudes and metaphors is another matter, and one for which I claim very little insight. The historical choices of major palaeontologists have been the theme of this paper. I would only reiterate my claim that attitudes are largely set by *a priori* predispositions, and that these predilections, in turn, are strongly influenced by cultural belief and social position. I think I understand why theistically inclined, nineteenth century, socially comfortable evolutionists reacted against the materialism of natural selection by asserting a DI theory of macro-evolution. In any case, we all have our biases. (I confess to a general SEP bias for micro-evolution, with a dose of DIP for an important aspect of macro-evolution — see below.)

Resolvable arguments about the three issues arise when proponents of one or another position extend their favourite metaphors beyond the domain of evident utility into the large middle ground of unresolved arguments or into the realm of the other camp itself. Intersections between attitudes occur at *domains* and *levels*. By domains, I refer to the competition between attitudes for an explanation of specific issues at a level. By levels, I refer to the more interesting situation in which one alternative is appropriate at one level (of a hierarchical system, for example) and another for what seems to be the same question at a higher level. I believe, for example, that the independence of macro- from micro-evolution resides in the propriety of DI thinking for a process that is invisible in ecological time (where the SE metaphor is appropriate). But the process is thoroughly Darwinian, and the error of most nineteenth century evolutionary palaeontology lay not in identifying the DI component in macro-evolution, but in thinking that it had no Darwinian basis.

I conclude, therefore, by illustrating the conflict of domains for the gradualist—punctuationist and internalist—environmentalist debates, and the interaction of levels for steady-statism and directionalism.

(1) G vs. P. I take it that no one would now deny the importance of rapid and major faunal crises in the history of life. The great Permian extinction has

been copiously documented (e.g., Newell, 1973), and we now have explanations that invoke sound ecological theory and the latest in plate-tectonic reconstruction (Schopf, 1974). On the other hand, no one would quarrel with the claim that our view of speciation in ecological time is more gradual than that proposed by such discredited theories as creationism and de Vriesian macromutationism.

There is, however, a vast territory between these two phenomena — the patterns of organic change in "normal" geological times, in particular. Traditional palaeontology accepted Darwin's view of phyletic gradualism — change accumulates by slow and steady transformation of entire lineages, or by slow separation, at about the same rate, of descendant species from their persisting ancestors. Eldredge and Gould (1972), however, have proposed the alternative of "punctuated equilibria." Species do not generally change after their successful origin. In geological terms, they arise very rapidly from very small populations peripherally isolated from the parental range (the model works equally well for sympatric or parapatric speciation, as long as diverging populations remain small and speciate rapidly). The geological record of descent in a local section should record the "sudden" replacement of ancestors by their fully formed descendants (the event actually recorded is the migration of the established descendant into the ancestral range, and the replacement of the ancestor by competition or emigration).

If Eldredge and I are generally right, then gradualism can only be important in ecological and micro-evolutionary time. The entire range of palaeontological or macro-evolutionary time (not just the rare event of mass extinction) falls into the domain of punctuation.

Gingerich (1976) has defended phyletic gradualism for the evolution of mammalian tooth size in Eocene sections of the Bighorn Basin; I believe that his data fit the model of punctuated equilibria equally well (Gould and Eldredge, in press).

(2) E vs. I. Environmental control for certain events would be denied by no one. We may be sure that Bermuda's endemic pulmonate land snail would no longer grace this earth if the island had ever been completely submerged during Pleistocene fluctuations in sea level. But physical scientists are forever trying to extend their "billiard ball" models to major events in the history of life. (By "billiard ball model", I refer to a habit of explanation that treats organisms as inert substances, buffeted by an external environment and reacting immediately to physical stress without any counteracting, intrinsic control or even temporary resistance.) Thus, in the Berkner—Marshall hypothesis (Berkner and Marshall, 1965), the Cambrian explosion merely represents the point at which atmospheric oxygen reached a sufficient level to provide ozone screening and effective respiration. No lag times, no recognition that metazoan *Baupläne* do not arise automatically from a procaryotic level just because external conditions now favour them. Likewise, four physical scientists (Reid et al., 1976) would tie mass extinctions to decay of the geomagnetic field during reversals — to "catastrophic depletions of stratospheric ozone caused by solar-proton irradiation over a reduced geomagnetic field" (Reid et al., 1976, p. 177). I have

labelled as "physicalist" these purely environmental explanations, based upon billiard ball models; i.e., stimulus leads to immediate and passive response (Gould, 1974c).

For all my general support of environmental control, I applaud the attempt of several palaeobiologists to counteract these physicalist explanations by asserting the independence and internal dynamic of biological processes in complex systems, particularly of ecological interaction and the genetic and morphological prerequisites of complexity. It is hard, for example, to see how the Cambrian "explosion" could have occurred before the eucaryotic cell evolved with its potential for enhanced genetic variability in sexual reproduction. And I doubt that any crude physicalism will explain the origin of the eucaryotic cell. Environmentalism will have to be tempered with a strong dose of internalism to explain these major events. I am, for this reason, strongly attracted to Stanley's (1973, 1976) ingenious ecological (and internalist) explanation of the Cambrian explosion. He argues that the Precambrian flora formed a stromatolitic "monoculture," with low diversity characteristic of uncropped ecosystems. The Cambrian explosion was the final result of a biogenic process mediated by positive feedbacks accompanying a single, key biological event — the evolution of a cropping herbivore (almost surely a protist). This cropper freed space for a rapid diversification among primary producers. These, in turn, provided a diversity of habitats for other herbivores and, eventually, carnivores. The ecological pyramid expanded rapidly in both directions.

(3) S vs. D. Directionalists of the last century did not assert their claims with the blindness of *a priori* social prejudice alone. They had quite a few facts going for them, especially the appearance of "progress" in vertebrate history, the body of data that eventually converted Lyell to progressionism. Yet Darwin's theory speaks, just as clearly, of adaptation to local environments alone. Since local environments fluctuate stochastically, with no directional trend through time, the basic Darwinian mechanism would seem to offer no rationale for a belief in progress. This apparent dilemma led most nineteenth century evolutionary palaeontologists to believe their own data and reject Darwin for a DI theory of evolution.

I believe that the concept of independent levels provides a resolution. Natural selection is undeniably steady-statist and environmentalist in ecological time. But once we discard the shackles of phyletic gradualism as an explanation for "trends", we see that the operation of natural selection in evolutionary time can yield direction. Sewell Wright (1967, p. 120) proposed a profound analogy: just as mutations are random with respect to the direction of selection in ecological time, so might speciation itself be random with respect to the direction of trends in evolutionary time. The primary events of speciation yield no direction, for they only adapt populations to local environments. But all speciations do not have an equal phyletic longevity or an equal opportunity for further speciation. Trends represent the differential success of subsets from a random spectrum of speciations (Fig. 4). Improved biomechanical efficiency, for example, represents one pathway to adaptation in local environments. The species that follow this path — rather than the acquisition of a limiting, mor-

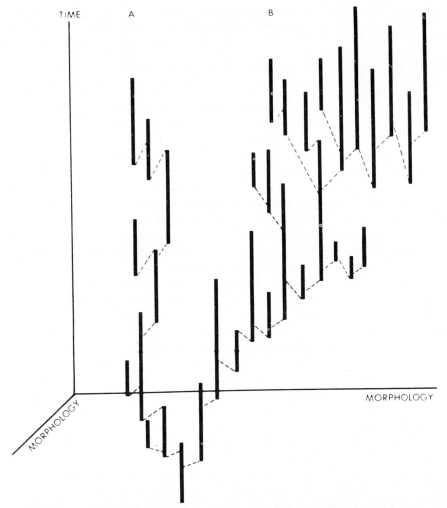

Fig. 4. An evolutionary trend in clade B according to Wright's rule (from Eldredge and Gould, 1972). There is no preferred direction of speciation, merely the differential success of those species defining the "trend". Note also the P feature of this "punctuated equilibrium" model — evolutionary change occurs during very rapid events of speciation.

phological specialization — might form the subset of a directional trend. (I would also regard such a trend as "internal" because an engineer might predict its path on structural principles alone.) Eldredge and Gould (1972, p. 112) write:

"A reconciliation of allopatric speciation with long-term trends can be formulated along the following lines: we envision multiple 'explorations' or 'experimentations' (see Schaeffer, 1965) — i.e., invasions, on a stochastic basis, of new environments by peripheral isolates. There is nothing inherently directional about these invasions. However, a subset of these new environments might, in the context of inherited genetic constitution in the ancestral components of a lineage, lead to new and improved efficiency. Improvement would be consistently greater within this hypothetical subset of local conditions that a population might invade. The overall effect would then be one of net, apparently directional change: but, as in the case of selection upon mutations, the initial variations would be stochastic with respect to this change. We postulate no 'new' type of selection."

Stanley (1975) has given a name to the phenomenon that Eldredge and I chose explicitly not to christen. He calls it "species selection", with a strong hint of its fundamentally non-Darwinian nature. But surely, the differential success of species is as Darwinian a process as their origin; there is nothing non-Darwinian about the extinction and expansion of species. The issue is not the scope of Darwinism, but the results of Darwinism at different levels — SE effects in the origin of each species as an adaptation to local environments, and potential DI effects in the differential success of species within larger clades through time. ("Species selection" furnishes an appropriate description, but it carries the unfortunate implication of a "new", non-Darwinian mechanism. If a key aspect of the phenomenon must have a name, I would prefer "Wright's rule of differential success", or — of this be judged too cumbersome — just "Wright's Rule", which is no vaguer than "Cope's Rule" or "Dollo's Law", and at least has the virtue of pleasant alliteration. Its précis need be no longer than: "Speciation is stochastic with respect to the direction of evolutionary trends". The rest follows from basic assumptions of the "punctuated equilibrium" model.)

In any case, Stanley (1975) correctly argues that Wright's rule "decouples" micro-evolution from macro-evolution, thus affirming the independence of palaeontology as an evolutionary subdiscipline. (Under phyletic gradualism, trends are merely extensions of directional selection within local populations, and macro-evolution has no separate status.) There can scarcely be a more important task for palaeontologists than defining the ways in which macro-evolution depends upon processes not observed in ecological time. In uniting palaeontology with neo-Darwinism, Simpson (1944, 1953) argued that macro-evolution had no such independence. This was an appropriate (and necessary) argument while palaeontology stood its distance from modern evolutionary theory by asserting principles that genetics could not support. Yet the greatest testimony I can offer to Simpson's success is my claim that the time has come to affirm the independence of macro-evolution by recognizing that Darwinian processes work in different ways at different levels. If Simpson had not buried so thoroughly the ghosts of vitalism and orthogenesis, I would be fearing their unintended resurrection in asserting such a claim.

I began this paper with a lament that all great thoughts had been developed in days of old. By now, I hope I have demonstrated that their modern contexts are not without interest (even fascination). I must also confess that I cheated a bit on the first quotation. Alexander's statement is usually cited as I gave it — as a lament for a world with no new directions (as by Wordsworth in his Supplementary Essay of 1815). But the first classical reference I can find cites it in an opposing context (from Plutarch's Morals):

"Alexander wept when he heard from Anaxarchus that there was an infinite number of worlds. . . 'Do you not think it a matter worthy of lamentation that when there is such a vast multitude of them, we have not yet conquered one?' "

A better statement, to be sure. Our ignorance is in no danger of ceding its general domination to our understanding.

References

Agassiz, L., 1857. Essay on classification. In: E. Lurie (Editor), Contributions to the Natural History of the United States, 1. Harvard University Press, Cambridge, Mass. (reprinted in 1962).

Berkner, L.V. and Marshall, L.C., 1965. On the origin and rise of oxygen concentration in the earth's atmosphere. J. Atmos. Sci., 22: 225—261.

Buckland, F., 1890. Curiosities of Natural History. Richard Bentley, London, 362 pp.

Buckland, W., 1836. Geology and Mineralogy Considered with Reference to Natural Theology. W. Pickering, London.

Clark, R.B., 1964. Dynamics in Metazoan Evolution. Clarendon Press, Oxford, 313 pp.

Cuvier, G., 1817. Essay on the Theory of the Earth. W. Blackwood, Edinburgh (translated by R. Jameson).

Darwin, C., 1859. The Origin of Species. John Murray, London, 490 pp.

Darwin, F. and Seward, A.C. (Editors), 1903. More Letters of Charles Darwin. John Murray, London.

Eldredge, N. and Gould, S.J., 1972. Punctuated equilibria: An alternative to phyletic gradualism. In: T.J.M. Schopf (Editor), Models in Paleobiology, Freeman, Cooper and Co., San Francisco, Calif., pp. 82—115.

Flessa, K.W. and Levinton, J.S., 1975. Phanerozoic diversity patterns: tests for randomness. J. Geol., 83: 239—248.

Gingerich, P.D., 1976. Paleontology and phylogeny: patterns of evolution at the species level in early Tertiary mammals. Am. J. Sci., 276: 1—28.

Gould, S.J., 1970. Private thoughts of Lyell on progression and evolution. Science, 169: 663—664.

Gould, S.J., 1971. D'Arcy Thompson and the science of form. New Lit. Hist., II (2): 229—258.

Gould, S.J., 1972. Allometric fallacies and the evolution of Gryphaea: A new interpretation based on White's criterion of geometric similarity. In: T. Dobzhansky et al. (Editors), Evolutionary Biology. Vol. 6. Appleton-Century-Crofts, New York, N.Y., pp. 91—118.

Gould, S.J., 1974a. Darwin's delay. Nat. Hist., 83 (10): 68—70.

Gould, S.J., 1974b. The evolutionary significance of "bizarre" structures: Antler size and skull size in the "Irish Elk," Megaloceros giganteus. Evolution, 28: 191—220.

Gould, S.J., 1974c. An unsung single-celled hero. Nat. Hist., 83 (9): 33—42.

Gould, S.J., 1977. Ontogeny and Phylogeny. Harvard University Press, Cambridge, Mass., in press.

Gould, S.J. and Eldredge, N., in press. Punctuated equilibria: the tempo and mode of evolution reconsidered. Paleobiology, 3.

Hallam, A., 1968. Morphology, palaeoecology and evolution of the genus Gryphaea in the British Lias. Philos. Trans. R. Soc. Lond., B 254: 91—128.

Hallam, A. and Gould, S.J., 1975. The evolution of British and American Middle and Upper Jurassic Gryphaea: a biometric study. Proc. R. Soc. Lond., B 189: 511—542.

Huxley, J.S., 1958. Evolutionary processes and taxonomy with special reference to grades. Uppsala Univ. Arsk., pp. 21—39.

Hyatt, A., 1880. The genesis of the Tertiary species of Planorbis at Steinheim. Anniv. Mem. Boston Soc. Nat. Hist. (1830—1880): 114 pp.

Jahn, M.E. and Woolf, D.J. (Editors), 1963. The lying stones of Dr. J.B.A. Beringer. University of California Press, Berkeley, Calif., 221 pp.

Lang, W.D., 1923. Evolution: a resultant. Proc. Geol. Assoc., 34: 7—20.

Lyell, C., 1842. Principles of Geology. Hilliard, Gray and Co., Boston, Mass., 6th ed.

May, R.M., 1973. Stability and Complexity in Model Ecosystems. Princeton University Press, Princeton, N.J., 235 pp.

Newell, N.D., 1973. The very last moment of the Paleozoic Era. Mem. Can. Soc. Pet. Geol., 2: 1—10.

Osborn, H.F., 1921. Adaptive radiation and classification of the Proboscidea. Proc. Natl. Acad. Sci., 7: 231—234.

Osborn, H.F., 1922. Orthogenesis as observed from paleontological evidence beginning in the year 1889. Am. Nat., 56: 134—143.

Ospovat, D., 1975. Charles Lyell's theory of the earth: the relationship between the theory of climate and the doctrine of non-progression in Lyell's Principles of Geology. Unpublished ms.

Raup, D.M., 1972. Taxonomic diversity during the Phanerozoic. Science, 177: 1065—1071.

Raup, D.M., Gould, S.J., Schopf, T.J.M. and Simberloff, D.S., 1973. Stochastic models of phylogeny and the evolution of diversity. J. Geol., 81: 525—542.

Reid, G.C., Isakson, I.S.A., Holzer, T.E. and Crutzen, P.J., 1976. Influence of ancient solar-proton events on the evolution of life. Nature, 259: 177—179.

Rudwick, M.J.S., 1972. The Meaning of Fossils. Macdonald, London, 287 pp.

Schaeffer, B., 1965. The role of experimentation in the origin of higher levels of organization. Syst. Zool., 14: 318—336.

Schopf, T.J.M., 1974. Permo-Triassic extinctions: relation to sea-floor spreading. J. Geol., 82: 129—143.

Simpson, G.G., 1944. Tempo and Mode in Evolution. Columbia University Press, New York, N.Y., 237 pp.

Simpson, G.G., 1949. The Meaning of Evolution. Yale University Press, New Haven, Conn., 364 pp.

Simpson, G.G., 1953. The Major Features of Evolution. Columbia University Press, New York, 434 pp.

Simpson, G.G., 1961. Three 19th century approaches to evolution. Am. Sch., 30: 238—249.

Simpson, G.G., 1974. The concept of progress in organic evolution. Soc. Res., pp. 28—51.

Stanley, S.M., 1973. An ecological theory for the sudden origin of multicellular life in the Late Precambrian. Proc. Natl. Acad. Sci., 70: 1486—1489.

Stanley, S.M., 1975. A theory of evolution above the species level. Proc. Natl. Acad. Sci., 72: 646—650.

Stanley, S.M., 1976. Fossil data and the Precambrian—Cambrian evolutionary transition. Am. J. Sci., 276: 56—76.

Steno, N., 1667 (1958 ed.). The earliest geological treatise (translated from *Canis carchariae dissectum caput* by A. Garboe). Macmillan, London.

Thompson, d'A.W., 1917. On Growth and Form. Cambridge University Press, London, 793 pp.

Thompson, d'A.W., 1942. On Growth and Form. Cambridge University Press, London, 1116 pp.

Valentine, J.W., 1973. Evolutionary Paleoecology of the Marine Biosphere. Prentice-Hall, Englewood Cliffs, N.J., 511 pp.

Wilson, L.G., 1970. Sir Charles Lyell's Scientific Journals on the Species Question. Yale University Press, New Haven, Conn., 572 pp.

Wright, S., 1967. Comments on the preliminary working papers of Eden and Waddington. In: P.S. Moorehead and M.M. Kaplan (Editors), Mathematical Challenges to the Neo-Darwinian Interpretation of Evolution. Monograph No. 5, Wistar Inst. Press, Philadelphia, pp. 117—120.

CHAPTER 2

GENERAL PATTERNS OF METAZOAN EVOLUTION

JAMES W. VALENTINE

Introduction

The history of life reflects the realization of some of the evolutionary potentials of organisms in response to environmental opportunities. Both the potentials and the environment change with the passage of time, and the evolutionary patterns are thereby altered. The processes which permit evolution to occur
are by no means completely understood. The very measurement of evolutionary activity is beset with grave difficulties; most investigators restrict themselves to a single evolutionary aspect and devise a measure which is useful for
their restricted purposes. The pattern of evolution perceived depends upon the
perspective from which it is viewed.

Evolutionary patterns within major groups of animal fossils are discussed in
detail in following chapters. In this chapter I will concentrate on some patterns
discerned from the interrelations among major animal groups, chiefly phyla,
and patterns generated by the metazoans as a whole. Three main evolutionary
perspectives are involved. One is from the standpoint of phylogeny. A second is
concerned with the taxa as adaptive solutions to environmental problems. A
third regards the taxa as components of ecosystems at the community and
provincial levels. Although the fossil record may be studied from any of these
perspectives more or less independently, each is really only a separate view of
the results of the same complex of processes, and all (and many more) must be
understood before a well-rounded evolutionary picture is available.

Phylogenetic Models

Metazoans may be defined as multicellular animals that share a common
ancestor with both the cnidarians and flatworms. The phylogenetic relationships among the metazoan phyla have long been of interest to evolutionists.
However, the features that characterize the phyla are very general ones such as
the number and quality of tissue layers, symmetry, body cavities, and developmental topology — the features that serve to make up the basic body plan or
ground plan of the phyla — and the interrelations among such characters have
never been made clear. Hyman noted in 1959 that on information then available
"anything said on these questions lies in the realm of fantasy". Kerkut (1960)
emphasized this famous phrase by devoting an interesting book to review the
many arguments and alternative hypotheses of phylogeny, and thus demonstrating the extent of our ignorance of the interrelations of higher taxa. The
situation has not improved spectacularly since that time. Such improvements

as have occurred have diverse sources, including studies of adaptation (discussed in the next section) and the insights of molecular biology. The fossil
record can also play a role.

For our purposes the phyla listed in Table I are adequate; they are those employed for the Eumetazoa by Hyman (1940) with the addition of Nematoda,
Pogonophora and Urochordata (tunicates). Four different phylogenetic
arrangements of these phyla are shown in Fig. 1. These arrangements are an
attempt to contrast the opinions of four separate authorities. Because they
have been drawn to a common plan and because only the phyla of Table I were
included, there are some small distortions of the author's original figures.

Hyman's phylogeny is conservative, based chiefly upon patterns of morphological resemblance among adult body plans but also leaning upon developmental resemblances during very early ontogeny. It features a stem leading
from cnidarians to acoelomates and then dividing into two main branches of
higher metazoans, the protostomes and deuterostomes. The scheme of Hadzi
(1963; Fig. 1B) differs chiefly from that of Hyman in that primitive flatworms
are the earliest Metazoa. They are believed to have evolved not from colonies
of unicells but from compound cells, probably ciliate syncytia. From the early
flatworm metazoans a number of non-coelomate groups evolved, together with
the molluscs, to form the Ameria. An advanced amerous form then gave rise
to coelomates with regionated or segmented coelomic cavities. Coelomic architecture thus plays an important role in Hadzi's arrangement. The placing of the
flatworms at the base of the Metazoa has raised a storm of controversy. The
arrangement of the coelomates, however, is more conventional.

Salvini-Plawen (1969; Fig. 1C), a molluscan specialist, has presented a
scheme resembling Hyman's for lower metazoans. Molluscs, however, arise
from flatworms, a feature common to Hadzi's model. Salvini-Plawen's work has

TABLE I

Living metazoan phyla employed herein [1]

Phyla	First appearances	Phyla	First appearances
Cnidaria	Late Precambrian*	Annelida	Late Precambrian
Ctenophora	No definitive record	Arthropoda	Early Cambrian*
Nemertea	Middle Cambrian*	Phoronida	?Devonian; Cretaceous
Entoprocta	No definitive record	Brachiopoda	Early Cambrian*
Priapuloidea	Middle Cambrian*	Ectoprocta	Early Ordovician*
Nematoda	Carboniferous*	Chaetognatha	Carboniferous
Other aschelminths	No definitive record	Pogonophora	Early Cambrian*
Platyhelminthes	No definitive record	Echinodermata	Early Cambrian*
Sipunculida	Middle Devonian	Hemichordata	Middle Cambrian*
Mollusca	Early Cambrian*	Urochordata	?Silurian; Permian
Echiuroidea	No definitive record	Chordata	Middle Cambrian

[1] Records of first fossil appearances are of uneven quality and should be treated with some
caution. The less reliable are preceded by a question mark, the more reliable are followed by
an asterisk.

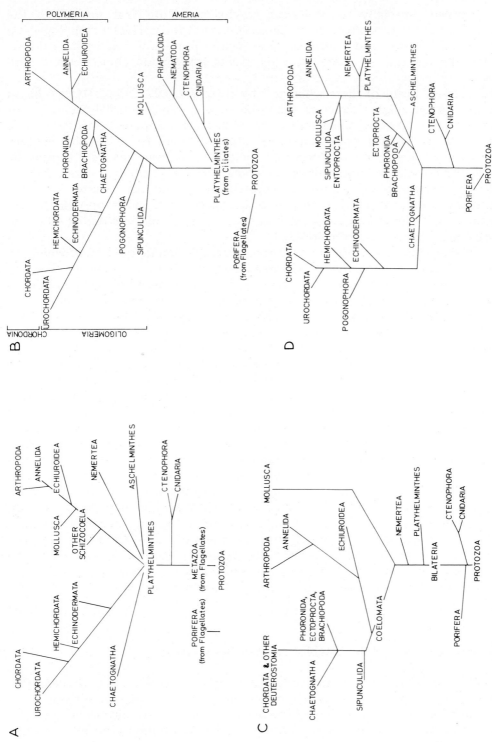

Fig. 1. Four models of the phylogeny of metazoan phyla. All are somewhat modified after schemes proposed by the following: A, Hyman, 1940; B, Hadzi, 1963; C, Salvini-Plawen, 1969; D, Jägersten, 1972.

contributed especially to molluscan phylogeny. Note that instead of the appearance of two phyla at the uppermost ends of the upreaching phylogenetic tree — arthropods and chordates, presumably represented at their peaks by butterflies and man — there are three peaks; arthropods, chordates and molluscs.

The most recent of these schemes is by Jägersten (1972; Fig. 1D). It is based much less on adult characters than the others, relying heavily on a set of assumptions regarding the significance of larval development and form. The phylogenetic conclusions are therefore rather different from the other models. For example, the ancestor from which the two main branches arose is regarded as a non-coelomate, non-seriated creeping benthic form. The Brachiopoda, being epifaunal benthic forms, are regarded as nearer this ancestor than the Phoronida. Another unusual feature is the association of Entoprocta, Sipunculida and Mollusca in a common branch. Flatworms play no part at all in the ancestry of the deuterostome phyla.

The existence of a considerable number of phylogenetic models, of which we have had only four examples, indicates how difficult it is to disprove many of the pathways of descent that can be suggested from morphological and developmental studies. The techniques of molecular biology may eventually prove decisive in establishing a metazoan phylogeny (Dayhoff, 1969; Fitch and Margoliash, 1970). This is possible through determining the sequence of amino acids in homologous proteins from different groups of organisms. The process is laborious, but it has now been accomplished for several proteins in a number of disparate groups. From a knowledge of differences in the sequences of amino acids in homologous proteins it is possible to calculate the minimum number of changes required in the genetic code necessary to derive one sequence from another. This is the "mutation distance" between the proteins, and represents a conservative estimate of their affinity. From these data a dendrogram may be constructed that indicates a possible phylogeny of the organisms containing the different protein homologues. A phylogeny based on any given protein may be somewhat in error owing to the occurrence of back mutations and other factors, but nevertheless provides another important criterion to supplement morphological and other data. When the sequences of large numbers of proteins have been investigated in representatives of all metazoan phyla, we should be nearly certain of their phylogenetic relationships. And when this is done our understanding of the origin and relations of body plans will stand closer to reality and farther from fantasy. The main work still lies ahead, however.

The Fossil Record of the Establishment of Body Plans

How much help is the fossil record in deciding between the possible phylogenetic patterns? In so far as it is viewed as a simple stratigraphic record, it is not very instructive. Undoubted metazoan remains have not yet been identified in rocks older than about 700 m.y., nearly four billion years after the formation of the earth and over two and a half billion years after the first indications that life had appeared. Absolute dating of the beds containing probable early metazoan fossils has proven difficult, for the beds are commonly not associated with rocks that are easily dated radiometrically. The problem is compounded by the difficulty of separating the traces of early animal activity from sedimentary structures that arise from inorganic causes (Cloud, 1973). We are therefore not yet sure of the precise sequence of events that these rocks display. However, it does appear that the oldest metazoan fossils known are traces and that they may possibly be as old as 800 m.y. but are more likely around 700 m.y. old (Glaessner, 1969, 1971; Banks, 1970; Webby, 1970; Crimes, 1974). Earlier rocks that have similar lithologies and palaeoenvironmental settings have been studied but they appear to be barren of metazoans (for example Banks, 1970; Germs, 1974). On the other hand, later rocks contain increasing numbers of traces, and during Cambrian time trace-fossil diversity becomes relatively high, essentially as high as or even higher than is found subsequently in shallow water environments (Crimes, 1974). The Late Precambrian traces include horizontal trails, horizontal burrows and vertical burrows; we are not yet sure which appeared first.

Sediments dating from the Late Precambrian interval that contain the trace fossils have also yielded a small variety of body fossils that lack mineralized skeletons. These are termed the Ediacaran fauna, after the South Australian rocks from which they are particularly well known (Glaessner and Daily, 1959; Glaessner and Wade, 1966). The Ediacaran metazoans are interpreted as being chiefly cnidarians, but there is also a number of worm-like organisms, and other types, of uncertain affinities. Some of these have annulations, but whether or not they are segmented is not certain. Conceivably all the Ediacaran forms could be diploblastic, but it is likely that some are triploblastic, possibly coelomates, though their assignment to living phyla is difficult. Most of the Ediacaran fauna appears to have been epifaunal or pelagic. The Ediacaran assemblage is known from several continents that are now widely dispersed.

Thus, by the start of Cambrian time we have certain records only of the Cnidaria, possible records of Annelida, and indirect (though convincing) evidence from burrows and trails that a variety of metazoans were present. Before discussing skeletonized forms, it is convenient to examine the next younger soft-bodied metazoan assemblage that is known: that of the Middle Cambrian Burgess Shale from British Columbia, Canada. This important assemblage is currently being restudied. Simon Conway Morris has graciously provided the following account of the major soft-bodied groups in advance of publication of all of the formal descriptions. Conway Morris writes:

Burgess Shale metazoan association

The Burgess Shale (Middle Cambrian) of British Columbia contains extraordinarily well-preserved fossils, many of which are entirely soft-bodied. This allows a unique glimpse into metazoan evolution and dynamics relatively shortly after the late Precambrian radiation of the Metazoa took place. C.D. Walcott, who discovered the Burgess Shale, issued a series of preliminary reports on much of the fauna and flora, and the entire biota is currently being reappraised (e.g. Conway Morris, 1976a,b, in press; Hughes, 1975; Sprinkle, 1973; Whittington, 1971a,b, 1974, 1975a,b) with the aid of additional collections obtained in 1966 and 1967 by a Geological Survey of Canada team led by Dr. J.D. Aitken.

The composition of the Burgess Shale fauna (total about 120 genera) is approximately as follows: arthropods, excluding trilobites 25.5%; trilobites 10%; arthropod-like invertebrates 3%; hemichordates and primitive chordates 3.5%; lophophorates, excluding brachiopods 1%; brachiopods 4%; echinoderms 4%; molluscs 3%; polychaetes 5%; priapulids 6%; nemerteans 1%; miscellaneous worms 11%; coelenterates 4%; and sponges 18% (Conway Morris, in press).

The majority of the fauna (about 80%) can be assigned to modern phyla, although most of the genera occupy extinct classes. This part of the fauna is providing invaluable evidence for separating primitive from advanced characteristics, which in turn indicates possible evolutionary pathways. For example, it is apparent that the priapulid worms underwent a considerable radiation in the Cambrian giving a fauna that, in some ways, was more diverse than the modern priapulid fauna. Primitive features can be discerned and their persistence in some modern priapulids may be recognized, so that a tentative phylogeny can be proposed (Conway Morris, 1976b). Many Burgess Shale genera are fairly specialized, although primitive forms that must be fairly close to their basal stocks have also been identified. They include the earliest crinoid (Sprinkle, 1973), and a laterally compressed worm which had a notochord and muscles arranged in myotomes suggesting that it was not far removed from the ancestors of the fish.

No fossil from the Burgess Shale appears to be ancestral to two or more modern phyla, and the fauna as a whole throws little light on the beginnings of the major phyla or the origin of metameric segmentation and the coelom. A few species have affinities with superphyla, but cannot be placed in any of the constituent phyla. *Odontogriphus omalus* (Conway Morris, 1976a), a probable conodontophorid, belongs to the superphylum Lophophorata, but it is quite distinct from the brachiopods, phoronids and ectoprocts. The Lophophorata may have formed a more diverse group than present evidence would suggest.

Other species cannot be accommodated in any known phylum. They represent ground plans that have not survived. Two of these species are *Banffia constricta* (Walcott, 1911) and *Canadia sparsa* (Walcott, 1911; Conway Morris, 1976b). In the former worm an annulated anterior section was separated from a sac-like posterior section by a prominent transverse constriction. The constriction may have been effective in damping down movements of the body cavity fluid caused by contraction of the anterior section. One disadvantage of an animal with a large continuous body cavity is that muscle movements in one part are transmitted by the body cavity fluid to all other parts of the body, thus necessitating the use of antagonistic muscles. The usual solution to this problem is to subdivide the body cavity into compartments separated by septa, as in the oligochaetes and polychaetes (Clark, 1964). The constriction, therefore, probably acted to more or less isolate the two sections. The posterior section was probably more passive and may have contained the urinogenital organs that presumably functioned more efficiently in a semi-isolated body cavity. *B. constricta* was probably a survivor of a primitive group that in attempting to subdivide its body cavity with a constriction could not compete with septate worms.

Canadia sparsa had a bizarre appearance. A horizontal, elongate and cylindrical trunk, possibly with a bulbous head, was supported by a row of seven pairs of spines, whilst seven tentacles and a posterior group of short tentacles arose from the dorsal surface. The seven tentacles had bifid tips and appear to have joined a central tube that was presumably the gut. *C. sparsa* appears to have been a scavenger, but the method of feeding is uncertain and the possibility that the tentacles acted as individual mouths cannot be dismissed.

Study of the palaeoecology of the Burgess Shale allows comparison with modern marine communities. The priapulids were numerically the dominant infaunal group and unlike modern communities the polychaetes were relatively unimportant. The priapulids were mostly predatory and even cannibalistic. The appearance of eunicid-like jawed polychaetes in the Ordovician suggests that it was at this time that the priapulids were relegated to a minor role. Many of the small arthropods (see Hughes, 1975; Whittington, 1971b, 1974) may have occupied the same niches as the modern isopods, amphipods and tanaidacean Crustacea. Some of the larger arthropods probably occupied similar niches to those of benthic fish today.

Considerable research on the Burgess Shale fauna and its ecology remains to be completed, but preliminary results are showing how complex early metazoan evolution must have been. (I am grateful to Prof. H.B. Whittington for advice and encouragement during this research, which was undertaken under a N.E.R.C. studentship, and now continues under a research fellowship at St. John's College, Cambridge.)

Of other occurrences of soft-bodied metazoans in younger rocks, the Lower Carboniferous deposits at Mazon Creek, Illinois are particularly notable. They have yielded soft-bodied animals with body plans that make them difficult to associate with living phyla (Richardson and Johnson, 1970). Nematodes are probably known from the Lower Carboniferous, and other soft-bodied phyla appear during the Mesozoic and Cenozoic (Table I). Five soft-bodied groups are not certainly known as body fossils: Ctenophora, Entoprocta, some aschelminths, Phoronida, and Echiuroidea. Trace fossils have sometimes been attributed to the Phoronida from the Late Precambrian onwards.

During the latest Precambrian, mineralized skeletons make their first appearance as fossils (Matthews and Missarzhevsky, 1975). These are minute remains described chiefly from Russia. They are of unknown affinities but may well represent metazoans. One form, *Protohertzina*, resembles conodonts. The first well-skeletonized fauna appears in the Lower Cambrian (Tommotian of Russia and equivalents). It includes coiled shells that are probably molluscan (but may not be gastropods; Yochelson, 1975), and hyolithids (which may or may not be molluscs; see Runnegar et al., 1975, and Marek and Yochelson, 1976). During the next stratigraphic stage a host of skeletonized representatives of modern phyla appear, including trilobites. By the close of the Lower Cambrian, skeletons of all phyla that are known to have representatives with well-mineralized hard parts have appeared, except for Cnidaria, ectoprocts and chordates, thus including the lowest and highest levels of organization among living metazoan phyla. Actually the Cnidaria and Chordata both seem to be represented by unskeletonized body fossils well before the end of the Cambrian, so among skeletonized phyla only the Ectoprocta are totally unknown by that time, and they appear almost immediately in the Early Ordovician.

In summary, then, the earliest metazoan body fossils are the cnidarians, but they are accompanied and probably even preceded by burrows that strongly suggest the presence of more advanced body plans capable of somewhat regular, sustained burrowing activities — pseudocoelomate or, more likely, coelomate animals (Valentine, 1973a,b). Of the remaining phyla, those that have the best chance for preservation nearly all appear within the next two stratigraphic stages, and soft-bodied living phyla appear in numbers when we are so fortunate as to obtain even a moderate local sample of an unskeletonized Cambrian fauna. There are numbers of both soft-bodied and skeletonized fossils of late Precambrian or Cambrian ages which are not assignable to living phyla. They do not seem to represent intermediates between the living phyla, nor can any of them yet be considered as ancestral to any specific living phylum. Rather they appear to be additional animal groups, separate phyla or just possibly different branches of some living phyla, which have arisen from some unknown ancestors.

Adaptive Models of Metazoan Body Plans

Even though their lines of descent have not been established, there has been much work done on the adaptive significance of phyla. In this the fossil record has proven useful, and the results help delimit the phylogenetic possibilities. These studies are aimed at understanding the primitive functions of the features that characterize each body plan and working out the adaptive pathways that led from preceding adaptations to these functions. Hyman (1940), Kerkut (1960) and Clark (1964) provide many references and cover this topic broadly. Here I shall briefly sketch adaptive pathways that seem likely on present evidence, and examine their implications in the light of the fossil record and phylogenetic models.

Origin of Metazoa

Multicellularity has arisen independently several times. The advantages of multicellularity seem to involve size, shape and the modular repetition of cellular machinery, from which may flow such useful properties as increased homeostasis, longevity and reproductive ability. By definition, the Metazoa originated when a multicellular lineage appeared that was to number among its descendants the cnidarians and flatworms. Metazoans probably descended from colonial, flagellated protozoa whose replicating cells remained in contact.

However, once multicellularity was under genetic control it could be exploited by selection in response to adaptive opportunities, and probably a variety of growth patterns and shapes were produced. Hummocky or dome-like shapes might be most advantageous for rather sessile forms in benthic sites, for they would increase surface area and promote turbulent flow, enhancing food and oxygen supply. In quiet waters, cup or cylindrical shapes, perhaps with extended cellular strands, might provide maximum benefits. Even such simple shapes could give rise to cellular differentiation. For example, in dome-shaped forms that rely on suspended particulate food, the higher cells would tend to receive more nourishment than cells in marginal or downwarped regions (Knight-Jones and Moyse, 1961, illustrate analogous cases among aggregated individuals). Cell differentiation to increase the feeding efficiency of the best-nourished cells would be favoured by selection; at the same time, reproduction might be promoted among better-nourished cells at neighbouring localities, while differentiation of marginal cells could favour supporting and strengthening functions. Cell colonies with other shapes and life modes would have distinctive patterns of differentiation.

Bottom-feeding forms, required to be vagile in order to locate fresh food sources, might develop locomotory rims with central cells differentiated for feeding, perhaps invaginating in larger organisms. Pelagic forms might become differentiated either in radial or bilateral patterns. In short, we can imagine early multicellular forms evolving a large number of shapes and cell patterns and becoming increasingly integrated. When the point was reached that members of some lineage ceased to be colonial and became truly multicellular individuals, the metazoan grade had appeared. Just which of the many possible early metazoan forms was the actual primitive metazoan may never be known; all metazoan lineages of this grade are extinct. Fossils are unlikely to occur and would not be conclusive if found. There is no shortage of speculative reconstructions, however, the most famous being the blastaea—gastraea pathway envisioned by Haeckel (1874).

Lower metazoan grades

The simplest metazoan architecture among living phyla is the diploblastic grade. An evolutionary pathway led from simple multicellular forms to forms with definite tissue layers. A dome-shaped or conical detritus feeder with invaginated or internal digestive tissues would essentially be diploblastic and might have been a cnidarian ancestor (somewhat as envisioned by Jägersten, 1955) though it would presumably have been bilateral. Radial or biradial symmetry might develop in a cup-shaped benthic suspension feeder, or in a similar pelagic form (Hand, 1959). A small, benthopelagic tetraradiate cnidarian ancestor with hollow tentacles has been suggested by Rees (1966); such a form might have developed from almost any primitive multicellular lineage. Thus, our problem is not in visualising a plausible adaptive pathway leading to diploblastic forms, but in deciding which of the alternative possibilities was the historical case.

A similar problem occurs when considering the origin of the triploblastic grade. The most primitive triploblastic phylum living is the Platyhelminthes. The only living forms that much resemble possible flatworm ancestors are the planula larvae of cnidarians. If planulae settled on the sea bed (perhaps originally as a prelude to a sessile adult hydroid generation) and reproduced via neoteny, the adult cnidarian would drop out of their life cycles and a sessile worm would result. The subsequent development of mesenchyme as a primitive mesoderm would be adaptive to the demands of benthic locomotion.

Other suggested flatworm ancestors include benthic diploblastic adults (Lang, 1884) and ciliate protozoa (Hadzi, 1963). It is also possible that flatworms evolved from an early multicellular metazoan independently of the cnidarian line; that is, that the cnidarian and flatworm ancestors diverged before attaining the characterizing features of either phylum.

A characteristic architecture developed among the flatworms, which seems to have been inherited by some later metazoan phyla. Flatworms lack circulatory systems and respire through body walls. In order to provide nourishment throughout the body interior, the digestive tract is elongated and frequently diverticulate, with gut outpockets occurring at intervals along the longitudinal axis; thus digestive spaces are furnished to lateral body regions. Excretion is provided by paired series of protonephridia situated between gut outpocketings, and other organs are similarly repeated, so that the basic architecture is one of seriated organs. A similar seriation occurs in the acoelomate Nemertea, some of which contain thick cores of mesenchyme.

Higher metazoan body plans

We shall neglect the acoelomate Entoprocta and the pseudocoelomates, which have poor fossil records and shed little light upon those metazoans that do have good records, namely coelomates. Many writers have argued that the coelomate phyla have arisen from flatworm stocks. Whether or not coelomates are monophyletic is obviously a matter of disagreement (Fig. 1). An outstanding discussion of the origin and architecture of the coelomic body plans is presented by Clark (1964). It seems likely that a primitive function of the coelom as such was as a hydrostatic skeleton, adaptive for peristaltic burrowing. There are five distinct coelomic plans among living phyla, and each may be interpreted as originally adaptive to a particular life mode.

The molluscs have body plans that in the more primitive groups (monoplacophorans, solenogastres) involve a serial arrangement of repetitive paired organs. Monoplacophora, for example, have seriated pedal muscles, nephridia, gills, gonads and other organs (Lemcke and Wingstrand, 1959). Thus they resemble flatworms in this aspect of their architecture. However, the molluscs also have coelomic spaces, chiefly around the heart but they are particularly extensive in Monoplacophora and Cephalopoda (probably including the more primitive and more advanced living members of the phylum, respectively). The coelomic spaces, though paired, are not divided into distinct longitudinal regions. This coelomic plan, with seriated organs but unregionated coelomic spaces, is termed pseudometamerous. It is reasonable that it was inherited from a flatworm ancestor (Clark, 1964; Vagvolgyi, 1967; Salvini-Plawen, 1969). The rise of the molluscs probably involved a size increase and created a need for a circulatory system as the coelom developed and as internal organs became remote from the exterior. Primitively, locomotion was provided by pedal waves as in some flatworms, but the viscera are removed from involvement with pedal motion and are placed dorsally in coelomic compartments. Gills were developed around the lateral and posterior body margin. A dorsal mineralized shell over the viscera provided attachment for pedal musculature.

Other types of metazoan coeloms (eucoeloms) surround the digestive tract. One such type of coelomic architecture is displayed by annelids — the familiar metamerous (or eumetamerous) body plan of serially repeated coelomic modules, most containing a number of paired organs separated by septa crossing the coelom and tied together by short bundles of muscles in the body wall. A strong case has been made that this segmented body plan is adaptive to prolonged peristaltic burrowing (Clark, 1964) and would therefore have been favoured in active infaunal lineages such as detritus feeders or predators. The segmentation acts to isolate the high coelomic fluid pressures generated during a peristaltic contraction promoting locomotory efficiency. Among annelid lineages that have become sedentary, such as suspension feeders, some septa are reduced and may have become obsolete since they are no longer required to isolate segments during prolonged burrowing. The internal repetition of organs is retained, however.

Still another coelomic body plan involves coelomic partitioning into only two or three distinctive regions, a trunk region and a tentacular crown, separated by a septum. This condition is common to the phoronids, brachiopods and ectoprocts. Each coelomic region is charged with a separate function. The trunk region presumably burrowed peristaltically (as in phoronids at present) but not for sustained periods, since it is unsegmented. This suggests a suspension-feeding or other sessile life mode. The tentacular coelom, antagonized by tentacle wall muscles, operated a lophophore in feeding. In brachiopods and ectoprocts the functional importance of the trunk coelom is reduced.

In fact a tri-regionated coelomic plan is displayed by the deuterostome phyla. Presumably the trunk coelom was employed primitively for burrowing and a tentacular coelom for feeding. A third, intermediate compartment may have been employed in locomotion within the burrow (Clark, 1964) or outside the burrow for short distances. In various deuterostomes the coelomic spaces have been much modified, presumably as required to serve many derived modes of life. Body plans with two or three longitudinal coelomic regions are termed oligomerous.

A final major style of coelomic architecture, amerous, is simply a longitudinally undivided tube without any associated seriation of organs. Sipunculids are amerous; they possess introverts for burrowing and feeding. Echiuroids resemble the amerous condition but display seriation during ontogeny and are probably directly derived from an annelidan or at least a metamerous stock.

Phyla which share a common coelomic architecture may be classed in a taxonomic category above that of the phylum; the superphylum is sometimes used and seems appropriate. If we assume that each superphylum is monophyletic then each must possess an ancestral stem which gave rise to the modern phyla via radiation and modification. Alternatively, a stem group itself may have survived and be represented among living phyla. A model of coelomate superphyla is depicted in Fig. 2.

The phyla which have radiated from the coelomate superphyla trunk stocks are diverse in habit of life and include all the well-skeletonized metazoans except for cnidarians. The primitive skeletonized forms seem to have one striking feature in common; they were all adapted to epifaunal habitats (Valentine, 1973a, 1975), with possible minor exceptions. Molluscs can best be interpreted as monophyletic, radiating from a monoplacophoran-like stem into the living classes, and probably into some extinct ones as well. Whether or not an infaunal coelomate worm preceded the epifaunal monoplacophoran is a key question; if so, then molluscs may be monophyletic with other coelomates.

The primitive metamerous stem presumably evolved as a burrowing and rather vagile infaunal worm (Clark, 1964) descended from a seriated protocoelomate and resembling an archetypal protoannelid. The Arthropoda, however, while evolving from annelidan or protoannelidan ancestors, developed as epifaunal animals with jointed appendages for locomotion in co-adaptation

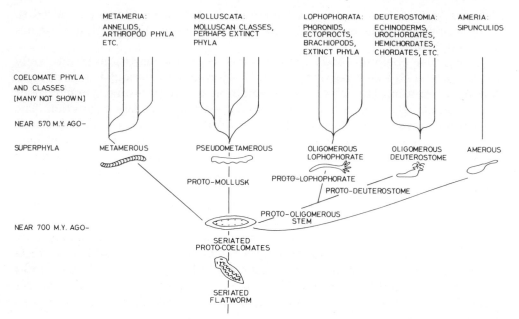

Fig. 2. Major coelomate superphyla, with radiations into phyla or classes sketchily indicated. The phylogenetic model is based chiefly on coelomic architecture. (After Valentine, 1973a.)

with the requisite integumental skeleton. Arthropodization probably occurred more than once from non-arthropod worm stocks (Teigs and Manton, 1958) so that the phylum should eventually be split up into monophyletic groupings.

The primitive oligomerous stems may have evolved as sedentary infaunal worms, but their skeletonized Cambrian descendants were, again, primarily and probably primitively epifaunal. From a phoronid or proto-phoronid complex (Hyman, 1959), several lineages of epifaunal lophophorates arose to form the Lophophorata (Fig. 2). Brachiopods may well have developed more than once from non-brachiopod worm stocks (Valentine, 1975), and if so should be split into the appropriate phyla. Brachiopod skeletons appear to be chiefly adaptations to harbour the feeding apparatus and to increase homeostasis in epifaunal habitats, though linguloids, which are infaunal (perhaps secondarily so), employ their skeletons to aid in burrowing (Thayer and Steele-Petrovic, 1975). Ectoprocts, which are not known to have skeletonized until the Ordovician, are also epifaunal and seem best interpreted as descendants of small epifaunal phoronid-like worms which adopted a colonial strategy (Farmer, Valentine and Cowen, 1973).

The Deuterostomia may have arisen from a proto-phoronid plexus which gave rise to the lophophorate phyla. If not, there does not appear to be any living analogue of the proto-deuterostome stem. At any rate the deuterostome invertebrate phylum with mineralized skeletons — the Echinodermata — appears to have been primitively epifaunal. The hemichordates, some of which have a good fossil record, were epibenthonic when they first appear (Rickards, 1975). The early chordates may have been pelagic, at least in part; their later mineralized dermal skeletons seem to be associated chiefly with protection and their still later mineralized axial skeletons with locomotion. Finally, the amerous worms did not produce any known skeletonized descendants, perhaps because they employ extensive introverts and require flexible body walls.

Among the coelomates, then, the Mollusca are most similar to a presumed flatworm ancestor, and the amerous ground plan of the sipunculids is farthest removed (Hadzi, 1963). If we base a coelomate phylogeny upon resemblances among the inferred primitive superphyla stocks, then we obtain a pattern resembling Fig. 2. In this model the distinctive deuterostome ontogenetic features arose partly in a generalized oligomerous ancestor, and partly within the primitive deuterostome worm lineage that became the trunk stock of living deuterostome phyla. There are numerous adaptive "peaks" to this phylogenetic tree, rather than two or three. Some other phylogenetic models are certainly possible, but some are most likely if we accept reasonable interpretations of the fossil record of adaptations. For example, if we accept that the eucoelom originated as an adaptation for burrowing, then some burrowing eucoelomates must have preceded epifaunal ones. As we have seen the fossil record does contain burrows that appear earlier than epifaunal, skeletonized metazoan fossils. To be sure the early burrows may have been formed in part by acoelomates or pseudocoelomates, but eucoelomates may be presumed to have been involved as well. The phylogenetic model of Jägersten (Fig. 1D) specifies that epifaunal eucoelomates such as articulate brachiopods and others arose before their infaunal relatives. This seems very doubtful indeed.

The Timing of the Evolution of Body Plans

There has certainly been controversy concerning the significance of the relatively sudden appearance of many living phyla, concentrated as it is around the beginning of Cambrian time. Some authors have supposed that the skeletonized phyla pre-existed for a long period as soft-bodied groups, and that the Early Cambrian radiation represents merely the acquisition of mineralized skeletons by these groups. For some this cannot be the case (Cloud, 1949, 1968) for their skeletons form critical components of their body plans. The brachiopods form the best example; an unskeletonized brachiopod would have to have such a different body plan that it would not be classed as a brachiopod (Cloud, 1949). The arthropod body plan also requires an exoskeleton, but since they are not normally mineralized heavily their fossilizability is doubtful and the significance of their early record is equivocal. It is also likely that primitive molluscs evolved dorsal skeletal caps in co-adaptation with pedal locomotion in a molluscan body plan, although a naked primitive mollusc can be imagined.

How can we best interpret all the evidence on the timing of evolution of the metazoan body plans? It is useful to consider two contrasting possibilities, one involving the earliest reasonable times of origin and a rather gradualistic development, the other involving the latest reasonable times and requiring rather sudden evolutionary pulses or explosions (Fig. 3).

In a model based on the longest reasonable time for the evolution (Fig. 3A), the ancestral metazoan may have appeared at any time after the development of appropriate eukaryotic ancestors, possibly well over a billion years ago. Cnidarians, Platyhelminthes and some other acoelomate phyla evolved during the following several hundred million years, in a gradual fashion, until the coelomates finally developed about 800 m.y. ago or more. These then gradually diversified into the living phyla over the next 300 m.y. or so.

In the other model (Fig. 3B), the major advances in metazoan evolution appear as a series of evolutionary bursts. The metazoans originated less than 800 m.y. ago, conceivably after 750 m.y. ago, and quickly diversified into primitive cnidarians, flatworms and other non-coelomates. The coelomates than arose considerably less then 700 m.y. ago and diversified into superphyla. Finally the living phyla arose, chiefly between about 600 and 550 m.y. ago. In neither of these models is there any reason to specify precise ancestral—descendant relations among the phyla. The contrast between the soft-bodied Ediaca-

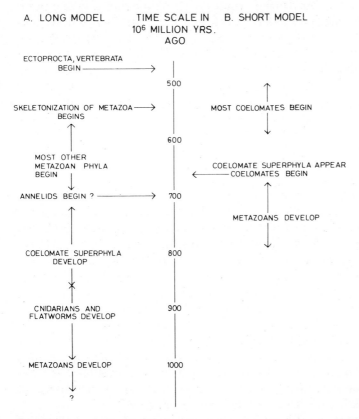

Fig. 3. Two models of the timing of appearance of major metazoan grades and ground-plans, respectively based on the longest and shortest reasonable histories in the light of the fossil record.

ran fauna, depauperate in higher metazoans, and the Burgess Shale fauna, teeming with higher metazoan phyla including novel ones, certainly suggests the second model, although this may be an artefact of preservation.

Some supporting evidence for the timing of the origin of major grades and body plans may come from molecular studies that attempt to establish molecular "clocks" by learning the rates of change in given proteins as they are traced through time. If the rates were constant and could be calibrated against well-dated evolutionary events, they could be used to date the last common ancestors of living organisms. Unfortunately the rates of evolution of proteins appear to vary through time (Fitch, 1976). When our knowledge of changes in numerous proteins becomes available, so that average rates of protein evolution can be calculated, then molecular evidence may indeed be employed to time phylogenetic events. As far as available molecular dates go, they can be interpreted as consistent with the idea of a major coelomate radiation in the late Precambrian, but should probably not be given stronger weight at present.

Models of Evolutionary Explosions

Adaptive models

Episodically during the Phanerozoic, the appearances and diversifications of large numbers of animal groups are clustered in time. These episodes are of two types or, more usually, are compounded of two elements. One is the increase in diversity at lower taxonomic levels, such as of species and genera, which involves a multiplication of lineages within body plans that are already extant. The other is the increase in the diversity of major body plans themselves. It is this second type that is considered in this section; the first type is considered later. The late Precambrian—Cambrian diversification is certainly the most spectacular example of the second type, even if the shortened model of evolutionary timing is too extreme. It has proved difficult to explain the processes underlying such rapid evolutionary bursts at high taxonomic levels.

An ecological—evolutionary hypothesis involving the invasion of new adaptive zones has been put forward to account for the sudden rise of novel body plans (Simpson, 1944, 1953). This hypothesis has proved to have much explanatory power and it has been widely accepted. Simpson observes that the environment contains many natural ecological barriers — the land-sea boundary is an extreme example — which pose difficult adaptive problems to any lineage attempting to evolve from life on one side to life on the other. The barriers may be thought of as thresholds separating two adaptive zones (Wright, 1932). For any given threshold it happens that from time to time lineages in one zone may evolve adaptations that, although selected as solutions to environmental problems in their own zone, happen by chance to prove useful in invading the other zone. These adaptations are pre-adaptations. Lineages which then proceed across the adaptive threshold to breach an ecological barrier may find little or no competition or predation awaiting them, and may manage to inhabit the new zone and endure even though they are in a low state of adaptedness that would ensure their extinction in their original zone. In time, selection co-adapts most of their characters to the key pre-adaptive features and to the new ecological requirements (post-adaptations). While crossing the adaptive threshold, the population size of the invading lineage may be small indeed (perhaps only tens or hundreds) and evolution rapid. Only after post-adaptive changes would the population ordinarily become large. At this time also, radiation may occur to exploit the potentialities of the novel adaptive type in an environment that is not heavily tenanted with competitors. With an increase in population size and with radiation, the new type is more likely to enter the fossil record than it would have done previously; it might well appear suddenly, already more or less distinctive and even well diversified.

Genetic models

This adaptive model, calling on short periods of high evolutionary rates, implies that many of the sudden appearances of diverse evolutionary novelties represent real evolutionary bursts. Neo-Darwinian evolution, however, is thought to depend upon the accumulation of small genetic changes and the recombination of these micromutations which may, under the influence of selection, create new types of organisms. Can this process operate rapidly enough to effect the evolution of whole new grades of animal organization in a relatively short time? Some workers have had their doubts, and some, such as Goldschmidt and Schindewolf, have postulated other evolutionary mechanisms. But the processes they have postulated have not been confirmed either in the laboratory or in theory. More recently, the rise of molecular genetics has focused much interest on the mechanisms of gene regulation, and it now seems likely that evolution of the regulatory apparatus of the genome may be responsible for the rise of truly novel organisms (Britten and Davidson, 1971) within relatively short periods of time (Valentine and Campbell, 1975).

Some of the regulatory systems in prokaryotic cells have been worked out, but mechanisms of the eukaryotic regulatory apparatus are not yet understood. Nevertheless it is clear that many developmental and evolutionary effects can largely be attributed to the regulatory as opposed to the structural gene complement. Structural genes are transcribed and then translated to produce polypeptides, forming proteins that act in development and metabolism. Regulatory genes control the activities of the structural genes. Their activities are made particularly clear by considering the large variety of cell types — perhaps 150 to 200 major types — present in an individual higher metazoan. Each of the cells has exactly the same genotype (with unimportant exceptions and neglecting gametes). Yet the cell types are very different in form and function, ranging from muscle cells to nerve cells, for example. This cellular differentiation must be based upon the activities of the regulatory apparatus, which must activate different groups and sequences of structural genes during the formation of different types of cells.

Examples of the effects of the evolution of the regulatory apparatus are most obvious in the development of different grades of complexity among groups of metazoan phyla. Early multicellular organisms required a more elaborate regulatory system than unicells, since genetic information is required to specify the cell interrelations. As cell differentiation appeared, more than one cell ontogeny had to be encoded within the same genome. Organs require still more genetic information to form and control the number and pattern of cells in each of their tissues, and the organs are themselves integrated into systems and into the whole organism, the form and function of which thus demands harmonious integration of myriad developmental and metabolic pathways at several levels. As evolution has developed each new grade of organization, or more generally as organisms have become more complicated, the functions of the regulatory genome have been enlarged.

Britten and Davidson (1971) have proposed a hierarchical regulatory model which contains those features essential to fulfil the requirements of the genetic regulatory apparatus in metazoans. As they point out, this particular model has not been verified in detail and will probably require some modification as new data appear, but it serves admirably to provide insight into the evolutionary significance of the regulatory genome. In their model (Fig. 4) the structural genes are activated by *receptors;* these may be visualized as DNA segments contiguous to the structural genes. The receptors are switched on by products of *integrator* genes; these products may be RNA transcriptions of the integrators. The integrators are themselves controlled by sensor genes which detect the need for certain gene products to meet some developmental or metabolic demand. Thus at some signal (perhaps carried by a hormone or protein) the sensors may cause a whole suite of structural genes to transcribe; their products would presumably be employed in integrated activity along some biochemical pathway or cycle. Some structural gene products are required in many biochemical reactions; the relevant structural genes would therefore be included in many polygene batteries (as the gene S_1 in Fig. 4). Batteries would not be mutually exclusive but would incorporate combinations of structural genes appropriate to their functions. Gene batteries could themselves be integrated by master sensors, and suites of gene batteries by even higher-level regulatory loci. The gene products of a given battery, or their derivatives, may often serve as stimuli for the sensors of other batteries. The workings of such integrated, cascading hierarchies of gene suites could be controlled so as to provide the proper combinations of enzymes and other materials in adequate and appropriate amounts to create and operate a harmoniously functioning multicellular organism. The form of the organism, particularly of its basic body plan, depends on the regulatory pattern far more than on the nature of the alleles found at the various loci, although these may also often be important.

Consider the results of mutations affecting the function of the regulatory units — sensors, integrators and receptors. Mutations at sites affecting the receptor function could have several consequences: they could remove an associated structural gene from some of the batteries to which it has belonged, or add it to others, or cause it to become silent. Mutations affecting the integrator function might change the composition of a gene battery to either add or elimi-

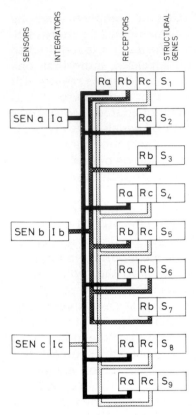

Fig. 4. The Britten—Davidson model of gene regulation to illustrate a hierarchical regulatory apparatus and non-exclusive, polygene structural gene batteries. See text for explanation.

nate one or many structural genes, or it might deactivate the battery entirely, or activate a previously silent battery. The effects of mutations to a sensor could range from switching of its gene battery from one ontogenetic or metabolic milieu to another (by responding to a different activating substance), or to activating or deactivating the battery. Mutations to master sensors that are progressively higher in the regulatory hierarchy would enable the timing and association of the products of vast gene batteries to be altered into novel patterns. Thus, even without any changes in the frequency of structural genes, major alterations in biochemical pathways and fundamental changes in the organization of developmental processes could arise. Since mutations that have large phenotypic effects are unlikely to increase fitness, the evolution of the regulatory structure, like that of structural gene frequencies, must take place in many small steps.

Still other patterns of gene regulation probably occur, and indeed the Britten—Davidson model was itself presented in alternative versions. Nevertheless, for the interpretation of major evolutionary patterns in the fossil record, the important point is that fundamental changes in body plans are likely to arise from changes in the regulatory genome.

Different classes of evolutionary change within the genome have different potentials for evolving novel body plans. One class is genome growth, which involves adding new DNA to the regulatory apparatus (and adding new structural genes as well). A second is genome repatterning, brought about by the reconstitution of batteries of genes and suites of such batteries, without enlarging the genome. A third is composed of the structural gene mutations and frequency changes that form the classic processes of micro-evolution. All three classes must be operating during the production of novel body plans, and indeed all three may operate during speciation. It is expected, however, that genome growth will predominate during the evolution of new grades or levels of complexity; that genome repatterning will predominate during the evolution of new adaptive types at a given grade, as when a phylum radiates into a number of primitive classes; and that structural gene evolution may predominate or at least be especially important during phyletic evolution within species lineages, fine-tuning the adaptive potential of the species to the ambient environmental regime (Valentine and Campbell, 1975).

If we look upon the appearance of animal superphyla and phyla in the light of its implications for the evolution of genetic regulatory systems, then a genetic model can be developed that renders plausible the rapid evolution of these new body plans. This is partly because the achievement of a new body plan is basically an addition to an old one. To achieve multicellularity does not require the re-invention of cells, but merely the use of cells to exploit their properties in aggregation. The development of the coelomates requires, not the invention of a new body-plan from the ground up, but (probably) the addition of a mesodermal body space to a flatworm-like plan that already existed. Accordingly, these developments involved, first of all, a growth in the regulatory genome, but the old structural genes and the old regulatory patterns could continue to function.

Primitive coelomates radiated into a number of coelomic plans, each trend representing an adaptive pathway leading to some different mode of life accessible to coelomate worms, as already discussed. Among the major genetic changes accompanying this radiation must have been the modification of growth gradients and patterns and the re-ordering of organ systems to create new geometries. In this case the early changes would involve the repatterning of the regulatory genomes. Co-adaptive additions and elaborations of structures that involved genome growth should be post-adaptive at least in large part. Thus, the radiation of coelomic plans could occur rapidly, involving chiefly the repatterning of the activities of gene batteries that were already formed and proven. The appearance of most epifaunal phyla probably signals another important episode of genome growth.

Another reason that the evolution of new ground plans appears to have been rapid is simply that taxonomists place them in different higher taxa. It is not clear, however, that the original differentiation of coelomate superphyla involved more evolutionary activity, when measured by some index of genome change, than did the development of the molluscan radula, or the hominid brain, or other features that are of less importance as systematic characters.

Thus it appears that the evolutionary bursts postulated during the Late Precambrian and the Cambrian do not require unbelievably rapid evolutionary rates. Furthermore, the staging of new body plans in a hierarchical sequence of adaptive novelties creates morphological distance in a very efficient manner. By the time epifaunal metazoan phyla had appeared, the similarities among their stem superphyla were masked by the next stage of modification, which naturally exploited and therefore emphasized such architectural differences as did exist. Extinction of the early coelomate lineages and of many of the soft-bodied precursors of the skeletonized phyla, coupled with their rarity or absence as fossils, has left us with a cryptogenetic record of distinctive modern phyla.

Evolutionary bursts

A more or less simultaneous rise of the coelomate superphyla, required by the shorter model of evolutionary timing, is in fact expected if they all originated in an adaptive radiation, in the classic sense, from a protocoelomate ancestor. The development of epifaunal phyla from several different superphyla during a short time interval cannot be explained in this manner, however. A large number of theories have been put forward to account for the more or less contemporaneous appearance of numerous phyla; space does not permit an extensive review here. Two of these theories that are commonly cited may well play important roles in the explosion but do not seem likely to have been decisive. One, argued by Nicol (1966) and previous writers, suggests that phyletic size increases brought about the need for skeletons. Size may well be among the factors that determined the early skeletonization of some lineages, such as brachiopods as opposed to ectoprocts, and may have contributed to the timing of skeletonization in other groups also; but there are arguments against this factor as a general explanation. It does not account for the correlation of the timing in separate superphyla, for example. Furthermore, Precambrian fossils suggest that metazoans had generally attained the sizes of early skeletonized forms well before the Cambrian.

A second suggestion by Hutchinson (1958) and others is that the origin and increase of predation created selection pressure for protective skeletons. It is certain that skeletons are employed widely for protection, and selection for this function would certainly influence their form. Probably we shall never be able to estimate closely the relative importance of such selection pressure. Perhaps the question to be asked in this regard is whether skeletonization would have occurred as epifaunal communities proliferated, even in the absence of predation pressure. It is my opinion that it would, but there is no certain answer at present.

If it was the development of taxa to function in epifaunal communities that caused skeletonization, then how can we account for the timing of this event? Two suggestions are current. One is that in the latest Precambrian, the concentration of free atmospheric oxygen (and therefore dissolved oxygen in the sea) reached a level capable of sustaining the high oxygen-demand functions required for mineralized skeletonization (Berkner and Marshall, 1964; Rhoads and Morse, 1971). The resulting opportunity for skeletonization may have created the possibility of epifaunal radiations in several superphyla. Another suggestion is that a stabilization of trophic resources permitted the elaboration of epifaunal diversity, and that skeletonization permitted the exploitation of such adaptive zones within this increasingly available environment (Valentine, 1973a,b). These suggestions are not necessarily mutually exclusive, nor does either of them entirely exclude the previous ideas, but we cannot assess their relative importance at present.

Phanerozoic Metazoan Patterns

Even after the metazoan phyla were well established, there were evolutionary events that affected the Metazoa as a whole. The more spectacular of these, after the diversification of the Cambrian fauna, are associated with episodic waves of appearances and disappearances of taxa in the fossil record. Although these have sometimes been interpreted as artefacts of a spotty fossil record, it seems likely that they are real episodes of diversification and extinction, since descendants of the taxa that may disappear are not found, and there is no obvious reason for the clustering of appearances not to be real, though the actual timing of all such events is certainly blurred in the record. For the well-skeletonized families of benthonic invertebrates, the changing rates of Phanerozoic appearances are indicated in Fig. 5. Similar figures based on different combinations of taxa can be found in Newell (1967 and references therein) and in Flessa and Imbrie (1973).

Extinctions

There are two fundamentally different approaches to the study of extinction. One is to search for the lethal weapons that caused mortalities or restricted birth rates and thus led directly to extinction. The other seeks the causes that underlay the ability of some such weapons, which are always present, to promote extinctions. The potential number of lethal weapons of extinction is enormous, and most possible causes of mortality have been invoked. Warming or cooling of the climate, salinity changes, oxygen depletion, habitat failure, competition, predation, disease, famine, cosmic radiation, natural pollution and poisoning of certain oceanic regions — even, at one time, old age — have been suggested as extinction causes, in various contexts and combinations.

An obvious result of extinction is to lower the number of living species below the level that would have occurred had the extinction not happened. Ecological theorists have considered at length the problems of the control of species diversity in modern ecosystems. Processes of speciation pervade the bio-

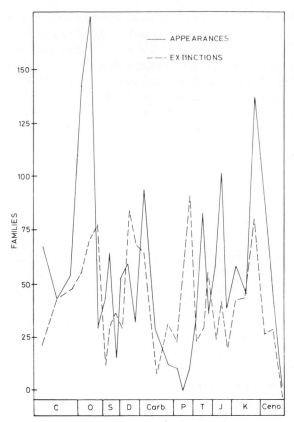

Fig. 5. Appearances and disappearances of families of well-skeletonized marine benthic taxa during the Phanerozoic, plotted by System (Valentine, 1973b).

sphere with a powerful machinery that is capable of creating vast arrays of new species at a high rate. If all the potential new forms survived, diversity would soon increase many times over. There are two principal ways in which diversity can be kept at some level; through suppression of speciation, and through extinction. Most of the proposed mechanisms of diversity regulation operate to suppress speciation. They assume that the biosphere can be filled up with species, and that when this limiting capacity is reached, the formation of new species is suppressed for as long as the relevant environmental conditions remain unchanged.

There are two classes of change that could upset such an equilibrium and lead to extinctions. One class involves environmental changes that do not affect the species capacity of the environment; these are diversity-independent changes (Valentine, 1973b). The extent of the extinctions that follow such changes does not depend upon the number of species within the ecosystems, but simply upon the number of species that happen to be intolerant of the change. As soon as such extinctions begin, the species-making machinery is free to evolve forms that are adapted to the new conditions, and thus to return the diversity to its equilibrium level. The other class of changes lowers the species capacity of the environment; it is diversity-dependent. Extinctions must follow because the environment has lost the ability to support all the living species. Factors that may lower the diversity capacity probably include the lessening of spatial heterogeneity (and thus a reduction in the opportunities for species to partition habitat resources spatially) and an increase in the seasonality of primary productivity or other trophic resources (and thus a reduction in the opportunity for species to partition energy supplies). The proximal causes of extinction that follow a reduction in diversity capacity might be rather varied but would chiefly be related to stresses, commonly competitive, associated with the increased difficulties of exploiting limited resources (space and food, primarily).

There is a great diversity of marine species living at present, and they are accommodated on two separate ecological levels. The first is the provincial level. Over thirty marine provinces exist on the present shelves, and each contains large numbers of endemic species. Additionally, many species range through more than one but not through many provinces. Thus, provinces represent a partitioning of environmental resources among species on a broad geographic scale. If provinciality were reduced, the numbers of species that could be accommodated by the partitioning would be reduced. The main factors that underlie provincial partitioning at present, and that serve to maintain provincial boundaries, are climatic barriers and geographic barriers. Marine species are nearly all poikilotherms, and thus their distributions are particularly sensitive to water-temperature regimes. If the climate changed so as to lower the temperature gradients in the seas, provinciality would decrease and diversity would fall. Furthermore, marine shelf species cannot often propagate across deep-sea barriers of any width, and are of course barred by land. If the topographic barriers were reduced — if seas overran land barriers or if some continental shelves now separated by deep seas were united — then provinciality would decrease also and diversity would fall. There is in fact good evidence that the patterns of provincial barriers were very different in the past. Temperature gradients often appear to have been far more gentle than today, and continents have been united or fragmented and ocean basins have been closed and opened by the processes of plate tectonics. It is therefore to be expected that these events would be a major cause of extinctions (Valentine and Moores, 1970).

A second level of accommodation of species diversity is in communities. Living communities with similar habitat ranges, but developed in different regions, may contain very different numbers of species. Arctic communities are less diverse than analogous temperate communities, which are in turn less diverse than analogous tropical communities. The causes of these differences are still in dispute, but they seem to be related to latitudinal variations in energy supply, perhaps to its seasonality. (For reviews of other hypotheses see Pianka, 1966, and Valentine, 1973b.) At any rate, if a larger proportion of provinces were brought under a regime that lowered species numbers within communities, diversity would obviously fall, requiring extinctions. Since there are many times in the past when fossil associations appear to have been less diverse than are living communities in comparable environments, this type of diversity accommodation may have been an important one at times. Diversity capacities lower than today's may have been caused by the presence of continents in higher average latitudes, their enclosure of small oceans, their inclusion in larger land masses, and their exposure by low sea levels. Any of these factors could serve to increase the temporal variability, especially the seasonality, of marine shelf environments above that of today.

Among the more interesting corollaries of plate-tectonic processes and of the sweeping changes in land—sea geography that accompany them are the varied and extensive changes that must have been created in the biotic environment. A better understanding of these changes and of their consequences in biotic history is urgently required, because they lead to just those conditions that are necessary to explain the sorts of continuous evolutionary activity, episodic in rate and various in outcome, displayed in the fossil record. Evolution must act continually to adapt the world's biota to ambient environmental conditions, but these are continually changing, first slightly, then greatly as both diversity-dependent and diversity-independent factors are altered. Explanations for extinctions and diversifications must take the ecological and taxonomic character of these events into account, and place them in the context of environmental changes that can be inferred from the physical evidence of the earth's history.

The Palaeozoic waves of extinction are not all equally well known; the Late Permian one has received the most attention. It certainly corresponds to a reduction in marine provinciality; there must have been about six or more Permian marine provinces (see Yancey, 1975), while in the Early Triassic (and throughout most of the Lower Mesozoic) there is evidently only a single one. Considerable species extinction had to accompany this deprovincialization. Detailed comparisons between Triassic and Permian communities have not yet been made, but their relative species richnesses may roughly be judged from systematic accounts of their faunas. Lower Triassic communities appear to have been especially depauperate in species, while Permian communities were fairly rich, especially in reef complexes where skeletonized species attained diversities comparable to or exceeding those of some living reefs. The disappearance of the reefs in the Late Permian must have swept away a significant proportion of Permian species. This event was presumably related to the lowering of the sea level and the associated rise in continentality that occurred near the Permian—Triassic boundary (see Valentine and Moores, 1972; Schopf, 1974). Thus the Permian extinctions seem to have involved a large diversity-dependent component, and indeed diversity seems to have remained low for tens of millions of years.

Earlier Palaeozoic extinction peaks are also associated with reductions in provinciality, and often with lowered sea levels (see the curves in Wise, 1974). The diversity changes within analogous communities at those times have not been studied, and they are not as obvious as those across the Permian—Triassic boundary. It is therefore difficult to assess the community-level contributions to these extinctions. The Late Cretaceous extinction seems not to be associated with any reduction in provinces at all, but there does seem to have been a short-lived lowering of sea level. Diversity may have decreased on the community level. Whether this was a cause or effect of the extinctions is not clear. Perhaps these extinctions were caused primarily by diversity-independent factors, such as climatic change, with extinct lineages tending to be replaced rather rapidly.

An interesting problem has been illuminated by Van Valen (1973), who made the observation that the duration frequencies of taxa, when plotted as cumulative curves on a log scale have a nearly straight-line relationship with time. Two examples are shown in Fig. 6. Not all taxonomic groups display a straight-line relationship, but many at least come close. In demography, such a straight survivorship curve indicates that an individual's chances of survival are independent of age — a young or an old individual have about the same chances of dying in a given interval (except at the extreme old-age end of the curve). The mortality rate for a cohort with this pattern of survivorship is essentially constant.

For the taxonomic survivorship curves such as Van Valen has constructed, a straight-line relationship suggests that the taxa became extinct independently of their ages; that recently evolved taxa and ancient taxa had the same chances of dying out within a given time interval. However, the curve does not necessarily mean that extinctions were stochastically constant, because the taxa do not form a cohort. To learn whether extinctions are stochastically constant requires a different analysis. The apparent correlation of extinction waves with key environmental events implies that they are deterministic. Even if extinctions were so inconstant that there was only one single extinction episode, a straight survivorship curve would be created as long as old and new lineages were carried off indiscriminately.

Such a result is not surprising in retrospect. It means that evolution has no foresight, that environmental changes are not anticipated by selection, and that

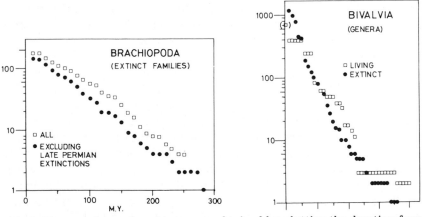

Fig. 6. Taxonomic survivorship curves obtained by plotting the duration frequencies of taxa on a log scale against time. (After Van Valen, 1973.)

when they occur they create problems for new and old lineages alike. Lineages do not "learn" to survive with time. As the environment changes, it tends to deteriorate for most taxa (Van Valen, 1973) irrespective of their ages. This is so plausible a consequence of Darwinian evolution that the problem now shifts to explaining the many taxonomic survivorship curves that are *not* straight lines. Actually, this may not prove too difficult, though the problem has by no means been solved. If bursts of diversification occur episodically through time, for whatever reason, then the standing crop of taxa at a given moment will have neither an even nor a random distribution of ages, but will contain large numbers of taxa of certain age classes — sorts of taxonomic cohorts that all originated during the same diversification peaks. Now if an extinction wave occurs, it will naturally carry off more taxa in the large age classes (in proportion to their sizes). The resulting survivorship curves will therefore contain steps at the duration represented by the larger age classes and thus may depart significantly from linearity.

Diversifications

The diversification peaks depicted in Fig. 5 are composed of representatives of numbers of phyla, and can be explained in very general terms. Evidently they are not all due to similar causes. The great Ordovician peak of diversification probably represents the first thorough exploitation of marine habitats by the coelomate phyla, especially by suspension-feeding lineages. Although provinciality seems to have been on the increase in Cambro-Ordovician times, it is evidently a rise in diversity within ecosystems that was chiefly associated with this wave of diversification. Studies of Cambrian communities are badly needed.

The next three peaks of diversification, in the Silurian, Devonian and Carboniferous, follow upon extinction peaks, and seem to represent the re-establishment of high levels of diversity following depletions in the marine biota. It is likely that the new taxa chiefly indicate the appearance of new provincial regions and regimes. The Mesozoic embraces three more peaks of diversification. The first is clearly a recovery of the marine fauna from the depauperation of Permian—Triassic extinctions after a long period when diversification was effectively suppressed (Fig. 5). Since provinciality is not on the increase, this Triassic diversification must have been accompanied by an enriching of communities, and indeed Upper Triassic fossil assemblages do appear to be richer than earlier Triassic ones, though again we do not have comparative community studies for these times as yet. The Jurassic peak, however, is probably due at least in part to the beginnings of the provincialization that was to characterize the Cretaceous and Cenozoic. Indeed the peak Upper Cretaceous levels, and the high levels during the early Cenozoic, must represent the unprecedented partitioning of the world's shelves into marine provinces. This came about as a result of the extensive and continued splitting up of Pangaea into separate continents, leading to extensive topographic barriers to faunal dispersal. These began to be significant in the Late Cretaceous. The cooling of high latitudes to create chains of latitudinal provinces within contiguous shelf regions has finally led through the Cenozoic to the high diversity levels of the present day. Whether the more diverse of living shelf communities, in tropical reef complexes, are richer in species than the reefs of, say, the Permian, is not certain, though it appears that they may be. But the important differences between Permian and Recent marine diversities are accommodated chiefly on the provincial level.

Summary of Major Metazoan Evolutionary Patterns

The patterns inferred for metazoan evolution at the largest scale, that of the origin of organizational grades and of body plans at the levels of subkingdom, superphylum and phylum, are similar to the familiar patterns described at much lower taxonomic levels, such as those interpreted among placental orders and families by Simpson (1944, 1953). Probably they have a similar significance. An evolutionary novelty is obtained by selection for fitness in the immediate environment, and then proves to be highly adaptive in many other environments or to many different modes of life. The novelty is therefore rapidly elaborated in a radiation, with each radiating clade enjoying special modifications as adaptations to the peculiarities of the environment and life mode that it follows. It is not at all clear that the initial radiations of taxa at high levels involved more evolutionary activity within the genomes of the lineages involved than have some radiations at lower levels. The morphological distance among the primitive lower metazoan ancestors of living phyla may not have been very great, and similarly the early coelomate worms may have differentiated in coelomic regionation but they retained many common features of tissue or organ structure, organized eventually into distinct anatomical plans.

Although the mechanisms of gene regulation have not yet been clarified in eukaryotes, it is of heuristic value to consider the various possible effects of regulatory evolution. Distinctive anatomical organization can be attained through additions to and modifications of developmental pathways via evolution of the regulatory genome, and this provides insight into the competence of evolution to rapidly produce novel body plans and adaptive modifications on given anatomical themes. The coelomate superphyla may be regarded as modifications of a basic coelomate theme, and the cluster of phyla and/or classes developed from each of these may each be regarded as modifications of the various superphyla themes. Since the clusters of phyla are by and large exploiting the potentials of their distinctive superphylum themes, the radiations of phyla and classes tend to emphasize the distinctiveness of the superphyla, functionally and morphologically.

By the time that fossils are common in the record, their radiations at higher levels are in the process of completion. The radiations appear to be adaptive responses to rather pervasive changes in the environment which provoke the development of those stem novelties from which radiations then depart. Such environmental parameters as oxygen concentration levels and regimes of primary productivity may be closely involved; these provide the major energy sources for animal life.

Throughout the Phanerozoic and doubtless before, environmental change must have been incessant though varying in rate and effect. In plate-tectonic processes alone we have powerful agents of environmental modification. The biosphere responded to changed conditions in both composition and structure. Species were multiplied when environmental partitioning was possible or necessary, and were extinguished when their adaptive zones or biospaces deteriorated. Some changes favoured lineages with certain body plans or life modes, and this altered the relative importance of higher taxa, sometimes leading to their extinction.

As extinction piled upon extinction, more and more of the earliest higher taxa were eliminated. Although the numbers of skeletonized phyla seem to have been fairly constant (Fig. 7), recent work on soft-bodied assemblages has revealed that previously unknown phyla have become extinct, though we are not yet sure of even the era in which such extinctions occurred. For classes and

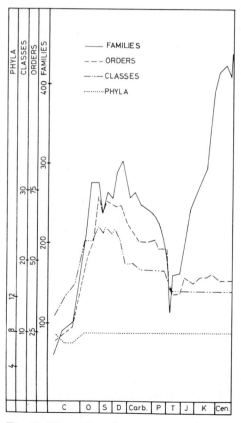

Fig. 7. Diversities of taxa in major categories during the Phanerozoic Systems. (After Valentine, 1973c.)

orders with mineralized skeletons, earlier Palaeozoic diversities were also greater than at present (Fig. 7), and we can trace the diminution of their numbers during the Middle and Late Palaeozoic. Today, classes remain at about their Late Palaeozoic level, but orders have recovered somewhat from that period. Families, however, have now surpassed their highest former levels that are recorded. Presumably genera and species have been represented by respectively higher levels of diversity in the Cenozoic, relative to the Palaeozoic. This pattern follows as a consequence of the patterns of extinction and radiation. Even though some environments and life modes were affected more than others by a given extinction wave, according to the underlying pattern of change, the species involved would not be entirely extirpated. When rediversification began, then, it was not necessary to invent (or re-invent) body plans; the surviving lineages, representing fewer high taxa than formerly, were nevertheless the natural sources of new taxa, modified so as to fit the requirements. Indeed after the Cambrian, natural selection could not easily develop a novel body plan, unless a wide adaptive zone (such as the terrestrial environment) was available for colonization. As phyla and classes passed from the biosphere they were replaced, not with architecturally novel taxa at their same distinctive levels, but with modifications of surviving architectural themes which are now chiefly recognized, and correctly so, as only new orders or families.

There is no reason to believe that this pattern of evolution acted to optimize progress. Had the sequence of environmental changes happened differently, quite different groups of extinction survivors would probably have been present, and subsequent radiations would have occurred from stocks, some of which are now long extinct. Whether alternate biospheres would have been superior to our own is an intriguing question, though at this point we leave the domain of science.

Acknowledgements

This article was reviewed by Dr. R.B. Rickards and Simon Conway Morris, Sedgwick Museum, Cambridge, and C.A. Campbell, King's College, Cambridge, to whom I am most grateful.

References

Banks, N.L., 1970. Trace fossils from the late Precambrian and Lower Cambrian of Finnmark, Norway. In: T.P. Crimes and J.C. Harper, Trace Fossils. Geol. J., Spec. Iss., 3: 19—34.

Berkner, C.V. and Marshall, L.C., 1964. The history of growth of oxygen in the earth's atmosphere. In: C.J. Brancuzio and A.G.W. Cameron (Editors), The Origin and Evolution of Atmospheres and Oceans. Wiley, New York, N.Y., pp. 102—126.

Bockelie, T. and Fortey, R.A., 1976. An early Ordovician vertebrate. Nature, 260: 36—38.

Boucot, A.J., 1975. Evolution and Extinction Rate Controls. Elsevier, Amsterdam, 425 pp.

Britten, R.J. and Davidson, E.H., 1971. Repetitive and non-repetitive DNA sequences and a speculation on the origins of evolutionary novelty. Q. Rev. Biol., 46: 111—133.

Clark, R.B., 1964. Dynamics in Metazoan Evolution, the Origin of the Coelom and Segments. Clarendon Press, Oxford, 313 pp.

Cloud, P., 1949. Some problems and patterns of evolution exemplified by fossil invertebrates. Evolution, 2: 322—350.

Cloud, P., 1968. Pre-metazoan evolution and the origin of the Metazoa. In: E.T. Drake (Editor), Evolution and Environment. Yale University Press, New Haven, Conn., pp. 1—72.

Cloud, P., 1973. Pseudofossils: a plea for caution. Geology, 1: 123—127.

Conway Morris, S., 1976a. A new Cambrian lophophorate from the Burgess shale of British Columbia. Palaeontology, 19: 199—222.

Conway Morris, S., 1976b. Worms of the Burgess Shale, Middle Cambrian, Canada. Thesis, Cambridge University, Cambridge.

Conway Morris, S., in press. The Burgess Shale. In: R.W. Fairbridge and D. Jablonski (Editors), Encyclopedia of Paleontology. Dowden, Hutchinson and Ross, Stroudsburg, Pa.

Cowie, J.W. and Spencer, A.M., 1970. Trace fossils from the late Precambrian/Lower Cambrian of East Greenland. In: T.P. Crimes and J.C. Harper, Trace Fossils, Geol. J., Spec. Iss., 3: 91—100.

Crimes, T.P., 1974. Colonisation of the early ocean floor. Nature, 248: 328—330.

Dayhoff, M.O., 1969. Atlas of Protein Sequence and Structure, 4. Natl. Biomed. Res. Found., Silver Springs, Md.

Durham, J.W., 1971. The fossil record and the origin of the Deuterostomia. N. Am. Paleontol. Conv., Chicago, Ill., 1969, Proc., H: 1104—1131.

Farmer, J.D., Valentine, J.W. and Cowen, R., 1973. Adaptive strategies leading to the ectoproct groundplan. Syst. Zool., 22: 233—239.

Fitch, W.M., 1976. Molecular evolutionary clocks. In: F.J. Ayala (Editor), Molecular Evolution. Sinauer Assoc., Sunderland, Mass., pp. 160—178.

Fitch, W. and Margoliash, E., 1970. The usefulness of amino acid and nucleotide sequences in evolutionary studies. Evol. Biol., 4: 67—109.

Flessa, K.W., and Imbrie, J., 1973. Evolutionary pulsations: evidence from Phanerozoic diversity patterns. In: D.H. Tarling and S.K. Runcorn (Editors), Continental Drift, Sea Floor Spreading and Plate Tectonics, 1. Academic Press, London, pp. 247—285.

Germs, G.J.B., 1974. The Nama Group in South West Africa and its relationship to the Pan-African geosyncline. J. Geol., 82: 301—317.

Glaessner, M.F., 1969. Trace fossils from the Precambrian and basal Cambrian. Lethaia, 2: 369—393.

Glaessner, M.F., 1971. Geographic distribution and time range of the Ediacara Precambrian fauna. Bull. Geol. Soc. Am., 82: 509—514.

Glaessner, M.F., 1972. Precambrian fossils — a progress report. Proc. Int. Paleontol. Union, Int. Geol. Congr., 23rd Sess., Prague, 1968. pp. 377—384.

Glaessner, M.F. and Daily, B., 1959. The geology and late Precambrian fauna of the Ediacaran fossil reserve. S. Aust. Mus. Rec., 13: 369—401.

Glaessner, M.F. and Wade, M., 1966. The late Precambrian fossils from Ediacara, South Australia. Palaeontology, 9: 599—628.

Hadzi, J., 1963. The Evolution of the Metazoa. Pergamon Press, Oxford, 499 pp.

Haeckel, E., 1874. The gastraea theory, the phylogenetic classification of the animal kingdom and the homology of the germ lamellae. Q. J. Microsc. Sci., N.S., 14: 142—165.

Hand, C., 1959. On the origin and phylogeny of the coelenterates. Syst. Zool., 8: 191—202.

Hughes, C.P., 1975. Redescription of Burgessia bella from Middle Cambrian Burgess shale, British Columbia. Fossils Strata, 4: 415—435.

Hutchinson, G.E., 1958. The biologist poses some problems. Am. Assoc. Adv. Sci. Publ., 67: 85—94.

Hyman, L.H., 1940. The Invertebrates: Protozoa Through Ctenophora. McGraw-Hill, New York, N.Y., 726 pp.

Hyman, L.H., 1959. The invertebrates: smaller coelomate groups. McGraw-Hill, New York, N.Y., 783 pp.

Jägersten, G., 1955. On the early phylogeny of the Metazoa. The bilaterogastraea theory. Zool. Bidr. Uppsala, 30: 321—354.

Jägersten, G., 1972. Evolution of the Metazoan Life Cycle. Academic Press, London, 269 pp.

Kerkut, G.A., 1960. Implications of Evolution. Pergamon Press, Oxford, 174 pp.

Knight-Jones, E.W. and Moyse, J., 1961. Intraspecific competition in sedentary marine animals. Symp. Soc. Exp. Biol., 15: 72—95.

Lang, A., 1884. Die Polycladen des Golfes von Neapel. Fauna Flora Golfes Neapel, Monogr. 11.

Lemche, H. and Wingstrand, K.G., 1959. The anatomy of Neopilina galathea Lemche, 1957 (Mollusca, Tryblidiacea). Galathea Rep., Copenhagen, 3: 1—71.

Marek, L. and Yochelson, E.L., 1976. Aspects of the biology of Hyolitha (Mollusca). Lethaia, 9.

Matthews, S.C. and Missarzhevsky, V.V., 1975. Small shelly fossils of late Precambrian and early Cambrian age: a review of recent work. J. Geol. Soc. Lond., 131: 289—304.

Newell, N.D., 1967. Revolutions in the history of life. Geol. Soc. Am. Spec. Pap., 89: 63—91.

Nicol, D., 1966. Cope's rule and Precambrian and Cambrian invertebrates. J. Paleontol., 40: 1397—1399.

Pianka, E.R., 1966. Latitudinal gradients in species diversity: a review of concepts. Am. Nat., 100: 33—46.

Rees, W.J., 1966. The evolution of the hydrozoa. In: W.J. Rees (Editor), The Cnidaria and Their Evolution. Symp. Zool. Soc. Lond., 16: 199—221.

Rhoads, D.C. and Morse, J.W., 1971. Evolutionary and ecologic significance of oxygen-deficient marine basins. Lethaia, 4: 413—428.

Richardson, Jr., E.S. and Johnson, R.G., 1970. The Mazon Creek faunas. N. Am. Paleont. Conv., Chicago, Ill., 1969, Proc., I: 1222—1235.

Rickards, R.B., 1975. Palaeoecology of the Graptolithina, an extinct class of the phylum Hemichordata. Biol. Rev., 50: 397—436.

Runnegar, B. and Pojeta, J., 1974. Molluscan phylogeny: the paleontological viewpoint. Science, 186: 311—317.

Runnegar, B., Pojeta, J., Morris, N.J., Taylor, J.D., Taylor, M.E. and McClung, G., 1975. Biology of the Hyolitha. Lethaia, 8: 181—191.

Salvini-Plawen, L. von, 1969. Solengastres und Caudofoveata (Mollusca, Aculifera): Organisation und phylogenetische Bedeutung. Malacologia, 9: 191—216.

Sanders, H.L., 1969. Benthic marine diversity and the stability-time hypothesis. Brookhaven Symp. Biol., 22: 71—81.

Schopf, T.J.M., 1974. Permo-Triassic extinctions: relation to sea-floor spreading. J. Geol., 82: 129—143.

Simpson, G.G., 1944. Tempo and Mode in Evolution. Columbia University Press, New York, N.Y., 237 pp.

Simpson, G.G., 1953. The Major Features of Evolution. Columbia University Press, New York, 434 pp.

Sokolov, B.S., 1972. Vendian and early Cambrian Sabelliditida (Pogonophora) of the U.S.S.R. Proc. Int. Paleontol. Union, Int. Geol. Congr., 23rd Sess., Prague, 1968, pp. 79—86.

Sprinkle, J., 1973. Morphology and evolution of blastozoan echinoderms. Spec. Publ. Mus. Comp. Zool. Harvard Univ., 283 pp.

Thayer, C.W. and Steele-Petrovic, H.M., 1975. Burrowing of the lingulid brachiopod Glottidia pyramidata: its ecological and paleoecological significance. Lethaia, 8: 209—221.

Tiegs, O.W. and Manton, S.M., 1958. The evolution of the Arthropoda. Biol. Rev., 33: 255—337.

Vagvolgyi, J., 1967. On the origin of the molluscs, the coelom, and coelomic segmentation. Syst. Zool., 16: 153—168.

Valentine, J.W., 1973a. Coelomate superphyla. Syst. Zool., 22: 97—102.

Valentine, J.W., 1973b. Evolutionary Paleoecology of the Marine Biosphere. Prentice-Hall, Englewood Cliffs, N.J., 511 pp.

Valentine, J.W., 1973c. Phanerozoic taxonomic diversity: a test of alternate models. Science, 180: 1078—1079.

Valentine, J.W., 1975. Adaptive strategy and the origin of grades and groundplans. Am. Zool., 15: 391—404.

Valentine, J.W. and Campbell, C.A., 1975. Genetic regulation and the fossil record. Am. Sci., 63: 673—680.

Valentine, J.W. and Moores, E.M., 1970. Plate-tectonic regulation of faunal diversity and sea level: a model. Nature, 228: 657—659.

Valentine, J.W. and Moores, E.M., 1972. Global tectonics and the fossil record. J. Geol., 80: 167—184.

Van Valen, L., 1973. A new evolutionary law. Evol. Theory, 1: 1—30.

Walcott, C.D., 1911. Middle Cambrian annelids. Cambrian geology and paleontology II. Smithson. Misc. Coll., 57: 109—144.

Webby, B.D., 1970. Late Precambrian trace fossils from New South Wales. Lethaia, 3: 79—109.

Whittington, H.B., 1971a. The Burgess shale: history of research and preservation of fossils. In: Extraordinary fossils. N. Am. Paleontol. Conv., Chicago, Ill., 1969, Proc., I: 1170—1201.

Whittington, H.B., 1971b. Redescription of Marrella splendens (Trilobitoidea) from the Burgess shale, Middle Cambrian, British Columbia. Bull. Geol. Surv. Can., 209: 1—24.

Whittington, H.B., 1974. Yohoia Walcott and Plenocaris n.gen., arthropods from the Burgess shale, Middle Cambrian, British Columbia. Bull. Geol. Surv. Can., 231: 1—27.

Whittington, H.B., 1975a. The enigmatic animal Opabinia regalis, Middle Cambrian, Burgess shale, British Columbia. Philos. Trans. R. Soc., B 271: 1—43.

Whittington, H.B., 1975b. Trilobites with appendages from the Middle Cambrian, Burgess shale, British Columbia. Fossils Strata, 4: 97—136.

Wise, D.U., 1974. Continental margins, freeboard and the volumes of continents and oceans through time. In: C.A. Burke and C.C. Drake (Editors), The Geology of Continental Margins. Springer-Verlag, New York, N.Y., pp. 45—58.

Wright, S., 1932. The roles of mutation, inbreeding, crossbreeding and selection in evolution. Proc. VI Int. Congr. Genetics, 1: 356—366.

Yancey, T.E., 1975. Permian marine biotic provinces in North America. J. Paleontol., 49: 758—766.

Yochelson, E.L., 1975. Discussion of early Cambrian "molluscs". J. Geol. Soc. Lond., 131: 661—662.

CHAPTER 3

STOCHASTIC MODELS IN EVOLUTIONARY PALAEONTOLOGY

DAVID M. RAUP

Introduction

Considerable efforts are now being made in palaeobiology to interpret groups of events in the fossil record in a probabilistic way instead of seeking purely causal (deterministic) explanations for specific events. This is not a fundamentally new approach to the fossil record; palaeontologists have long searched for statistical laws or generalizations that can be applied to classes of events such as adaptive radiations, periods of extinction and ontogenetic relationships. However, in the last few years, more rigour has been applied to the search for generalizations. Also, the development of computers has made possible the application of a wider variety of analytical methods.

The development of stochastic approaches to palaeobiology follows logically from advances in other areas of geology and biology. In geomorphology, Shreve (1966, 1967, 1974) and others made substantial breakthroughs in the study of drainage patterns by noting that probabilistic (= stochastic) models could be used to describe the development of real-world patterns in many cases. Krumbein (1969a) and others have successfully applied such models to sedimentary sequences. On the biological side, the use of probabilistic models in population genetics (Wright, 1964; Frazer and Burnell, 1970; Nei, 1975) and in mathematical ecology (MacArthur and Wilson, 1967) has had a powerful influence on the development of these fields. On a broader scale, various aspects of physics and chemistry were revolutionized by the successful application of probabilistic models, as exemplified by the formulation of the gas laws, quantum physics, and so on.

Evolutionary palaeontology is basically an historical science and its problems are somewhat different from those in the fields just mentioned. In the fossil record we have only an historical sequence and thus we share with historians the problem that experimentation in the usual sense is impossible. But this does not disqualify the fossil record as an area for the application of many probabilistic approaches. It does mean, however, that certain kinds of models and certain approaches are more appropriate than others in palaeontology.

In this chapter I will discuss some of the basic principles of random models and attempt to show what kinds of random processes are most likely to be represented in the evolutionary record.

Concepts of Randomness

To most people, randomness implies chaos and a process which is haphazard, ill-defined or, at least, totally disorganized. It is commonly thought that an interpretation which relies on random processes implies substantial ignorance of the underlying mechanism and is probably unscientific. But at certain levels, models based on an *assumption* of randomness can be extremely effective and anything but ill-defined or disorganized. The reviews by Anderson (1974), Krumbein (1969a, b) and Watson (1969) are particularly helpful in providing the perspective on this subject.

Radioactive decay serves as a good example of the application of randomness to a scientific problem. The example is also appropriate in a palaeobiologic context because Van Valen (1973) has discussed a situation in which extinction rates may descriptively be quite similar to radioactive decay. All the atoms in a sample of a radioactive isotope are unstable in the sense that sooner or later they "decay" to form atoms of other isotopes. There is no way of predicting when a given atom will decay, although presumably there is a cause for each decay event. It has been shown empirically that the frequency of decay does not change with time. Thus, it is possible to predict statistically what proportion of the atoms in a sample will decay in a given period of time. If the sample is large enough, this relationship can be used quite deterministically in geologic age determination. The important point is that the fate of individual atoms is not known and is never predicted. One is not concerned with when or why a particular atom decays but only with the behaviour of a group of decay events.

In population genetics, the use of the Hardy—Weinberg principle illustrates another successful application of random or stochastic models. The Hardy—Weinberg principle says simply that in the absence of mutation and selection, it can be assumed that allelic frequencies will not change significantly from generation to generation. However, this prediction is true only in a statistical sense. There is nothing in the system to prevent a disproportionate number of the carriers of one allele from dying or otherwise failing to produce offspring. In fact, in very small populations this often happens (genetic drift), but in large populations the probability of substantial departures from expectation is small and can generally be ignored. Using the Hardy—Weinberg principle, valid predictions can be developed in a deterministic way. By the same token, if a population is subject to mutation favouring one allele, it is possible to predict from the frequency of this mutation how the frequency of the favoured allele will increase over time. As in the case of radioactivity, we can do this without any real knowledge of the process of mutation and without knowing which particular individuals in the population undergo mutation. Selection in interbreeding populations is treated in the same way.

Harbaugh and Bonham-Carter (1970) define a random variable as "one that varies in an uncertain manner", meaning that it is not possible to predict with accuracy how the variable will change. By this definition, the decay of a radioactive isotope and the behaviour of allelic frequencies qualify as random variables under the Hardy—Weinberg principle.

It is debatable whether anything in our universe is truly random in the sense of having no cause. It can be argued, for example, that if the basic processes were fully understood we would find that true randomness does not exist. Lucas (1964) used the term "pseudorandom" to describe variables which *behave* in a random fashion whether or not the fundamental process is random. *We either cannot know or do not choose to know the actual deterministic basis of the process we are dealing with.* Thus, if a process is termed random it is not necessarily truly random but it satisfies a mathematical definition of random behaviour.

The implications of scale are extremely important in the present context. It is often true that a process can be described as a random process (or pseudo-random in the sense of Lucas) only on a large scale (a large group of similar events). The approach to extinction used by Van Valen (1973) is an example. No one, least of all Van Valen, would suggest that extinctions in the evolutionary record occur without any cause. Extinction results when all members of a taxon die without issue. In the real world, the death of each individual must have had a reason. For the taxon as a whole, a great many different causes may have contributed to death: predation, starvation, ageing, genetic malformation, and so on. It is technically invalid, therefore, to say that a given individual died by chance alone. But if frequency of death is looked on at the population or taxon level, it may well be reasonable and mathematically valid to describe the frequency of death as being governed by a random process. The observed *frequency* of death can be used to compute the *probability* of extinction if the population or taxon size is known.

A final, non-palaeontologic example may serve to clarify and emphasize the points just made. Stock-market averages may behave as a random walk yet many stock-market fluctuations have clearly identifiable causes. Events such as the assassinations of political leaders and sharp changes in interest rates influence the stock market in a fairly predictable way. Yet the occurrence of such events is usually not predictable and when a combination of these events operates over a long period, the market averages show many random properties. Thus, individual fluctuations can often be explained in terms of cause and effect — after the event — even though the pattern as a whole is random.

Path-Dependent Versus Independent Events Models

A *path-dependent process* is a fully determined one, an example of which is represented by the following equation:

$$Y = Y_0 e^{-aX} \tag{1}$$

If Y_0 and a are constants, then the path followed by Y is fully determined and fully dependent on the path followed by X. Thus, if Y_0 and a are known, Y can be determined precisely for any value of X. If, on the other hand, a is subject to variability or uncertainty, Y can be determined only in a statistical or probabilistic sense. Let us suppose, for example, that a fluctuates in a uniformly random fashion between a high value and a low value. This means that for a given value of X, Y can be predicted to fall within a range determined by the possible extremes of a. If the process is "run" many times, the average value of Y will be that that would be predicted by assuming that a is constant and located at its observed mean value. This is, in fact, the case in the analysis of radioactivity. Eq. 1 is used in the theory of radioactivity: a is the "decay constant", though it is constant only in a statistical sense and it can be considered a constant only when dealing with very large samples (large numbers of atoms). When one of the "constants" in the process behaves as a random variable, we have departed somewhat from the truly path-dependent process and we have a process which contains some probabilistic elements (the variable a) and some deterministic elements (the constant Y_0).

Van Valen's formulation of his Law of Extinction (1973) is algebraically identical to the radioactivity equation: a is the extinction frequency of taxa within an adaptive zone. According to Van Valen, a is stochastically constant, meaning that there is unpredictable variation around some mean or expected value.

An *independent-events process* is one which is fully probabilistic; that is, it contains no path-dependent elements. Coin tossing is a classic example. In a series of coin tosses, the outcome of any one toss is unpredictable and is in no way affected by the record of previous results. The process has no memory. Again, however, if sufficiently large samples are used, rather accurate predictions can be made about the properties of a large group of coin tossing events.

Between the truly path-dependent process and the independent-events process lie an infinite number of intermediates: situations with combinations of deterministic and probabilistic elements. Most evolutionary processes are examples of the intermediate condition.

Markov Processes

The concept of the Markov process is applicable to evolutionary palaeontology. A Markov process produces a sequence of events such that each event is partly dependent on the outcomes of preceding events and partly dependent on a random (= pseudorandom) process acting at the time of the event itself. The sequence of events thus produced is commonly called a *Markov chain*. In geology, Markov processes have been applied with particular success to the study of sedimentary sequences, especially in cases of clear-cut cyclic sedimentation (such as cyclothems).

A *random walk* is a type of Markov chain which is applicable to an evolutionary record. Fig. 1 shows an example of a classic random walk. It was produced in this case by a computer generated sequence of random numbers, but it could equally well have been produced by a series of coin tosses. Starting from the origin, the line has an equal probability of going up or down for each increment on the horizontal (time) axis. Thus, its position at any point will largely be dependent on the sum of all previous independent events. At point A in Fig. 1 the line is 3 units above the horizontal axis. At point B, it is 4 units above the axis because the random process involved at that point happened to call for a rise of 1 unit. The only other alternative was a fall of 1 unit. Given the record up to point A, the position of the random walk at point B was

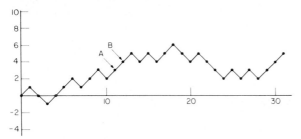

Fig. 1. A classic random walk. The line has the same probability of going up as down with each increment on the horizontal (time) axis. The walk was terminated after 31 steps.

limited to two possibilities: 2 and 4. Thus, the random walk at any point is largely, though not completely, constrained by the position of the preceding point and that point is constrained by the sum of all preceding events.

The topology of the random walk is often appropriate for studies of evolutionary sequences. If we think of the vertical dimension in Fig. 1 as being a morphological trait in an evolving lineage, then it is clear that the value of the trait at any point in time (such as the point B) is highly dependent on the historical sequence of events leading to that point.

In an evolving lineage, the changes that take place with each increment of time are relatively small. Each new form is dominated genetically by inheritance from its immediate ancestor. The same is generally true in trans-specific evolution. In relatively simple systems, the change may be positive or negative; that is, the new form may continue an earlier trend or reverse the direction. Therefore, the basic framework of change is markovian and is approximated by the random-walk model. By contrast, a truly independent-events model is decidedly not appropriate in this context. If each step in an evolving lineage were totally unconstrained by its ancestral steps, the system would be one without heredity and would be totally chaotic. We need not be concerned now whether evolutionary change actually contains random elements; we are only concerned with finding the most appropriate framework.

Another kind of Markov chain is known as the *branching process* and is readily familiar (topologically at least) to any student of evolution. The concept of the branching process is commonly applied to such problems as nuclear chain reactions, survival of family names, and even to problems of genetic drift (Feller, 1968). In the simplest type of branching process, a particle existing at a point in time (equivalent perhaps to a species in geologic time) has a certain probability of producing one or more other particles. If it does not produce other particles, it becomes extinct. If new particles are produced, they are each subject to extinction or splitting in the subsequent time interval. Estimates can be made of the number of particles that will be present after many iterations of the process. This is clearly applicable to a variety of problems in the analysis of the evolution of diversity. Elaboration of the basic model is possible. For example, one can provide for the persistence of any of the particles so that they coexist with their offspring. This is analogous to the coexistence of a species with its ancestor.

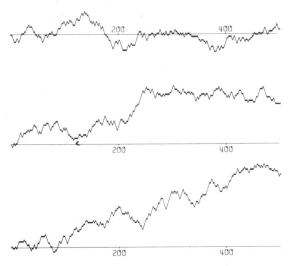

Fig. 2. Three computer-generated random walks. Each is 500 steps long. The three were selected from a sample of twenty (see text for selection procedure).

Random walks and branching processes have been the subject of rigorous theoretical study by statisticians and mathematicians and they are surrounded by a well-developed body of scholarship. Many of the results of the mathematical analysis of these Markov processes are strikingly counterintuitive! Because random walks and branching processes are partially path-dependent, it is indeed dangerous to think of them only in terms of independent-events models. Some examples from random walks will serve to illustrate this point. Fig. 2 shows three relatively long random walks. The ground rules for constructing them are the same as those applied in Fig. 1: that is, each record shows a sequence of a great many independent decisions. This is analogous to the commonly cited gambling situation wherein players have an equal chance of winning and losing but where their net gain or loss is shown by a random walk. When a line crosses the horizontal axis, the number of positive and negative moves leading to that point are equal. To the layman and even to statisticians not familiar with Markov processes, it is intuitively reasonable to expect that a random walk will fluctuate back and forth across the horizontal axis fairly frequently.

In palaeontology, many so-called evolutionary trends are established because the path followed by a morphological character through time clearly does not fit the preconceived idea of the random walk just described. In other words, the interpretation of the evolutionary trend is based on the assumption that if the observations were literally random in character, the line would have many reversals in direction and would tend to stay close to the horizontal axis.

The random walks in Fig. 2 were computer generated and were terminated after 500 steps. The vertical scales are all the same, the maximum deviation from the zero axis being 49 units (bottom diagram). In studying random walks graphically, it is difficult to select "typical" results because one is inevitably influenced by what one is looking for. In this case, twenty random walks were generated and the three shown here were selected from the twenty. The top one was chosen because it matches most closely what most people expect from random walks; it contains quite a bit of noise and it fluctuates back and forth across the axis. The lower two were chosen because they appear to be decidedly non-random in that they display long periods of monotonic change and few crossings of the axis.

Theoretical analysis of the random-walk phenomenon (particularly by Feller, 1968) indicates that the lower two examples in Fig. 2 are not unusual. In fact, the upper example is probably the least typical. It can be shown, for example, that in any reasonably long random walk the probability that the line will cross the horizontal axis in the second half of the random walk is only 0.5. In 20% of random walks, the line will stay on one side of the axis 97.6% of the time (Feller, 1968).

Strategies For Applying Stochastic Models

Fig. 3A shows data on change in proloculus diameter for the fusulinid *Lepidolina multiseptata* through part of the Permian. The plot is redrawn from Hayami and Ozawa (1975). Fig. 3B is a second example from the Hayami and Ozawa paper and shows the change in frequency of a discrete phenotype in the scallop *Cryptopecten vesiculosus* in the late Cenozoic. Hayami and Ozawa interpreted both examples as showing unidirectional, genetic change resulting from a single cause (orthoselection). The viability of this biological interpretation depends on ruling out the possibility that the indicated trends are due just to a chance combination of other factors. There are at least three general alternatives to the orthoselection interpretation.

(1) *The observed trends may result from incomplete sampling of highly variable populations.* The implication of this is that if the study were redone with

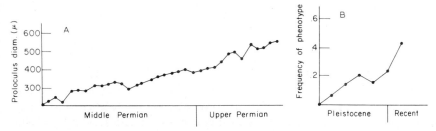

Fig. 3. Fossil time series data expressed in the format of a random walk (redrawn from Hayami and Ozawa, 1975). A. Size variation through time in the fusulinid *Lepidolina multiseptata*. B. Variation in frequency of one phenotype in the bivalve *Cryptopecten vesiculosus*. In the *Cryptopecten* data, the right-hand point is an average of five living populations. Also, data given by Hayami and Ozawa for six Pliocene and Early Pleistocene populations are not shown because the frequency of the phenotype was uniformly zero (that is, the phenotype had not yet appeared in the species).

new samples, the trends might disappear (or be reversed) merely because of the vagaries of sampling. This is a common problem and a host of statistical techniques designed to measure sampling error are familiar to most palaeontologists. In the present cases, Hayami and Ozawa give data on the standard errors of the plotted means which provide convincing evidence that the trends are "real" in the sense of being reproducible.

(2) *The biological causes of change may be so many and varied that change behaves as a random variable.* By this alternative, the observed trends would be due to chance in the same sense that random walks generated by coin tossing often show regular trends. This interpretation does not deny genetic change through natural selection but it does deny that the series of changes has a single interpretation (orthoselection).

(3) *The morphological traits may be adaptively neutral and the changes caused by genetic drift.* By this alternative, the trends are also seen as the result of a random-walk phenomenon but the biological mechanism is different (selection is absent).

Alternatives (2) and (3) make the claim that the observed trends may be approximated by a random process, specifically a random walk. Both the fusulinid and scallop examples satisfy the basic topological conditions of the random walk, as presented earlier. In both cases, we have a time series of numerical observations and there is every reason to believe that the process is markovian: the morphological increments from one horizon to the next are consistently small relative to the total range in values. Also, in these examples at least, the variation in the size of increments is small so that the classic, coin-tossing random walk is an appropriate model.

The key to evaluating the random walk idea is in formulating the proper null hypothesis. If it can be shown that the observed trends could have occurred by chance, we cannot reject the hypothesis that the underlying process is random or pseudorandom. And if we cannot reject the hypothesis of randomness, it is not legitimate to make a biological interpretation of orthoselection. Given a random model (such as the random walk), there are two testing strategies that can be followed: (1) analytical and (2) Monte Carlo.

Analytical methods

In simple situations it is often possible to test the null hypothesis by direct, analytical means. That is, it is often feasible to develop equations that yield a direct answer to the question: What is the probability that the observed trends could have occurred by chance? Following conventional standards, if the calculated probability is greater than 0.05, we cannot reject the null hypothesis and we are forced to abandon the non-random, biological interpretation. If more conservatism is desired, then the 0.01 probability level can be used as a cut-off.

For random walks, Feller (1968) provides some elegant theorems for assessment of the relevant probabilities. For example, his theorem III.7.1 states that in a random walk of length n, the probability that the highest point m equals some constant c is:

$$P_r[m = c] = \left(\frac{n + \frac{n}{c}}{2}\right)2^{-n} \text{ or } \left(\frac{n + \frac{n}{c} + 1}{2}\right)2^{-n} \tag{2}$$

whichever yields a positive value. (This choice depends on whether the quantity $(n + c)$ is odd or even.) The right-hand side of eq. 2 is equivalent to:

$$\frac{n!}{\left(\frac{n + c}{2}\right)!\left(n - \frac{n + c}{2}\right)!2^n} \text{ or } \frac{n!}{\left(\frac{n + c + 1}{2}\right)!\left(n - \frac{n + c + 1}{2}\right)!2^n}$$

The probability that the maximum value is less than or equal to c is:

$$P_r[m \leqslant c] = \sum_{i=0}^{c}(P_r[m = i]) \tag{3}$$

The maximum cannot be less than zero because the random walk starts at zero.
    The probability of equalling or exceeding a given c is simply:

$$P_r[m \geqslant c] = 1 - P_r[m \leqslant (c - 1)] \tag{4}$$

This form of Feller's theorem is the most useful in the palaeontological case.
    The Hayami and Ozawa data (Fig. 3) can be viewed in the context of Feller's theorem. The starting point of each plot (lower left) can be taken as the start of a random walk (zero ordinate). Each upward move may be thought of as a positive increment of one unit and each downward move as a negative increment of one unit. In Fig. 3A, there are 25 positive steps and 8 negative steps. Thus, the end point has a value of +17 and this also happens to be the highest point m. By the same reasoning, the highest point in Fig. 3B is 4 (5 minus 1). The question regarding eq. 4 is whether the highest points (17 and 4, respectively) could reasonably have been expected to occur by chance in random walks of these lengths (33 and 6, respectively).
    For a random walk of 33 steps, computation using eq. 4 shows that the probability of m being at least as large as 17 is only 0.003. Thus, for the fusulinid case we are justified in rejecting the hypothesis of randomness and the interpretation of Hayami and Ozawa is shown to be plausible.
    For the scallop case, however, the probability of the maximum being at least as large as 4 is 0.125. Thus the null hypothesis cannot be rejected and the suggestion of a phyletic trend in the sense of Hayami and Ozawa cannot be sustained.
    It is intuitively obvious that the problem with the scallop case is that the time series is very short; so short that an apparent trend can easily be produced by chance alone. In less obvious situations, eq. 4 is useful in providing a rigorous assessment of probabilities. Table I shows calculated probabilities for several arbitrary values of c for the 33-step random walk.
    In many palaeontological time series, the attribute being studied does not change during some time intervals. This does not fit the random-walk model being used here because the line must always go up or down. In actual cases

TABLE I

Calculated probability that the maximum value m encountered in a 33-step random walk will be at least as large as some designated value c (Computations made using eq. 4)

c	$P_r[m \geqslant c]$
5	0.392
10	0.080
15	0.009
20	0.001

where a time series has level segments, eq. 4 can still be used, however. The simplest procedure is just to ignore the level segments. For computation purposes, the number of steps n in the series is reduced by the number of level segments ignored.

Other direct tests of the fusulinid and scallop data are available. Feller gives a variety of other characteristics of random walks that may be assessed rigorously. Also, more conventional statistical tests can be applied. For example, if the fossil data are random, the number of increases (positive increments) in the time series should equal the number of decreases, within the limits of sampling error. The χ^2 statistic can be used. In the fusulinid case, there were 25 positive increments and 8 negative increments. This produces a χ^2 high enough to reject the hypothesis of randomness and thus the results agree with the random-walk test presented above. In the scallop case, the χ^2 value is too low to reject the null hypothesis and this result is also in agreement with the earlier one.

It is also intuitively reasonable to apply the correlation coefficient to palaeontological time series. In the search for evolutionary trends, the question is often asked: Is the attribute significantly correlated with time? The attribute can be a morphological trait (as in Fig. 3) or some measure of taxonomic diversity, extinction rate, or whatever. While formal statistical tests are rarely applied to evolutionary time series, most palaeontologists evaluate such data as if they were comparable with the bivariate scatters of points commonly dealt with in biometrics. The difficulty with using this approach is that the underlying statistics (specifically the correlation coefficient) require assumptions that are patently false in the present context. The correlation coefficient calls for an independent-events model, whereas evolution is clearly markovian. In most evolutionary sequences, each point is constrained to be close to its ancestor (relative to the range shown by all the points).

The danger of using the correlation coefficient — or intuition based on it — can be shown by computing the correlation coefficients for a series of artificially generated random walks. The results of such a test are shown in Table II. Several hundred random walks were generated by computer. They varied in length from 10 to 30 steps. For each random walk, the coordinates of all points (including the start at 0,0) were used to compute a correlation coefficient r by standard means. Table II shows the somewhat astonishing result that well over half the random walks yield statistically significant correlation coefficients at

TABLE II

Levels of significance of correlation coefficients calculated from random walks of various lengths (n)

Length n of random walk	Number of random walks	Number not significant	Number significant at 5% level	Number significant at 1% level
10	100	41	59	46
20	100	32	68	56
30	100	25	75	64

the 5% level and most of these are also significant at the 1% level. The longer random walks have a higher proportion of significant correlation coefficients than the short ones.

Table II demonstrates, therefore, that a purely random process will appear to produce statistically significant trends in the majority of cases if the conventional independent-events models of biometrics are used. How many supposed instances of Cope's Rule are only illusions resulting from improper evaluation of time series data?

Monte Carlo methods

Stochastic models can also be tested by stimulation using automatically generated random numbers. In order to find the expected maxima for random walks of various lengths, large numbers of imaginary random walks can be constructed and the probabilities estimated empirically. This has been done for 2000 random walks of 10 steps each and in Table III the results are compared with the analytical method.

TABLE III

Comparison of analytical and Monte Carlo methods of assessing the probability of the maximum value m of a 10-step random walk being at least as large as some designated value c

c	Analytical: probability calculated from eq. 4 $P_r[m \geqslant c]$	Monte Carlo: number of cases where $m \geqslant c$ in artificial random walks	
		first 1000	second 1000
1	0.754	782	759
2	0.549	583	547
3	0.344	356	345
4	0.227	246	218
5	0.109	129	108
6	0.065	83	64
7	0.021	23	31
8	0.012	11	15
9	0.002	3	3
10	0.001	2	2

It is clear that the Monte Carlo method yields results comparable with those of the direct method using eq. 4. It should be noted that eq. 4 contains no approximations; it is derived by combinatorial methods and is precise. The differences seen in Table III are due entirely to inevitable sampling errors in the Monte Carlo results. In this case, the direct method is obviously preferable because it is faster and more accurate. In many situations the Monte Carlo method is the only one available either because the model being used is mathematically intractable or because the necessary mathematical theory has not been developed.

In the brief treatment of modelling strategies just presented, the random-walk model was used as an example partly because it is amenable to both analytical and Monte Carlo treatment and partly because it is appropriate to many studies of evolutionary trends and patterns using fossil data. A wide variety of other models (both markovian and non-markovian) are appropriate for other types of evolutionary problems.

Applications

The use of stochastic models in evolutionary palaeontology has two main objectives.

(1) *To test for randomness in order to avoid the pitfall of erecting deterministic explanations where none are justified.* The random-walk examples discussed earlier illustrate the importance of testing for randomness. If an array of real-world data does not depart significantly from chance expectations, it is folly to attempt a specific biological interpretation of the data. Tests of randomness are inherently negative in philosophy: they tell what cannot be done in the interpretive phase of a study.

(2) *To search for statistical generalizations which have predictive value in the fossil record.* Given a probabilistic hypothesis (or hunch!) about the behaviour of some aspect of the evolutionary process, a stochastic model can be used as a vehicle for testing the hypothesis. The model is used to make certain predictions and these predictions are compared with real-world data. If the match is good, then a strong though circumstantial argument can be made in favour of the hypothesis.

Papers by Anderson (1974) and Anderson and Anderson (1975) illustrate the joint use of the two objectives just described. These authors were concerned with the evolution of patterns of taxonomic diversity: species per genus, species per family, and so on. Fig. 4A shows cumulative frequency distributions of taxon size for Recent mammals. If raw frequencies are plotted (Anderson, 1974, fig. 2), the so-called "hollow curve" of Willis (1922) results. The hollow curve and the various other ways of expressing the same frequency data have been the subject of a variety of interpretations (see Anderson, 1974, for review). Some interpretations deal only with evolutionary phenomena (origination and extinction of lineages in clades) while others take into account aspects of taxonomic procedures (splitting and lumping). Some of the interpretations

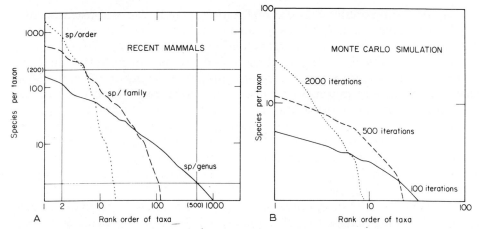

Fig. 4. Cumulative frequency distributions of species per taxon (redrawn from Anderson and Anderson, 1975). A. Genera, families, and orders of Recent mammals. B. Monte Carlo simulation based on random origination and extinction of species. [Explanation: in Fig. 4A, the largest genus has about 160 species (left end of solid curve) and the next largest genus has about 110 species. There are about 1000 genera in all (right end of solid curve), of which 500 have 2 or more species. Therefore, about 500 genera are monotypic.]

are highly deterministic while others are stochastic.

Anderson and Anderson (1975) developed three simple stochastic models based on branching processes and used the Monte Carlo method to find out whether real-world data (such as Fig. 4A) could be simulated. In all three models, 100 monotypic higher taxa were established at the start. Then, through an iterative procedure involving a sequence of imaginary time units, species were either removed from the taxa (simulating species extinction) or added (simulating speciation). Extinction and speciation were determined by random numbers and the two kinds of events had equal probabilities. Thus, the total number of species fluctuated as a random walk, as did the sizes of the several higher taxa. Inevitably, some higher taxa became extinct while others grew very large. The three models differed in relatively minor regards which are not of concern to us in the present context.

Fig. 4B shows some of the results of computer runs using one of the models. The three curves show the simulated frequency distributions after 100, 500, and 2000 iterations. As the simulation ran, the total number of higher taxa decreased (the right-hand end of the curve shifted to the left), the size of the largest taxon increased (the left-hand end of the curve shifted upward), and the relative number of monotypic taxa decreased (the curve became steeper). The shapes and spatial relationships of the curves in Fig. 4B compare well with those in Fig. 4A. Furthermore, the effect of increasing taxonomic rank from genus to order in the mammals appears to be simulated by the increase in number of iterations in the computer model.

Anderson and Anderson have thus shown that the real-world diversity patterns could have been produced by a simple stochastic process and they have thereby carried out objective (1). By showing that a simple random model can produce results comparable with the real world, they have demonstrated that a deterministic explanation (perhaps involving exploitation of an adaptive zone by some taxa at the expense of others) is neither necessary nor justified.

With regard to objective (2), Anderson and Anderson suggest that their models may in fact describe what was going on in the evolution of mammalian diversity. This is only a circumstantial argument, of course, because the success of the simulation does not prove the model; it proves only that the simulation represents one way in which the observed diversity patterns could have developed. The logical next step is to use the model to make other kinds of predictions that can be tested with real-world data. If these tests are successful, the model can be elevated to the status of a generalization or statistical law — which is the ultimate purpose of objective (2).

Another example of the application of stochastic models to evolutionary problems is the continuing series of papers using the so-called MBL computer program (Raup et al., 1973; Raup and Gould, 1974; Schopf et al., 1975; Gould and Raup, 1975). The MBL program originated at a series of conferences convened by T.J.M. Schopf at the Marine Biological Laboratory, Woods Hole. The methodology is entirely Monte Carlo and is similar in concept to the simulations of Anderson and Anderson (1975) discussed earlier.

The MBL program produces an entirely probabilistic simulation of the evolutionary branching process and includes the following steps.

(1) A single lineage is established at time $t = 1$. It has a morphology described by up to twenty imaginary traits; each trait is expressed as an integer with an initial value of zero.

(2) At time $t = 2$, the starting lineage either persists without branching, persists with branching to form a second lineage, or does not persist (i.e., becomes extinct). The choice between the three alternatives is signalled by computer-generated random numbers, with the probabilities of the alternatives being specified in advance. If the starting lineage becomes extinct by action of the third alternative, the simulation aborts and starts again. If a new lineage is formed by branching , its morphology is subject to change. That is, each character may shift from its immediately ancestral state; the presence or absence of change is signalled by random numbers, as is the direction of change (positive or negative). All traits are treated independently and all changes are increments of one unit on the integer scale. The probability of morphological change is set at a relatively low value so that the morphology of the new lineage is close to that of its ancestor.

(3) From time $t = 3$ onwards, all surviving lineages are treated in a manner identical to that of the starting lineage. Branching events form new lineages which are in turn subject to branching, extinction, or simple persistence as described above. Each new lineage inherits the morphology of its immediate ancestor subject to randomly determined change in its morphological traits.

(4) The simulation is stopped at some designated number of time iterations (such as 120). At this point, a branching "tree" of lineages has been produced. The number of branches depends on the probability of branching. By this time

the morphology of the surviving lineages has diverged from the starting condition. The amount of divergence depends on the probabilities that were designated for morphological change at branch points and on the number of branching events.

(5) The lineages are then automatically grouped into "taxa". The taxonomy may be based entirely on the branching pattern; monophyletic groups within certain size limits are defined as higher taxa. Alternatively, the grouping can be done entirely on morphological information using standard techniques of numerical taxonomy. If many traits are used (such as 20), the taxonomy based on morphology is virtually identical to (or parallel to) the taxonomy based on branching patterns (Schopf et al., 1975). In other words, classification errors caused by convergence and parallelism are rare when as many as 20 traits are used.

The MBL program contains certain other constraints and procedures. The probability of branching may be set equal to the probability of extinction and, if so, the number of coexisting lineages (standing crop) will behave as a random walk. As an alternative to this, the two probabilities may be allowed to vary in a controlled fashion such that when the standing crop goes down (by chance), the branching probability is automatically increased and the extinction probability decreased so as to push the standing crop upwards. If random increases in standing crop are similarly constrained, the result is a fluctuation in the number of lineages about some mean value. Lineage diversity is then in dynamic equilibrium.

The MBL program may be thought of as a giant null hypothesis to describe phylogeny. The model makes only a few biological assumptions about the evolutionary process. The important ones are the following.

(1) *Monophyly*. Once separated, lineages cannot come back together and no new lineages are formed by confluence of lineages. Thus, introgressive hybridization is ruled out.

(2) *The process is markovian*. All new lineages come from a pre-existing lineage. Each new lineage inherits most of the morphology of its immediate ancestors.

(3) *Morphology changes by small steps*. This rules out large changes such as might be called macromutations.

(4) *All morphological change takes place at branch points*. This implies an extreme expression of the punctuated equilibrium model of Eldredge and Gould (1972). The program can be modified easily to avoid this assumption and this is being done by M.K. Hecht (personal communication, 1975).

Therefore, the model used for the MBL program makes no assumptions about adaptation or natural selection; morphology is allowed to drift in a totally random fashion subject only to the constraints common to all Markov processes. No assumptions are made about competition between lineages or higher taxa unless standing crop is constrained as outlined earlier. Even in this

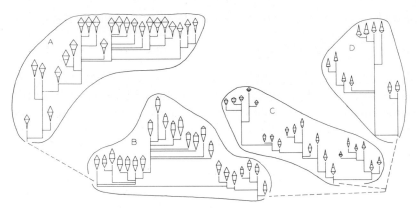

Fig. 5. Part of an MBL simulated phylogeny. Vertical axis is time with the "Recent" at the top. Each glyph ("triloboid") expresses five morphological traits for one lineage. Four higher taxa (A—D) are distinguished. See text for further explanation.

case, competition is non-specific, because all lineages "living" at the same time are subject to the same probabilities of branching and extinction. The morphological traits are independent of each other so that the evolution of a functionally integrated set of characters is impossible.

The MBL program has been used primarily as an exploratory tool. If a pattern commonly observed in the real world can be simulated readily by the program, then causes not included in the MBL model need not be called on for explanation and interpretation. Fig. 5 shows an example. Here, part of an MBL simulated phylogeny is shown. Each of 77 lineages (which may be considered as populations, species, or genera) is indicated by a vertical line, the length of which corresponds to the time duration of the lineage. The several lineages at the top of the diagram were extant at the end of the computer run. Ancestry of the lineages is shown by horizontal lines. Four higher taxa (A—D), defined by the computer, are delineated and their phylogenetic relationships are given by dashed lines. Some intermediate lineages (belonging to other taxa) are not shown. Only a few such lineages have been removed, however, and the four taxa may be considered to be "closely related". The computer run used for Fig. 5 is the one described by Raup and Gould (1974); for comparison, taxa A—D are the same as clades 88, 1, 9, and 91, respectively, of the 1974 paper.

Above each lineage in Fig. 5 there is a glyph that expresses the states of five morphological traits. The glyphs are similar to those used by Raup and Gould and are called "triloboids". Two of the traits determine the dimensions of the upper triangle (head of the triloboid), one determines the length of the mid-section (thorax), and the remaining two determine the dimensions of the lower triangle (tail). If these triloboids are compared with those in the earlier paper (Raup and Gould, 1974, fig. 13), it will be noticed that glyphs for the same lineages differ slightly. The conversion from the numerical scale of the computer to the glyph was modified slightly in the preparation of Fig. 5 (a constant was added to all values) in order to avoid the complete loss of the tail at extreme values of the fifth trait.

The taxonomy for Fig. 5 was done entirely on cladistic evidence; that is, morphology was not considered. Nevertheless, there is considerable order in the pattern of triloboid evolution. Higher taxa are quite distinct morphologically and changes in one trait tend to correlate with changes in others. There are also fairly regular patterns in triloboid size; size is the sum of the magnitudes of the five traits.

The patterns shown in Fig. 5 are remarkably like real-world patterns. If the MBL data were real data, boundaries between higher taxa would probably be placed differently because total cladistic information would not be available. Also, because monophyly was the principal taxonomic criterion, other taxonomic schemes could be established which would be just as valid as that shown in Fig. 5.

If Fig. 5 were found in the real world of evolutionary palaeontology, some interpretations would be virtually automatic. It would be conventional to suggest that the triloboids of taxa A and B became large because of a selective advantage of large size in their particular ecological setting. The small triloboids in taxon C would indicate a different ecology or adaptive strategy. The characteristic shape of most forms in a given taxon would be said to result from the adaptive success of that particular set of traits: these would be examples of integrated sets of characters. The interpretations are almost limitless. Cases of parallelism and convergence can be found. Examples of Cope's Rule are even evident, depending on how Cope's Rule is applied.

But all of these interpretations require evolutionary mechanisms which are known not to have been at work in the simulation! How then can the MBL results be interpreted in terms of the null hypothesis? The data in Fig. 5 (plus other MBL data described by Raup et al., 1973, and Raup and Gould, 1974) could be interpreted as evidence that Darwinian evolution does not obtain or at least that Darwinian theory is not necessary to explain the real-world data. This is largely unacceptable because there is every reason to believe that adaptation through natural selection does work in many instances. Not only have population biologists provided ample evidence of the credibility of the mechanism but also many structures are known to have clear functional purposes not explicable by chance. Furthermore, the integrated traits seen in Fig. 5 make no sense functionally because the imaginary triloboids have no functions. To emphasize this, Fig. 6 shows the same phylogeny (from the same computer run) but with triloboids constructed on the basis of a second set of traits. The cladistically defined taxa are still quite coherent morphologically but the forms are different. For example, the triloboids of taxon A have a small head and a large, wide thorax, in contrast to the large head and small, tapered thorax seen in Fig. 5. If real trilobite characters were made to evolve using the MBL program, correlations between characters would be produced but they would be nonsensical with regard to the known functional morphology of trilobites.

The MBL results do say, however, that trends and patterns of the sort illustrated in Figs. 5 and 6 cannot by themselves be used as evidence for Darwinian theory. A phylogenetic increase in size, for example, may or may not be adaptive; evidence other than the regularity of the trend and the homogeneity of

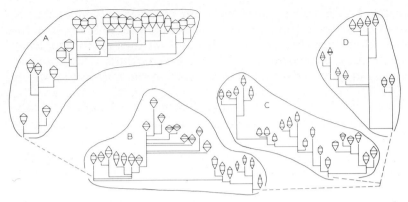

Fig. 6. The same simulated phylogeny as in Fig. 5 but with triloboids based on a different set of five traits.

higher taxa must be found, because regular size increase (or decrease) is a common outcome of the simple random process. In fact, in the absence of independent evidence, we have no reason to reject the hypothesis that some morphological traits are adaptively neutral and have evolved in a strictly random-walk fashion. The MBL simulation therefore opens the way for interpreting some aspects of morphology as being adaptively neutral even though they show considerable regularity and order in their phylogeny.

The foregoing conclusions apply equally well to a variety of taxonomic diversity and extinction patterns not discussed here (see Raup et al., 1973). Some aspects of the MBL results, on the other hand, do not resemble the real world. For example, dramatic mass extinctions of the Permo-Triassic type do not occur. Such phenomena are thus not explicable in terms of the model and for them the null hypothesis of randomness is clearly rejected. One of the benefits of the stochastic simulation is to pin-point in this way those aspects of the evolutionary record which require more study and a fuller explanation — perhaps in a completely deterministic mode.

Summary

Stochastic models serve to separate those features of the evolutionary record which are amenable to a deterministic explanation from those where the search for specific causes is not warranted. If treated rigorously, the models provide a necessary test of the investigator's intuition. And as we have seen, many of the results of stochastic modelling are counter-intuitive. If, in a given case, the null hypothesis of randomness cannot be rejected, randomness has not been proven; it has been shown only that the data are not distinguishable from random data.

The application of stochastic models in palaeontology has many potential pitfalls. One of the most serious is that the models may contain unrecognized assumptions; these "cryptic" assumptions may be quite deterministic. Monte Carlo simulations are especially prone to this problem (Schull and Levin, 1964). Also, there is an unfortunate tendency to make models too complex. For example, several of the discrepancies between the MBL simulations and the real world could be eliminated by clever doctoring of the model; parameters and constraints could be added until the match between the simulations and at least some real-world situations was perfect. In the right (or wrong) hands, Monte Carlo simulation becomes an almost unbeatable system. When this happens, the null hypothesis being tested becomes too narrowly defined, the number of cryptic assumptions increases, and the results become virtually meaningless. The best safeguard is to maintain utter simplicity in the underlying models.

Where a choice is possible, analytical methods are preferable to the Monte Carlo approach. The principal advantage of analytical testing (such as was illustrated with random walks) is that the result contains an accurate assessment of the probabilities attached to the null hypothesis. Thus, the MBL simulation results should be backed up by rigorously derived analytical tests.

Acknowledgements

This research was supported by the Earth Sciences Section, National Science Foundation, NSF Grant DES75-03870, and by the University of Rochester.

References

Anderson, S., 1974. Patterns of faunal evolution. Q. Rev. Biol., 49: 311—332.
Anderson, S. and Anderson, C.S., 1975. Three monte carlo models of faunal evolution. Am. Mus. Novit., No. 2563: 6 pp.
Eldredge, N. and Gould, S.J., 1972. Punctuated equilibria: an alternative to phyletic gradualism. In: T.J.M. Schopf (Editor), Models in Paleobiology. Freeman, San Francisco, Calif., pp. 82—115.
Feller, W., 1968. An Introduction to Probability Theory and Its Applications. Wiley, New York, N.Y., 509 pp.
Frazer, A. and Burnell, D., 1970. Computer Models in Genetics. McGraw-Hill, New York, N.Y., 206 pp.
Gould, S.J. and Raup, D.M., 1975. The shape of evolution: a comparison of real and random clades. Geol. Soc. Am., Abstr. Progr., 7: 1088.
Harbaugh, J.W. and Bonham-Carter, G., 1970. Computer Simulation in Geology. Wiley—Interscience, New York, N.Y., 575 pp.
Hayami, I. and Ozawa, T., 1975. Evolutionary models of lineage zones. Lethaia, 8: 1—14.
Krumbein, W.C., 1969a. The computer in geological perspective. In: D.F. Merriam (Editor), Computer Application in the Earth Sciences. Plenum, New York, N.Y., pp. 251—275.
Krumbein, W.C., 1969b. Deterministic and probabilistic models in geology. In: P. Fenner (Editor), Models of Geologic Processes. Am. Geol. Inst., Washington, D.C., pp. WCK-B-1—WCK-B-14.
Lucas, H.L., 1964. Stochastic elements in biological models; their sources and significance. In: J. Gurland (Editor), Stochastic Models in Medicine and Biology. University of Wisconsin Press, Madison, Wisc., pp. 355—383.
MacArthur, R.H. and Wilson, E.O., 1967. The Theory of Island Biogeography. Princeton Univ. Press, Princeton, N.J., 203 pp.
Nei, M., 1975. Molecular Population Genetics and Evolution. Elsevier, New York, N.Y., 288 pp.
Raup, D.M. and Gould, S.J., 1974. Stochastic simulation and evolution of morphology — towards a nomothetic paleontology. Syst. Zool., 23: 305—322.
Raup, D.M., Gould, S.J., Schopf, T.J.M. and Simberloff, D.S., 1973. Stochastic models of phylogeny and the evolution of diversity. J. Geol., 81: 525—542.
Schopf, T.J.M., Raup, D.M., Gould, S.J. and Simberloff, D.S., 1975. Genomic versus morphology rates of

evolution: influence of morphologic complexity. Paleobiology, 1: 63—70.

Schull, W.J. and Levin, B.R., 1964. Monte carlo simulation: some uses in the genetic study of primitive man. In, J. Gurland (Editor), Stochastic Models in Medicine and Biology. University of Wisconsin Press, Madison, Wisc., pp. 179—196.

Shreve, R.L., 1966. Statistical law of stream numbers. J. Geol., 74: 17—37.

Shreve, R.L., 1967. Infinite topologically random channel networks. J. Geol., 75: 178—186.

Shreve, R.L., 1974. Variation in mainstream length with basin area in river networks. Water Resour. Res., 10: 1167—1177.

Van Valen, L., 1973. A new evolutionary law. Evol. Theory, 1: 1—30.

Watson, R.A., 1969. Explanation and prediction in geology. J. Geol., 77: 488—494.

Willis, J.C., 1922. Age and Area. Cambridge University Press, Cambridge, 259 pp.

Wright, S., 1964. Stochastic processes in evolution. In: J. Gurland (Editor), Stochastic Models in Medicine and Biology. University of Wisconsin Press, Madison, Wisc., pp. 199—241.

CHAPTER 4

BRACHIOPOD EVOLUTION

ALWYN WILLIAMS and JOHN M. HURST

Introduction

Brachiopods are one of the few groups of living metazoans to be represented by distinctive and complex skeletal remains more or less continuously throughout the Phanerozoic record. Their bivalved shells have nearly always been composed of such durable components that ultrastructural differences in the original fabric are as easily determinable in Cambrian specimens as in Recent ones (Williams, 1968b). Both valves, even of the oldest stocks, are usually rich in internal impressions and sufficiently variable in morphology to provide a means for inferring evolutionary changes in gross anatomy as well as in shell shape, ornamentation and devices responsible for the articulation and movement of the valves. The geological history of brachiopods, as measured by generic diversity and commonness of occurrence, is also revealing. The phylum attained its evolutionary climax by Devonian times, and notwithstanding subsidiary peaks of proliferation during the Permian and Jurassic Periods, has been in decline ever since (Fig. 7). Brachiopods, therefore, are equipped with the right sort of skeleton and have lived through a long enough segment of geological time to afford a unique illustration of the processes of evolution affecting marine invertebrates.

There are, none the less, obstacles to providing an objective account of the evolutionary history of the phylum. A complete anatomy can never be reconstructed from skeletal impressions alone, yet there are basic differences in soft parts, like the loss of the anus or the rearrangement of filaments into pairs during lophophore growth, that mark significant changes in phyletic history. Furthermore, the ordinal classification of the phylum reflects breaks in the fossil record that are disconcertingly real even if they are the consequences of evolution by punctuated equilibria. The main problem, however, is to determine the most appropriate taxon in which to process the data and measure the effects of evolution. Within a quantitatively controlled polythetic classification, the species is the ideal unit, but most fossil and living species of brachiopods are too subjectively and vaguely drawn to be reliable indicators of phyletic differentiation. Genera are better because, with few exceptions, the same character complexes with about the same order of weighting have been used in their definition throughout the phylum; and deficiencies in the circumscription of species tend to be reduced when they are assembled into genera. Consequently the use of genera does, at least, allow meaningful comparisons to be drawn between different groups, especially when these correspond to families and superfamilies which are normally founded on generic clusterings. It is noteworthy that the frequency of superfamilies relative to the number of genera

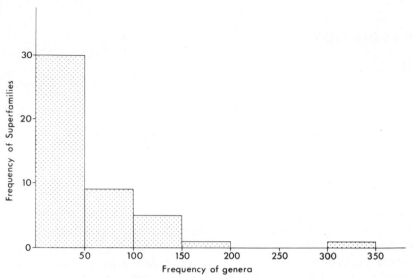

Fig. 1. Variation in the size of brachiopod superfamilies according to the frequency of genera assigned to them.

assigned to them (Fig. 1) is negatively skewed in a distribution generally conceded to be the pattern of apportionment of species among genera. This is different from the diffuse scatter of ordinal size in respect of generic content and, as already suggested, groupings corresponding to orders seem to require a different scale of assessment. This inference will be explored later in the chapter.

A final point to be noted by way of introduction is the basis for our estimates of absolute time. We have used the radiometric age determinations compiled for the Phanerozoic aeon by Harland et al. (1964), and adapted them to the chronostratigraphic divisions of the Geological Column as set out in the *Treatise on Invertebrate Palaeontology*. Neither classification is entirely satisfactory but each probably constitutes the most widely used tabulations of relative and absolute time. Both tables are divided into series; but in view of the wide margin of error attending radiometric age determinations and the imprecise range of most brachiopod genera, our estimates of absolute time have not been calibrated more finely than those representing subsystems.

Nature of Numerical Data

In 1884, Davidson's index of brachiopod species, published posthumously, recorded 68 genera recognised by him during compilation of his classic monographs. Ten years later Hall and Clarke accepted no fewer than 323 genera as valid taxa. Since then the number of new genera appearing in the literature, including synonymies, has increased linearly ($r = 0.80$) at a steady rate of about 23 per annum up to the latest count of nearly 2400 by mid-1975. Some of the newly described taxa represent discoveries arising from access to new exposures or the use of improved retrieval techniques. However, a sampling estimate (Williams, 1957) suggesting that such genera are proposed at about one-half the rate of those created by continual taxonomic revision, still appears to hold good. Estimates of brachiopod evolutionary rates may therefore be more affected by subjectively derived opinions of taxonomists than by the fortuitous nature of rock exposure and fossil preservation. The bias can be demonstrated by exploring the proliferation and morphological foundation of genera within their superfamilial context.

If the rate of generic evolution is a function of time one would expect to find some relationship between the longevity and size of superfamilies. Despite the dispersion shown in Fig. 2, there is a positive correlation ($r = 0.381$) between the absolute time range of the 46 superfamilies composing the phylum in 1965 (Williams et al.) and the number of genera they contained in 1975. A reduced major axis fitted to the scatter suggests that the normal rate of proliferation was about five genera per 10 m.y. and was characteristic of some inarticulates (Acrotretacea, Paterinacea, Siphonotretacea), Strophomenida (Lyttoniacea, Triplesiacea), Pentamerida (Porambonitacea, Pentameracea), Rhynchonellida (Stenoscismatacea), Spiriferida (Cyrtiacea, Dayiacea, Spiriferinacea) and

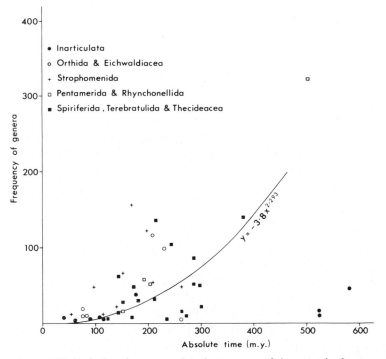

Fig. 2. The relation between the size measured in generic frequencies and longevity (expressed in absolute time) of brachiopod superfamilies.

Terebratulida (Terebratulacea, Zeilleracea). Bearing in mind the idiosyncrasies of the systematists who have recently erected genera assigned to these superfamilies — Amsden (1964; 1968), Boucot and Johnson (1966; 1967), Cooper (1956), Cooper and Grant (1969), Dagys (1974), Havlíček (1967, 1971), Johnson (1973, 1974), Rowell (1966) and Williams (1974) — the alignment of so many unrelated groups along the fitted axis suggests at least a broad concurrence on the taxonomic weighting of morphological characters. The axis may also be regarded as a boundary between two sets of superfamilies with different rates of evolution. One is a low diversity, slowly evolving group which included especially the inarticulates Craniacea, Discinacea and Lingulacea, while the other consists of high diversity, rapidly evolving stocks which seem to have been prevalent among the Strophomenida (Productacea, Strophalosiacea, Strophomenacea), and characteristic too of the Orthacea and Spiriferacea. In this distribution the most anomalous superfamily is the Rhynchonellacea. It is so large, as is shown by its position at the end of the histogram in Fig. 1, that the 325 genera assigned to it more than outweigh a longevity approaching 500 m.y. to make it the fastest evolving group of brachiopods ever to have existed. This conclusion contradicts the relatively restricted range of rhynchonellacean morphology and another explanation has to be considered.

Since the phyletic history of brachiopods is here measured in genera, each of which is assumed to be morphologically unique, the rate of evolution may be a function of the number and complexity of those characters used to distinguish genera. Schopf et al. (1975) have considered the effects of this function on comparisons between phyla but we are concerned with the bias it may have introduced within relatively lowly taxa of familial or superfamilial rank. It is, for example, theoretically possible for a brachiopod stock with simple shells but rapidly evolving unsupported lophophores to be regarded as more conservative than a group with variably ornamented shells enclosing a stable lophophore because only evidence of skeletal differentiation survives fossilization. Even within the range of skeletal features there are ranks of complexity which favour some groups more than others in the promotion of genera. This relationship has been tested by estimating the number of character complexes used in defining the genera of the 46 superfamilies belonging to the phylum. By a character complex we mean a feature expressed in two or more states which has been used to distinguish genera. Ninety-four were identified during our review of generic diagnoses and include: shell dimensions, the attitude and dimensions of the cardinal areas, the nature of the various types of ornamentation and each of the several attributes related to the delthyrium and notothyrium, cardinalia, muscle scar complexes, mantle impressions and lophophore support systems.

There is a high positive correlation between the number of genera belonging to a superfamily and the range of characters used in their discrimination ($r = 0.74$ when the Rhynchonellacea are excluded). The relationship is predictably exponential and the line of best fit is curvilinear (Fig. 3) with the Paterinacea, Craniacea, Spiriferinacea, Clitambonitacea, Athyridacea, Atrypacea, Plectambonitacea and Terebratellacea forming a progressive series on, or close to, the curve. The scatter about the curve is noteworthy. Some terebratulide (Stringo-

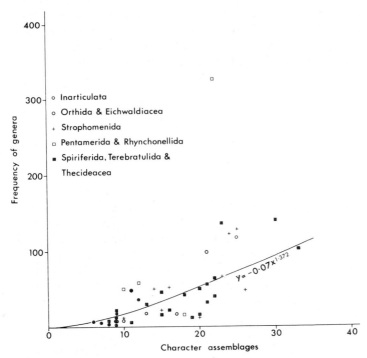

Fig. 3. The relation between the size of superfamilies and the number of characters used in the erection of genera assigned to the superfamilies.

cephalacea, Zeilleriacea), spiriferide (Cyrtiacea, Dayiacea, Suessiacea) and strophomenide (Richtofeniacea, Strophalosiacea) superfamilies contain fewer genera than expected in relation to their large number of character complexes. The converse is true of the Lingulacea, Pentameracea, Porambonitacea, Productacea, Rhynchonellacea, Spiriferacea and Strophomenacea. These deviations stem from different taxonomic attitudes. Students of the former groups tend to be "lumpers", those of the latter "splitters", although neither practice differs significantly from the more normal systematic procedure, except that governing the erection of genera assigned to the Rhynchonellacea.

In relation to its longevity and morphological variability, therefore, the rhynchonellacean superfamily is much too big. The fitted curves calculated for both these features (Figs. 2 and 3) suggest that it should contain 220 and 60 genera, respectively. These estimates fall too far short of the 325 genera assigned to the superfamily to be dismissed as the sort of error that can arise in the preparation of statistics of this nature. They are much more likely to indicate that rhynchonellacean genera are more finely drawn than in any other superfamily. This does not necessarily signify that systematists responsible for rhynchonellacean classification are using ill-founded criteria for generic discrimination. Only time will tell whether they and not the students of other superfamilies are adopting more realistic taxonomic standards. It should, however, be borne in mind that no attempt has yet been made to regroup the 325 genera

into more than one superfamily which suggests that the variability of rhynchonellacean morphology is unanimously admitted to be much less than among a
like number of genera belonging to any other articulate order. For this reason,
the quantitative data in the Rhynchonellida have been treated circumspectly.

General Aspects of Brachiopod History

Two aspects of brachiopod history require some attention before discussing
how the main groups, approximating to orders, originated. They are the rate
and diversification of phyletic evolution.

Quantification of evolutionary rates was first rigorously explored by Simpson. His use of "survivorship graphs" to illustrate the structure of phyletic
extinction and of "reciprocals of longevity" to estimate quantile rates of evolution constituted a major advance in the presentation of numerical data. These
devices were used by one of us (Williams, 1957) to demonstrate changes that
had taken place in estimates of rates of brachiopod proliferation and extinction
based on data compiled at different periods during the last century. Since 1957
the number of valid genera included in the phylum has almost doubled and the
stratigraphic ranges of the great majority are also better known. Consequently
it has been possible to calculate the mean longevity of 2037 extinct genera, in
relation to radiometric conversions of their chronostratigraphic ranges taken to
the nearest subsystem, as 36 m.y. The data used for these calculations have
been arranged in histograms (Fig. 4).

The larger histogram, based on a 5 m.y. interval, gives a reasonably faithful
picture of the extreme changes in generic frequencies which conform to an
overall positively skewed distribution relative to absolute time. It shows that
the shortest-lived genera were not less than 10 m.y. old and that the modal longevity was between 16 and 20 m.y. with subsidiary peaks between 46 and 50
and 56 and 60 m.y. These statistics, however, do not necessarily reflect the true
time ranges of extinct brachiopod genera even after allowance has been made
for approximations introduced by clustering at 5 m.y. intervals. Thus the
shortest-lived genera are those restricted to one of the three subsystemic divisions of the Silurian, each of which was estimated to have lasted for only 14
m.y. The modal longevity is shared by genera restricted to subsystemic divisions of the Devonian (16 m.y.), Triassic (18 m.y.), Cambrian (20 m.y.) and
Jurassic (20 m.y.). The subsidiary peak between 46 and 50 m.y. is compounded
of genera ranging throughout the Devonian or the Upper Devonian and Lower
Carboniferous (48 m.y.) or two-thirds of the Ordovician (50 m.y.); while that
between 56 and 60 m.y. includes genera spanning the Permian (56 m.y.), the
Cambrian (60 m.y.) and the Jurassic (60 m.y.). The only means of reducing this
disjunctive effect of basing calculations on subsystemic divisions of unequal
time spans is to plot generic frequencies for intervals of 32 m.y. which, except
for the Palaeogene, is the duration of the longest subsystemic divisions, the
Lower and Upper Carboniferous. The regrouping is shown as the inset histogram in Fig. 4. It conceals the fact that all genera must have existed for a deter-

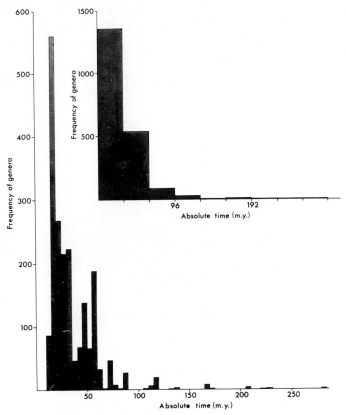

Fig. 4. The distribution of genera according to their longevity estimated in intervals of 5 and 32 m.y. for the larger and smaller histograms, respectively.

minable amount of time before becoming extinct but it does approximate to a die-away curve which has long been considered by many palaeontologists to be the normal survival pattern for an evolving group.

These positively skewed distributions are represented by similarly structured cumulative curves. Two have been prepared, one for extinct and another for living brachiopod genera, to facilitate comparison with other phyla (Fig. 5). We have not used the reciprocals of these curves to construct Simpson's decile distributions of evolutionary rates which seem an unnecessary elaboration. Notwithstanding the taxonomic and stratigraphic deficiencies in the data, rather because of them, it appears more appropriate to relate evolutionary rates to the more tangible statistics of generic frequency as a function of time expressed either as histograms or regression lines. Within this amended context the positively skewed brachiopod distribution, dominated as it is by short-lived genera, constitutes Simpson's horotelic pattern. His tachytelic and bradytelic patterns, respectively, represented by frequency distributions approximating to a positive skew rectangle and a gaussian curve, are rarely represented among brachiopod phylogenies. Productaceans, identified as a high diversity group in Fig. 2,

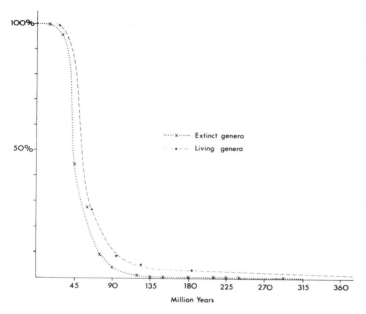

Fig. 5. Survivorship curves for extinct genera and for living genera with a geological record. The tail of the latter cumulative curve should be extended for another 75 m.y. to accord with the full geological record of *Lingula*.

have a mean longevity as high as 40 m.y. because they are mainly restricted to the Permo-Carboniferous with subsystemic divisions of greater than average duration. However, the cumulative curve for the superfamily climbs sharply and approximates to a tachytelic positive skew rectangle (Fig. 6). Low-diversity lingulaceans, in contrast, with a mean longevity of 50 m.y. for extinct genera are distributed on a less skewed curve than for the phylum as a whole although as a frequency distribution it is still more horotelic than bradytelic (Fig. 6).

Attempts to obtain reliable estimates of the phyletic diversity at any instant of geological time are constantly frustrated by the imprecise nature of the fossil record. This is especially so for periods when generic turnover is high which might equally well represent replacement within a group of constant size as successive expansions and contractions in the size of the group dependent on whether recruitment or extinction is dominant. Harper has recently (1975) discussed the problem and described a formula to meet some of the difficulties. We do not subscribe to his view that calculations can safely ignore taxa restricted to the interval of time for which information is required. For estimates based on geological periods or even epochs we prefer the formula:

$$D_{mi} = \frac{E}{2} + \frac{N}{2} + T_1 + \frac{T_r}{n}$$

where D_{mi} is the diversity at the mid-point in a stated interval of time; E, the

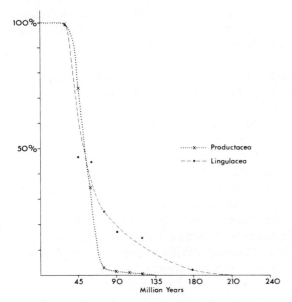

Fig. 6. Survivorship curves for productacean genera and extinct lingulacean genera.

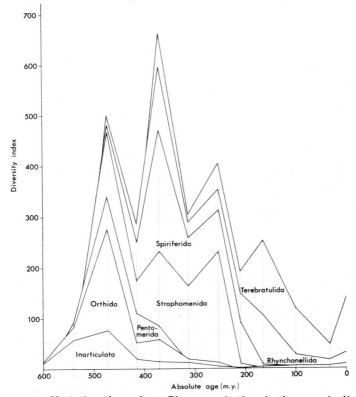

Fig. 7. Variation throughout Phanerozoic time in the generic diversity of the Inarticulata and the six orders (excluding the Thecideidina) making up the Articulata. The vertical dotted lines represent estimated mid-points for the Phanerozoic periods.

number of taxa becoming extinct during the interval; N, the new taxa appearing during the interval; T_1, the taxa ranging through the interval; T_r, the taxa restricted to the interval; and n, the interval of time measured in units of 10 m.y.

Fig. 7 shows the variation throughout the Phanerozoic of generic diversity among the Inarticulata (including the Kutorginida) and the six articulate orders recognized in this paper. The Devonian peak coincides with the rhynchonellide and spiriferide maxima and is further boosted by chonetidine and stringocephalacean climaxes, although these do not represent the fullest diversification of their respective orders. Ordovician diversity, on the other hand, is sustained by the maxima for the Inarticulata, Orthida (Clitambonitidina and Enteletacea as well as Orthacea), Pentamerida (almost exclusively Porambonitacea) and the Strophomenidina (mainly Plectambonitacea). Indeed with respect to the ordinal grouping of brachiopods only the peaks of strophomenide and terebratulide diversity culminate in other periods. The maximum for the former order embraces a proliferation of davidsoniaceans, strophalosiaceans, richthofeniaceans and lyttoniaceans and occurred during Permian times, while the Terebratulida is most widely represented by living terebratellaceans. The only other noteworthy maxima are those of the reticulariaceans and productaceans within the Carboniferous, and the terebratulaceans, dielasmataceans and pentameraceans within the Jurassic, Permian and Silurian, respectively.

Major Events in Brachiopod Evolution

The Brachiopoda conform to a pattern characteristic of most phyla in embracing a number of groups, mainly ordinal in taxonomic rank but also including aberrant subordinal and superfamilial units, which are morphological discontinuities in time and space. They are distinguished by certain persistent features which appear to be much more exclusive in phylogenetic development than was thought to be so a decade ago. There are, however, enough clues in other character combinations to delineate the broad path of phyletic descent, and the origin of the more important discontinuities is best illustrated by tracing the evolution of the phylum from its hypothetical prototype.

The prototypic brachiopod

The Precambrian lophophorate which was ancestral to a monophyletic brachiopod phylum was probably a triple-segmented coelomate with a poorly developed protosome undergoing atrophy, a pair of metanephridia and a recurved gut with the anus outside the feeding apparatus (lophophore) consisting, as in the juvenile trocholophe in living species, of a single row of filaments disposed in a ring around a centrally situated mouth (Fig. 8). It was attached to or within the substrate by a muscular extension of the ventral body wall (pedicle) which was capable of secreting a cementing mucopolysaccharide. The animal was protected by a pair of unarticulated valves secreted by a pair of

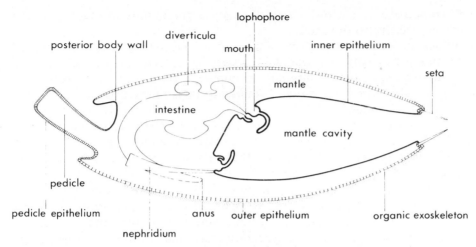

Fig. 8. The sagittal section of the inferred prototype of the Brachiopoda.

epithelial folds (mantles) which developed subsequent to the cleavage of a ring-like flap just anterior to the rudimentary pedicle. Initially the valves would have been entirely organic with an impersistent external film of mucopoly-saccharide underlain by a fibrillar periostracum consisting essentially of a triple-unit membrane of lipids and mucoproteins. However, towards the end of Pre-cambrian times the valves probably became stiffened by chitinous layers con-comitantly with the development of a musculature capable of mobilising the valves into more efficient protective/feeding attitudes. The chitin, or a muco-polysaccharide associated with it, then induced the epitactic growth of hydroxyapatite crystals which, initially, was a convenient way of getting rid of sporadic excesses of calcium and phosphate ions in the coelomic fluid circu-lating in an open system throughout the mantle canals.

Origin and phylogeny of the Inarticulata

The evolution of inarticulate brachiopods has been outlined by Williams and Rowell (in Williams et al., 1965) and our views on the history of the class do not differ much from that version except for amendments necessitated by recent studies of inarticulate shell structure and acrotretacean functional mor-phology. Since the class includes many of the earliest appearing, as well as the great majority of the most primitive, brachiopod stocks, it is possible that some were independent offshoots from a rapidly radiating and evolving prototypic group. This would explain why all four orders unequivocally assigned to the Inarticulata are represented in the Lower Cambrian rocks. On morphological grounds, the Lingulida and Acrotretida are comparatively closely related (an affinity confirmed by the ontogeny and anatomy of living species), the Obolel-lida much less so, and the Paterinida only remotely so (see Fig. 17). The close resemblance between modern lingulides and acrotretides may be partly attrib-utable to convergence; but they have always compared closely in musculature and shell composition, and have been distinguishable mainly in the disposition of the pedicle relative to the shell.

The valves of both orders have always been closed by a broken ring of adductor muscles usually segregated into two pairs, and slid or rotated by a group of interspersed oblique muscles. In contrast, the pedicle of acrotretides typically protruded through a notch, foramen or tube in the pedicle valve while that of lingulides emerged between the valves and was accommodated by median grooves in the thickened posterior arcs of the valves (pseudointerareas). Chuang (1971) has recently challenged this functional interpretation of acrotretide morphology preferring to believe that the foramen may have been an opening for an exhalant current from the mantle cavity and that the pedicle emerged between the valves as in the lingulides. We find this interpretation less satisfactory than the traditional one. It is noteworthy that the oldest stock assigned to the Acrotretida, the botsfordiids, were like contemporaneous lingulide obolids in possessing a grooved pseudointerarea, and only the morphology of the brachial valve suggests that the two families belong to different orders. Thereafter the acrotretide pedicle was accommodated by modifications of the posterior part of the pedicle valve or lost among craniaceans and the acrotretacean *Undiferina* (Rowell and Krause, 1973), so that their pedicle valves were almost invariably cemented to the substrate. The main evolutionary trend in the Lingulida was the atrophy of the pseudointerareas among the lingulids, including living species.

Of the two remaining inarticulate orders, the Obolellida which are restricted to the Lower and Middle Cambrian were calcareous-shelled and had a musculature similar to those of the Lingulida and Acrotretida. The pedicle emerged either between grooved pseudointerareas as in lingulides (e.g., *Obolella*), or through a foramen (e.g., *Trematobolus*), or an apical tube (e.g., *Alisina*) as in the acrotretides. The Paterinida, on the other hand, were typically inarticulate only in the chitinophosphatic composition of the skeleton and in the lack of articulation between the valves. Their elongate, divergent muscle tracks and pseudointerareas, which are median arched covers instead of grooves, are more reminiscent of articulate brachiopods.

The Kutorginida may also be conveniently referred to here as a calcareous-shelled Lower Cambrian order which is allocated to the Articulata only because of its elongate muscle tracks and shelf-like posterior sectors (cardinal area) in both valves. Both the Kutorginida and especially the Paterinida have been identified in the past as ancestors of Phanerozoic brachiopods. This viewpoint is no longer tenable; indeed a contrary interpretation is more feasible and serves to emphasize the continual recombining of characters that must have gone on among prototypic brachiopods before the main successful lines of descent emerged.

Another aspect of evolution, which is well illustrated by changes in the structure and composition of the inarticulate skeleton, is its re-iterative nature (Fig. 9).

The inarticulate skeleton is typically chitinophosphatic with layers of sclero-protein and chitin interspersed with lenses of apatite. Jope (1967) has identi-

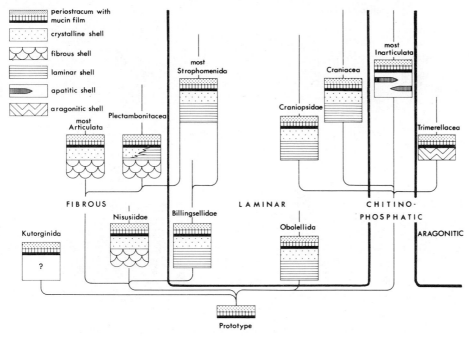

Fig. 9. The inferred relationships among the different types of shell structures found in the Brachiopoda.

fied hydroxyproline in chitinophosphatic shells suggesting that a collagen-type protein might contribute to the organic constituent of that kind of shell. Kelly et al. (1965) have also demonstrated that the *c*-axes of the apatite needles which are up to 100 nm long are alined in the plane of the shell parallel with the fibrillar lineation of the chitin. Thus chitin and subordinate collagen could have acted as organic frames for epitactic apatite seeds. In living lingulids both the outer mucopolysaccharide film and the proteinous periostracum developed external triple-unit membranes; among the Acrotretida other types of periostraca seem to have prevailed. The protegulum of many acrotretaceans appears to have been covered by a highly vesicular periostracum which served as a bubble-raft mould during deposition of the underlying shell (Biernat and Williams, 1970), although Ludvigsen (1974) found a pattern of overlapping circular depressions in the Devonian *Opsiconidion* which he preferred to ascribe to selective skeletal resorption. The entire surface of the siphonotretacean shell, in contrast, showed no superficial elaboration suggestive of periostracal differentiation (Biernat and Williams, 1971).

Although the chitinophosphatic shell is regarded as typical of inarticulates throughout the Phanerozoic, 26% of all genera belonging to the class had skeletons composed of scleroproteins and calcium carbonate (Williams and Wright, 1970; Schumann, 1970). They are not monophyletic (Fig. 9) but have independently evolved from a prototypic brachiopod stock, as in the case of the Obolellida, or from the Acrotretida (Craniacea), or the Lingulida (Craniopsidae

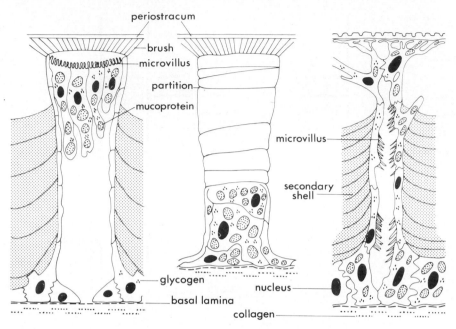

Fig. 10. Stylized sagittal sections of the caeca penetrating the shells of terebratellaceans (left), thecideidines (middle) and craniaceans (right).

and Trimerellacea). Biochemical changes which resulted in the emergence of all four groups must have repeatedly included not only secretion of calcium carbonate instead of apatite but exudation of a specific scleroprotein with comparatively high proportions of glycine and less alanine and suppression of collagen-type protein and chitin secretion (Jope, 1967). The carbonate constituent of the shell is also identical for the Obolellida, Craniacea and Craniopsidae because it consisted of a granular or acicular crystallite primary layer and a laminar secondary layer which grew by perpetuating screw dislocations. The Trimerellacea appear to have been unique among brachiopods in possessing an aragonitic shell (Jaanusson, 1966).

Finally it is worth remarking that the craniacean shell (Fig. 10) is pierced by fine canals (puncta) which accommodate mantle papillae (caeca) acting as storage centres for muco- and glycoproteins (Williams and Wright, 1970). The shells of some articulate groups are also permeated by storage caeca (Owen and Williams, 1969; Williams, 1973), although these with their microvillous brushes connecting with the periostracum are structurally different from the craniacean caeca with their sparsely branched, blind terminations (Fig. 10).

Origin of the Articulata

The earliest articulate brachiopods are Lower Cambrian orthides including the billingsellaceans, *Nisusia* and *Eoconcha*, and the orthacean *Eorthis*. These differ from inarticulates in the presence of posterior planar cardinal areas (interareas) with their growing edges (hinge lines) in contact. The ventral inter-area bears a median triangular notch (delthyrium) flanked by a pair of pro-cesses (teeth); the dorsal interarea, a median notch (notothyrium) bounded by a pair of hollows (sockets) fronted by flat-lying socket plates. The teeth and accommodating sockets formed the pivoting articulatory device which was operated by a different muscle system from that found in typical inarticulates. The adductors were attached to a posteromedian muscle scar in the ventral umbo and a quadripartite scar in the brachial valve. The diductors, which opened the shell, flanked the ventral adductors in the pedicle valve but were attached to the notothyrial floor or outgrowths of it (cardinal process) in the brachial valve. In addition, adjustors, controlling the contraction of the pedicle, were normally inserted on either side of the ventral diductors and on the socket ridges.

These differences in the articulation of the valves and the arrangement of the muscles operating them are hinted at in the Paterinida and Kutorginida, which suggests that they were viable character assemblages, probably neotenously derived and promoted during the prototypic brachiopod radiation. As a foun-dation for a new class of organisms, they were supplemented by a basic distinc-tion within the orthides and by another novel development affecting the skele-tal fabric.

The principal morphological distinction among the Orthida (Fig. 11) is between those groups with delthyria and notothyria that were primarily open as in *Eorthis* and those in which the delthyria and notothyria are primarily cov-

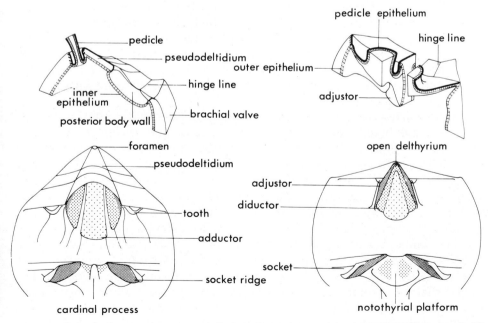

Fig. 11. The morphology of the ventral and dorsal interiors of billingsellaceans (left) and orthoceans (right) with the inferred relations between the pedicle and shell shown above.

ered by continuous arched strips of shell (pseudodeltidia and chilidia, respectively). Williams and Rowell (in Williams et al., 1965) believed that persistence
of an entire pseudodeltidium throughout ontogeny and the concomitant
restriction of the pedicle foramen within the pedicle valve meant that the
posterior junction between the two valves was underlain by a strip of inner epithelium as in inarticulates, and that the pedicle was an outgrowth of the ventral
body wall. In contrast, the open delthyrium and notothyrium of *Eorthis* and all
other orthaceans and related enteletaceans indicated that the pedicle developed
from a rudiment which was differentiated quite early in the ontogeny of the
embryo as in living articulates. Such differences could have been promoted by
paedomorphic changes in the ontogeny of a billingsellacean stock leading to the
emergence of the orthaceans which, in turn, were ancestral to the Pentamerida,
Rhynchonellida, Spiriferida, Terebratulida and Thecideidina. The billingsellaceans also persisted and gave rise to the Clitambonitidina and Strophomenida
which, in our opinion, became extinct in Late Permian or Early Triassic times.

Differences in the skeletal fabric also suggest a basic division of the Articulata more or less along the lines indicated by the presence or absence of pseudodeltidia (Fig. 10). The differences derive from the secretion of an entirely new
kind of carbonate layer during formation of the post-protegular skeleton. All
articulate and inarticulate carbonate shells, with the exception of the trimerel-
lacean, consist of external organic layers, an underlying finely crystalline or
granular primary layer, and a secondary layer composed of discrete carbonate
units segregated from one another by proteinous membranes. In the craniaceans, craniopsids, obolellides and some articulates shortly to be identified,
these units consists of tabular or blade-like laminae which grew spirally or vec-
torially by accretion at the edges. In the great majority of articulates, however,
each carbonate unit is a long fibre secreted by the posterior part of an outer
epithelial cell on a seeding membrane exuded by the anterior arc of the same
cell; and, because the cells are regularly arranged in alternate rows, the fibres
form a corresponding mosaic of keeled structures separated from one another
by interconnected protein sheets.

Such a fibrous secondary shell was already characteristic of the earliest
known articulate brachiopods, the nisusiids *Nisusia* and *Kotujella* (Williams,
1968b), and in view of its occurrence in other Cambrian orthaceans, like
Oligomys and *Orusia* (Williams, 1970), it probably was also part of the skeletal
succession of *Eorthis*. The skeletal structure of post-Cambrian species is now
sufficiently well known to identify the essentially fibrous groups. They are: the
Nisusiidae and all other Orthida except the Billingsellidae; the Pentamerida,
Rhynchonellida, Spiriferida, Terebratulida and Thecideidina; and the strophomenide Plectambonitacea. Some of these stocks are punctate including even
the Early Cambrian *Kotujella*, but, since at least two different types of caeca
are known (Fig. 10), this condition is almost certainly polyphyletic. The
secondary shell of plectambonitaceans and orthide gonambonitaceans is penetrated by calcitic rods (taleolae) which are secreted simultaneously with the
fibres (pseudopunctate). As will be shown later, this condition is also polyphyletic.

A monophyletic fibrous secondary shell could only have arisen as a caeno-genetic novelty. Its accretion involves a different type of secretion from that giving rise to spirally growing tabular laminae. Consequently it may have evolved from a prototypic chitinophosphatic brachiopod independently of other Early Cambrian stocks with carbonate shells (i.e., the Obolellida and Kutorginida). A morphological comparison between the billingsellids and the older nisusiids supports the view that the former were derived from the latter. Yet the secondary shell fabric of at least *Billingsella* is laminar although the units are arrays of blades not tablets as in inarticulates and, therefore, possibly derived by a flattening of the inner and outer surfaces of nisusiid fibres (Williams, 1970).

The discovery that the billingsellid secondary shell consists of flat-lying blades confirms long-held suspicions that the Strophomenida are likely to have been polyphyletic (Fig. 9). The strophomenaceans are morphologically close to the earlier appearing plectambonitaceans; their invariably pseudopunctate secondary shell, consisting of arrays of lath-like laminae disposed at acute angles to one another in successive layers (cross-bladed condition of Armstrong, 1969), could have been derived from laminae introduced paedomorphically into an indefinitely growing primary layer in plectambonitaceans following the suppression of a fibrous secondary layer. This branch of the order appears, then, to have descended from the Nisusiidae. The cross-bladed laminar secondary shell of the Triplesiacea, on the other hand, is invariably impunctate and that of the Davidsoniacea became pseudopunctate only during Devonian times. This discrepancy in shell structure and other features suggest that neither superfamily is closely related to contemporaneous strophomenaceans but both could have descended from the laminar-shelled *Billingsella*. The convergence in the skeletal ultrastructure among Upper Palaeozoic strophomenides is as striking as the initial divergence among the billingsellaceans, which gave rise to both ancestral sources of the order.

Orthide phylogeny

The flat-lying socket plates, characteristic of Early Cambrian orthides, persisted in the Clitambonitidina which were mainly remarkable for the development of spondylia supporting ventral muscle bases and a lengthening of the ventral interareas. But by mid-Cambrian times, the socket ridges of many orthacean species had rotated into the submedian plane to project ventrally as a pair of rods or blades (brachiophores) secreted within invaginations of outer epithelium (compare Figs. 12 and 13). These continued to provide attachment areas for the dorsal adjustors. They also underwent elaboration in several familial groups including the development of plates to form elevated sockets and supporting struts reaching to the floor of the valve which were disposed in various attitudes and exceptionally were greatly enlarged to form a platform (cruralium) bearing the dorsal adductor bases, as in *Mystrophora*. The most significant development involving the brachiophores, however, resulted from their distal growth in a ventro-anterior direction so that in a minority of orthi-

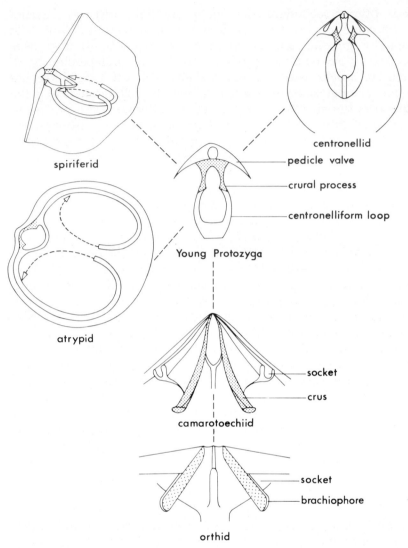

Fig. 12. A morphological series showing the derivation of some brachiopod brachidia from crura and brachiopores (after Williams et al., 1965).

des, like *Phragmorthis*, they must have been long enough to support the lophophore on either side of the mouth sector. This elaboration of the brachiophores anticipated the growth of crura performing the same function in other articulate orders and were the precursors of more complicated lophophore supports like loops and spiralia (Fig. 12).

During their phylogeny, the orthides underwent a wide range of external and internal modifications. These included the growth of: deltidial plates over the delthyrium (*Barbarorthis*); spondylial (skenidiids) and shoe-lifter (*Parenteletes*) structures supporting ventral muscle bases; blade-like dorsal septa dividing the mantle cavity into two compartments (*Phragmorthis, Kayserella*); the reduction of interareas (*Productorthis*); and the adoption of every basic brachiopod shape and outline including that of the bilobed *Dicoelosia* (Wright, 1968). Such stocks were relatively short-lived but their evolution anticipated similar divergences among later, unrelated articulate species. The only important features found in other orders which did not appear during orthide radiation were: cementation of the shell following loss of the pedicle; the persistent growth of surface spines; and the development of elaborate lophophore supports. With respect to the last feature we agree with Jaanusson (1971) that the aberrant Devonian *Tropidoleptus*, previously accepted as an enteletacean (Williams and Wright, 1961), is unlikely to be an orthide on account of its complexly shaped teeth (cyrtomatodont). Provisionally the genus has been assigned to the Terebratulida. Yet even with the exclusion of this stock, the orthides must have occupied, at some time or other during their existence, as many ecological niches as did the remaining articulate brachiopods.

Strophomenide phylogeny

The Strophomenida constitute the largest ordinal grouping of brachiopods and are so distinctive even in their full diversity as to be immediately identifiable. Yet the order is far from homogeneous and, apart from its probable polyphyletic origin already referred to, its contents and phylogenetic relationships are controversial. In 1965 Muir-Wood (in Williams et al., 1965) described *Chonostrophia* as having descended from the strophomenaceans, the rest of the chonetidines from the plectambonitaceans, and the productidines from the strophomenacean *Leptaenisca*. The lyttoniaceans were believed to be productide in origin. In 1970 Rudwick derived the productidines exclusively from the chonetidines, the latter from the strophomenaceans, and the lyttoniaceans from the davidsoniaceans. More recently Brunton (1972), mainly by comparisons of shell fabric, confirmed Havlíček's identification (1967) of the plectambonitacean aegiromenids as likely ancestors of the chonetidines and suggested that productidines were linked to the latter. Dr. Brunton (personal communication, 1976) remains sceptical about the inclusion of *Chonostrophia* among the chonetidines rather than the strophomenaceans. Grant (1972) in a survey of newly discovered lophophore supports within certain productidines, continued to identify *Leptaenisca* as ancestral to the suborder. We are satisfied that the chonetidines were paedomorphically derived from a precocious sowerbyellid like *Aegiromena* equipped with hinge spines and that, on balance, the productidines arose paedomorphically from a strophomenacean like *Leptaenisca*, again through the acquisition of spines, although in this stock scattered over the shell surface.

Certain post-Permian groups including the spire-bearing koninckinaceans (with *Cadomella*) and *Thecospira*, and the Thecideidina and *Bactrynium*, have also been classified as strophomenides by various researchers mainly because of certain morphological similarities like shell shape and lophophore supports, the presence of "pseudodeltidia", misinterpretations of shell structure, and especially the cementing habit of many of the species in question. However, it is now known that the pedicle foramen of koninckinaceans is not supra-apical in position as in strophomenides (Brunton and MacKinnon, 1972), and that, together with *Thecospira*, members of that superfamily have an orthodoxly fibrous secondary layer (Williams, 1968a; MacKinnon, 1971). Furthermore, as will be shown later, the thecideidines (including *Bactrynium*) are more likely to have evolved from a spiriferide or terebratulide than a strophomenide (Williams, 1973). With these changes in mind it is possible to identify the main trends in strophomenide phylogeny more succinctly than previously.

The most diverse modifications of the primitive strophomenide shell with its semicircular, concave—convex shape, small supra-apical foramen, umbonal ventral muscle scar, and simple notothyrial platform were undoubtedly promoted by the loss of the pedicle. The presence of a pedicle affords a brachiopod a certain amount of mobility which is probably critical to its survival under normal conditions. In terebratulides, for example, contraction of the pedicle not only tilts the anterior margin upwards away from the substrate but also rotates the shell through 270°. Both movements assist in the adoption of feeding postures and in precluding settlement by sedimentary particles or microbenthos. Yet atrophy of the pedicle can be compensated for by certain changes in skeletal growth so that it was not a rare event in brachiopod evolution; what was unique in the case of early strophomenides was its loss in flat-shelled species which were consequently liable to burial by rapid sedimentation.

The least response to such a selection pressure was in the first appearing groups, the plectambonitaceans and strophomenaceans, among which a functional pedicle sporadically persisted in young shells and even in some adults of the Carboniferous *Leptagonia*. In this leptaenid the anterior shell margin was elevated above the substrate by a secondary biconvex growth supplemented by a sharp concentric bend (geniculation) in the shell. A similar geniculation directed dorsally or ventrally (resupination) occurred during the growth of many early strophomenide stocks and contributed to their survival. Species with flat or gently convex shells also continually evolved but these normally displayed very large, muscle scars and might have been capable of limited movement by a snapping action of the valves as in the bivalve *Pecten* (Williams, 1953). In the light of the subsequent history of the Order, however, the most significant response was that of the strophomenid *Leptaenisca* and the stropheodontid *Liljevallia*. Both were attached to the substrate during early stages of growth by cementation of the pedicle valves.

The other two groups, the triplesiaceans and davidsoniaceans, which descended directly from a billingsellacean ancestor were basically biconvex in shape and therefore less affected by pedicle atrophy. Triplesiaceans probably retained a functional pedicle throughout their history irrespective of a repeated elongation of the pedicle valve which was always matched by a complementary lengthening of the forked cardinal process. The earliest davidsoniacean also had a functional pedicle, but the Late Ordovician *Fardenia* was free-lying, and the majority of Devonian davidsoniacean species were either cemented to the substrate by the ventral umbonal area throughout life, or at least during the early stages of growth. Umbonal cementation accompanied by accelerated growth of the ventral interarea led to subconical forms with a lid-like brachial valve as in *Meekella*, one of many trends conducive to high-level suspension feeding by cemented strophomenides during late Palaeozoic times.

Among the later appearing strophomenides, the development of spines occurred at least three times in response to selection pressures, which must have militated against the survival of immobilized free-lying shells. The spines were not superficial but were secreted by persistent cylindroid evaginations of the mantle, each tipped by a generative zone responsible for the exudation of an external organic cover and for the proliferation of generative epithelium to keep pace with the lengthening spine (Williams, 1956). These extensions of the mantle did not always secrete carbonate tubes (e.g., some chonetacean daviesiellids) although presumably they could have extended beyond the shell surface within cylinders of stiffened periostracum. In any event all types of spines must have been capable of exuding an outer mucopolysaccharide film which could have assisted in the adhesion of the animal to the substrate.

The spines of chonetidines are restricted to the posterior edges of the ventral interarea and probably acted as anchors for adult forms subsequent to the loss of a pedicle which was functional in the juveniles of many Silurian and Devonian species (Muir-Wood in Williams et al., 1965). In view of the evidence for the derivation of the group from an *Aegiromena*-like stock, it is noteworthy that the Late Ordovician *Eochonetes*, also a sowerbyellid, developed oblique canals piercing the ventral interarea. These canals must have accommodated evaginations of the mantle identical to those occupying chonetidine spines, although *Eochonetes* appears to have been aberrant and without descendants.

The arrangement of productidine spines was much more complex and involved a degree of specialization unrivalled in any other group of brachiopods. A functional pedicle existed in certain juvenile Carboniferous productaceans (Brunton, 1965) and may have persisted in far more productidine stocks, although traces of it are likely to have been obliterated by the ventral cicatrix of attachment. In the juvenile stages of growth, attachment to cylindroid objects like a crinoid stem or a bryozoan branch was further facilitated by the development along the cardinal margin of a pair of clasping spines (rhizoid) which converged to form a ring around the substrate. As the animal grew, further rhizoid spines appeared, as well as others (halteroid) scattered over the surface of one or both valves, either as sets of symmetrically disposed strut-like spines (*Muirwoodia*) or as dense brushes of long thin spines (*Waagenoconcha*). These halteroid spines served to buoy up and anchor on the soft substrate those adults which broke away from the elevated attachment areas they occupied as juveniles (Grant, 1966). Mats of fine short spines also developed and must have served as protective devices.

Regrouping of halteroid spines to form efficient floating platforms was rife and led to the appearance of bizarre forms. This was especially so when trends maintaining species as high-level suspension feeders resulted in a disproportionate growth of the pedicle valve into a cone-like structure stabilized by proximal halteroid spines. In such stocks, the soft parts were usually restricted

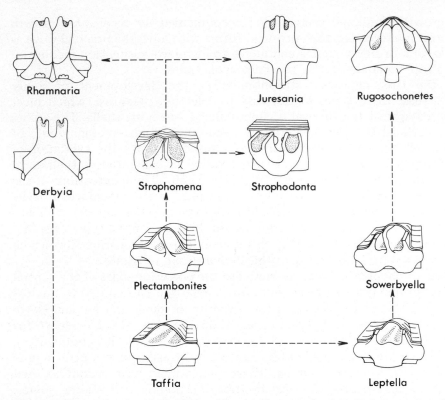

Fig. 13. The main morphological grades of the notothyrial platform and cardinal process found in the Strophomenida (after Brunton, 1968; Muir-Wood and Cooper, 1960; Williams et al., 1965).

to the distal zone of the pedicle valve by cystose plates and the brachial valve was either like an external lid, as in *Scacchinella*, or so deeply inserted in the cone as to have operated like a trap-door beneath a carbonate meshwork made up of ventral apertural spines, as in the richthofeniacean *Hercosestria*.

A number of other trends repeatedly occurred during strophomenide evolution. One of the most orderly, affecting all major groups, involved the fashioning of a cardinal process out of the notothyrial platform (Fig. 13). In the most primitive stocks, like the plectambonitacean *Leptella*, the dorsal ends of the paired adductor muscles were implanted directly on the notothyrial floor and covered by an arched canopy of shell (chilidium). In many plectambonitid descendants, a cardinal process appeared as a median partition extending to the chilidium or discrete chilidial plates, and must have separated the adductor ends from each other. In most later plectambonitaceans, concomitant posterior growth of the notothyrial walls tended to form a trilobed structure with the median partition; and, when all three projected ventrally as in *Sowerbyella*, they delineated a pair of slot-like troughs for the insertion of the adductor ends. The chonetacean cardinal process, although variably lobed ventrally and posteriorly, seems to have been derived from the sowerbyellid trilobed structure by a superficial cleavage of what was originally the median partition.

Even among the most primitive strophomenaceans, davidsoniaceans and triplesiaceans the dorsal ends of the adductors were attached to a pair of outgrowths from the notothyrial floor, which constitute the typical strophomenidine bilobed cardinal process. Only in some primitive strophomenaceans (*Rafinesquina, Leptaena*) are there traces of a vestigial median partition between the lobes (Fig. 13). Generally the trend in all strophomenidines (e.g., *Strophodonta*) was for the lobes to become discrete structures growing posteroventrally away from the floor of the brachial valve where all signs of the notothyrial platform and chilidium are lost. The productidine cardinal process, although basically a bilobed structure, was subject to much more differential growth and is extremely variable in morphology. In many stocks (e.g., *Juresania*), fusion of the submedian sides of the two lobes produced a trilobed design comparable with that of the chonetidines (*Rugosochonetes*); while in forms like *Rhamnaria* the distal ends remained discrete like those found in triplesiaceans and later davidsoniaceans (*Derbyia*). In *Titanaria*, atrophy of the lateral walls of an ankylosed pair of lobes resulted in a simple myophore and shaft as in orthaceans.

The spread of complementary processes and hollows along the hinge-lines (denticulation) was also a recurrent trend. It occurred three times among plectambonitaceans (plectellids, *Plectodonta* and *Sampo*): spreading laterally of persistent teeth and sockets. Among the stropheodontids, denticles first appeared on the teeth and dental plates and did not spread along the hinge-line until each tooth had fused with its supporting dental plate to form a single ledge-like feature known as the denticular plate (Williams, 1953).

The most spectacular trend in its culminative effect was promoted by the weak development or absence of socket ridges or any other outgrowths from the dorsal hinge-line which could have evolved into apophyses supporting the mouth segment of the lophophore. In the absence of such supports and as an outcome of the general narrowness of the mantle cavity, ridges and platforms arising from the floor of the valve were repeatedly evolved to accommodate the lophophore (Fig. 14).

This trend became apparent early in the phylogeny of the order because bilobed platforms or rows of small spines, which are assumed to have supported a schizolophe with a single row of filaments, appeared in several Ordovician stocks including the plectambonitaceans *Leptellina, Bilobia* and *Sericoidea*, and the strophomenacean *Christiania*. In two younger strophomenaceans, *Leptaenisca* and *Leptodonta*, a pair of ridges curving into one or two convolutions occur on the valve floor (brachial ridges) and probably gave support to an attached lophophore. On configuration alone, one would expect the lophophore to be disposed as a pair of plano-spiral lobes or a convoluted ptycholophe, which is how Grant (1972) has interpreted the disposition of the lophophore in relation to similar brachial ridges found in productidines. Yet in some stocks like *Gigantoproductus*, the ridges are associated with subconical impressions and the ensemble suggests a plectolophe with side arms attached to the valve floor and disposed in the same plane as a pair of low spires representing

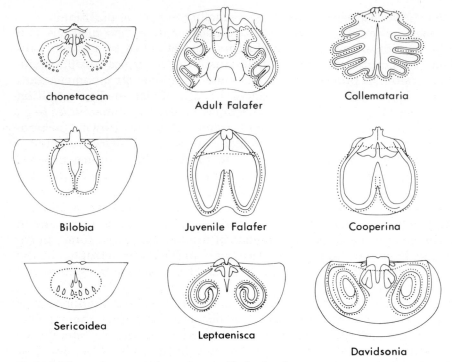

Fig. 14. The main types of strophomenide lophophore supports with the inferred disposition of the lophophore indicated by broken lines.

the younger parts of the lophophore. Coiled, ridge-like impressions found on the floor of the pedicle valve of *Davidsonia* suggest that the davidsoniacean lophophore was spirolophous, while the chonetidine schizolophe (or spirolophe) was supported not only by brachial ridges but also sporadically by a pair of processes (anderidia of Brunton, 1968) which extended anteriorly from the base of the cardinal process to the inferred position of the mouth segment of the lophophore. The anderidia must, therefore, have functioned like crura.

Diverse as these inferred lophophore supports are, they are not as spectacular as those discovered by Grant (1972) in certain productaceans, nor as bizarre as those of the oldhaminidines.

The interior of strophalosiacean cooperinids like *Falafer* is dominated by a pair of lobate, crenulate outgrowths directed posteriorly from the floor of the brachial valve. This structure is like an exaggerated version of the platform found in some plectambonitaceans like *Bilobia* and, as Grant has demonstrated, could only have served as a support for a ptycholophe. Following his discovery Grant concluded that the prevailing attitude of the productidine lophophore was ptycholophous, although it was normally unsupported.

The oldhaminidines represent the climax to strophomenide divergence. The suborder embraces a small number of Permo-Carboniferous genera which have been so greatly modified by paedomorphosis that their antecedent can only be broadly identified as an aspinose productidine (the spines which prompted

Sarycheva to found *Spinolyttonia* apparently belong to a productidine attached to a normal specimen of *Lyttonia*). The pedicle valve varies in shape from a free-lying or cemented cone reflecting an essentially holoperipheral growth (Cooper and Grant, 1974) to a flat attached plate which was formed when part of the valve everted itself as a posterior flap and became cemented to the substrate. The internal floor of the adult pedicle valve is ridged by septa disposed laterally or, more rarely, radially. The skeletal succession consists orthodoxly of a thin cryptocrystalline primary layer and a laminar pseudopunctate secondary layer.

The "brachial valve" is flat and lobate in outline so that it fits within the septa on the floor of the pedicle valve. On morphological grounds (Termier and Termier, 1949) and from a study of shell structure (see Williams, 1973) it was independently concluded that this lobate skeletal piece was not a normal valve but an internal plate supporting a ptycholophe. Study of the shell structure shows that only the posterior triangular area behind the bilobed cardinal process represents a vestigial valve (Williams in Williams et al., 1965). The remaining lobate part is composed exclusively of laminar secondary shell with pseudopuncta directed externally and internally (Williams, 1973). This arrangement could only have come about if the entire structure had been a plate secreted within a fold of epithelium in the same way as the lobate lophophore supports in *Falafer*.

Thus the oldhaminidines, although remaining functional bivalves, were the only brachiopods ever to approximate to structural univalves. Their appearance represented the most extreme departure from the main line of descent in the history of the phylum.

Pentameride phylogeny

The pentamerides were never a dominant group of brachiopods and were, moreover, restricted to the Lower and Middle Palaeozoic successions. None the less, they are usually regarded to be among the least difficult to identify although the principal diagnostic features are the culmination of a series of trends which affected the Order.

The earliest pentamerides, like the Middle Cambrian porambonitacean *Cambrostrophia*, are comparable with contemporaneous orthaceans in their open delthyrium, discrete dental plates and rudimentary brachiophores. They differed from what must have been an orthacean ancestor only in the presence of a dorsal fold (uniplicate condition) and a reduction in the size of interareas. The uniplicate condition persisted in all porambonitaceans which, with the exception of *Anastrophia*, are restricted to the Cambro-Ordovician, but it is found in only a minority of the younger pentameraceans (e.g., *Clorinda*) which range from Late Ordovician to Late Devonian in age and are mainly sulcate or evenly convex. The reduction of the interarea and hinge-line is more comprehensive. All late porambonitaceans and most pentameraceans are rostrate, and the few pentameraceans, like *Stricklandia* and *Aliconchidium*, which are not, probably acquired a hinge-line secondarily.

The two most significant trends affecting the shell interior involved modification of the plates associated with the articulatory devices of the valves. In the earliest porambonitaceans the ventral muscle bases were attached to the postero-median segment of the floor of the pedicle valve and to the inner surfaces of a pair of discrete dental plates bounding that segment. This condition persisted in a few Ordovician stocks like *Porambonites*, but Late Cambrian huenellids and clarkellids display every stage in the elevation of the ventral muscle field from the development of a raised median pad of secondary shell (pseudospondylium) to a trough-like feature formed by the convergence of dental plates (spondylium). This structure is the dominant mode of support for the ventral muscle field among post-Cambrian porambonitaceans and the exclusive one for pentameraceans among which it may be supported by a median septum as in *Holorhynchus*. A distinction is frequently drawn between a spondylium "simplex" and "duplex" in the belief that a dividing partition of prismatic calcite found in the supporting septum of the latter indicates an incomplete fusion of dental plates. It is unlikely that this shell differentiation, which is sporadic in taxonomic distribution, is diagnostic of an imperfect ankylosis of dental plates (Williams and Wright, 1961).

Differentiation of the cardinalia also occurred during pentameride phylogeny. By Late Cambrian times brachiophores supported by divergent to convergent bases and subtending socket plates had become well defined. In a few porambonitaceans, like *Camerella*, and in all pentameraceans, the brachiophores were long enough to have functioned as lophophore supports and almost certainly acted in that capacity in the manner of crura. The climax to the trend whereby skeletal support was given to the lophophore is the apparatus of the Middle Devonian *Enantiosphen*. This is an impunctate rostrate form with typically pentameracean spondylium, cruralium and cardinalia, except that the crura support a pair of wide blades linked by a transverse plate. The entire structure so simulates a terebratulide loop that there is still some controversy about the affinities of the genus (Jaanusson, 1971), although the general morphology, the impunctate shell and the nature of the loop itself confirm that it is more likely to be a pentameracean than a centronellidine terebratulide. In that event the apparatus probably supported a schizolophe, although a spirolophe may have been more typical of other pentameraceans (Williams and Wright, 1961).

The supporting bases of the crura were also subject to phylogenetic changes. In some porambonitaceans and the majority of pentameraceans they remained discrete, but in most of the younger porambonitaceans they converged to form a septalium and in a few pentameraceans (e.g., *Pentameroides*) they were large enough to receive the dorsal adductor bases (cruralium). In contrast, the bases atrophied more or less synchronously in British and American stricklandiid lineages during Llandovery times (Williams, 1951; Boucot and Ehlers, 1963).

The pentameride shell succession, which basically consists of crystalline primary and fibrous secondary layers, is frequently made more complicated in pentameraceans by the presence of prismatic calcite either as sporadically occurring lenses to the secondary layer or as the sole constituent of a continu-

ous tertiary layer. Whether the appearance of a tertiary layer follows a phylogenetic pattern is not yet known despite attempts by Gauri and Boucot (1968)
to use pentameracean shelly fabrics systematically.

In retrospect, pentameride evolution is notable for the emergence of an articulate brachiopod which, with its rostrate shell and functional crura, approximated to the modern form. Both characters occurred rarely in the Orthida, but
in the Pentamerida the combination was relatively successful and apparently
provided the ancestors of extant orders.

Rhynchonellide phylogeny

Although rhynchonellides were never a dominant group of brachiopods, they
constitute one of the most successful orders because, having first appeared in
Early Ordovician times, they still survive in a wide range of marine environments today. Their ancestral connections can be inferred from the modal characteristics of the earliest stocks like *Ancistrorhynchia*, *Oligorhynchia* and
Rostricellula. Species of these genera were normally uniplicate, non-strophic
forms with greatly reduced ventral cardinal areas and an open delthyrium occasionally restricted by deltidial plates. Internally, divergent dental plates support
laterally set teeth while sockets were delineated by concave plates bounded by
ridges and separated from functional crura by outer hinge plates. The preserved
skeletal successions include a finely crystalline primary layer and a secondary
layer composed of orthodoxly stacked fibres. Many of these features independently evolved among orthaceans but as a combination they are decidedly
porambonitacean and the prototypic rhynchonellide probably descended neotenously from that superfamilial group by the precocious development of crura
and suppression of a tendency to form a spondylium.

The lack of a spondylium among the earliest rhynchonellides may have been
linked to a forward shift of the ventral muscle field. In the Orthida, Strophomenida and Pentamerida, the ventral muscle bases occupied the umbonal area
and part of the valve floor; the pedicle base must have been superficial. In the
Rhynchonellida the umbonal area has always accommodated a sunken pedicle
base and the muscle field is inserted forward of that area. This condition is also
typical of the Spiriferida and Terebratulida and has been interpreted (Williams
and Wright, 1963) as an adult expression of mantle reversal which occurs
during the ontogeny of all living articulates. Mantle reversal may, therefore,
first have taken place only with the emergence of the Rhynchonellida.

Very few morphological changes underwent divergent trends. Deviations
from the basic morphological facies appear from time to time, like the secondary growth of cardinal areas in the Ordovician *Orthorhynchula* and the Triassic
Hallorella, the frequent recurrence of a cardinal process, and the presence of
puncta in *Rhynchopora* and *Tretorhynchia* (Williams, 1968b; Brunton, 1971);
but none of them is phylogenetically significant except for muscle platforms
which distinguish the stenoscismataceans from rhynchonellaceans. These platforms include a trough-shaped receptacle (camarophorium) for the dorsal
adductor muscles and a spondylium for the ventral muscle field. Both struc-

tures are fully developed in the earliest known stenoscismatacean, the Devonian *Atribonium*, and evolved by convergence of the dental plates and as outgrowths of the dorsal median septum respectively (Grant, 1965). The presence of these structures prompted Rudwick (1970) to assign the stenoscismataceans to the Pentamerida, but he ignored the facts that the spondylium evolved, sometimes repeatedly, in every other order and that the camarophorium is structurally distinct from a cruralium. We therefore regard his proposed systematic revision as untenable.

Rhynchonellide conservatism is well typified by the almost invariable uniplicate condition of the shell. Indeed the chief feature of this history of the Order is the stability of a basic design which may, through its very simplicity, have been the key to survival.

Spiriferide phylogeny

Over 500 genera of spire-bearing brachiopods constitute the second largest ordinal grouping within the phylum after the Strophomenida. Their classification, however, is controversial and, although this is readily admitted by the cognoscenti, every scheme proposed in place of that used by Pitrat et al. in the *Treatise* suffers the same handicap of having been erected without due regard for the full range in time of all the characters of taxonomic importance. Both Copper (1967) and Rudwick (1970), for example, advocate the separation of the atrypidines from the spiriferidines by the erection of a new order for the former group. This proposal may prove worthwhile, but not necessarily for the reasons advanced by them. Copper maintained that the well-known links between early atrypidines and contemporaneous rhynchonellides, and the absence of any connecting forms between atrypidines and spiriferidines called for an atrypide order closely associated with the Rhynchonellida, while Rudwick went so far as to consider the orthides as a possible source of the spiriferidine ancestor.

These arguments do not take into account the extreme variability of the earliest spire-bearers like the atrypidine zygospirids. All zygospirids are distinguishable from contemporaneous rhynchonellides at least in the presence of a loop or spiralia, and usually by the muscle scars and the form of the shell. Both groups share obsolescent cardinal areas, incipient deltidial plates and crura with well-developed inner socket ridges. The distinction is no less profound than that between other orders and is impressive when one considers that the atrypidines were almost certainly neotenously as well as paedomorphically derived. This is demonstrated by the fact that the average length of 46 Ordovician spire-bearing species was less than one-half the modal length of 1113 contemporaneous species belonging to other articulate orders (Williams and Wright, 1961). Thereafter, the descent of Lower Silurian spiriferidines from a zygospirid stock like *Catazyga*, with its secondarily developed cardinal areas, involves fewer morphological changes than the derivation of the atrypidines from the rhynchonellides and precludes further speculation on alternative ancestors for the spiriferidines.

A study by MacKinnon (1974) of the evolution of spiriferide shell structure provides guide-lines to the use of differences in skeletal fabric in determining intra-spiriferide affinities. MacKinnon showed that the earliest spiriferides were characterized by orthodoxly developed impunctate primary and secondary shell successions and that tertiary prismatic layers appeared independently in the Athyridacea, Atrypacea, Dayiacea, Koninckinacea and Reticulariacea. He also found (1971) that the puncta of Jurassic *Spiriferina* must have accommodated caeca with terminal microvillous brushes perforating distal canopies of primary shell as in living terebratulides and thecideidines (Fig. 10). Now that the microvillous brush is known to reflect the condition of secreting epithelial cells at the mantle edge and that caeca have been shown to be storage centres (Owen and Williams, 1969), it is reasonable to infer that these complex structures could have evolved repeatedly during the phylogeny of *related* stocks. Emphasis must be laid on there having been a close relationship between the stocks under scrutiny because the caeca of craniaceans are different from those of articulates, and perforated canopies have yet to be found covering the puncta of orthides. In this context, we reject Ivanova's thesis (1972) that all punctate spiriferides are monophyletic and should be included in a new suborder "Spiriferinidina", and symphathize with Rudwick's suppression of the Athyrisinacea and Suessiacea on the grounds that their systematic segregation is based solely on punctation.

Spiriferides are normally easily distinguishable externally despite a fundamental contrast between strophic spiriferidines with their well-developed interareas and wide hinge-lines and the remaining non-strophic spiriferides with their rostrate shells. If the spiriferidines descended from atrypidines, the strophic condition was a secondary development accompanying the eversion of the lophophore and the rotation of its arms and their brachidial supports. Recurrent elongation of the ventral interarea, leading to the repeated appearance of a pyramidal pedicle valve among cyrtiids, syringothyridids, cyrtospiriferids, etc., was a logical consequence of such a change. The persistence of a well-defined dorsal fold among most spiriferides, except for a few later stocks like *Ambocoelia*, *Brachythyris* and *Martinia*, was consistent with the strong partitioning of the filter current systems by laterally directed spiral arms of the lophophore whether the median confluent current be inhalent or exhalant. In other spiriferides with variably disposed lophophores, the dorsal fold, although common, was frequently subdued or even reversed into a sulcus.

Outgrowths from the valve floors to support soft tissues other than the lophophore appeared sporadically. The commonest support for ventral muscle bases was a narrowly arched platform (shoe-lifter process) which evolved independently in some Lower Palaeozoic stocks like *Aulidospira* and *Merista*. The spondylium (as in the reticulariid *Bojothyris*) and the median septum of the spiriferinids, which are the normal supports for ventral muscles in other orders, very rarely occur.

The most diagnostic feature to have evolved among the spiriferides was the skeletal apparatus supporting the lophophore (Fig. 12). For reasons already discussed we prefer to include the aberrant spire-bearing koninckinaceans and thecospirids among the Spiriferida. This revision of many current classifications emphasizes our view that all spire-bearers shared a common ancestry and that, except for the leptocoeliids which apparently lost their spiralia (Nikiforova and Andreeva, 1961), the spiriferides are an extraordinarily compact group by normal phyletic standards.

The spiralia do not, as is still fairly commonly believed, represent a calcification of lophophore tissue. They were secreted within growing sheaths of outer epithelium extending from the crura, which presumably grew in phase with the lophophore. Vectors of growth and zones of resorption in relation to secondary fibres making up the calcitic ribbons can therefore be delineated (e.g. Mac-Kinnon, 1974) and should in future assist in reconstructing filtering current systems within the mantle cavity as has already been done by Rudwick (1960).

The number of attitudes adopted by the spiralia is limited and all can be found in the Ordovician atrypids. The earliest known spiriferide, *Protozyga*, is essentially a loop-bearer, because the main part of the lophophore support is a loop formed by the anteromedian fusion of a pair of prongs which grew forward from the crura (Fig. 12). Resorption fashioned the median part into a transverse band (jugum) and differential growth of the antero-lateral corners resulted in the appearance of a pair of coiled ribbons (primary lamellae) lying parallel to the median plane of the shell. In contemporaneous *Zygospira*, however, the primary lamellae were disposed in low cones more or less at right angles to the median plane. These two patterns became the basis for subsequent evolution of the spiralia. The zygospirid pattern was the model from which atrypidine spiralia, with their apices directed dorso-medially ventrally or laterally, were derived. The protozygid pattern was the model for the spiralia of the remaining spiriferides with their apices mainly directed laterally. Some stocks like the cyrtinids underwent reversion to the zygospirid pattern by rotation of the connections between the spiralia and crura (umbonal blades) into the plane of the brachial valve. The most extraordinary development was a recurrent trend consequential to the growth of a posteriorly directed prong from the jugum (jugal stem). In the Devonian *Bifida* and the Triassic *Diplospirella* and *Koninckina*, the jugal stems continued to grow and bifurcated to form accessory lamellae co-extensive with the primary spiralia.

The most surprising aspect of spiriferide phylogeny is the extinction of the order in contrast to the survival of the Rhynchonellida, Terebratulida and Thecideidina. Throughout spiriferide history, a number of stocks repeatedly arose which, by recombination of morphological characters alone, should have survived to the present day. Their failure to do so may have been linked with the elaborate mimicry by a skeletal apparatus of the disposition of the lophophore thereby reducing the necessary flexibility of the feeding organ to a degree which invited extinction.

Terebratulide phylogeny

The terebratulides are the youngest ordinal group to emerge during brachiopod evolution because the earliest known genera, *Brachyzyga*, *Mutationella* and

Podella, occur in basal Devonian successions. The ancestry of the order is revealed in the basic features of these early species which include: a non-strophic punctate shell with variably developed deltidial plates; cyrtomatodont teeth; a hinge plate that may be united medially; and crura prolonged antero-ventrally into strong processes and supporting a centronelliform loop. The adult centronelliform loop was variable in shape and size due to differential secretion and resorption but early in its ontogeny it was like the lanceolate loop from which was fashioned the spiralia of spiriferides. Both types of loop were also comparable in proportionate growth. The modal length of adult loops, expressed as a proportion of the length of brachial valves, was 40% for 25 terebratulide and 45 spiriferide species of Devonian age. This array of diagnostic features implies that the terebratulides were neotenously derived from retziidines by suppression of the growth of calcareous spires (Williams and Wright, 1963).

Terebratulide phylogeny is overwhelmingly expressed in the evolution of the loop (Fig. 15). The most primitive apparatus, the oval to subcircular centronelliform loop, must have supported a trocholophe or schizolophe even when it attained diameters approaching 5 cm as in *Rensselandia.* But the adult loop of other early Devonian terebratulides, like *Cryptonella,* was a narrow folded structure so that the calcareous ribbon consisted of a pair of subparallel segments pendant from the crura (descending branches) which were bent back on themselves (ascending branches) and distally connected by a transverse band. A long loop of this nature must have supported a plectolophe, consisting of a pair of folded side-arms and a medially situated pair of tips coiled in the plane of symmetry of the shell, which is characteristic of most living terebratulides. Yet the short loop found in the later Palaeozoic dielasmataceans might also have supported a plectolophe. In *Dielasma,* for example, the young centronelliform loop became modified during growth into a curved transverse band joining a pair of descending branches at acute angles. These corner angles could have accommodated side-arms.

The early diversification of the centronellacean loop and the ultimate survival of short-looped terebratulaceans and long-looped terebratellaceans throughout the Mesozoic and Cainozoic are classic illustrations of radiation and selection. The apparently profound differences in the loop development of Palaeozoic and Cainozoic terebratulides have led to the recognition of a number of superfamilies, like Dielasmatacea, Cryptonellacea and Zeilleriacea, to include many Permo-Triassic terebratulides with poorly known loop ontogenies. This taxonomic fragmentation has tended to exaggerate the differences between living and extinct species, but recent work, like that of Dagys (1974), has clarified relationships among older and younger terebratulides and indicates that distinctions may be less fundamental than was once believed.

The dielasmataceans with their juvenile centronelliform loops form a link between the terebratulaceans and the centronellidines. Terebratulaceans are likely to have descended from a *Dielasma*-like ancestor by suppression of the first-formed phase involving the appearance of a lanceolate loop; thereafter the only noteworthy trend has been the exaggerated growth of the crural processes

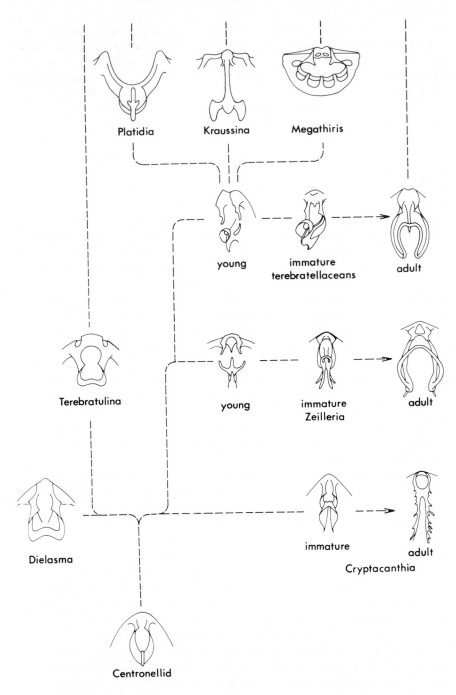

Fig. 15. A phylogeny and selected ontogenies of the terebratulide brachidia (after Baker, 1972; Richardson, 1975; Williams et al., 1965).

to form a more or less continuous ring with the transverse band supporting the dorso-posterior part of the plectolophe, as in *Terebratulina.*

The long-looped terebratellaceans must have descended from another group of Palaeozoic terebratulides because their loop development involves the growth of a dorsal medium septum which contributes to the formation of various elements of the apparatus. Unfortunately very little is known of the early ontogenetic development of the superficially similar long-looped cryptonellaceans. However, they are unlikely to have been directly ancestral to the terebratellaceans because their loops probably developed from a juvenile centronelliform structure in the same way as did the long loop of the centronellidine *Cryptacanthia* (Cooper, 1957). On the other hand, ontogenetic studies of *Aulacothyris* and *Zeilleria* by Babanova (1965) and Baker (1972), respectively, demonstrated that the zeilleriacean loop did include elements derived from a median septum. The terebratellaceans, therefore, probably descended from the zeilleriaceans which, on the basis of general internal morphology including a spinose condition of the loop (Baker, 1972), are believed to have evolved out of the cryptacanthiids.

The phylogeny of the terebratellacean loop is now fairly well understood, largely through the researches of Atkins (1959), Elliott (1953), Owen (1970) and Richardson (1975). The last-named, in particular, has sorted out relationships among long-looped Cainozoic terebratellaceans belonging to the Dallinidae, Laqueidae and Terebratellidae. She has shown (Fig. 15) that in these families, the first structure to appear has always been the dorsal median septum; then a hemiconical tube (hood) over the posterior edge of the septum, subsequently resorbed into an open ring; and finally a pair of lamellae, one on either side of the septum, which grow posteriorly to fuse with extensions from the crura (descending branches). At this stage in ontogeny, the attachments to the septum of the ring and descending branches fuse with each other. Thereafter the ring becomes differentiated into ascending branches joined by a transverse band. Concomitantly, resorption of the septum can result in freeing the loop from any connection with the floor of the brachial valve in all three families, although the route to this state differs according to the nature of resorption of the ring and the timing of its release from the septum. The process results in the development of: vertical connecting bands among the laqueids which Richardson regards as the ancestral group; doubled descending branches among the dallinids; and of neither feature among the terebratellids.

Early stages in terebratellacean loop development have been described because the most significant trend within the superfamily is the neotenous elimination of these metamorphoses and a simultaneous simplification of the lophophore. The trend is recurrent and progressively affected different stocks in the manner we shall now outline (Fig. 15).

In Cainozoic platidiids, the most elaborate apparatus supporting the lophophore, which is invariably schizolophous, consists of short descending branches and short discrete ascending branches arising from a small median septum. In Recent *Amphithyris* this ensemble is reduced to a median septum only; while in Recent *Thaumotosia,* described by Cooper (1973) as warranting a separate familial status although possibly related to the platidiids, the schizolophe is unsupported by any kind of skeletal structure.

Among Cainozoic kraussinids, a schizolophe might be supported by a septal ring and later developing descending branches. However, the latter structures might be absent and only a septum with a pair of divergent projections occurs in Recent *Pumilus*.

The megathyrids which first appeared in Late Cretaceous times have an even simpler arrangement because the apparatus consists exclusively of descending branches attached to the brachial valve and supporting a trocholophe. In Recent *Gwynia* even these descending branches may be absent and the trocholophe is unsupported.

These trends are especially striking for their modernity but they are not unprecedented. Baker (1970) has shown that a similar simplification characterized the apparatus of the Middle Jurassic terebratellacean *Zellania*. In this micromorphic form, the support of what must have been a schizolophe consisted solely of a median septum and a pair of curved ridges representing adnate descending branches.

Two other aspects of terebratulide evolution require comment. The shell has always been perforated by puncta accommodating caeca with distal brushes (Owen and Williams, 1969). The mineral skeletal succession is mainly two-layered, but in the terebratulids a tertiary layer was normally well-developed and, although it probably arose gerontomorphically, it may have been subjected to repeated neotenous suppression (MacKinnon and Williams, 1974).

Terebratulides are also characterized by the existence of true endoskeletons in many living species. These consist of meshworks of calcareous spicules found in the mesoderm of the mantle and lophophore (Williams, 1968c). They have been systematically studied by Schumann (1973) who has demonstrated how different types of spicules are specific to different genera and how their secretion has led to a stiffening of the lophophore base in all groups studied. Mesodermal spiculation is also known in the thecideidines and has been recorded in Mesozoic species belonging to that group and to the Terebratulida. As Schumann has pointed out, there is no obvious reason for the occurrence of an endoskeleton of this nature, but it is noteworthy that, when fully developed, it must have bestowed the same order of rigidity on the lophophore as the spiralia did in the spiriferides.

In conclusion it can be said of the terebratulides that, in many respects, they represent the climax of brachiopod evolution. Like the rhynchonellides they are typified by an economically ovoid shape to a shell articulated by a device that was securely interlocked, without any reduction in its efficiency, by complementary protuberances and depressions on the teeth and sockets. The loop support of the lophophore was, initially at least, an improvement on the more complex skeletal apparatus of spiriferides. There are admittedly indications that endoskeletal spiculation is re-introducing a rigidity, comparable with that afforded by spiralia, into the lophophore of some species; whether this is equivalent in effect to atavistic gerontomorphism remains to be seen. There are also signs that the more complex metamorphoses of the terebratellacean loop

are being shed. The appearance within Recent times of micromorphic, neotenously derived genera, like *Amphithyris, Gwynia, Pumilus* and *Thaumotosia*,
with only vestigial skeletal supports for the lophophore, may be the prelude to
entirely new types of brachiopods in the future.

Thecideidine phylogeny

The history of the last group to be considered, the Thecideidina, is currently
the most controversial aspect of brachiopod evolution. All post-Triassic species
assigned to the suborder are unambiguously related to one another and are
clearly monophyletic. The origin of the group is, however, obscured partly by
our imperfect knowledge of Triassic faunas but mainly by the neotenous and
caenogenetic changes that gave rise to the thecideidines. Consequently their
ancestry can only broadly be determined by attempting to discount the effects
of caenogenesis. For that reason aberrant Permo-Triassic stocks, with no more
than a superficial resemblance to thecideidines, tend·to get identified as close
relatives or even included in the Suborder, thereby sustaining confusion.

The present balance of opinion is as follows. The majority of brachiopod
researchers follow Schuchert (1893) and Elliott (1953) in maintaining that
thecideidines were descended from strophomenides. In particular, Pajaud
(1970) and Grant (1972) believe they originated from a strophalosiacean like
Cooperina; while Rudwick (1968) and Baker (1970) favour the davidsoniaceans
as likely ancestors. Many morphological and structural features, however,
which must have been characteristic of the prototype refute a strophomenide
relationship. The include uniquivocally spiriferide or terebratulide characters
like the cyrtomatodont teeth; the inner socket ridges fused into a cardinal process; the fully functional crura; and a shell composed of a crystallite—granular
primary layer and a fully developed fibrous secondary layer which were perforated by puncta containing caeca with distal brushes. Indeed the only features which, on balance, point to a strophomenide affinity are attachment by
cementation and the bilobed ridge arising from the floor of the brachial valve
which supports a schizolophe (peribrachial ridge). The impersistent tubercles
permeating the shell are different from the persistent taleolae penetrating the
strophomenide shell.

Attachment by cementation involves two prerequisites. The first, the loss
of a functional pedicle, occurred many times in brachiopod history. The
second is the exudation of a bonding glue which could not have differed very
much from the mucopolysaccharide secreted externally to the periostracum in
all brachiopods. Moreover, attachment by cementation is not unique to the
strophomenides. Even if the punctate spira-bearing *Thecospira* is still considered an aberrant strophomenide (Rudwick, 1968; Baker, 1970; Dagys,
1974), contrary to the findings of Williams (1968c) and MacKinnon (1974)
who regard it as a spiriferide, the inarticulate craniaceans may be cited as nearly
always having been cemented to the substrate.

Similarly, the growth of structures like the peribrachial ridge is not restricted
to the strophomenides or thecospiraceans like *Hungaritheca* (Dagys, 1972).

Among terebratulides, the mid-Jurassic *Zellania* bore an almost identical structure (Baker, 1970), and the emergence during the Holocene of *Gwynia* with a trocholophe supported only by traces of a loop continuous with the valve floor illustrates how neoteny can change the basic morphology of a stock. There are also many orthides which must have been served by a schizolophe supported by a median septum, so that lophophore supports arising directly from the floor of the brachial valve appeared many times during brachiopod evolution.

In attempting to evaluate the relative importance of these various features, the irreversibility of evolution as well as the effects of neoteny and deviation should be borne in mind. A strophomenide ancestry implies that the secondary laminar shell fabric found ubiquitously among late Palaeozoic strophomenides suddenly reverted to an orthodoxly fibrous condition in the emerging thecideidine prototype. This is as improbable as the concomitant acquisition of other non-strophomenide features like the microvillous caeca found in the punctate thecideidine shell, and the system of articulation and posterior lophophore support which is structurally comparable with that found in contemporaneous spiriferides and terebratulides. It is surely more likely that a neotenous offshoot from a spiriferide (or terebratulide) lost a functional pedicle and spiralia (or loop) and in so doing became the thecideidine prototype (Williams, 1973).

Of the two orders cited as the credible source of thecideidine ancestry, the Spiriferida is preferred because of certain differences in the thecideidine mantle compared with the terebratulide. These include a periostracum originating within a slot in the outer mantle lobe and caeca occupying the proximal parts of puncta, the distal parts being sealed off by proteinous partitions (Williams, 1973).

Contrary to his own conclusions, we think that the researches carried out by Dagys (1972, 1974) on Triassic brachiopods corroborate this link between thecideidines and spiriferides. Dagys has shown that *Thecospira* is closely related to *Hungaritheca* and *Thecospirella* (Fig. 16). Both these genera are characterized by the presence of a pair of subperipheral curved ridges in the brachial valves which are reminiscent of the peribrachial ridge and identical with the adnate descending branches of *Zellania*. In *Thecospirella* the descending branches are prolonged into a pair of double ascending branches each terminating in a convolution; but in *Hungaritheca* they are associated only with a high median septum. Dagys has derived the thecideidines from *Hungaritheca* and all three thecospiraceans from a mid-Triassic prototype equipped only with adnate descending branches. Yet the phylogenetic series:

Thecospira → *Thecospirella* → *Hungaritheca* → *Thecideidina*

could represent a neotenous shedding of various detached elements of what was originally a spirally disposed brachidium. In any event *Hungaritheca* is nearer the inferred thecideidine ancestor than any brachiopod so far described, and the main difference between the interpretations offered by Dagys and ourselves now rests on whether the thecospiraceans are strophomenides or spiriferides.

Whatever their origin, the thecideidines were unequivocally represented in the Upper Triassic (Norian) by *Thecidella* (Patrulius and Pajaud, 1974) and in

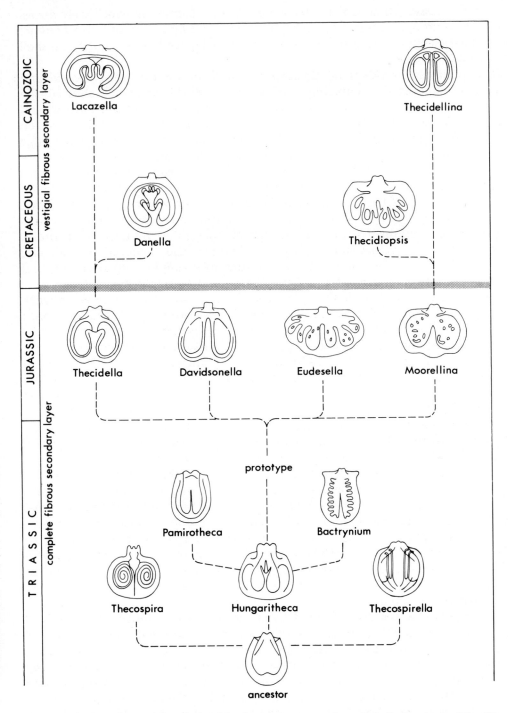

Fig. 16. A phylogeny of the skeletal lophosphore supports and shell structure of the Thecideidina and inferred related brachiopods (after Dagys, 1974; Pajaud, 1970; Williams, 1973).

the Late Triassic (Rhaetian) by *Davidsonella* and *Moorellina* which were ancestral to two groups culminating respectively in living *Lacazella* and *Thecidellina*.
Both lineages underwent parallel changes in their lophophore supports and shell
structure (Fig. 16).

The peribrachial ridges of *Davidsonella* and *Moorellina* were bilobed and
must have supported a schizolophe. However, as Backhaus (1959) showed, the
median ridge of the former was broad while that of the latter was narrow, and
further enlargement of the schizolophe was brought about by a folding of the
lophophore accommodated by a simultaneous lobation of the peribrachial ridge
(ptycholophe) and/or the median ridge (thecidiolophe). In general, the more
complex lophophore supports of the *Lacazella* group accommodated thecidio-
lophes, and those of the *Thecidellina* group, ptycholophes. Pajaud's detailed
study (1970) of the thecideidines has demonstrated how intricate the processes
of evolution can be. During the Jurassic the dominant thecideidine, irrespective
of group, was equipped with a bilobed peribrachial ridge. By Cretaceous times,
ptycholophous and thecidiolophous thecideidines had achieved ascendancy;
while during the Cainozoic there was a return to a simpler schizolophous and
thecidiolophous arrangement as in *Thecidellina* and *Lacazella*, respectively.

Pajaud also identified both palingenetic and neotenous processes at work
during lophophore evolution. The juvenile lophophore supports that grew into
the complex lobate structures of adult *Praelacazella* and *Thecidiopsis* from the
Cretaceous are identical to adult supports of Jurassic *Thecidella* and *Rioultina*,
respectively. In contrast to this example of palingenesis, adult supports of
Lacazella and *Thecidellina* are indistinguishable from those of immature
Danella and *Backhausina* and confirm the neotenous derivation of the former
genera from the latter.

The only aberrant stock that seems to us at present to merit inclusion in the
Thecideidina is the Late Triassic *Bactrynium* (Rudwick, 1968; Williams, 1973;
Dagys, 1974). If the relationship between *Bactrynium* and thecideaceans is as
close as this classification suggests, the folding of the peribrachial ridge of
Bactrynium into paired lobes disposed at right angles to the median ridge is
noteworthy. This folding was a function of size so that up to ten pairs of lobes
are found in adults which are three or four times as big as the earliest thecideaceans. Such a ptycholophous condition is a precocious development within the
phylogenetic framework of the suborder; it anticipates the folding of the peribrachial ridges of such thecideaceans as the Lower Jurassic *Eudesella* and the
Lower Cretaceous *Thecidiopsis*, except that the lobes of these genera are disposed parallel to the median ridge.

Neotenous changes also affected thecideidine skeletal fabric. The shells of
the oldest genera of both groups (*Thecidella*, *Davidsonella*, *Moorellina* and
Eudesella) were lined by a continuous secondary layer of fibres. In all descendants, so far as is known, the fibrous secondary layer became neotenously
reduced irrespective of any increase in species size or shell thickness (Williams,
1973). In the *Lacazella* group only a few secondary fibres survive on the teeth
of living species, and no fibres have been found in the Late Cretaceous *Eolaca-
zella*, which may indicate accelerated neoteny in an aberrant stock.

Among post-Jurassic members of the *Thecidellina* group, secondary fibres survive in the teeth and inner socket ridges of Cretaceous *Bifolium* and living *Thecidellina*, and in the ventral muscle support of *Thecidiopsis*. Some members of this group, like *Thecidellina hedleyi* with secondary fibres restricted to a few spinose outgrowths from the floor of the pedicle valve, may have undergone accelerated neoteny.

It seems that the change from secretion of a fully developed secondary layer to exudation of vestigial patches of fibres probably occurred within the *Thecidellina* group during Late Jurassic or Early Cretaceous times. The *Lacazella* group is less well documented but what is known suggests that a similar change also occurred in this group at about the same time. If this is so, evolution of the fabric of the thecideidine shell is a remarkable example of synchronous as well as parallel neoteny.

In retrospect, the phylogeny of the thecideidines provides a number of striking contrasts in the efficacy of evolutionary processes. Most characters, like the strophic hinge-line, the loss of pedicle and cementation of the pedicle valve, the entire pseudodeltidium, the teeth and cardinalia, and the development of postero-laterally situated adductors were immutable. Changes in shell structure and the peribrachial ridges, however, were numerous and complex. They were neotenously induced in shell fabric but palingenetically as well as neotenously controlled in the elaboration and simplification of the lophophore supports. They were also manifest as parallel trends in both of the two principal thecideidine lineages, but whereas those affecting the skeletal structure were more or less isochronous, those modifying the peribrachial ridges were heterochronous.

Conclusions

An outline of brachiopod history, like the one just given, inevitably reveals some of the complex phylogenetic relationships that existed within the phylum. Some of these relationships may have been rendered more complicated than they were by gaps in the geological record; others appear to be less intricate because they involved anatomical or physiological changes that were not recorded in the exoskeleton. Most, however, are expressions of repeated interventions of the same trends in independently evolving stocks, like the development of muscle platforms which took place in all articulate groups as well as in the Lingulida and Acrotretida. Such trends are, in themselves, fascinating and worthy of study but they tend to obscure fundamental changes that must have occurred during phylogeny. They are best identified by determining the basic differences between living articulates and inarticulates bearing in mind that both groups may have significantly diverged from a prototypic common ancestor. They are: shell composition; muscle arrangement; development of articulation and lophophore supports; presence or absence of an anus; development of pedicle; and mantle reversal during larval stages of growth. It may be argued that this comparison presupposes brachiopod monophyly. We accept the outcome of this supposition because all data concerning living and extinct species

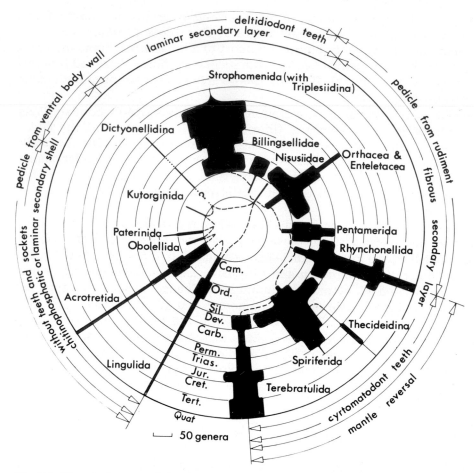

Fig. 17. The phylogeny of the Brachiopoda with an indication of the range of some of the more important morphological features developed during evolution (modified from Williams, 1965).

point to a basic homogeneity within the phylum. Living articulates may therefore be regarded as the culmination of a divergence from Late Precambrian prototypic brachiopods which, as already indicated, were similar to soft-bodied primitive inarticulates (Fig. 17). Such a culmination was achieved with only five basic changes in the prototypic design (Williams and Rowell in Williams et al., 1965), which may be summarized as consecutive stages of increasing complexity as follows:

(1) Acquisition of a solid mineral exoskeleton by the Acrotretida, Lingulida and Obolellida. In most species the shell has always been chitinophosphatic but that of others is composed of protein and aragonite (Trimerellacea) or calcite which occurs as laminae in the secondary layer of the Craniacea, Craniopsidae and Obolellida.

(2) Regrouping of muscle attachment areas in the median parts of the valves

of Paterinida and Kutorginida. These might have been adductor sites only because "articulation" in both orders was too primitive to require diductors to open the shell yet too advanced to allow the sliding of valves, so that oblique muscles probably atrophied.

(3) Appearance of deltidiodont teeth and sockets in the Billingsellacea, Clitambonitidina, Triplesiidina and Strophomenida. These structures acted as a fulcrum for the opening of the valves by newly developed diductors attached to the umbonal interiors of the shell. The loss of the anus may have independently taken place during this stage in evolution because its continued presence in the posterior region of the shell was incompatible with the development of articulation and closely fitting delthyrial and notothyrial covers. The exoskeleton was a fully differentiated carbonate shell consisting of a finely crystalline or granular primary layer and a fibrous or bladed laminar secondary layer.

(4) Development of a pedicle from a larval rudiment instead of the ventral body wall in the Orthacea, Enteletacea and Pentamerida. This change was probably accompanied by loss of the posterior body wall, and atrophy of the anus may also have been delayed until this stage in development. Socket ridges became elaborated into crura (or brachiophores) supporting the lophophore in later stocks. The secondary shell was invariably fibrous, or fibrous and prismatic as in the Pentamerida which may also have secreted a prismatic tertiary layer.

(5) Reversal of the mantle rudiment during larval growth indicated by the anterior shift of the ventral muscle scars in the Rhynchonellida, Spiriferida (? with Thecideidina) and Terebratulida. Lophophore supports as crura, loops and spires were almost invariably developed, as were cyrtomatodont teeth. Both the fibrous secondary and the prismatic tertiary layers may be neotenously lost.

References

Amsden, T.W., 1968. Articulate brachiopods of the St. Clair Limestone (Silurian), Arkansas, and the Clarita Formation (Silurian), Oklahoma. Paleontol. Soc. Mem. 1, J. Paleontol., 42 (Suppl.): 117 pp.

Armstrong, J., 1969. The cross-bladed fabrics of the shells of *Terrakea solida* (Etheridge and Dun) and *Streptorhynchus pelicanensis* Fletcher. Palaeontology, 12: 310—320.

Atkins, D., 1959b. The growth stages of the lophophore and loop of the brachiopod *Terebratalia transversa* (Sowerby). J. Morphol., 105: 401—426.

Babanova, L.I., 1965. New data on Jurassic brachiopods. Int. Geol. Rev., 7(8): 1450—1455.

Backhaus, E., 1959. Monographie der cretacischen Thecideidae (Brach.). Hamburg Geol. Staatsinst. Mitt., 28: 5—90.

Baker, P.G., 1970. The morphology and microstructure of *Zellania davidsoni* Moore (Brachiopoda), from the Middle Jurassic of England. Palaeontology, 13: 606—618.

Baker, P.G., 1972. The development of the loop in the Jurassic brachiopod *Zeilleria leckenbyi*. Palaeontology, 15: 450—472.

Biernat, G. and Williams, A., 1970. Ultrastructure of the protegulum of some acrotretide brachiopods. Palaeontology, 13: 98—101.

Biernat, G. and Williams, A., 1971. Shell structure of the Siphonotretacean Brachiopoda. Palaeontology, 14: 423—430.

Boucot, A.J. and Ehlers, G.M., 1963. Two new genera of stricklandi brachiopods. Mich. Univ. Mus. Paleontol. Contrib., 18: 47—66.

Boucot, A.J. and Johnson, J.G., 1966b. *Lissocoelina* and its homeomorph *Plicocoelina* n. gen. (Silurian, Pentameracea). J. Paleontol., 40: 1037—1042.

Boucot, A.J. and Johnson, J.G., 1967. Silurian and Upper Ordovician atrypids of the genera *Plectatrypa* and *Spirigerina*. Nor. Geol. Tidsskr., 47: 79—101.

Brunton, C.H.C., 1965. The pedicle sheath of young productacean brachiopods. Palaeontology, 7: 703—704.

Brunton, C.H.C., 1968. Silicified brachiopods from the Visean of County Fermanagh (II). Bull. Br. Mus. Nat. Hist. (Geol.), 16(1): 1—70.

Brunton, C.H.C., 1971. An endopunctate rhynchonellid brachiopod from the Visean of Belgium and Britain. Palaeontology, 14: 95—106.

Brunton, C.H.C., 1972. The shell structure of Chonetacean brachiopods and their ancestors. Bull. Br. Mus. Nat. Hist. (Geol.), 1, 21(1): 1—26.

Brunton, C.H.C. and MacKinnon, D.L., 1972. The systematic position of the Jurassic brachiopod Cadomella. Palaeontology, 15: 405—411.

Chuang, S.H., 1971. New interpretation of the morphology of Schizambon australis Ulrich and Cooper (Ordovician siphonotretid inarticulate). J. Paleontol., 45: 824—832.

Cooper, G.A., 1956. Chazyan and related brachiopods. Smithson. Misc. Collect., 127: 1—1245.

Cooper, G.A., 1957. Loop development of the Pennsylvanian terebratulid Cryptocanthia. Smithson. Misc. Collect., 134(3): 1—18.

Cooper, G.A., 1973. New Brachiopoda from the Indian Ocean. Smithson. Contrib. Paleobiol., 16: 1—27.

Cooper, G.A. and Grant, R.E., 1969. New Permian brachiopods from West Texas. Smithson. Contrib. Paleobiol., 1: 1—20.

Cooper, G.A. and Grant, R.E., 1974. Permian brachiopods of West Texas, II. Smithson. Contrib. Paleobiol., 15: 1—793.

Copper, P., 1967. Pedicle morphology in Devonian atrypid brachiopods. J. Paleontol., 41: 1166—1175.

Dagys, A.S., 1972. Ultrastructure of Thecospirid shells and their position in brachiopod systematics. Paleontol. J., 6: 359—369.

Dagys, A.S., 1974. Triassic Brachiopods (Morphology, Classification, Phylogeny, Stratigraphical, Significance and Biogeography). Nauka, Novosibirsk, 386 pp. (in Russian).

Davidson, T., 1884. Supplement to the fossil Brachiopoda. Palaeontogr. Soc. Monogr., 5(3): 243—476.

Elliott, G.F., 1953. Brachial development and evolution in terebratelloid brachiopods. Biol. Rev., 28: 261—279.

Gauri, K.L. and Boucot, A.J., 1968. Shell structure and classification of Pentameracea M'Coy, 1844. Palaeontographica, 131: 79—135.

Grant, R.E., 1966. Spine arrangement and life habits of the productoid brachiopod Waagenoconcha. J. Paleontol., 40: 1063—1069.

Grant, R.E., 1972. The lophophore and feeding mechanism of the Productidina (Brachiopoda). J. Paleontol., 46: 213—248.

Hall, J. and Clarke, J.M., 1892—94. An introduction to the study of the genera of Palaeozoic Brachiopoda. N.Y. State Geol. Surv., Palaeontol., N.Y., 8, (1), 1892, i—xvi, 1—367, 41 pl.; (2) 1894, i—xvi, 1—394, pls. 21—84.

Harland, W.B., Smith, A.G. and Wilcock, B. (Editors), 1964. The Phanerozoic Time-Scale. The Geological Society of London, London, 458 pp.

Harper, C.W., 1975. Standing diversity of fossil groups in successive intervals of geologic time: a new measure. J. Paleontol., 49: 752—757.

Havlíček, V., 1967. Brachiopoda of the suborder Strophomenidina in Czechoslovakia. Rozpr. Ustred. Ust. Geol., Praha, 33: 1—235.

Havlíček, V., 1971. Brachiopodes de l'Ordovicien du Maroc. Notes Mem. Serv. Mines Carte Géol. Maroc., Rabat, 230: 7—69.

Ivanova, Y.A., 1972. Main features of spiriferid evolution (Brachiopoda). Paleontol. J., 6: 309—320.

Jaanusson, V., 1966. Fossil brachiopods with probable aragonitic shell. Geol. Fören. Förh., 88: 279—281.

Jaanusson, V., 1971. Evolution of the brachiopod hinge. In: J.T. Dutro Jr. (Editor), Paleozoic Perspectives: A Paleontological Tribute to G. Arthur Cooper. Smithson. Contrib. Paleobiol., 3: 33—46.

Johnson, J.G., 1973. Some North American Renoselandiid brachiopods, Part 2. J. Paleontol., 47: 1102—1107.

Johnson, J.G., 1974. Affinity of Dayracean brachiopods. Palaeontology, 17: 437—439.

Johnson, J.G. and Boucot, A.J., 1967. Gracianella, a new late Silurian genus of atrypoid brachiopods. J. Paleontol., 41: 868—873.

Jope, M., 1967. The protein of brachiopod shell — I. Amino acid composition and implied protein taxonomy. Comp. Biochem. Physiol., 30: 209—224.

Kelly, P.G., Oliver, F.G.E. and Pautard, F.G.E., 1965. The shell of Lingula unguis. In: Proceedings of the Second European Symposium on Calcified Tissues, pp. 337—345.

Ludvigsen, R., 1974. A new Devonian acrotretid (Brachiopoda, Inarticulata) with unique proregular ultrastructure. Neues Jahrb. Geol. Paläontol. Mh., 1974: 133—148.

MacKinnon, D.I., 1971. Perforate canopies to canals in the shells of fossil Brachiopoda. Lethaia, 4: 321—325.

MacKinnon, D.I., 1974. The shell structure of Spiriferide Brachiopoda. Bull. Br. Mus. Nat. Hist. (Geol.), 25(3): 189—261.

MacKinnon, D.I. and Williams, A., 1974. Shell structure of terebratulid brachiopods. Palaeontology, 17: 179—202.

Muir-Wood, H. and Cooper, G.A., 1960. Morphology, a classification and life habits of the Productoidea (Brachiopoda). Geol. Soc. Am. Mem., 81: 1—447.

Nikiforova, O.I. and Andreeva, O.N., 1961. Stratigrafiya Ordovika i Silura sibirskoy Platformy i ee Paleontologicheskoe Obasnovanie (Brakhiopody). Tr. Vses. Nauchno-Issled. Geol. Inst., 56: 412 pp.

Owen, E.F., 1970. A revision of the brachiopod subfamily Kingeninae Elliott. Bull. Br. Mus. Nat. Hist. (Geol.), 19(2): 27—83.

Owen, G. and Williams, A., 1969. The caecum of articulate Brachiopods. Proc. R. Soc. B, Lond., 172: 187—201.

Pajaud, D., 1970. Monographies des Thécidées (Brachiopodes). Mem. Soc. Géol. Fr. (N.S.), 49(112): 1—349.

Patrulius, D. and Pajaud, D., 1974. Présence de Thécidées et de thécospires dans les dépôts détritiques du norien des monts apuseni (Roumanie). Docum. Lab. Géol. Fac. Sci. Lyon, 62: 129—135.

Richardson, J.R., 1975. Loop development and the classification of terebratellacian brachiopods. Palaeontology, 18: 285—314.

Rowell, A.J., 1966. Revision of some Cambrian and Ordovician inarticulate brachiopods. Paleontol. Contrib. Univ. Kansas, 7: 1—35.

Rowell, A.J. and Krause, F.F., 1973. Habitat diversity in the Acrotretacea (Brachiopoda, Inarticulata). J. Paleontol., 47: 791—800.

Rudwick, M.J.S., 1960. The feeding mechanisms of spire-bearing fossil brachiopods. Geol. Mag., 97: 369—383.

Rudwick, M.J.S., 1968. The feeding mechanisms and affinities of the triassic brachiopods *Thecospira* Zugmayer and *Bactrynium* Emmrich. Palaeontology, 11: 329—360.

Rudwick, M.J.S., 1970. Living and Fossil Brachiopods. Hutchinson University Library, London, 199 pp.

Schopf, T.J.M., Raup, D.M., Gould, S.J. and Simberloff, D.S., 1975. Genomic versus morphologic rates of evolution: influence of morphologic complexity. Paleobiology, 1: 63—70.

Schuchert, C., 1893. A classification of the Brachiopoda. Am. Geol., 11: 141—167.

Schumann, D., 1970. Inäquivaler Schalenbau bei *Crania anomala*. Lethaia, 3: 413—421.

Schumann, D., 1973. Meodermale Endoskelette terebratulider Brachiopoden. 1. Paläontol. Z., 47: 77—103.

Termier, H. and Termier, G., 1949. Sur la classification des brachiopodes. Soc. Hist. Nat. Afr. Nord, Bull., 40: 51—63.

Williams, A., 1951. Llandovery brachiopods from Wales with special reference to the Llandovery district. Q. J. Geol. Soc. Lond., 107: 85—136.

Williams, A., 1953. North American and European stropheodontids: their morphology and systematics. Mem. Geol. Soc. Am., 56: 1—67.

Williams, A., 1956. The calcareous shell of the Brachiopoda and its importance to their classification. Biol. Rev., 31: 243—287.

Williams, A., 1957. Evolutionary rates of brachiopods. Geol. Mag., 94: 201—211.

Williams, A., 1968a. Shell structure of the billingsellacean brachiopods. Palaeontology, 11: 486—490.

Williams, A., 1968b. Evolution of the shell structure of articulate brachiopods. Spec. Pap. Palaeontol., No. 2: 1—55.

Williams, A., 1970. Origin of laminar-shelled articulate brachiopods. Lethaia, 3: 329—342.

Williams, A., 1973. The secretion and structural evolution of the shell of thecideidine brachiopods. Philos. Trans. R. Soc. Lond., Ser. B., 264: 439—478.

Williams, A., 1974. Ordovician Brachiopoda from the Shelve District, Shropshire. Bull. Br. Mus. Nat. Hist. (Geol.), 11: 1—163.

Williams, A. and Wright, A.D., 1961. The origin of the loop in articulate brachiopods. Palaeontology, 4: 149—176.

Williams, A. and Wright, A.D., 1963. The classification of the "Orthis testudinaria Dalman" group of brachiopods. J. Paleontol., 37: 1—32.

Williams, A. and Wright, A.D., 1970. Shell structure of the Craniacea and other calcareous inarticulate Brachiopoda. Spec. Pap. Palaeontol., No. 7: 1—51.

Williams, A. et al., 1965. Brachiopoda. In: R.C. Moore (Editor), Treatise on Invertebrate Paleontology, H. University of Kansas Press, Lawrence, Kansas, 927 pp.

Wright, A.D., 1968. The brachiopod *Dicoelosia biloba* (Linnaeus) and related species. Ark. Zool., 20: 261—319.

CHAPTER 5

EVOLUTION OF PRIMITIVE ECHINODERMS

C.R.C. PAUL

Introduction

Whatever theories of evolution one may entertain, evidence to support them must come from the fossil record. Consideration of the nature of the fossil record of echinoderms, particularly the early primitive ones, should precede interpretation. Echinoderms possess compound skeletons which usually disintegrate very rapidly after death. Their skeletal fragments form a significant proportion of many Palaeozoic rocks, especially limestones. To be preserved intact or nearly so, echinoderms must be buried immediately after death without any subsequent disturbance. Indeed many of the best preserved primitive echinoderms seem to have been buried alive. Such preservation is relatively rare and primitive echinoderms do not normally occur in continuous evolutionary sequences. Those which are preserved often yield a detailed picture of their palaeoecology particularly since many were gregarious. The fossil record of primitive echinoderms is analogous to a collection of detailed frames from a ciné film: the individual pictures may be clear but the story may not be, nor can one be certain of the sequence. The fossil records of other groups (e.g., the ammonites with a single shell which cannot disarticulate but which is almost never buried *in situ*) may be likened to a more continuous film in which the sequence and story are more obvious, but all the frames are somewhat blurred.

Important taxonomic and evolutionary consequences arise from the type of fossil record of primitive echinoderms. Recognition of the limits of fossil taxa is rarely difficult because the nearest related forms are usually widely separated in space and/or time and intermediates have just not been preserved. This is true at all taxonomic levels and may contribute to the fact that so many (21) echinoderm classes are recognized. Conversely, since all echinoderm classes, and many lower taxa, appear abruptly in the fossil record, construction of a phylogenetic tree for the phylum is very difficult. Taken at face value, the echinoderm type of fossil record inevitably supports "saltative" theories of evolution, although the nature of the record is due more to the processes of preservation than to any evolutionary process. Indeed the fossil record of echinoderms would seem of little use in testing evolutionary ideas. However, probably the best example of an evolutionary lineage, that of *Micraster*, involves echinoderms. Also, if competition and selection played important roles in evolution, one must interpret the fossil record in terms of function and functional efficiency. Although surprisingly few such attempts have been made, echinoderms are particularly suited to functional studies. In terms of numbers of skeletal parts, they have the most complex morphology of any phylum and they are second only to the vertebrates in the intimacy of the relationship between soft and hard anatomy. Thus I believe that provided one acknowledges its bias, the fossil record of echinoderms can contribute significantly to the study of evolution.

With this in mind, I shall discuss the fossil record of echinoderms in three sections. The first considers the origin and relationships of the phylum. Echinoderms probably arose from a minor coelomate phylum, such as the Phoronida (Nichols, 1967) and gave rise to the chordates (Jefferies, 1967, et seq.). The second section compares the pattern of echinoderm evolution at class level with computer generated models of phylogeny to determine which features cannot be explained by chance alone. The explosive radiation of short-lived minor classes early in the record and the subsequent decline in numbers of classes are interpreted as being due to a lack of competition early in the Phanerozoic followed by the elimination of less efficient designs. This idea is tested in the third section which reviews fossil evidence for the vital functions necessary for survival (protection, feeding, respiration, etc.); comparing extant and extinct forms and interpreting trends in terms of functional efficiency. By and large the fossil record supports the hypothesis and Cambrian echinoderms are shown to have been less efficient than their descendents.

The evolution of primitive echinoderms has not been studied extensively. The following sections not only summarize current knowledge, but raise questions for which, as yet, no answers exist.

Origin, Characteristics and Relationships of Echinoderms

Comparative anatomy and embryology indicate that echinoderms are primitive bilateral, coelomate metazoans with a mesodermal skeleton. They undergo radial cleavage and coelomic cavities arise by enterocoely. Most of these characters are shared with chordates, hemichordates, ectoprocts, phoronids and brachiopods. Echinoderms characteristically possess a mesodermal skeleton composed of many plates or spicules which have a meshwork structure of calcite rods (trabeculae) intimately interwoven with soft tissue. Two other typical features are pentameral symmetry and an extensive hydrovascular system. Evidence for all three features can be found in the fossil record, although the first is by far the most conspicuous. Probably the water vascular system (? Precambrian) preceded the appearance of a calcite skeleton (Lower Cambrian) which in turn preceded pentamery. The latter apparently affected the ambulacra first (Lower Cambrian) and later spread to the theca (Middle Cambrian) and, when present, the stem (Ordovician). Stephenson (1977) has recently suggested a plausible explanation for the appearance of pentamery which agrees with this progressive extension of its effects, but which would only have involved immovable filter-feeders.

Fig. 1 illustrates the ranges, sizes and supposed relationships (the latter often

Fig. 1. Range chart for all echinoderm classes and calcichordates to show sizes (numbers of genera) and suggested relationships.

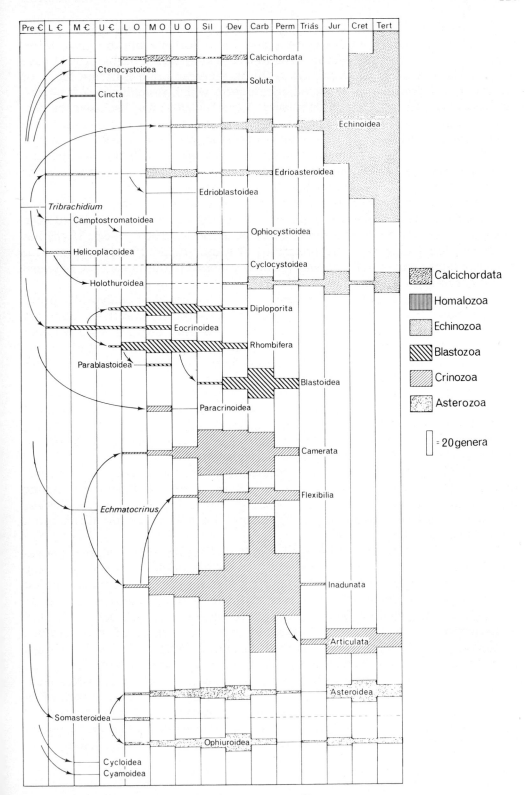

| Pre € | L € | M € | U € | L O | M O | U O | Sil | Dev | Carb | Perm | Triás | Jur | Cret | Tert |

Calcichordata
Ctenocystoidea
Soluta
Cincta
Echinoidea
Edrioasteroidea
Edrioblastoidea
Tribrachidium
Camptostromatoidea
Ophiocystioidea
Helicoplacoidea
Cyclocystoidea
Holothuroidea
Diploporita
Eocrinoidea
Rhombifera
Parablastoidea
Blastoidea
Paracrinoidea
Camerata
Flexibilia
Echmatocrinus
Inadunata
Articulata
Asteroidea
Somasteroidea
Ophiuroidea
Cycloidea
Cyamoidea

Calcichordata
Homalozoa
Echinozoa
Blastozoa
Crinozoa
Asterozoa

= 20 genera

highly speculative) of all classes and subclasses. Four classes appear in the Lower Cambrian which implies considerable Precambrian evolution prior to the acquisition of a calcite skeleton as Durham (1964) and Ubaghs (1971) have argued. Possibly *Tribrachidium*, from the late Precambrian Ediacara fauna of Australia, is a primitive pre-skeletal echinoderm which resembles a soft-bodied edrioasteroid with only three ambulacra. Nichols (1967) has discussed the possible origins of echinoderms from a zoological standpoint. He concluded that they were derived from a phoronid-like ancestor, with the water vascular system derived from the phoronid lophophore. This agrees with the idea that the water vascular system was acquired before the calcite endoskeleton which was affected progressively by pentamery during the Cambrian.

Although the divergence of phyla is generally held to have taken place in the Precambrian, which therefore implies that a search for fossil evidence of relationships between phyla will prove fruitless, Jefferies (1967, 1968, 1975) has produced convincing new interpretations of some "carpoid" echinoderms, his calcichordates, which, in my view, strongly confirm the close link between echinoderms and chordates.

There is no single feature of morphology which could unequivocally prove the affinities of calcichordates. The alternative hypotheses, that these fossils are primitive chordates or unusual echinoderms, can only be judged on whether they are internally consistent, particularly from a functional point of view. I shall discuss briefly Ubaghs', Caster's and Jefferies' different views of three "carpoid" groups, the solutes, mitrates and cornutes.

Ubaghs (1962) interpreted the "stem" or "stele" of the mitrates and cornutes (= calcichordates) as a feeding organ, the aulacophore. He reorientated the whole animal with the aulacophore anterior. Subsequently (Ubaghs, 1967) he maintained them as echinoderms and regarded the tripartite stele of the solutes as quite distinct from the tripartite aulacophore.

Caster (1967, 1971) alternatively regarded the tripartite stems of calcichordates and solutes as homologues, accepted those of the former as feeding organs but interpreted that of the latter as having lost its primitive feeding function. A separate feeding organ (a brachiole) occurs in solutes at the opposite pole to the stem.

Functionally both interpretations run into difficulties. If the solute stele was ever a feeding organ, as Caster proposes, then solutes originally had two feeding organs (the brachiole and stele) which entered the theca at opposite poles. Fitting a gut into such a theca is very difficult. If the posterior stele lost its feeding function, what happened to the gut to which it was connected? Equally if the brachiole is a new development, how did a new branch of the gut evolve to connect with it? Ubaghs apparently realises these difficulties and has always maintained that the calcichordate aulacophore was quite different from the solute stele. There is thus no need to suggest that the solute stele lost its feeding function with the attendant problems of internal anatomy. However, of all possible echinoderm groups only the cornutes, mitrates and solutes possess a tripartite stele and I believe Caster is correct in arguing that this organ is homologous in all three groups. If the solute stele did not gather food, it seems most unlikely that the aulacophore of the cornutes and mitrates did. It is just as difficult to develop a new branch of the gut to connect with the calcichordate aulacophore, if it was derived from a non-feeding solute stele.

In contrast to these echinoderm interpretations, Jefferies has proposed very detailed analyses of the calcichordates as chordates, providing functional explanations for virtually all anatomical features. A complex brain is present, divisible into fore, mid and hind parts, giving rise to cranial nerves which feed homologous structures to the numbered cranial nerves of craniates. Minor asymmetries are found (e.g., the anus always opens on the left) which lack obvious functional significance but have counterparts in the asymmetries of larval tunicates, *Amphioxus* and other related groups. The usual explanation for the occurrence of such apparently homologous non-functional features is common ancestry and I think it applies in this case as in others. Whatever criticisms of these interpretations may be advanced, functionally they are consistent and form the first interpretations of these fossils which in any way explain the vast amount of preserved internal anatomy.

In summary, both Ubaghs and Caster stress the distinctiveness of the homalozoans from other echinoderms and present interpretations of the calcichordates which raise serious functional problems. Jefferies presents detailed interpretations of the calcichordates as chordates which do not raise such functional problems but require a reappraisal of conventional ideas on the origin of vertebrates. In my view calcichordates make more sense as chordates with echinoderm affinities.

Thus echinoderms probably arose from one of the minor coelomate phyla, such as the Phoronida, and gave rise to the chordates as outlined by Jefferies.

Patterns of Echinoderm Evolution

Fig. 1 presents the most up-to-date information available on the stratigraphic ranges and sizes (as estimated by numbers of described genera) of all echinoderm classes and subclasses, and the Calcichordata. At class level there was a rapid early radiation: 15 classes in the Cambrian, 19 in the Ordovician; followed by a decline to the present seven classes and subclasses. Early in the record a number of very small (some with only a single genus), very short-lived classes occur. Indeed all classes confined to one geological period, or part of a period, occurred in the Cambrian and Ordovician. The Blastoidea (Silurian—Permian) is the only class to arise since the Ordovician, although two subclasses, articulate crinoids (Trias—Recent) and euechinoid sea-urchins (Trias—Recent; not shown separately on Fig. 1), appeared still later. Even so, at class level, virtually all echinoderm innovation was over by the end of the Ordovician. Subsequent evolution involved the refinement of successful, established lineages.

Fig. 1 may be compared with theoretical evolutionary patterns generated by computer, such as Raup et al. (1973) have presented. Such a comparison enables the distinction of those aspects of the fossil record of echinoderms which are likely to have occurred by chance from those aspects which are unique to echinoderm evolution. Raup et al. have already compared the actual fossil record of seventeen reptile clades, or lineages, with their computer models. They identified four important differences. The numbers of genera in the reptile clades varied more widely than in the computer clades, largely due to the "rules" of the computer program under which very large clades would have been subdivided. The other three features had no counterparts in the computer-generated diagrams: (1) the simultaneous extinction of five reptile clades; (2) very rapid evolution (explosive radiation) within one reptile clade; (3) the survival of very small lineages for very long periods of time — the "coelacanth effect". Examples of all three features occur in the echinoderm fossil record. Fig. 1 also differs from the computer-generated phylogenetic diagrams in the following additional respects: (4) a large number (nine) of very small, very short-lived clades occur in the early part of the record; (5) simultaneous maxima occur in the Middle Ordovician for six clades and in the Carboniferous for three clades; (6) several clades have "gaps" in their record. The last point could not occur under the "rules" of the computer program and is, of course, due to the incompleteness of the fossil record and our knowledge of it. Points (4) and (5) are more fundamental and also require consideration when evaluating echinoderm evolution. All five points will now be discussed. The early occurrence of small, short-lived classes and the early explosive radiation of echinoderms at class level are obviously linked. The other points do not correlate well and will be dealt with first.

Simultaneous extinctions

Fig. 2 shows the overall size (total genera) of the phylum compared with the number of classes and with the exception of the Permian extinctions, the two correlate poorly. Six classes became extinct in the Cambrian, four in the Ordovician, none in the Silurian, and six in the Devonian. Despite these extinctions, the total number of genera increased steadily to an all-time maximum (living genera excepted) in the Carboniferous. A rapid decline in overall numbers followed, accompanied by one class extinction in the Carboniferous and one class and two subclasses at the close of the Permian. The simultaneous extinction of numerous small classes early in the echinoderm fossil record was partly due to the larger number available (hence it was not repeatable later) and had no effect on the success of the phylum as a whole. In my view it was due to the elimination of small inefficient groups of echinoderms which could not compete with their more successful contemporaries (see later). The Permian extinctions, however, are different; they represent a real life-crisis for echinoderms, and coincide with numerous extinctions in other phyla.

Simultaneous maxima

Five small echinoderm classes and the calcichordates have their maximum diversity in the Middle Ordovician; three larger groups in the Carboniferous,

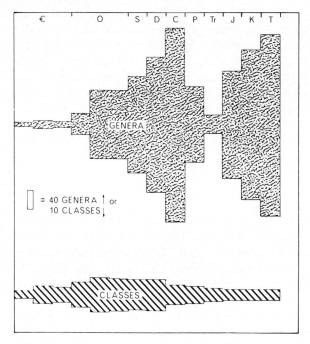

Fig. 2. Numbers of genera and classes of echinoderms throughout the Phanerozoic.

which is the period of maximum diversity for the phylum. The classes with maxima in the Ordovician — calcichordates, solutes, diploporites, rhombiferans, parablastoids and paracrinoids — are very diverse and it is difficult to advance a general explanation for their simultaneous success. Geographical location will not account for this entirely. It is true that the paracrinoids occur almost exclusively in North America where Middle Ordovician echinoderm faunas are prolific and echinoderm workers are active, but the Diploporita are almost entirely absent from the Ordovician of North America. The last four groups have various respiratory pore-structures. The Ordovician undoubtedly saw the greatest diversity, both taxonomic and morphological, of pore-structures, but respiration apparently has little bearing on the simultaneous success of the calcichordates and solutes. This simultaneous success is probably partly an accidental result of the large number of Middle Ordovician classes (18) and also partly due to the fact that they occur in rich, well-described faunas.

The "coelacanth effect"

The somasteroids form an excellent example among echinoderms of the "coelacanth effect". They first appear in the Tremadoc and are known as fossils until the Devonian. Thereafter they have no fossil record but the living *Platasterias* is accepted as one and ranks with the monoplacophoran *Neopilina*, or the coelacanth *Latimeria*, as a "living fossil". A general explanation for these exceptional survivors of typically ancient groups is difficult either in terms of adaptation or competition. Nevertheless, examples are known among verte- brates (*Sphenodon*, *Latimeria*), invertebrates (*Neopilina*, *Platasterias*) and plants (*Ginkgo*, *Araucaria*) and are thus too widespread to be ignored. Clearly there is some pervasive effect which allows the survival of small groups for long periods, but what it is remains unknown. The occurrence of 'living fossils' is partly due to our better knowledge of the living world compared with the fossil record. Small groups have a relatively low preservation potential, but this only explains why small groups are not preserved, not why they manage to sur- vive for long periods.

Small, short-lived classes

Nine very small classes are recognized in the Cambrian and Ordovician, and are confined to parts of these systems. For this reason Fig. 1 distinguishes Lower, Middle and Upper divisions of these two systems while treating the rest as single units. Five small classes — Ctenocystoidea, Parablastoidea, Campto- stromatoidea, Cyamoidea and Cycloidea — are founded on a unique type genus. The largest, the Paracrinoida, contains perhaps eight valid genera (Parsley and Mintz, 1975). The cycloids and cyamoids look quite similar to me, but even if one amalgamates these there is still a considerable number of small early echinoderm classes. Following Sprinkle (1973) I have rejected the class Lepido- cystoidea, incorporating it within the Eocrinoidea; and Fig. 1 ignores *Bolbo- porites*, an enigmatic Ordovician fossil which is undoubtedly part of an echino- derm but quite different from all known contemporary classes. Thus no attempt has been made to over-emphasize the numbers of bizarre, early echino- derm groups. The incredible variety of basic designs in early echinoderms is real and undeniable. To my mind the evolution of the echinoderm classes was genuinely explosive and requires explanation.

Explosive radiations

Explosive radiations are not uncommon in the fossil record. They occur not only when a group of organisms colonizes a new environment (e.g., the early radiation of Amphibia or birds), but also when adaptations to novel modes of life are involved (e.g., the radiation of "irregular" echinoids in the Jurassic is largely correlated with adaptation to an infaunal, deposit-feeding mode of life; that of angiosperms is related to a mode of reproduction involving a new type of plant—insect symbiosis; that of early reptiles to the evolution of the amnio- tic egg). Early in the colonization of a new environment or the development of a new mode of life, competition will be low because few organisms will have adapted to the new conditions. Relatively inefficient organisms may develop, but as more and more life forms become adapted to the new conditions the inefficient ones are likely to become extinct due to increased competition.

This general hypothesis would apply most strongly to the early stages of the evolution of Metazoa, since almost all marine environments and modes of life would have been open to colonization or adaptation. Initially there would have been very little competition between organisms and almost any type, however bizarre and inefficient, would have been able to survive. As faunal diversity increased competition would have eliminated the less efficient clades which would have a short fossil record early in the Phanerozoic. This idea explains many features of the echinoderm fossil record. Virtually all radiation of echinoderm classes occurred in the Cambrian and Ordovician giving rise to a number of short-lived classes all of which are confined to these two geological periods. Post-Ordovician evolution of echinoderms involved a reduction in the number of classes and subclasses to the present seven, accompanied by an expansion in the number of genera, to a peak in the Carboniferous. Thus, while small inefficient classes were being eliminated, the more efficient echinoderms were flourishing. Only the Permian life-crisis, which affected most phyla, reversed this trend and brought about a concomitant decline in both classes and genera.

In summary, the overall pattern of echinoderm evolution involves a rapid early diversification of basic designs followed by a continuous decline to the present seven classes and subclasses. This early radiation produced a number of small, short-lived classes which I interpret as inefficient groups that could only survive under conditions of low competition such as one might expect early in the evolution of Metazoa. The extinction of these and other early classes had no effect on the total number of echinoderm genera, which rose to a maximum in the Carboniferous. The succeeding Permian life-crisis affected echinoderms as much as other phyla, and was followed by a Jurassic radiation of 'irregular' echinoids. The phylum as a whole is probably as abundant now as at any time in the past. The general idea that the least efficient echinoderm classes arose first and were short-lived is open to test in terms of their efficiency in performing the vital functions necessary for survival. The next section is devoted to an analysis of the idea.

Vital Functions in Fossil Echinoderms

All living animals must perform a relatively small number of vital functions in order to survive, as all fossils must also have done when they were alive. Animals need to eat (nutrition) and avoid being eaten (protection). Energy from food is released by metabolism which requires a constant supply of oxygen (respiration) and by-products must be eliminated (excretion). Surplus energy is absorbed in individual growth (ontogeny) or the perpetuation of the species (reproduction). All animals have a more-or-less precise ecology. By considering each vital function in turn, together with palaeoecological information from preservation, burial attitude, inclosing sediments, etc., a detailed picture of the modes of life of fossils can be deduced. In this section I shall review the fossil evidence for vital functions in echinoderms to see if there is any relationship between survival and efficiency, considering first the general pattern of echinoderm evolution, then I shall make a more detailed analysis of primitive echinoderms, and finally discuss any well-documented trends. In all cases it is the first appearance of more efficient structures which is considered important.

Protection

All living things require protection, not only from the physical energy of their environment, but also from predators and parasites. Protection may be afforded by morphological, physiological or behavioural features and it is convenient to treat these in turn, even though combinations occur in nature. The most obvious morphological features involved with protection in echinoderms are the calcified endoskeleton which appears first in the Lower Cambrian, and accessory spines which appear in the Lower Ordovician. Fossil evidence for physiological protection (poisonous and distasteful secretions) is naturally lacking but there is evidence for the most obvious protective behaviour, adopting an infaunal mode of life, which first occurred in the Lower Ordovician. Almost all evidence for protection in primitive echinoderms relates to the theca which, in effect, forms a protective envelope.

The paradigm, or ideal design, for a protective envelope would be as strong as possible and completely cover all vital organs without any gaps. However, in marine organisms like echinoderms, such a paradigm would isolate the internal organs from the ambient sea water which contains the food and oxygen necessary for life. It might also pose problems in growth. All paradigms should carry with them the qualification that the performance of one vital function should not interfere with the performance of any other. The conflicting requirements of protection on the one hand, and of nutrition and respiration on the other, are sometimes overcome by temporal alternations of function: brachiopods and bivalves gape to feed and breathe, but snap shut on the approach of danger. Alternatively, specialized orifices may develop in the protective envelope and these may need further protection themselves. In this section on protective morphology I shall first consider the evolution of protective envelopes, then accessory structures, and lastly the protection of orifices.

The first and most obvious requirement of a protective envelope is a hard skeleton. The earliest plated echinoderms occur in the lowest Cambrian, but the theca was generally flexible as in helicoplacoids (Fig. 3, *4*), imbricate eocrinoids (Fig. 3, *10*), camptostromatoids and possibly the edrioasteroid, *Stromatocystites*. Indeed all Lower Cambrian thecae appear to have been composed mainly of imbricated plates, with a small oral surface of weakly tesselated plates in the eocrinoids. *Stromatocystites*, from the Lower Cambrian, are not well enough preserved for one to be certain whether the oral surface was rigid or flexible.

Fig. 3. Evolution of protective features in echinoderms. Calcichordata and echinoderm subphyla indicated by same shading as in Fig. 1. Note changes from flexible thecae in the Lower Cambrian (*4, 10*) to thecae with marginal rims (*1—3*), or to tesselated plating (*11—17*), accompanied by thecal reduction in the crinozoa (*18—22*). Spines first appear in the Asterozoa (*23*) but are most obvious in the echinoids (*7*).
1, Chauvelicystis; *2, Ctenocystis*; *3, Trochocystites*; *4, Helicoplacus*; *5, Volchovia*; *6, Isorophus*; *7, Bothriocidaris*; *8, Lepidopsolus*; *9, Holothuria*; *10, Kinzercystis*; *11, Lichenoides*; *12, Gogia*; *13, Rhopalocystis*; *14, Aristocystites*; *15, Eucystis*; *16, Lepadocystis*; *17, Orophocrinus*; *18, Ecmatocrinus*; *19, Aethocrinus*; *20, Platycrinites*; *21, Homalometra*; *22, Ilycrinus*; *23, Villebrunaster*; *24, Asterias*; *25, Ophiothrix*.

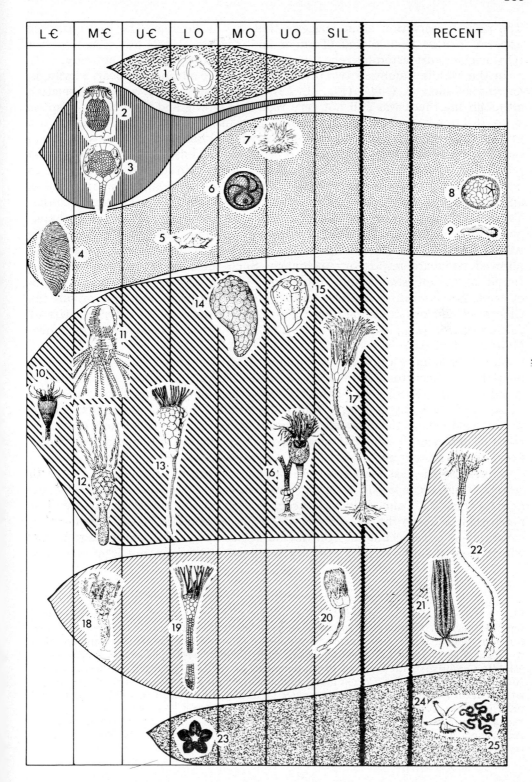

| L€ | M€ | U€ | LO | MO | UO | SIL | | RECENT |

The poor preservation alone suggests that it was flexible. The lower surface appears to have been plated (Termier and Termier, 1969) but certainly later (Ordovician) edrioasteroids relied on the substrate for protection aborally.

In the Middle Cambrian two different protective advances appear. Firstly, in eocrinoids such as *Gogia* (Fig. 3, *12*), most plates were weakly tesselated, although the theca was still not very rigid and strong and respiratory epispires were still interposed along most plate margins. *Lichenoides* (Fig. 3, *11*), a Middle Cambrian eocrinoid from Bohemia which I believe floated upside down, represents the first truly rigid theca with a small number of strongly tesselated plates, but even this genus still has epispires along plate sutures. The alternative development, already weakly initiated in Lower Cambrian *Stromatocystites*, and more fully developed in the succeeding Middle Cambrian edrioasteroid, *Cambraster*, was the appearance of a theca with a strong, heavily plated marginal rim which would resist deformation better than an imbricating theca. The weakly plated, almost membranous, thecal wall within the rim presumably allowed adequate respiratory exchange. Such thecae with marginal rims are found in *Ctenocystis* (Middle Cambrian; Fig. 3, *2*), the only known ctenocystoid, *Trochocystites* and *Gyrocystis* (Middle Cambrian; Fig. 3, *3*) among the Cincta, all cyclocystoids (Middle Cambrian—Devonian), and several genera of calcichordates (e.g., *Chauvelicystis*, Fig. 3, *1*) and eocrinoids (Tremadoc—Middle Ordovician).

Many echinoderms possess an "ornament" of thickened ridges or folds connecting plate centres. These form a rigid triangulated girder system which greatly strengthens the cup. The first suggestions of this "ornament" occur in *Gogia? radiata* Sprinkle (Fig. 4e) from the Middle Cambrian of Canada and it is well developed in the Tremadoc rhombiferans *Macrocystella* and *Cheirocystella*. Thereafter it is a common "ornament" in many echinoderm groups, down to the present day.

Spines are the last protective morphological feature to appear in the fossil record. They occur in the Asterozoa and two genera of calcichordates, *Chauvelicystis* (Fig. 3, *1*) and *Amygdalotheca*, as early as the lowest Ordovician, but reach their acme in the sea urchins which first appear high in the Middle Ordovician. The earliest echinoid, *Bothriocidaris*, had spines adjacent to its tube-feet which were themselves protected by small calcified rings, but later echinoids

---

Fig. 4. Protective features in echinoderms. a. *Helicoplacus* (Lower Cambrian) with entirely flexible theca. b. *Kinzercystis durhami* (Lower Cambrian) with imbricate lower theca and tesselate oral surface. c. *Gogia palmeri* (Middle Cambrian) with tesselate theca and epispires. d. *Gogia longidactylus* (Middle Cambrian) with less extensive epispires. e. *Gogia?radiata* (Middle Cambrian) with the first development of strengthening ridges on thecal plates. f. *Homocystites alter*, and g. *Coronocystis angulatus* (Middle Ordovician) with well-developed triangulation ridges. h. *Eucidaris tribuloides* (Recent) camouflaged in *Thalassia* grass with primary spines producing a disruptive outline. (b,d,f,g × 1.6; c × 1.8; e × 1.4.)
Fig. a is reproduced from Durham and Caster (1963, Science, 140, p. 820, copyright by the Am. Assoc. Adv. Sci.); figs. b—e from Sprinkle (1973); fig. h from Kier (1975). Permission to reproduce these photographs is gratefully acknowledged.

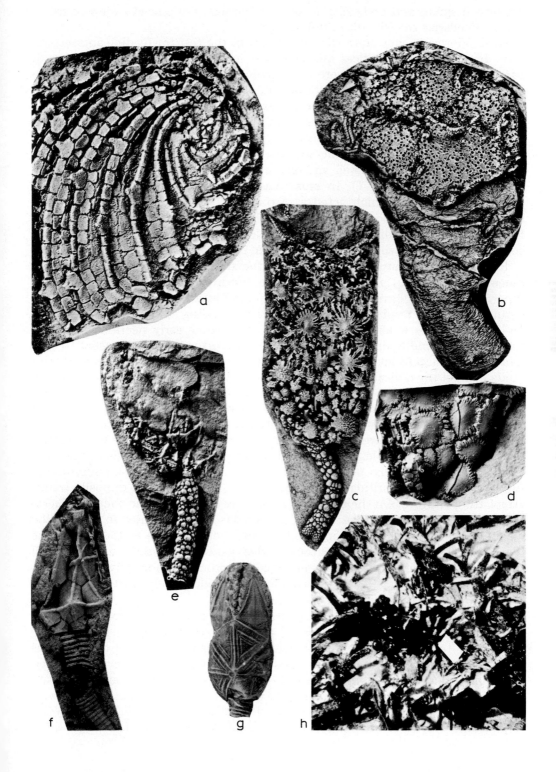

developed spines and pedicellariae all over the test. For a brief review of spines and pedicellariae see Nichols (1969, p. 126).

Echinoderms usually have special orifices in the test for feeding and respiration. In many, these orifices are large and might be entered by predators or parasites if not protected. Where large orifices are covered with flexible membranes, the membranes are usually plated, as in the peristomial and periproctal membranes of recent echinoids or the extensive periproctal membrane of the Ordovician—Devonian rhombiferan family Pleurocystitidae (Paul, 1967b, fig. 2). In the diploporite families Sphaeronitidae and Holocystitidae (Upper Ordovician—Devonian) the peristome was covered by a palate of six plates; food entered by four or five small ambulacral orifices at the margins of the peristome (Paul, 1971, fig. 1). In most blastozoans the anus was covered by a plated anal pyramid which acted as a one-way valve opening outwards only. Even the small gonopore of some Rhombifera had a similar gonal pyramid. Food particles and radial extensions of the coelomic systems were protected by cover plates in blastozoans and by lappets in crinozoans.

Respiratory pore-structures weakened the theca, and the evolution of the theca in the crinoid family Porocrinidae (Ordovician) and the rhombiferan superfamily Glyptocystitida may be interpreted as a compromise between the requirements of respiration and protection (Kesling and Paul, 1968; Paul, 1972b). Epispires weakened plate sutures as do some early pectinirhombs and fissiculate blastoid hydrospires. Accessory ridges often develop adjacent to conjunct pectinirhombs and counteract their weakening effects. The spines of echinoids protect the delicate uncalcified tube-feet which are thus much better protected than the similar diplopores of Palaeozoic diploporites. This protection may partly explain the success, in terms of survival and diversity, of respiratory tube-feet in living echinoids compared with the extinct diploporites.

Trends do occur in protected orifices but are less obvious than those affecting the theca as a whole. The peristome enlarges with time in the Sphaeronitidae but is adequately protected by the palatal plates. In the rhombiferan superfamily Glyptocystitida (Tremadoc—Upper Devonian) two opposite trends affect the periproct. On the one hand it enlarges significantly in the line Macrocystellidae — Cheirocrinidae — Pleurocystitidae, but decreases in size in the lines Cheirocrinidae — Glyptocystitidae and Cheirocrinidae — Callocystitidae (Paul, 1972b, p. 25). In the Pleurocystitidae, which would seem to be less well protected than their ancestors, the enlarged periproct was normally held closely against the substrate and may well have been adequately protected by the strongly plated dorsal surface of the theca.

In summary, if one ignores those forms with a spicular skeleton which have a poor fossil record, the evolution of protective morphological features was from poor to better protection and the most weakly protected echinoderms survived for the shortest periods. Those forms with the most easily distorted and disarticulated tests, the helicoplacoids, camptostomatoids and imbricate eocrinoids, appear first and are confined to the Lower Cambrian. Thecae with thickened peripheral plates appear in the Middle Cambrian (*Ctenocystis*, Cincta) and survive to the Ordovician (eocrinoids) or Devonian (cyclocystoids).

Tesselated eocrinoids which still have weakened sutures bearing epispires appear in the Middle Cambrian and survive to the Ordovician. Rigid cups without sutural pores and with strengthening triangulation structures appear in the Tremadoc and survive to the present day. Finally, echinoderms with protective spines appear last in the Ordovician and also survive to the present day.

Physiological protection is afforded by poisonous spines and pedicellariae and by generally repellant substances such as must occur among holothurians. The majority of this latter group has all but abandoned any skeletal protection. A few holothurians have no calcified structures whatsoever and most have only spicules and sclerites buried in the body wall. Needless to say their fossil record is meagre. Echinoid spines with similar morphology to extant poisonous spines occur as far back as the Cretaceous but there is, as yet, no way of knowing whether they were poisonous or not. Pedicellariae are reported from the Upper Carboniferous but there is some doubt about their age (Kier, 1974, p. 4). Otherwise pedicellariae are Mesozoic and later features.

Behavioural protection involves a variety of acts from simply dwelling in crevices, through cryptic camouflage and infaunal behaviour, to dymantic display. Again fossil evidence is not easy to substantiate except for infauna. Many modern crinoids and filter-feeding ophiuroids dwell in crevices or under stones and extend their arms to feed at night. Presumably fossil crinoids and ophiuroids did also. The large primary spines of some cidaroids are used to wedge these echinoids into crevices in rocks, producing a characteristic wear. I have seen this type of abrasion on cidarid spines from the Corallian (Upper Jurassic) of England. The outline and appearance of recent cidaroids is often modified by the large primary spines (Fig. 4h) especially when these bear growths of calcareous algae. These algal growths are potentially preservable but I know of no fossil examples. They should be sought. Many recent echinoids and holothuroids camouflage themselves by carrying sediment, shell, or marine plant fragments all over their dorsal surfaces. This in turn leads to infaunal modes of life where the animal is completely camouflaged by being buried in the sediment. Evidence for an infaunal mode of life is abundant among echinoids from the Lower Jurassic onwards. Many sea-urchin tests depart from the paradigm of a protective envelope. They are often thinly plated (e.g., many spatangoids), have flattened tests (e.g., many clypeasteroids), and have small, ineffectual spines. Among living echinoids, almost all of these forms are infaunal and presumably selection for a protective envelope or protective spines is relaxed since the enclosing sediment affords considerable protection. *Inter alia*, these infaunal echinoids have petaloid ambulacra for specialized respiratory tube-feet and such features can be recognized readily in fossil echinoids. One major question about echinoid evolution is why did it take so long (from the Middle Ordovician to the Lower Jurassic) to exploit an infaunal mode of life, especially as the subsequent explosive radiation of "irregular" echinoids demonstrates how successful infaunal sea urchins have become. Perhaps this is related to respiratory problems. The necessary deposit-feeding apparatus, suckered tube-feet, were already highly developed. This is perhaps the biggest puzzle of echinoderm evolution.

Finally some recent sea-urchins, e.g., *Diadema antillarum*, actively rotate and wave their spines when approached. This is the only example of dymantic or threatening behaviour that I have come across among echinoderms but others probably exist. Incidentally this simultaneously involves morphological, physiological and behavioural features but is unlikely to be detectable in fossils.

Feeding

Four broad types of feeding are found among living and fossil echinoderms: filter-feeding (Fig. 5), deposit-feeding, predation (including scavenging) and herbivorous browsing. In addition to these modes of ingesting particulate food, living echinoderms absorb dissolved organic nutrients directly through the body wall. Undoubtedly this forms a useful supplementary energy source and was probably utilized by fossil echinoderms but there is, of course, no preserved evidence of this. All known Lower Cambrian echinoderms were probably filter-feeders. The first likely deposit-feeders, ctenocystoids (and calcichordates)

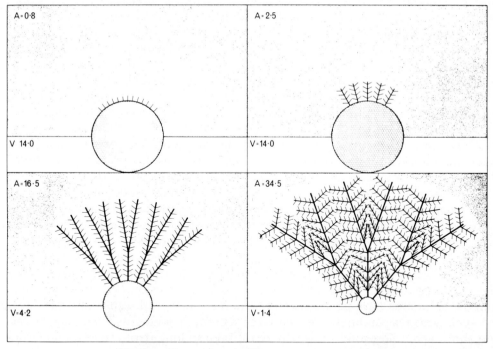

Fig. 5. Evolution of filter-feeding structures in echinoderms, showing a progressive increase in filter area and hence food-gathering ability with a concomitant decrease in thecal volume and therefore food requirements. Upper right: helicoplacoid grade (Lower Cambrian) with tube-feet arising directly from the theca. Upper left: blastozoan grade (Lower Cambrian—Permian) with brachioles. Lower left: primitive crinozoan grade (Lower Ordovician) with branched arms. Lower right: advanced crinozoan grade (Middle Ordovician—present) with pinnate branched arms. A = area of filter; V = volume of theca (in arbitrary units).

appear in the Middle Cambrian. Starfish probably represent the first appearance of true predators (Lower Ordovician) and echinoids the first herbivorous browsers (Middle Ordovician). Certainly passive microphagous feeding preceded the more complex, active macrophagous types and will be considered first.

The paradigm for a filter has as large an area as possible, an equidimensional mesh, a means of maintaining a current through the mesh and of separating filtered from unfiltered water to prevent recycling. Broadly speaking, the food gathering capacity of a filter-feeder will be proportional to the area of filter and the food requirements to the volume of soft tissue. A potential "starvation effect" occurs during growth since food requirements increase faster than the capacity to meet them. In fossil echinoderms the volume of the theca is a rough measure of food requirements and the area of the subvective system a measure of food-gathering capacity.

Crinoids probably represent the most complex filter-feeders among living echinoderms. The histology and activities of the tube-feet and water vascular system in feeding activity have been described by Nichols (1960, 1972) while *in situ* observations of crinoid feeding behaviour have been published by Magnus (1963, 1967), Meyer (1973) and Macurda (1973). All branches of the arms and all pinnules bear extensions of the radial water vessels. In the pinnules these give rise alternately to groups of three tube-feet: one long, one short and one intermediate in length. The tube-feet bear sensory papillae which trigger mucous glands to trap food particles when stimulated. The tube-feet bend rapidly towards the central food groove and deposit trapped food particles which are then carried towards the mouth along the food groove in a bolus.

In situ observations suggest that modern crinoids fall broadly into two feeding categories: rheophilic (current seeking) and rheophobic (current avoiding). Rheophilic crinoids form filtration fans, which compare closely with the paradigm of a filter in many cases (Fig. 6a, 6c), by orientating their arms and pinnules in a plane perpendicular to the current with the food-grooves downstream. When a complete fan is formed, tips of the long tube-feet on adjacent pinnules just touch (Fig. 6a) and the tips of pinnules on adjacent arm branches also touch (Fig. 6c). Not all rheophilic crinoids form complete fans, however (Fig. 6d). Rheophobic crinoids do not form a fan, extend their arms much less regularly, and hold the pinnules in four radial series along each arm (Fig. 6b). Rheophilic deep-water stalked crinoids apparently also form parabolic filtration fans (Figs. 6e-g).

Filter-feeding occurred (and occurs) in all Lower Cambrian echinoderms, virtually all Blastozoa and Crinozoa, and some Asterozoa. A progressive increase in the ability to meet food requirements may be inferred from fossil evidence by comparing thecal volume with the size and complexity of the subvective system (Fig. 5). Of the earliest echinoderms helicoplacoids were apparently most inefficient since they possessed a large theca with a single ambulacrum from which tube-feet arose directly. Even so, the effective length of the ambulacrum is increased by the spiral arrangement and the single branch. The earliest edrioasteroid *Stromatocystites* had five short ambulacre, but the tube-feet still arose directly from them. The earliest blastozoans, *Kinzercystis* and *Lepidocystis*,

a

b

c

d

e

f

g

show the first increase in complexity of filter and are typical of later Blastozoa. Both genera had recumbent ambulacra off which three to eight brachioles arose. Although Sprinkle (1973, p. 21) has questioned the existence of radial branches of the water vascular system in the ambulacra of blastozoans, most authors have assumed these to be present and Breimer and Macurda (1972) present convincing evidence for their presence in blastoids. I infer, therefore, that in blastozoans tube-feet arose from the brachioles which in turn arose from the recumbent ambulacra (diagrammatically shown in Fig. 5, upper right). Among blastozoans only members of the rhombiferan superfamily Hemicosmitida evolved any more complex food gathering apparatus. In *Caryocrinites*, and probably other hemicosmitid genera, "pinnate" free arms developed but a large theca was still retained (Sprinkle, 1975). This group of blastozoans independently evolved a crinozoan type of food-gathering apparatus in which erect free arms bore biserial brachioles which in turn bore tube-feet.

Crinozoans typically have complex and efficient subvective systems and arose in the Middle Cambrian, that is, if *Ecmatocrinus* is a crinoid, but first flourished in the Lower Ordovician. Most crinozoans possess free arms which may branch several times. Each branch bears a series of uniserial pinnules which in turn bear series of tube-feet (Fig. 5, bottom). Even so Palaeozoic camerate crinoids usually had large heavy cups and hence high food requirements. The Palaeozoic inadunates and their living descendents, the articulates, have gone a stage further and reduced the cup to an insignificant structure at the junction of the stem and arms (e.g., Fig. 3, 22). The majority of the living articulates, the comatulids, lose the stem in the adult stage. Thus crinozoa have the most efficient filtering organs, and modern crinoids have reduced food requirements to a minimum.

The overall evolution of filter-feeding echinoderms shows that the least efficient (helicoplacoids) appeared first and had the shortest stratigraphic range (Lower Cambrian). The next most efficient (edrioasteroids and blastozoans) appeared almost as early (high Lower Cambrian), reached a peak in the Ordovician, and survived to the close of the Palaeozoic. The most efficient forms arose last (?Middle Cambrian, certainly Lower Ordovician) and survived to the present day.

Even though modern comatulid crinoids can and do move about, when feeding they, like all filter-feeders, are essentially passive. The other modes of feeding found in echinoderms involve locomotion of necessity: to find new sedi-

---

Fig. 6. Filter-feeding in living crinoids. a and c. The rheophile *Heterometra savignyi*, showing filtration fan. Note that the tips of adjacent pinnules (c) and of adjacent tube-feet (a) just touch to form an equidimensional mesh. b. Rheophobic feeding posture in *Nemaster discoidea*. Note pinnules arranged in four series along arm. d. *Democrinus* sp., a probable rheophile with incomplete filter. e—g. *Cenocrinus asternis*, a stalked rheophile which forms a complete parabolic filter.

Figs. a and c from Magnus (1963); fig. b from Meyer (1973); figs. d—g from original photographs kindly supplied by D.B. Macurda Jr., taken at 300 m (d) and 150—250 m (e—g) depth off north coast of Jamaica. Permission to publish these photographs is gratefully acknowledged.

ment particles, fresh unbrowsed algae, prey or uneaten carcasses, etc. Food is presumably located by taste and/or smell and hence the oral surface is opposed to the substrate, an inversion of the animal with respect to normal filter-feeding orientations. The water vascular system becomes at least as much a locomotor system as a feeding system. Tube-feet lack papillae but develop suckers for more efficient holding in both feeding and locomotion.

The earliest probable deposit-feeders were ctenocystoids and possibly other homalozoans (Middle Cambrian). Evidence for a water vascular system is scanty to say the least in this group which also lacks any trace of pentameral symmetry. Deposit-feeding stelleroids appear in the Lower Ordovician along with truly predaceous forms, while herbivorous echinoids appear in the Caradoc (Middle Ordovician). Even the echinoids' complex feeding organ, the Aristotle's lantern, appears in a slightly modified form in the earliest known sea urchin *Bothriocidaris*. Although some Palaeozoic echinoids may have been deposit-feeders, the real development of this mode of feeding among echinoids occurred very late in their evolution during the Jurassic and was associated with the adoption of an infaunal mode of life (Kier, 1974).

Respiration

Respiration is a complicated process which involves, *inter alia*, gas exchange through a respiratory surface. Only this aspect will be considered since fossil evidence for internal circulation and metabolism is extremely limited. Exchange of oxygen and carbon dioxide takes place by diffusion and Farmanfarmaian (1966) and I (Paul, 1972a) have shown that oxygen will penetrate into an echinoderm 1—3 mm by diffusion alone. Thus echinoderms of more than 3 mm radius require special respiratory surfaces. As with feeding (which is quantitatively linked to respiration rates) respiratory requirements are approximately proportional to thecal volume and exchange capacity proportional to the area of respiratory surface. Hence with growth there is a potential "suffocation effect". With regard to primitive echinoderms, the early Palaeozoic forms generally had large thecae and small, inefficient exchange surfaces, while modern crinoids have virtually no theca and all their arms, etc. are thin enough to be oxygenated directly by diffusion from surrounding sea water. Modern echinoids, with large thecae, gain most of their oxygen through the tube-foot/ampulla systems and Fenner (1973) has reviewed respiratory adaptations of echinoid tube-feet and ampullae. Infaunal echinoids have specialized respiratory tube-feet in the petaloid ambulacra.

Even in echinoderms where the water vascular system is associated with respiration and hence both the internal and external fluids may be sea water, an exchange surface is essential for efficient exchange. If equal volumes of saturated and deoxygenated sea water are mixed, the maximum resulting saturation is 50%. Respiratory surfaces with a counter-current system (Paul, 1968, fig. 22) can theoretically achieve 100% saturation of internal fluids. A respiratory surface is only one type of exchange surface. The paradigm for an exchange system has a surface of large area, is as thin as is compatible with its strength, and maintains a concentration gradient across the surface. The most efficient means of maintaining a concentration gradient is by a counter-current system. Evidence for counter currents is difficult to substantiate in fossils but separate entrances and exits indicate the potential for a one-way current in one half of the system.

Pore-structures may be endothecal or exothecal, where exchange takes place within and outside the theca, respectively. Endothecal pore-structures are protected by the theca and tend to be thin, extensive and delicate. However, sea water flowed through them and the dangers of fouling by extraneous particles or of recycling are greater than in exothecal pore-structures. Protective devices at the entrances of endothecal canals take the form of narrow slits or fine meshes (Paul, 1968), while devices to prevent recycling include ciliated ridges and pores in chimney-like raised tubercles. Excurrent apertures were also generally smaller than incurrent apertures, thereby generating stronger excurrents (e.g., Macurda, 1966b). Effective exchange was limited by the rate at which sea water could pass through the canals. Exothecal pore-structures are not protected, tend to have thicker walls and to be much less extensive than endothecal pore-structures. Fouling and recycling are impossible and the pore-structure is bathed in an infinite ocean of oxygenated sea water. Again effective exchange is controlled by the rate at which body fluids can pass through the canals.

Many echinoderm pore-structures are calcified and are hence frequently preserved and measurable in fossils. Echinoderm calcite is a meshwork; gases diffuse through the soft tissue portion. Echinoderms, above all other phyla, yield excellent fossil evidence of respiration.

Evidence of respiratory surfaces is found in the earliest calcified echinoderms. The Lower Cambrian helicoplacoid *Waucobella* had sutural pores along each side of the ambulacrum which probably served as a connection between an external podium and an internal ampulla (Fig. 7, *5*) as Durham (1967, fig. 2) has suggested. Oxygen would have diffused into the podium and would be carried thence into the ampulla where it would have diffused into the internal coelomic fluids. The pores are single, and circulatory currents such as occur in the echinoid tube-foot/ampulla systems (Fig. 7, *6*), and are inferred in fossil diplopores, were not possible. Similarly, Lower and Middle Cambrian eocrinoids have epispires which were probably simple extensions of coelomic cavities like the papulae of modern starfish. These were potentially more efficient than helicoplacoid tube-feet since there would have been only one soft-tissue layer through which gases had to diffuse (Fig. 7, *1*). Again, with a single sutural pore, circulatory currents were impossible. Some epispires spread out over the external surface of the thecal plates as in the eocrinoid *Lichenoides* (Middle Cambrian; Fig. 7, *2*; Fig. 8a) and the inadunate crinoid *Carabocrinus* (Middle Ordovician), thus increasing the area of exchange surface; but a major advance in respiratory structures occurred in the Tremadoc with the first appearance of diplopores (Fig. 7, *3*; Fig. 8b). In diplopores two canals open in an external depression, the peripore, which was covered with soft tissue in life. Circulation was now possible (Fig. 7, *3*) with body fluids rising up one canal, being

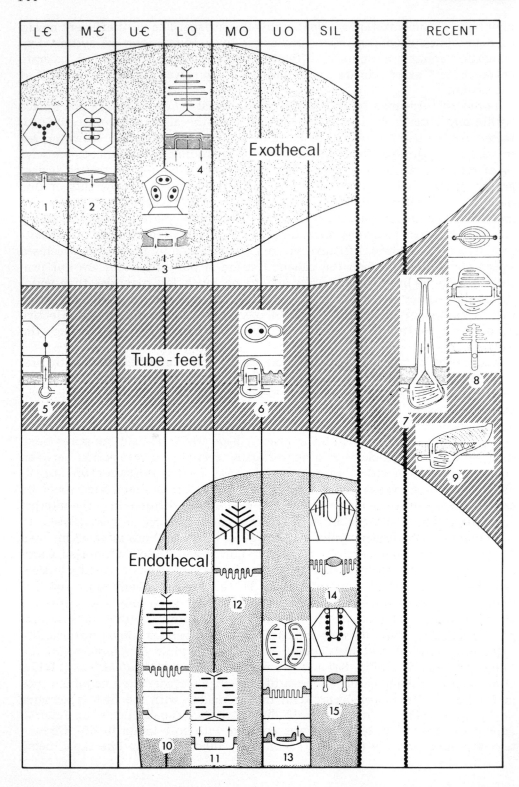

oxygenated within the peripore, and descending down the other canal. These are potentially highly efficient pore-structures. Computer programs developed to simulate exchange indicate efficiencies approaching 100% in a diplopore 1 mm long. Actual fossil diplopores only rarely exceed this length (Paul, 1977). Although apparently highly efficient, diplopores were external, soft-tissue structures not protected by spines as are sea-urchin tube-feet; and diploporites became extinct in the Middle Devonian. Lack of external protection may explain the brief success of calcified exothecal pore-structures (Fig. 7, 4) which occur in fistuliporite Rhombifera (Ordovician) and the diploporite family Holocystitidae (Upper Ordovician—Middle Silurian).

The Tremadoc also saw the appearance of the first endothecal pore-structures in the pectinirhombs of the rhombiferan *Cheirocystella*. These first pectinirhombs were relatively inefficient, having conjunct slits and discrete dichopores (Fig. 7, 10; Fig. 8h). Disjunct slits, with a separate entrance and exit, arose in the Arenig (Fig. 7, 11; Fig. 8i) and confluent dichopores in the Middle Ordovician (Fig. 7, 13; Fig. 8j—l). During the Ordovician, pore-structures reached a peak of taxonomic and morphological diversity and examples are found in groups which do not normally possess them, as for example, the inadunate crinoid *Porocrinus* (Fig. 7, 12). The last major type of endothecal pore-structure, hydrospires, evolved with the first true blastoids in the Middle Silurian and survived to the end of the Palaeozoic. Both fissiculate and spiraculate blastoids appear and disappear simultaneously (Figs. 7, 14 and 7, 15). Echinoids, with their tube-foot/ampulla systems, first appear in the Middle Ordovician and show little change until the development of petaloid ambulacra in the Lower Jurassic. Petaloid ambulacra contain specialized respiratory tube-feet and ampullae with enlarged surfaces and internal divisions to enhance circulatory efficiency (Figs. 7, 8 and 7, 9). Counter-currents occur in both the tube-feet and ampullae of the petals of living sea-urchins.

Trends in respiratory systems have been investigated, not least in the petaloid ambulacra of *Micraster*. I have summarized (1968, 1972b) the general evolution of pectinirhombs and shown a repeated trend from discrete to confluent dichopores in all families at about the Middle/Upper Ordovician boundary. Associated with this trend is another to reduce the number of pectinirhombs per theca from twenty or more in the Lower Ordovician to three or less by the

Fig. 7. Evolution of respiratory structures in echinoderms. *1*, Epispires of *Kinzercystis* (Lower Cambrian), surface view above, section below, currents indicated by small arrows; *2*, enlarged epispires of *Lichenoides* (Middle Cambrian); *3*, diplopores (Tremadoc—Middle Devonian); *4*, humatirhombs (Lower—Upper Ordovician); *5*, tube-feet with single pore, *Waucobella* (Lower Cambrian); *6*, primitive echinoid tube-feet, *Bothriocidaris* (Middle Ordovician); *7*, dorsal tube-feet and ampullae of *Strongylocentrotus* (Recent); *8*, specialized respiratory tube-feet of *Brissaster* and *9*, *Dendraster* (Recent); *10*, conjunct pectinirhomb with discrete dichopores (Tremadoc); *11*, disjunct pectinirhomb (Arenig); *12*, goniospires of *Porocrinus* (Middle Ordovician); *13*, disjunct pectinirhomb with confluent dichopores and vestibule rims (Middle Ordovician—Upper Devonian); *14*, fissiculate and *15*, spiraculate blastoid hydrospires (Middle Silurian—Permian).

Figs. 7—9 after Fenner (1973).

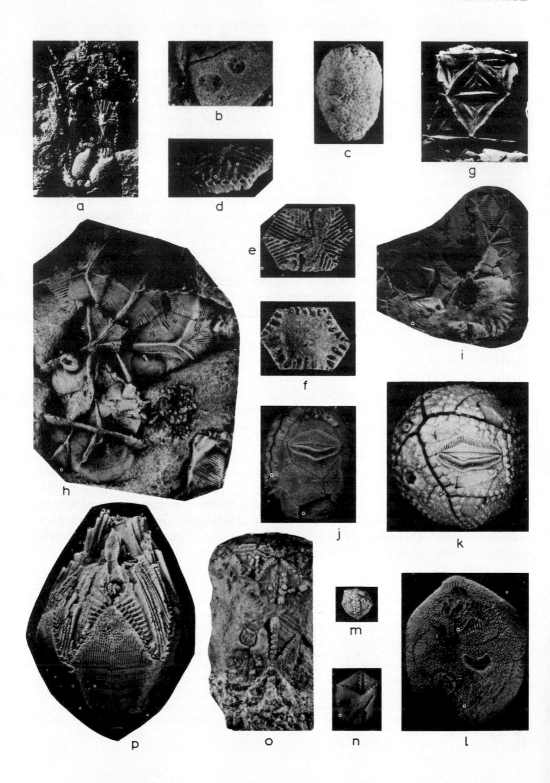

Middle Silurian. In the diploporite *Sphaeronites*, I have shown (1972a, 1973) a trend to reduction in diplopore density which has been interpreted (Bockelie and Paul, 1977), using computer programs to generate quantitative estimates of exchange rate, as a trend towards fewer, more efficient, diplopores that achieved the same general exchange rate in all species of *Sphaeronites*.

Rowe (1899) noted that, among many other features, the paired petaloid ambulacra of the chalk sea-urchin, *Micraster*, increased in length with time, i.e., phylogenetically as well as ontogenetically, but he made no mention of the numbers of respiratory tube-feet in the petals. Rowe stressed the stratigraphic importance of changes in the morphology of the paired petaloid ambulacra and recognized five types which he termed "smooth", "sutured", "inflated", "subdivided", and "divided". Nichols (1959) used regression analysis to show that the numbers of respiratory tube-feet in the paired petals increased significantly phylogenetically. His fig. 39 (Nichols, 1959, p. 414) also shows that the number of respiratory tube-feet increased more rapidly during growth in the later forms (*M. cortestudinarium* and *M. coranguinum*) than in the early form (*M. corbovis*). Nichols interpreted the morphological changes from "smooth" to "divided" petals as producing stronger ciliary currents associated with a deeper burrowing mode of life. The increase in the number of respiratory tube-feet presumably increased respiratory capacity in response to the more restricted oxygen supply available to deep-burrowing *Micraster*. The more rapid increase in the number of tube-feet during ontogeny in the later deep-burrowing forms would offset the "suffocation effect" mentioned early in this section. Even so, the shallow-burrowing *M. corbovis* regularly attained a much larger size than the later deep-burrowing forms. Perhaps the size of the later species was limited by their respiratory capacity. However, the late semi-infaunal *M.* (*Isomicraster*) *senonensis* has a similar maximum size to other late species of *Micraster* s.s., yet it has more respiratory tube-feet at all sizes. Nichols suggested that the individual respiratory tube-feet of *Isomicraster* may have been smaller than those of *Micraster* s.s.

Even if one cannot relate the increase in the numbers of respiratory tube-feet to depth of burrowing in *Isomicraster*, nevertheless throughout the evolu-

---

Fig. 8. Respiratory structures in echinoderms. a. Epispires in *Lichenoides priscus* (Middle Cambrian). b. Diplopores in *Eucystis* sp. (Upper Ordovician). c. Humatipores in *Holocystites scutellatus* (Middle Silurian). d—f. Humatirhombs in *Ulrichocystis eximia* (Middle Ordovician); oblique, external (weathered surface) and internal views, respectively. g. Folds in plates of *Macrocystella azaisei* (Tremadoc). h. Conjunct pectinirhombs with discrete dichopores in *Cheirocystella languedocianus* (Lower Ordovician). i. disjunct pectinirhombs with discrete dichopores in *Cheirocystis anatiformis* (Middle Ordovician). j—l. Disjunct pectinirhombs with confluent dichopores in *Lipsanocystis traversensis* (Middle Devonian), *Strobilocystites calvini* (Middle—Upper Devonian) and *Osculocystis monobrachiolata* (Lower Silurian), respectively. m—n. External and internal views of cryptorhombs of *Caryocrinites ornatus* (Middle Silurian). o. Cryptorhombs of *Hemicosmites* sp. (Middle Ordovician). p. Cataspires of *Meristoschisma hudsoni* (Middle Ordovician).
Figs. a and p reproduced from Sprinkle (1973). Permission to publish is gratefully acknowledged. (a × 1.7; b × 8; c × 0.8; d × 4.8; e,f,l × 3.2; g—k, m—o × 1.6; p × 1.8.)

tion of *Micraster* s.s. a progressive increase in respiratory capacity appears to have taken place. This is in direct contrast to the trend in *Sphaeronites* where the same respiratory capacity was achieved by progressively fewer diplopores. However, all *Sphaeronites* were epifaunal and did not have to overcome the problems of the deep-burrowing *Micraster*.

In summary, once again we find that in general the earliest respiratory structures were inefficient and short-lived while later structures were more efficient, and those which are still extant seem to be the most efficient. Again the helicoplacoids seem to have been the least efficient group. There are, however, some questions which remain unanswered: Why did endothecal pore-structures become extinct at the close of the Palaeozoic when they appear to be highly efficient? How did camerate crinoids, with large, heavily plated cups and no respiratory structures at all, survive? In addition the general trends outlined here need more detailed analysis before they can be accepted as established fact.

Excretion

Excretion is a puzzle in living echinoderms which have no circulatory or excretory systems. Soluble nitrogenous wastes are thought to diffuse through respiratory surfaces along with carbon dioxide. It has been suggested repeatedly that insoluble wastes may be eliminated by phagocytes, wandering cells which are said to burst through the body wall and escape, but the evidence is meagre (Endean, 1966). With such poor understanding of excretion in living echinoderms, very little can be said about the fossil record of excretion.

Reproduction

All living echinoids and crinoids normally have separate sexes. Less than 100 species of living echinoderms are known to be hermaphroditic (Hyman, 1955): 30—40 species of ophiuroids, about 30 holothuroids and some asteroids. Three ophiuroids are suspected to reproduce parthenogenetically. Most echinoderms reproduce by shedding eggs and sperm directly into the surrounding sea water. The chances of successful fertilization are enhanced by a chemical stimulant, the shedding substance, which issues with the genital products and hence the first echinoderm to shed triggers simultaneous shedding in adjacent mature individuals of the same species. Accessory sexual organs and complex courtship behaviour are generally absent in echinoderms. The fertilized eggs develop into free-swimming larvae which eventually settle and metamorphose, development following a distinctive pattern in each living class.

A number of modern echinoderms develop directly, without a free-swimming larval stage, and brooding is known in all living classes. Brooding is typical of, but not confined to, cold water (polar) species. All hermaphroditic and some dioecious ophiuroids brood their young in the genital bursae, which also have a respiratory function. Fertilization in dioecious species probably takes place when sperm brought in by respiratory currents reach the eggs in the bursae. Because of their currents, respiratory and filter-feeding organs are good sites for brood pouches. For example the gills of some bivalve molluscs and the pinnules of some Antarctic comatulid crinoids are modified as brood pouches. The starfish *Asterina* broods its young with no morphological modification, but pterasterid starfish have brood pouches in both males and females. These latter structures are probably also respiratory. Cidaroid echinoids carry young on their surface, effectively brooding them, again without any morphological modification.

Other recent sea-urchins, however, have distinct brood pouches in the females only and form the best evidence for sexual dimorphism in the echinoderm fossil record. Philip and Foster (1971) illustrate and describe an extensive fauna of marsupiate echinoids from the Tertiary of Australia. The earliest known marsupiate echinoids are from the Upper Cretaceous, but brooding without morphological modification may have arisen much earlier. Fay (1967, p. 281), among others, has suggested that the hydrospires of blastoids were brood pouches, rather than, or as well as, being respiratory structures. This interpretation is distinctly plausible given the known association of respiratory structures and brood pouches; however, several lines of evidence argue against it. The position of the gonopore in blastoids was unknown when Fay wrote. Breimer and Macurda (1972, p. 153) have described a *single* gonopore in the anal inter-radius of fissiculate blastoids. There are eight to ten hydrospire fields in all blastoids, two per radius, except that those adjacent to the anal inter-radius may atrophy. Thus, when the number of hydrospires is reduced, they are lost from precisely the area where one would expect to find brood pouches, in the anal inter-radius adjacent to the gonopore. Secondly, currents in blastoid hydrospires flowed from the aboral pole towards the mouth. Thus currents in the hydrospires of the anal inter-radius (and any other hydrospires) would have carried genital products out of, not into, the hydrospires. Finally, as no sexual dimorphism has been described in blastoids and the solitary gonopore does not correspond to the number of hydrospires, it seems likely that the latter were purely respiratory structures (see postscript on p. 158).

Apart from brooding, two other important changes occurred in echinoderm reproductive systems during their evolution. A change from a single gonad, as evidenced by a single gonopore, to multiple gonads, which probably appeared first in the Lower Ordovician in the asteroids; and the development of external gonads in crinoids. The latter is difficult to date precisely since the skeletal ossicles of the basal pinnules, which bear the reproductive tissue, do not differ from those of normal pinnules.

Evidence for internal gonads in Lower Cambrian echinoderms is scant, but *Kinzercystis durhami* has a possible gonopore (Sprinkle, 1973, p. 74; pl. 6, figs. 1, 3). A gonopore definitely occurs in the rhombiferans *Macrocystella* and *Cheirocystella* (Tremadoc). There is no unequivocal evidence for internal gonads in fossil crinoids, although a gonopore-like orifice occurs in the posterior oral plate of *Hybocrinus*, *Porocrinus*, *Palaeocrinus* and other Palaeozoic inadunates. Haugh's (1973, 1975a, b) extensive and detailed reconstructions of the internal anatomy of Mississippian Camerate crinoids do not include any internal gonads and these may well have been absent in camerates. If so, external gonads may have arisen with the first crinoids in the Lower Ordovician or possibly the Middle Cambrian.

It is difficult to measure efficiency in reproduction. Ultimately efficiency means survival and the extinction of groups need not necessarily result from any specific reproductive strategy. The fossil record shows, however, that most of the reproductive strategies known in living echinoderms could well have originated early in their evolution. Both multiple and external gonads could have evolved by the Lower Ordovician and brooding in specialized pouches was certainly present by the Pennsylvanian. Paradoxically external gonads may have had a protective effect. The most nutritious parts of an adult sea-urchin are the large internal gonads. If a predator can find a way through the protective armour of spines, there is a rich source of food to be exploited. Man preys on sea-urchins in many parts of the world for precisely this reason. In contrast, having the reproductive tissues dispersed among several tens (perhaps hundreds) of pinnules, each with a relatively small amount of nutritious tissue and a relatively large amount of hard calcite skeleton, makes the effort of predation too great for the benefit gained. It is true that some fish nibble at modern crinoids (Macurda, personal communication, 1975), but they do no serious damage. No animal is reported as preying exclusively, or even regularly, on living crinoids. External gonads also have the advantage that they can be oxygenated directly from sea water. Thus, the high ratio of skeleton to soft tissue in modern crinoids may help explain their success in surviving to the present, compared with other "pelmatozoan" groups. They are less palatable to predators and have simpler respiratory and feeding requirements than their Palaeozoic predecessors with large thecae and internal gonads.

Growth

Growth involves an increase in size and complexity. Although larval blastoids have been reported (Croneis and Geiss, 1940) and some microcrinoids may be larval forms, only post-larval development can be considered in the vast majority of fossil echinoderms. Even so it is impossible to review all aspects of growth in all echinoderms because of the variety of styles of growth and diversity of growing organs and organisms. Growth involves d'Arcy Thompson's (1917) principle of similitude or proportionality, and is thus related to all other functions. As has already been mentioned, food and oxygen requirements increase in proportion to *volume*, while the capacity to meet them increases in proportion to the *areas* of filter and respiratory surface. Strength of skeletal ossicles is similarly related to cross-sectional area, but loading requirements to weight, albeit reduced by immersion in sea water. Similitude explains many examples of allometric growth, where an organ increases in size at a different rate to the whole body.

As usual in considering fossils, attention will be directed primarily at skeletal growth, which is largely achieved by accretion in echinoderms. Although they have the ability to resorb their skeletons, echinoderms only rarely do so; certainly there is no wholesale remodelling such as occurs in vertebrate bones. Growth of plates will be considered first, since all parts of the skeleton are made of plates, then growth of the stem, subvective system, theca and orifices, in turn.

Plate growth

Echinoderm plates are a composite material, as are almost all organic skeletal materials. Composites (e.g., fibreglass) overcome the brittle nature of many strong materials, such as glass fibres, by embedding them in a flexible matrix (e.g., rubber, resin, protein) to produce a material which is stronger and more flexible than either component alone. Composites may be granular (tar Macadam), fibrous (fibreglass, wood), or laminar (plywood, laminated plastics) depending on the nature of the strong component. Examples of all three types occur in the plates of echinoderms, which are composed of calcite rods (trabeculae) and collagen fibres, set in a soft-tissue matrix. Echinoderm plates differ from most other biogenic calcite composites in having a much coarser meshwork, in only rarely being granular, and in not having a pure protein matrix. The calcite meshwork (called stereom by most palaeontologists) varies in density and coarseness in different parts of the skeleton (Fig. 9g, h) depending on function. In sea-urchin test plates, for example, there is usually a laminar inner region of alternating sheets of calcite trabeculae and soft tissue; a fibrous outer region in which both trabeculae and collagen fibres are set perpendicular to plate sutures, the collagen fibres being most densely arranged adjacent to the sutures (Moss and Meehan, 1967); and the external surface may be a granular composite in the articulation surfaces of the spine mamela. Furthermore, the galleries within the stereom may be aligned for the insertion of collagen fibres, as in the ligamental articulations of crinoids (Fig. 9a), or more random in areas of muscle attachment (Fig. 9c, above right). Thus, echinoderm plate structure is complex and capable of adaptation to a variety of functions. The effectiveness of the plates of living echinoderms as composites is indicated by the fact that when damaged in life they do not break along either suture lines or calcite cleavage planes. In fossils, however, with the soft-tissue matrix replaced by secondary calcite, cleavage is the most characteristic indication of an echinoderm plate.

Growth of plates is initiated at numerous centres and soon the small calcite spicules coalesce to form a single porous sheet. Subsequent growth is usually holoperipheral with new calcite added to existing trabeculae all around the plate. Plate growth is not identical in all classes. In diploporites and echinoids plates consist of two regions. The outer fibrous region grows by the extension of fibres perpendicular to plate margins (sutures); the inner laminar region by the addition of sheets of calcite on the internal surface. Growth lines are seen only exceptionally because they lie on the surface between the inner and outer regions. They can be revealed in Recent sea-urchins by shining light through the test or by immersion in suitable fluids (Jensen, 1969). In fossils the inner laminar part of the plates is occasionally preserved in a different manner from the outer region and subsequent solution of the original calcite may reveal

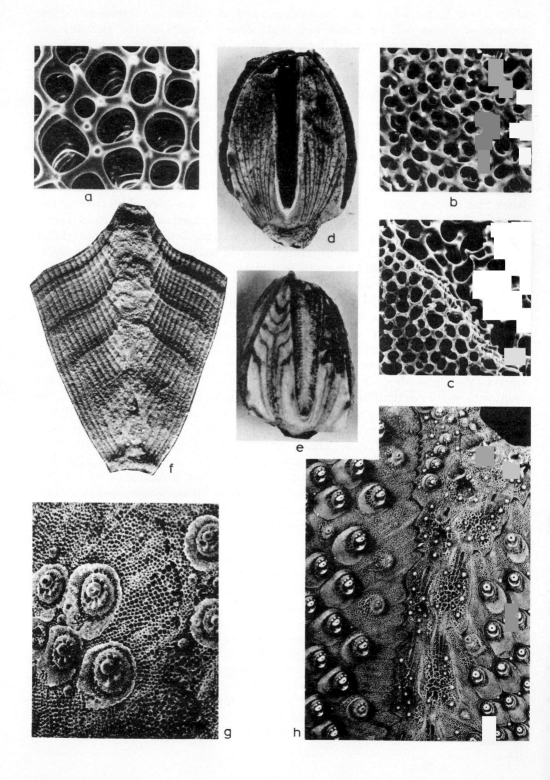

growth lines. I have (1971) figured examples of this among diploporites. In contrast to this plate structure, blastoids regularly have growth lines on the external surface (e.g., Macurda, 1966b). Meyer (1975) has figured growth lines on the internal surface of plates of the camerate crinoid *Platycrinites*. In diploporite and echinoid plate growth, the thickness is maintained across the plate by the addition of sheets of calcite on the internal surface. In blastoid and crinoid plates the new layers presumably continue across the inner and outer surfaces, respectively.

Stem

This term is restricted here to include only true columns with columnals. Polyplated "holdfasts" of early echinoderms grew in a manner analogous to that of the theca. Stems grow by both enlargement of existing columnals and the addition of new columnals. Enlargement is effectively the same as plate growth and is holoperipheral, although it may require special modification of the articulation surfaces (Roux, 1974). In columnals composed of a single piece, enlargement of the central canal was effected by resorption: one of the few undeniable cases of resorption in echinoderm growth.

Addition of new columnals may take place at one or two defined locations in the stem, or at more or less regular intervals along its length between primary ossicles (nodals). In many stemmed echinoderms new columnals were added just below the cup. In those with xenomorphic stems, new columnals of distinctive morphology may be added at the proximal (thecal) end of each section. The most rapid increase in length of stem occurs where secondary, tertiary and sometimes even quaternary columnals are added between nodals. The points of insertion of new columnals are usually the most flexible parts of the stem, so changing requirements of both flexibility and strength can be met during growth.

Subvective system

For purposes of discussing growth, I am restricting this to food-gathering structures that are free of the general test. The ambulacra and food grooves of echinoids and starfish are therefore excluded. Blastozoan and crinozoan subvective systems include main branches, arms, which may be erect (most crinozoans) or recumbent on the theca (many blastozoans), off which free pinnules

---

Fig. 9. Plate structure and growth in echinoderms. a. Galleried stereom in *Endoxocrinus parrae* (ligamental articulation). b. Internal surface of centro-dorsal plate of *Analcidometra armata*. c. Stereom of muscular (above right) and ligamental (below left) articulation surface in *Nemaster rubiginosa*. d. *Pentremites conoideus* (Carboniferous), and e. *Hyperoblastus alveata* (Devonian) showing colour bands. f. Basal plate of *Meristoschisma hudsoni* (Middle Ordovician) showing ?annual colour bands. g—h. Variation in surface stereom of *Plagiobrissus grandis*.
Figs. a—c reproduced from Macurda and Meyer (1975), figs. d—f from Sprinkle (1973); and figs. g—h from Kier (1975). Permission to publish these photographs is gratefully acknowledged. (a × 560; b × 390; c × 840; f × 5.25; g × 37; h × 10.)

or brachioles arise. In crinozoans the main arm branches are primitively uni-serial, although some become secondarily biserial; the pinnules are always uni-serial. In blastozoans the main ambulacral structures are usually composed of two biseries of plates, alternately large and small on either side of the main food groove. The brachioles are always biserial. Growth of all types is peripheral, with new arm, pinnular, or brachiolar plates being added at the tips of these structures. Ontogenetic series show that the total number of brachioles in the Middle Silurian rhombiferans, *Staurocystis quadrifasciatus* and *Pseudocrinites bifasciatus*, increased in proportion to thecal diameter (Paul, 1967a). Brower (1974) has considered ontogeny in camerate crinoids quantitatively. His table 8 (Brower, 1974, p. 38) shows that the mean length of the subvective system per cm^3 of thecal volume declines with growth from 8660 cm in the smallest to 2720 cm in the largest examples of six Upper Ordovician camerates; amply illustrating the "starvation effect" mentioned in the section on feeding. Macurda (1966b, 1975) shows a similar relationship between the number of brachioles and thecal volume in two Mississippian blastoids, *Globoblastus norwoodi* and *Pentremites elongatus*.

Theca
The theca, here interpreted in the broadest sense, grew by enlargement of existing plates coupled with the addition of new plates in many cases. Addition seems to predominate in the earliest (Lower Cambrian) echinoderms and was random in time and place. Nearly all plates remained quite small (less than 5 mm maximum dimension); distinct generations of plates (Paul, 1971) cannot be recognized nor are the points of insertion of new plates regularly arranged as happens, for example, in camerate crinoids and in echinoids. The eocrinoid *Lichenoides* (Middle Cambrian) is the earliest known echinoderm whose theca grew by enlargement alone. It apparently had a fixed number and arrangement of plates throughout growth, with the possible exception of the minute "basal" plates. Distinct generations of plates can be recognized in the Middle Silurian diploporite *Holocystites*, especially in species with a cylindrical theca. It is curious, though, that the contemporary and sympatric species *H. cylindricus*, *H. abnormis* and *H. alternatus* should have different growth patterns. *H. cylindricus* added no plates during growth; *H. abnormis* had the beginnings of a second generation; and *H. alternatus* had three fully developed generations (Paul, 1971). All three reach approximately the same size with *H. abnormis* a little smaller than the other two.

Blastoids were extremely conservative in their growth, virtually all the theca of all species being composed of the same eighteen plates. Only the minute anal deltoids vary in number and position. Thus all blastoid growth was achieved by these eighteen plates. Many inadunate and articulate crinoids are also characterized by a fixed number of cup plates. Camerates, on the other hand, regularly added plates during growth, but did so at specific locations. Arm plates (brachials) and inter-radial plates were added on top of the usual primary cup plates in five or ten definite series (e.g., Brower, 1974, figs. 8—12).

Among extant groups, echinoids add ambulacral and interambulacral plates along the margins of the five ocular plates in the apical disc. All extant echinoids have twenty columns of plates, a pair of columns in each ambulacrum and each interambulacrum; but Palaeozoic echinoids had much more variable arrangements (Kier, 1965). Even so, apparently all but *Bothriocidaris* and *Neobothriocidaris* added new plates along the margins of the oculars. In *Bothriocidaris* normal oculars are absent and may be represented by the single columns of plates without tube-feet. In *Neobothriocidaris* new ambulacral plates appear to have been added along the margins of these columns of imperforate plates (Paul, 1967c).

Orifices

Enlargement of orifices poses special problems during growth. In the vast majority of cases echinoderms have small orifices developed across a plate suture and large orifices surrounded by several plates. As the plates grew, the orifices could be enlarged by a slower secretion of calcite in the adjacent plates, or by no secretion at all. In either case resorption is unnecessary. Indisputable cases of resorption do occur where an orifice lies within a single plate as in the gonopore of some Aristocystitidae (Diploporita) and the central canal of columnals as already noted.

Of respiratory structures, all rhombs, hydrospires, epispires and the cataspires of parablastoids grew at plate margins. The canals, pores or slits were formed by modifications of the growing edge of two contiguous plates. Diplopores, however, grew by resorption; but the tube-feet of echinoids form at the same time as the ambulacral plates and resorption is unnecessary or reduced. The goniospires of the inadunate crinoid *Porocrinus* had a very complex mode of growth (Kesling and Paul, 1968).

Longevity

Growth lines and seasonal colour bands are preserved in some fossil echinoderms. These allow estimates of age, although the evidence is too scant to suggest evolutionary trends. I reported (Paul, 1967a) ages of 3—4 years for two Silurian rhombiferans, *Pseudocrinites bifasciatus* and *Tetracystis oblongus*. Reimann (1961) and Sprinkle (1973) figure colour bands in blastoids from the Devonian and Carboniferous (Fig. 9d,e). The latter author also figures colour bands in the Middle Ordovician parablastoid *Meristoschisma*, and suggests ages of 3—7 years for most blastozoans.

Living echinoids apparently have similar lifespans. Jensen (1969) describes a method of age determination in recent sea-urchins. Some echinoderms do live longer than this and Buchanan (1967) estimated lifespans up to 15 years for the ophiuroid *Amphiura chiajei* and the spatangoid sea-urchin *Echinocardium cordatum*. In the same area of the North Sea he found that the closely related *Amphiura filiformis* and *Brissopsis lyrifera* lived for only three or four years.

With such complex growth patterns it is difficult to recognize clear trends during evolution. However, from available information it seems that plates were uniformly small in Lower Cambrian echinoderms, but grew significantly larger in the Middle Cambrian. This may be associated with the change from imbricate to tesselate plates and is certainly related to the first appearance, in *Liche-noides*, of a theca with a fixed plate arrangement which grew by the enlargement of existing plates alone. Where plates were added during growth, the pattern was initially random but became more organized in Ordovician and later echinoderms. Fell (1963, 1967) has argued that changing growth gradients allowed the development of free-living asterozoans from fixed crinozoans, but the details are by no means settled. Again it seems that the basic growth patterns and lifespans of living echinoderms were first established early in the history of the phylum. For further studies of growth in fossil echinoderms see the following papers: blastoids (Macurda, 1966a,b, 1975; Breimer and Macurda, 1972); crinoids (Macurda, 1968; Brower, 1973, 1974); diploporites (Paul, 1971); edrioasteroids (Bell, 1976); rhombiferans (Paul, 1967a).

Summary of vital functions

Two general conclusions can be reached from the above review of vital functions in fossil echinoderms. First, almost all the features of the phylum arose in the Cambrian and Ordovician. The echinoids are perhaps the only exception in developing advanced types of teeth in the Aristotle's lantern and pedicellariae (probably in the Trias) and adopting an infaunal mode of life (Jurassic) relatively late in their evolution. All three developments are important in terms of living faunas. Secondly, at least with regard to three functions (protection, feeding and respiration), echinoderms of the Cambrian and Ordovician were significantly less efficient than their descendents. Both these points are compatible with the idea that there was a genuine explosive radiation of echinoderm classes in the Cambrian and Ordovician and that the subsequent decline in the numbers of classes was due to the elimination of inefficient groups by competition.

References

Bell, B.M., 1976. A study of North American Edrioasteroidea. Mem. N Y. State Mus. Sci. Serv., 21: 447 pp. (63 pls., 65 text-figs.).

Bockelie, J.F. and Paul, C.R.C., 1977. The Ordovician cystoid *Sphaeronites* (Diploporita) in Britain and Scandinavia. (in prep.).

Breimer, A. and Macurda Jr., D.B., 1972. The phylogeny of the fissiculate blastoids. Verh. K. Ned. Akad. Wet., No. 26: 390 pp. (34 pls.).

Brower, J.C., 1973. Crinoids from the Girardeau Limestone (Ordovician). Palaeontogr. Am., 7: 263—499 (pls. 59—79).

Brower, J.C., 1974. Ontogeny of camerate crinoids. Palaeontol. Contrib. Univ. Kansas, Pap. 72: 53 pp.

Buchanan, J.B., 1967. Dispersion and demography of some infaunal echinoderm populations. Symp. Zool. Soc. Lond., 20: 1—11.

Caster, K.E., 1967. Homoiostelea. In: R.C. Moore (Editor), Treatise on Invertebrate Paleontology. S. Echinodermata 1. Geological Society of America and University of Kansas Press, Lawrence, Kansas, pp. S581—S627.

Caster, K.E., 1971. Les echinodermes carpoides de l'Ordovicien inférieur de la Montagne Noire (France) by G. Ubaghs. (Review). J. Paleontol., 45: 919—921.

Croneis, C. and Geis, H.L., 1940. Microscopic Pelmatozoa: part 1, ontogeny of the Blastoidea. J. Paleontol., 14: 345—355.

Durham, J.W., 1964. The Helicoplacoidea and some possible implications. Yale Sci. Mag., 39: 24—25.

Durham, J.W., 1966. Evolution among the Echinoidea. Biol. Rev., 41: 368—391.

Durham, J.W., 1967. Notes on Helicoplacoidea and early echinoderms. J. Paleontol., 41: 97—102 (pl. 14).
Durham, J.W. and Caster, K.E., 1963. Helicoplacoidea: a new class of echinoderms. Science, 140: 820—822.
Endean, R., 1966. The coelomocytes and coelomic fluids. In: R.A. Boolootian (Editor), Physiology of Echinodermata. Wiley, New York, N.Y., pp. 301—328.
Farmanfarmaian, A., 1966. The respiratory physiology of echinoderms. In: R.A. Boolootian (Editor), Physiology of Echinodermata. Wiley, New York, N.Y., pp. 245—265.
Fay, R.O., 1967. Evolution of the Blastoidea. In: C. Teichert and E.L. Yochelson (Editors), Essays in Paleontology and Stratigraphy. R.C. Moore Commemorative Volume. University of Kansas Press, Lawrence, Kansas, pp. 242—286.
Fell, H.B., 1963. Phylogeny of the sea-stars. Philos. Trans. R. Soc. Lond., B246: 381—435.
Fell, H.B., 1965. The early evolution of the Echinozoa. Breviora, 219: 17 pp.
Fell, H.B., 1967. Echinoderm ontogeny. In: R.C. Moore (Editor), Treatise on Invertebrate Paleontology. S. Echinodermata 1. Geological Society of America and University of Kansas Press, Lawrence, Kansas, pp. S60—S85.
Fenner, D.H., 1973. The respiratory adaptations of the podia and ampullae of echinoids (Echinodermata). Biol. Bull., 145: 323—339.
Haugh, B.N., 1973. Water vascular system of the Crinoidea Camerata. J. Paleontol., 47: 77—90 (3 pls.).
Haugh, B.N., 1975a. Digestive and coelomic systems of Mississippian camerate crinoids. J. Paleontol., 49: 472—493 (5 pls.).
Haugh, B.N., 1975b. Nervous systems of Mississippian camerate crinoids. Paleobiology, 1: 261—272.
Hyman, L.H., 1955. The Invertebrates. Volume 4. Echinodermata. McGraw-Hill, New York, N.Y., 763 pp.
Jefferies, R.P.S., 1967. Some fossil chordates with echinoderm affinities. Symp. Zool. Soc. Lond., 20: 163—208.
Jefferies, R.P.S., 1968. The subphylum Calcichordata (Jefferies, 1967), primitive fossil chordates with echinoderm affinities. Bull. Br. Mus. Nat. Hist. (Geol.), 16: 241—339 (10 pls.).
Jefferies, R.P.S., 1975. Fossil evidence concerning the origin of the chordates. Symp. Zool. Soc. Lond., 36: 253—318.
Jensen, M., 1969. Age determination in echinoids. Sarsia, 37: 41—44.
Kesling, R.V. and Paul, C.R.C., 1968. New species of Porocrinidae and brief remarks upon these unusual crinoids. Contrib. Mus. Palaeontol., Univ. Mich., 22: 1—32 (8 pls.).
Kier, P.M., 1965. Evolutionary trends in Paleozoic echinoids. J. Paleontol., 39: 436—465 (pls. 55—60).
Kier, P.M., 1974. Evolutionary trends and their functional significance in the post-Paleozoic echinoids. Paleontol. Soc. Mem., 5: 95 pp.
Kier, P.M., 1975. The echinoids of Carrie Bow Cay, Belize. Smithson. Contrib. Zool., 206: 45 pp. (12 pls.).
Macurda Jr., D.B., 1964. The Mississippian blastoid genera Phaenoschisma, Phaenoblastus, and Conoschisma. J. Paleontol., 38: 711—724 (pls. 117—118).
Macurda Jr., D.B., 1966a. The functional morphology and stratigraphic distribution of the Mississippian blastoid genus Orophocrinus. J. Paleontol., 39: 1045—1096 (pls. 121—126).
Macurda Jr., D.B., 1966b. Hydrodynamics of the Mississippian blastoid genus Globoblastus. J. Paleontol., 39: 1209—1217.
Macurda Jr., D.B., 1966c. The ontogeny of the Mississippian blastoid Orophocrinus. J. Paleontol., 40: 92—124 (pls. 11—13).
Macurda Jr., D.B., 1968. Ontogeny of the crinoid Eucalyptocrinites. Paleontol. Soc. Mem., 2: 99—118.
Macurda Jr., D.B., 1973. Ecology of comatulid crinoids at Grand Bahama Island. Hydro-Lab J., 2: 9—24 (2 pls.).
Macurda Jr., D.B., 1975. The Pentremites (Blastoidea) of the Burlington Limestone (Mississippian). J. Paleontol., 49: 346—373 (5 pls.).
Macurda Jr., D.B. and Meyer, D.L., 1975. The microstructure of the crinoid endoskeleton. Paleontol. Contrib. Univ. Kansas Pap. 74: 28 pp. (30 pls.).
Magnus, D.B.E., 1963. Der Federstern Heterometra savignyi im Roten meer. Nat. Mus., 93: 355—368.
Magnus, D.B.E., 1967. Ecological and ethological studies and experiments on the echinoderms of the Red Sea. Stud. Trop. Oceanogr., 5: 635—664.
Meyer, D.L., 1965. Plate growth in some platycrinid crinoids. J. Paleontol., 39: 1207—1209.
Meyer, D.L., 1973. Feeding behaviour and ecology of shallow water unstalked crinoids (Echinodermata) in the Caribbean Sea. Mar. Biol., 22: 105—129.
Moss, M.L. and Meehan, M.M., 1967. Sutural connective tissue in the test of an echinoid Arbacia punctulata. Acta Anat., 66: 279—304.
Nichols, D., 1959. Changes in the chalk heart-urchin Micraster interpreted in relation to living forms. Philos. Trans. R. Soc., Lond., B242: 347—437.
Nichols, D., 1960. The histology and activities of the tube-feet of Antidon bifida. Q.J. MIcrosc. Sci., 101: 105—117.
Nichols, D., 1967. The origin of the echinoderms. Symp. Zool. Soc. Lond., 20: 209—229.
Nichols, D., 1969. Echinoderms. Hutchinson, London, 4th ed., 192 pp.
Nichols, D., 1972. The water-vascular system in living and fossil echinoderms. Palaeontology, 15: 519—538.
Parsley, R.L. and Mintz, L.W., 1975. North American Paracrinoidea: (Ordovician: Paracrinozoa, new, Echinodermata). Bull. Am. Paleontol., 68: 115 pp. (13 pls.).

Paul, C.R.C., 1967a. The British Silurian cystoids. Bull. Br. Mus. Nat. Hist. (Geol.), 13: 297—356 (10 pls.).

Paul, C.R.C., 1967b. The functional morphology and mode of life of the cystoid *Pleurocystites* E. Billings 1854. Symp. Zool. Soc. Lond., 20: 105—121.

Paul, C.R.C., 1967c. New Ordovician Bothriocidaridae from Girvan and a reinterpretation of *Bothriocidaris* Eichwald. Palaeontology, 10: 525—541 (pls. 84—85).

Paul, C.R.C., 1968. The morphology and function of dichoporite pore-structures in cystoids. Palaeontology, 11: 697—730 (pls. 134—140).

Paul, C.R.C., 1971. Revision of the *Holocystites* fauna (Diploporita) of North America. Fieldiana, Geol., 24: 166 pp.

Paul, C.R.C., 1972a. Morphology and function of exothecal pore-structures in cystoids. Palaeontology, 15: 1—28 (pls. 1—7).

Paul, C.R.C., 1972b. *Cheirocystella antiqua* gen. et sp. nov. from the Lower Ordovician of western Utah, and its bearing on the evolution of the Cheirocrinidae (Rhombifera: Glyptocystitida). Geol. Stud. Brigham Young Univ., 19: 15—63 (7 pls.).

Paul, C.R.C., 1973. British Ordovician cystoids. Palaeontogr. Soc. Monogr. part 1: 64 pp. (11 pls.).

Paul, C.R.C., 1977. Respiration rates in primitive (fossil) echinoderms. Thalassia Yugosl. (in press).

Philip, G.M. and Foster, R.J., 1971. Marsupiate Tertiary echinoids from southeastern Australia and their zoogeographic significance. Palaeontology, 14: 666—695 (pls. 124—134).

Raup, D.M., Gould, S.J., Schopf, T.J.M. and Simberloff, D.S., 1973. Stochastic models of phylogeny and the evolution of diversity. J. Geol., 81: 525—542.

Reimann, I.G., 1961. A color-marked Devonian blastoid. Okla. Geol. Notes, 21: 153—157 (2 pls.).

Roux, M., 1974. Observations au microscope électronique à balayage de quelques articulations entre les ossicules de squelette des crinoïdes pédonculés actuels (Bathycrinidae et Isocrinina). Trav. Lab. Paléontol. Univ. Paris, Fac. Sci. d'Orsay, 10 pp. (4 pls.).

Rowe, A.W., 1899. An analysis of the genus *Micraster*, as determined by rigid zonal collecting from the zone of *Rhynchonella cuvieri* to that of *Micraster coranguinum*. Q. J. Geol. Soc., Lond., 55: 494—547 (pls. 35—39).

Sprinkle, J., 1973. Morphology and evolution of blastozoan echinoderms. Spec. Pap. Mus. Comp. Zool. Harvard, 284 pp. (43 pls.).

Sprinkle, J., 1975. The "arms" of *Caryocrinites*, a rhombiferan cystoid convergent on crinoids. J. Paleontol., 49: 1062—1073 (1 pl.).

Stephenson, D.G., 1977. On the origin of the five-fold pattern of the echinoderms. Thalassia Yugosl. (in press).

Termier, G. and Termier, H., 1969. Les Stromatocystitioïdes et leur descendence. Essai sur l'évolution des premiers échinodermes. Geobios, 2: 131—156 (pls. 7—12).

Thompson, d'A.W., 1917. On Growth and Form. Cambridge University Press, Cambridge, 793 pp.

Ubaghs, G., 1962. Sur la nature de l'organe appelé tige ou pédoncule chez les Carpoïdes Cornuta et Mitrata. C. R. Acad. Sci., Paris, 253: 2738—2740.

Ubaghs, G., 1967. Stylophora. In: R.C. Moore (Editor), Treatise on Invertebrate Paleontology. S. Echinodermata 1. Geological Society of America and University of Kansas Press, Lawrence, Kansas, pp. S495—S565.

Ubaghs, G., 1969. General characteristics of the echinoderms. Chem. Zool., 3: 3—45.

Ubaghs, G., 1971. Diversité et spécialisation des plus anciens échinodermes que l'on connaise. Biol. Rev., 46: 157—200.

Ubaghs, G., 1975. Early Paleozoic echinoderms. Annu. Rev. Earth Planet. Sci., 3: 79—98.

Postscript

In proof stage Katz and Sprinkle's convincing article on fossilised eggs in a Pennsylvanian blastoid (Science, N.Y., 192: 1137—1139) appeared. Their evidence satisfies criteria for the anal hydrospires being brooding structures in this one species, and records the first occurrence of sexual dimorphism in echinoderms. Even so, only the anal hydrospires are dimorphic and contain eggs in one specimen. I remain convinced that hydrospires were basically respiratory structures.

CHAPTER 6

PATTERNS AND THEMES OF EVOLUTION AMONG THE BRYOZOA

THOMAS J.M. SCHOPF

General Considerations

The aim of this section is threefold: (1) I will argue that we are justified in isolating evolutionary patterns for bryozoans from the whole of the epifaunal community of which they are a part. To oppose this, it might be stated, for example, that to isolate one part of that community so distorts any general patterns that any notion of generality is a delusion. (2) I will present a model of speciation which seems to me to be of greatest applicability to Bryozoa. This model is largely influenced by recent data on genetic and morphologic differentiation, and emphasizes that both allopatric and phyletic evolution may be important for these animals. (3) I will outline the classification of Bryozoa adhered to in this paper.

Constraints on the system

Although this chapter is chiefly directed toward Bryozoa, many of the concepts discussed cannot be considered in isolation from other groups of animals and plants as long as they have similar biological and physical requirements. Bryozoan individuals within the colony are on the order of 1 mm long and somewhat less than that in width, and individual colonies are commonly on the order of cm². These animals obtain food chiefly by filtering phytoplankton from sea water. The natural competitors for food and space are similarly sized animals and plants with similar size-dependent physical constraints on circulatory, excretory and respiratory systems, etc., and methods of obtaining energy. These groups especially include, in modern seas, the hydroids, barnacles, tunicates and sponges.

The initial question is why are there a certain number of species of this size range in a given area, rather than simply just considering the bryozoans. Models of speciation patterns for organisms in this size range should be developed with this whole "adaptive zone" in mind. There is no reason to believe that living soft-bodied forms did not also exist in this community in the past. There is even extensive Tertiary evidence of tunicates from their spicules (reviewed by Monniot and Buge, 1971).

We can examine whether the species abundance of bryozoans in the epifaunal community follows the distribution for the community as a whole. In Fig. 1, the abundance of the whole epifaunal community, and the bryozoan component, is plotted against the species rank for material from settling plates put out at Woods Hole, Mass. (Osman, 1977). For the whole fauna, and the bryozoan component, the slopes of the curves are similar to each other for

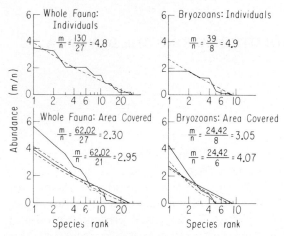

Fig. 1. Relative abundance of species of epifaunal community and of bryozoan component of that community from 103 cm² slate settling plate (m = number of individuals; n = number of species). Area covered is in cm² by the individuals (i.e., 1 colony = 1 individual). For the bottom two graphs, upper solid line is for whole population sampled; upper dashed line is for whole population minus those species so rare that their individual area covered is each less than 0.01 cm². The lower solid and/or dashed lines at the intercept in each graph are for distributions according to the MacArthur "broken-stick" model. Note that the bryozoan component behaves much as the total fauna. Data obtained and plotted by Richard W. Osman. See Osman (1977).

both (1) individuals (1 colony = 1 individual) or (2) area covered. Since a few colonies can control a large amount of area, the area-species ranking also shows dominance. [Note that the epifaunal community comes close to the distribution predicted by the MacArthur "broken-stick" model (MacArthur, 1957; King, 1964).]

Direction of speciation

The initial question is what set of general physical and biological conditions would lead, on the average, to the origin of new species? The approach which I find most useful in framing an answer to that question is derived from an optimization model previously used to make predictions about the extent of specialization between castes or polymorphs of animal colonies (Schopf, 1972, 1973). I would like to generalize the basic elements of that model to consider specialization and the origin of bryozoan species.

The basic assumption is that local populations of bryozoans (or other animals) will, on the average, continuously tend toward specialization of the use of resources to the extent that that resource is held sufficiently constant so that it is within the potential for specialization for a species. By reducing competition between similar species, specialization is of advantage to an individual, and to its local population, and (other things equal) should result in an increased proportion of viable offspring. The notion is simply that species should be maintained to the extent that continuing partitioning of environmental resources is possible. Saturation is simply defined as the situation when there is an equilibrium between species origination and extinction. This equilibrium is modulated by the capacity for specialization for environmental resources (Schopf, 1971).

Specialization should be enhanced by reinforcement of the selective regime in different local populations that are subject to somewhat different environments, as shown graphically for two local populations in Fig. 2. The optimum mix of individuals of the local population is the point at which there is a minimum number of individuals capable of utilizing each of the various resources. This presupposes that competition for resources continually presses populations downward toward this minimum number, and that below this number, populations rapidly tend to go extinct. Fig. 2 represents two different local populations. The important point is that specialization is continuously favoured in individuals comprising these populations. For various resources, a different optimum number of individuals may exist and so the ultimate population size would be a trade-off of the various competing influences.

This view basically assumes that ecologic restrictions regulate species diversity. This idea was discussed by Hutchinson (1968, p. 184) when he wrote: ". . .we seem forced to conclude that complex environments somehow subdivide naturally in one of a number of predetermined possible ways, as if they were intersected by cleavage planes." In this way, he suggested that one may pass from niche space to taxonomic space, and one should expect, in time, to have a 1 : 1 correspondence between them. We should come to have as many species in a given region as there were ways for a given group to partition the environment. The number of species in a specific adaptive zone is considered to be most directly a function of the ways to partition the environment within that zone.

This view of ecologic saturation can also be considered with the notion of an

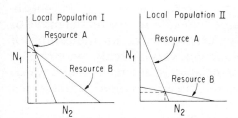

Fig. 2. Graphic illustration of principle of optimization for individuals within a local population capable of utilizing two different resources (say two different substrates). Resource A could be algal substrates, and B could be shell substrates. N_1 is the number of individuals especially suited to utilize resource B which are needed to utilize the available resources. N_2 is the number of individuals especially suited to utilize resource A which are needed to utilize the available resources. The point of intersection of the two resource lines marks the minimum number of individuals in local populations which are able to utilize both resources. In general, species are continually pressed toward this minimum number through competition for these resources. N_1 plus N_2 is the total number of individuals in any given local population. Note that the mix of N_1 and N_2 individuals changes as resources change.

equilibrium number of species in island biogeographic theory (MacArthur and Wilson, 1967). That is, for a fixed adaptive zone (k, in the expression $S = kA^z$), the probability increases that specialization will occur, and that the number of species (S) will increase as total area (A) increases. Terborgh (1973) has shown that within habitat speciation may well be a function of habitable area [and the importance of area is later emphasized (p. 189)].

We then have an operational definition of saturation: a substrate (or area) can be said to be efficiently utilized (saturated) if it holds as many species as would be anticipated from island biogeographic theory for substrates (or areas) of that particular environment. For example, for the epifaunal community on rocks, shells, and pilings, etc., in the Woods Hole region, this is about 13 species per 10 cm^2, doubling to 26 species as area is increased in order of magnitude to 100 cm^2, as theory predicts (Osman, 1977). The reason why the equilibrium number is about 13 per 10 cm^2 has not been analytically derived.

This view of evolution sees species continuously evolving in adaptive radiations, away from the unspecialized (r-selected) toward the specialized (K-selected). The common short-hand of r and K represent (note Table I) investment in early, rapid and high productivity (r is the intrinsic rate of increase from the logistic growth equation) versus late, slow and low productivity (K being the carrying capacity of the environment). The values r and K are the contrasting extremes in the game of assuring reproductive success to the next generation. This seems to me to be of surpassing importance in studies of evolution, at least for the Bryozoa.

Morphologic and life-history adaptations then can be viewed on the one hand as being few and chiefly concerned with very rapid growth and rapid reproduction, or on the other hand as being directed toward achieving stable populations, often especially through defence. Morphologic features can be graded from the point of view of which end of the spectrum they tend to occur. Table I summarizes some of the correlates of r and K selection, and the concepts are well discussed by MacArthur (1972, pp. 34, 226), among others.

Any species is of course a compromise among many factors, but, in general, bryozoan adaptations in the r versus K "game" include few or no defensive or reproductive polymorphs versus many polymorphs; few or no indications of colonial integration of growth and development versus highly coordinated behaviour; high larval production perhaps associated with large colonies versus very limited larval production, especially as seen in small colonies; free-swimming larvae versus brooded larvae.

Seen through time, as in Fig. 3, evolution is viewed as a continuing process of K-selected speciation along various structural or physiological "axes", diverging from r-selected, generalist, ancestral stocks. One way in which this general trend has been well documented for various taxa is in Cope's Rule of phyletic size increase (see Stanley, 1973).

In general, in this chapter, deterministic explanations which are dependent upon a particular and special ecologic or taxonomic history are de-emphasized. Rather, the point is that there is at any given time a spectrum of r versus K characters, and a spectrum of taxa that are dominantly r versus K oriented (as

TABLE I

Some of the correlates of r and K selection (directly from Pianka, 1970)

	r selection	K selection
Climate	variable and/or unpredictable; uncertain	fairly constant and/or predictable; more certain
Mortality	often catastrophic, nondirected, density-independent	more directed, density-dependent
Survivorship	often Type III (Deevey, 1947)	usually Type I and II (Deevey, 1947)
Population size	variable in time, nonequilibrium; usually well below carrying capacity of environment; unsaturated communities or portions thereof; ecologic vacuums; recolonization each year	fairly constant in time, equilibrium; at or near carrying capacity of the environment; saturated communities; no recolonization necessary
Intra- and interspecific competition	variable, often lax	usually keen
Relative abundance	often does not fit MacArthur's broken-stick model (King, 1964)	frequently fits the MacArthur model (King, 1964)
Selection favours	(1) rapid development (2) high r_{max} (3) early reproduction (4) small body size (5) semelparity: single reproduction	(1) slower development; greater competitive ability (2) lower resource thresholds (3) delayed reproduction (4) larger body size (5) iteroparity: repeated reproductions
Length of life	short, usually less than one year	longer, usually more than one year
Leads to	productivity	efficiency

shown graphically in Fig. 3). Since, on the average, relative uniformity of the environment reinforces specialization (as reviewed above), and leads to speciation, any given character complex, or lineage, will, on the average appear through time to go from the r to the K end of the spectrum. This result is viewed as a consequence of the stochastic process of speciation without any deterministic specification of a given morphology, or time of occurrence, or taxonomic composition. This conclusion — that r versus K selection can be viewed as a consequence of the stochastic model of speciation — is the main point of theoretical interest in this article.

This situation has been modelled in Monte Carlo simulations to examine trends in morphology where natural selection cannot have any significance at

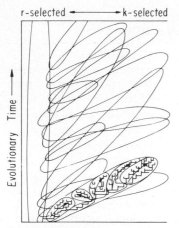

Fig. 3. Diagram of representation of the inferred importance of r versus K selection in bryo-zoan evolution. Each enclosed area represents a "foray" into specialization. In practice, the general problem is to identify the patterns of adaptive radiation, and then to work "back-wards" towards the "r-selected" ancestral groups. Some simplification occurs during evolu-tion, but this is considered of much lower frequency of occurrence than progressive speciali-zation. The enclosed area at the bottom with trellis patterns might represent a family with smaller circles genera, and individual branching species.

all (Raup et al., 1973). Descendent lineages almost always appear "specialized" with respect to ancestral stocks. As mean values of characters move stochasti-cally further and further from initial conditions, K selection seems to be occur-ring — except that no selection was operating in the stochastic model. Indeed, the problem is not how to account for morphological trends toward specializa-tion, but rather, how to account for the repeated occurrence (persistence?) of "general" types. Perhaps in the real world, mechanisms regarding changes in time of development, especially neoteny, may be significant, although this is not yet recognized to be significant for bryozoans.

Factors inhibiting speciation

From a physical point of view, environmental instability counteracts the trend toward specialization of the use of local resources (and therefore acts to inhibit speciation). An adequate biological response to an unstable environment requires such a generalist biology that specialization (thus leading to partition-ing of the environment and speciation) has a very low probability of occurring.

From a biological point of view, two factors seem to be important in inhibit-ing speciation.

(1) Competition among species may lead to dominance by one and even-tually to extinction of the weaker competitor. If a single species is able to achieve dominance, then this is interpreted to mean that the environmental conditions are sufficiently non-uniform so as to prevent further specialization with respect to the resource being competed for. As environmental instability

increases, so does dominance. A common way in which this relationship is expressed is in the higher species to individuals ratio of tropical marine areas relative to a lower species to individuals ratio of temperate areas.

(2) Gene flow may greatly restrict opportunities for specialization to local ecologic conditions. Gene flow is chiefly dependent upon larval dispersal. Marine Bryozoa are all similar in size, but are divided into two groups with regard to dispersal ability and potential gene flow. Probably 95% of the cheilostomes brood their young and these produce a non-feeding larva of short-lived duration (hours) capable of only slight dispersal. For this type of larva, I will introduce the term Barrois type (after Jules Barrois whose work on bryozoan development forms much of the basis of what we know today in this field). The remainder of the Bryozoa have a larva capable of feeding in the plankton, the Cyphonautes type of larva.

Schopf and Dutton (1976) document that morphologic differentiation in the Bryozoan *Schizoporella errata* parallels genetic differentiation in this same species. Gene flow via larval dispersal must act, in general, in the direction of reducing genetic differentiation. Thus it appears to me that in evaluating the probability of speciation in Bryozoa we must normalize for dispersal ability (= gene flow). In general, for any fixed distance, the probability of genetic isolation should be higher for animals that brood their young than for those with a feeding, long-lived larva.

Summary of speciation model

My general view of the process of speciation in Bryozoa is summarized in Fig. 4. If a species can extend its range 100 times that of its larval travel distance, then the time scale of gene flow between end members is an estimated 10^2-10^3 years (estimate based on judgement of Schopf, 1974, 1977a). At present, we have knowledge of one instance in which a distance of 30—40 km is sufficient to result in a significant change in both allele frequencies and in morphology (Schopf and Dutton, 1976). Hence the boundary between "local population" and "genetic and morphologic differentiation" is drawn at about that distance.

I believe it would be of great predictive value to be able to construct curves of the type shown in Fig. 5 in which the probability of gene exchange is related to the probability of speciation, each as a function of distance. Since biogeographic information is available in the fossil record, such a graph could be applied through geologic time. This type of graph must always be expressed in terms of probabilities since the percentage of the genome held in common by two populations will never be a precise measure of "speciation distance". Genetic changes do not have equal weight in determining whether or not genetic isolation obtains.

Distance *per se*, as discussed above, must be normalized for dispersal (i.e., gene flow in the present context). But in addition, distance must be normalized for environmental variability. Although distance and changes in the environment are positively correlated in general, the extremes range from rapid change

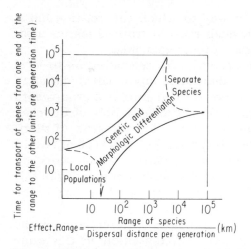

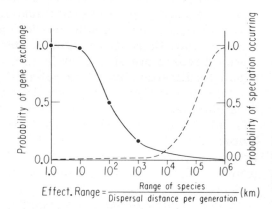

Fig. 4. Relationship between length of the time necessary for transport of genes, and the effective range of a species.

Fig. 5. Relationship between probability of gene exchange and the effective range of a species. Points are hypothetical.

in a single salt marsh to imperceptible change in the deep sea. Perhaps one way to express this relationship would be in terms of how far away one had to go from a given place before the mean value of a variable, such as temperature, changed by two standard deviations. This distance would be large for most variables in the tropics and smaller in temperate latitudes. Terborgh (1973) has shown the tremendous excess of area between 1° isotherms in the tropics compared with higher latitudes. In summary of this point, it appears to me that for bryozoans, the probability of genetic isolation is a function of the relative constancy of selective regime and of the distance between populations, when normalized for dispersal ability (gene flow) and environmental variability.

In attempting to account for patterns of evolution in the bryozoans, I think that explanations should be given in terms of the speciation process itself. Therefore I have tried to give greatest importance in these introductory remarks to those aspects of bryozoan biology and ecology which seem to me to most critically influence the speciation process. I think one can normalize for gene flow, and (in a broad way) for relative environmental variability. Thus, I see promise that illustrations such as Fig. 5 will be of value in predicting the probability of speciation, and more than an intellectual exercise. Naturally such diagrams represent probabilities of a stochastic process, and are not meant to be used deterministically for any given species. In this respect, such a diagram is akin to a probability distribution of a gas law, rather than as a time-table for a particular molecule.

I offer no strong evaluation of the possibility that speciation may occur *within* the range of a species. Well-known morphologic differences, and not so well known (but equally real) genetic changes, occur at physiographic barriers in the coastal region (e.g., at Cape Cod and Cape Hatteras). This suggests to me that geographic factors are of strong (and by themselves of sufficient?) importance to speciation. I know of no reason to suggest that sympatric speciation may be occurring in marine Bryozoa.

Bryozoan classification

The most significant development in approaches to the classification of bryozoans since the 1960's has been the explicit treatment of as *many* characters as possible and then using various statistical measures of association to help one to see "natural" groupings. The key paper in this approach is that of Boardman et al. (1970) which sets forth the basic philosophy and examples (this is also the approach to be used in the forthcoming five-volume revision of Bryozoa for the *Treatise on Invertebrate Paleontology*).

A thorough and detailed analysis of the topic of ectoproct—entoproct—bryozoan relationships that has appeared since Hyman's (1959) broad summary is Nielsen's (1971) monographic study of entoproct embryology and his detailed and sensitive consideration of the results of the nearly 100 species of ectoprocts whose development has now been examined. Nielsen writes (page 318) "...from the present knowledge of morphology and development of the entoprocts and ectoprocts it seems reasonable to interpret the body cavity of the ectoprocts as a specialized larval pseudocoele." And he goes on: "The first conclusion is that the Entoprocta and the Ectoprocta together must represent a separate phylum, the Bryozoa." I think Nielsen's carefully discussed conclusion should be accepted.

In this chapter, only the ectoproct bryozoans will be considered (Table II). And among the ectoprocts, greatest emphasis will be given to the orders Cheilostomata and Cyclostomata. There are abrupt discontinuities between the classes, but transitional forms exist among some of the orders.

Patterns of Evolution

In this section I will discuss patterns of evolution observed in (1) morphologic features, (2) behaviour; and (3) levels of colonial integration. A fourth type of pattern (biogeographic) is discussed elsewhere (Schopf, 1977b). Evidence has been chosen so that it is possible to ask for a given pattern what characterizes the r versus K ends of the spectrum of possibilities. For some patterns, such as the development of castes, and colony integration, the end members are relatively clear. For other patterns, such as calcification, the nature of end members is less easy to decide upon in some cases. Explanations for particular patterns are partially considered here, and partially in the next section on themes of evolution.

Morphological patterns

Calcification and support
Calcification is regulated by the coelom in bryozoans. The coelom is the

TABLE II

A present classification of Bryozoa to the ordinal level, with a few additional taxa mentioned in the text (for summary of geological ranges and composition of groups see Bassler, 1953, and Larwood et al., 1967)

Classification	Geological range	Places to begin if interested in phylogenetic relationships
Phylum Bryozoa		
Subphylum Entoprocta		
Subphylum Ectoprocta	Pleistocene to Recent	Reguant, 1959; Ryland, 1970
Class Phylactolaemata		Toriumi, 1956
Class Gymnolaemata		
Order Ctenostomata	Lower Ordovician to Recent	Jebram, 1973a, b
Order Cheilostomata	Upper Jurassic to Recent	Banta, 1975; Silén, 1942b
Suborder Cribrimorpha	Middle Cretaceous to Recent	
Suborder Anasca	Upper Jurassic to Recent	
Suborder Ascophora	Late Cretaceous to Recent	
Class Stenolaemata		Ross, 1964
Order Cryptostomata	Lower Ordovician to Lower Triassic	Blake, 1975; Tavener-Smith, 1975
Order Trepostomata	Lower Ordovician to Lower Triassic	
Order Cystoporata	Lower Ordovician to Permian	Utgaard, 1973
Order Cyclostomata	Lower Ordovician to Recent	Brood, 1972, 1975

pathway by which nutrients and waste products can be transported, both within a zooid, and within the colony. It seems reasonable, owing to the absence of a circulatory system, that the hormonal and growth-regulating substances would also be transferred through the coelom. By utilizing the colony-wide coelom, colonial control over growth patterns can also be maintained.

The importance of colony coordination of growth is that specialization for defence, reproduction and for other reasons can be achieved. High coordination of colony activity leads to efficiency and is a K-selected trait. In contrast, colonies in which each zooid performs all functions independent of other zooids leads to high productivity, an r-selected trait. It seems no accident that the two cheilostome families (Membraniporidae and Electridae, *sensu* Mawatari, 1974) with a plankton-feeding larva persisting for weeks to months are, in fact, the same families with very little calcification, little specialization, and a very large number of larvae. All other cheilostome families brood their young (and produce far fewer offspring) and usually have a strong investment in calcification as well as in defensive polymorphs.

The evolution of a colonial coelom, existing at least at the growing margin, occurs in the great majority of species of the extant order Cyclostomata, and is present in at least some species in the extant order Cheilostomata, and is inferred to have existed in all three orders known only from the fossil record [the Trepostomata, Cryptostomata, and Cystoporata (modern discussion of the colonial coelom of these five orders is given by Tavener-Smith, 1969; Brood, 1972; and Boardman and Cheetham, 1973)].

The particular manifestations of the colonial coelom differ in various groups. The Cheilostomata are organized so that the orifice through which the feeding polypide extrudes is placed along one side of a zooid; in the other four orders it is typically at the distal end of a zooid. The importance of this is that in all groups except the Cheilostomata, the zooid is therefore already completely calcified on all sides except in the very small opening through which the polypide is extruded. In the "primitive" Cheilostomata, the whole of one long side is almost entirely uncalcified, and many of the subsequent patterns of evolution within the Cheilostomata can be viewed as simply various answers to the theme of how to calcify (and thereby protect) that open side.

Among the four orders of non-cheilostome bryozoans, colonial control of total growth is always in evidence at the growing margin with the subdivision of the so-called common bud. In the "single-walled" forms, the common bud is continuous only at the growing margin and the calcareous walls occur only beneath the exterior cuticule. In the "double-walled" forms, the common bud continues to exist over a large portion of the colony, and an additional mineralized wall exists.

The cyclostome *Diplosolen* is a highly integrated colony that is polymorphic not only in space over the colony surface, but in time as zooid function changes while the colony increases in size. Silén and Harmelin's account (1974) is the first instance of time-dependent changes in caste reported in a bryozoan colony. This astounding example shows the extraordinary degree of coordination among colony members despite the fact that the animal lacks a full colony-wide coelom.

If the degree of colony coordination observed in *Diplosolen* occurs in forms whose colonial coelom is limited to the growing margin, and which is "single-walled", then what may be the additional advantage for colony coordination among double-walled Cyclostomata, Trepostomata, Cystoporata, and Crypto-stomata? What double-walled forms have in common is the capability for increased calcification, thereby attaining increased strength of support and increased size for the colony as a whole. The advantage of size may be chiefly in the number of brood chambers that can be supported.

In terms of reproduction, polyembryony, which is widespread in the Cyclo-stomata, occurs in *Diastopora*, a genus of the same family as *Diplosolen*. The function of polyembryony is to produce by cloning a large number of off-spring. It means that larvae released from a given parent share all their genes in common. In cheilostomes (all of which *lack* polyembryony), only the maternal genes will be in common among the larvae of a given colony. Since there is a high degree of genetic relatedness within a colony with polyembryony, kin-selection within the offspring would be very advantageous. *Diplosolen* is one of the genera in which one might assay for the existence of a pheromone released by the larvae to indicate good (or poor) places for settlement of one's fellow offspring.

That the colonial coelom functions importantly in calcification and thereby to strengthen the colony is indicated by cheilostome evolution (Cheetham, 1971). Six general morphotypes have been found to occur (Fig. 6). At the *r* end of the spectrum, types I and II have membranous frontals. Types III and IV have partial cover by a carbonate shield, and at the *K* end of the spectrum, types V and VI have full cover by a carbonate shield (developed together with a colonial coelom above the shield). Calcareous structures of the frontal surface may develop by joining across the frontal surface, by thickening the frontal surface, and ultimately by the binding of one zooid to another. Only morpho-types V and VI permit binding, and these same morphotypes are the most effective ones for permitting thickening and joining (Cheetham, 1971, p. 16, following his earlier suggestion: 1969, pp. 229).

Among cheilostomes the analysis of structural support has so far been directed at the implications of increased strength for resistance to physical stress, especially current strength, but quantitative or experimental approaches to indicate if sheer physical support is important have not been carried out. As indicated later (p. 177), the implications for reproduction in large versus small colonies should also be considered an immediate advantage for colonies with considerable vertical growth from the substrate.

Communication and sharing of nutrients on a colony-wide level is yet another key feature in the evolution of colonial specialization. A very high degree of colonial coordination can be achieved within colonies lacking the extensive colonial coelom, and whose only connection among zooids consists of a few pores (plugged with tissue) in zooid walls. Almost all species classified as cyclo-stomes from the Palaeozoic have non-porous skeletal walls (Brood, 1975). Nye (1968, 1976, pp. 36—38) therefore suggested that the development of porous walls by very similar forms in the Mesozoic permitted much greater colony

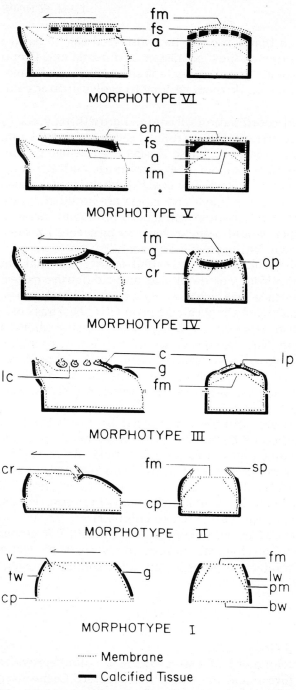

MORPHOTYPE VI

MORPHOTYPE V

MORPHOTYPE IV

MORPHOTYPE III

MORPHOTYPE II

MORPHOTYPE I

······ Membrane

━━ Calcified Tissue

Fig. 6. Zooid morphotypes I, II, III, IV, V, and VI. The relations of skeletal tissues to membranes are shown in diagrammatic sagittal (left) and transverse (right) sections (*bw* = basal wall; *cp* = communication pore; *cr* = cryptocyst; *fm* = frontal membrane; *g* = gymnocyst; *lw* = lateral wall; *pm* = parietal muscle; *sp* = spine; *tw* = transverse wall; *v* = vestibule beneath orifice; *c* = costa; *lc* = lacuna between costae; *lp* = lumen pore; *op* = opesiule; *a* = ascus; *em* = epifrontal membrane; *fs* = frontal shield). Lophophore, alimentary canal, and other organs have been omitted for simplicity; the direction of colony growth (distal) is indicated by the arrows. From Cheetham (1971).

circulation and integration, and in that sense can be considered the "adaptive breakthrough" which led to the tremendous expansion of Jurassic and Cretaceous cyclostomes. Since cyclostomes have been extant since the Early Palaeozoic, the question may well be what took them so long to evolve greater colonial coordination!

Among cheilostomes, increased coordination of colonial growth may act to decrease colony size as well as to free a colony from its reliance on substrate attachment (as in the lunulitiform colonies). These K-selected free-living gumdrop-shaped colonies have a colonial coelom (Tavener-Smith and Williams, 1972; Håkansson, 1973), and they also have highly evolved defensive structures known as vibracularia. The colony size is fixed, and larvae are brooded. These forms would seem to have about as much an investment in producing larvae, a relatively large percentage of which would seem to have to be successful since relatively few are produced, as occurs among living bryozoans.

The Cribrimorpha are of particular interest evolutionarily in that an extremely high degree of polymorphism was developed in the defensive castes. The protection of the frontal surface was achieved by overarching of spines, and subsequent fusion, rather than by skeletal calcification (Fig. 7). Thus colonies did not achieve the strength necessary for bearing an erect highly calcified growth, and remained encrusting. The various stages of development in the fusion of spines and of their overgrowths are said to occur in the series from *Membranipora* to *Membraniporella* to *Cribilina* according to Hincks (1890, p. 96), who wrote that "No doubt they are terms in an evolutionary series, connected by many transitional links. . .".

Other patterns of calcification are strongly emphasized by Ross (1964) in her review of morphologic groupings of Early Palaeozoic Bryozoa. At present, budding patterns are known for nine types of Palaeozoic tubular Bryozoa (McKinney, 1975, p. 69, figure 1) and six types of post-Palaeozoic cyclostomes (Brood, 1972, p. 33). One of the post-Palaeozoic types includes at least five variants in the family which includes the well-known genus *Idmonea* (Hinds, 1975). Some of these growth forms can be related to colony strengthening on the assumption of an erect mode of growth (Hinds, 1975, p. 897). Whether erect growth is chiefly owing to mechanical support to "withstand" water currents, or to an increase in reproductive potential (p. 177) should be evaluated.

Polymorphism: Avicularia and vibracularia
The extent and mode of development of castes within a colony provides another major indication of the investment in r versus K characters. Castes have been identified as defensive, reproductive, and as supporting (like tendrils). As presented by Schopf (1973), evolution in Bryozoa occurs at the colony level of organization, and anything to improve the reproductive potential of the *colony as a whole* should be favoured by natural selection. Thus, the evolution of separate castes by kin-selection should follow directly to the extent that the colony

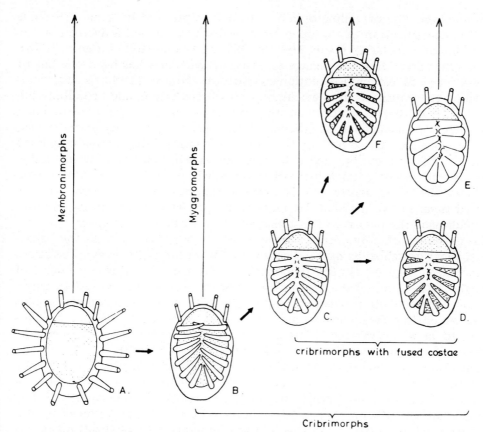

Fig. 7. Diagrams of single zooecia of a membranimorph and of derived Early Cretaceous cribrimorphs to show variations in the structure of costate secondary frontal walls. Note the transition from overarching spines unfused at the mid-line (*B*), to fused at the mid-line (*C*). Spines may grow together to form a shield (*E*), or spaces between spines may be partly (*F*) or wholly filled in (*D*). From Larwood (1969, p. 174).

is benefited. And the colony is benefited to the extent that a consistent environmental stimulus (such as a potential predator) is successfully handled, often by the evolution of a particular caste.

Polymorphs identified as probably defensive in function are abundant in the Cheilostomata (Silén, 1938) [see also statistical summaries showing increase in occurrence in tropical areas relative to the polar regions, in Schopf (1973), and Moyano (1975)]. Defensive polymorphs are rare in the Cyclostomata [they are known only in Levinsen's (1912) "Cyclostomata Operculata", later the sub-order Salpingina (see Brood, 1972), and are well seen in the genus *Melliceritites* of Cretaceous age but are not known younger than the Palaeocene; defensive polymorphs owing to a defensive secretion have been suggested in the single-tentacled polymorph of *Diplosolen* (Harmer, 1930, p. 102)].

Avicularia are simply modified normal zooids, with a "mandible" in place of the operculum. The prevailing opinion, strongly supported by Canu (1900), is that "the specialization of the operculum probably preceded a decrease in the dimensions of avicularia" (Dzik, 1975, p. 397; also supported by Banta, 1973).

Among cheilostomes, the change in shape of avicularia has been one line of evidence used to estimate evolutionary patterns. Hincks (1882, p. 22) commented on the transition from avicularium to vibracularium, and (1880b, p. 82, 1881, p. 6) on the transition from the encrusting *Membranipora* to the erect *Bugula*, by the occurrence of articulated birds-head type of avicularia (which "exactly resemble" those of *Bugula*) along the margin of the encrusting species *M. carteri*. Of very curious and still unexplained significance is the incredible occurrence of *internal* functional avicularia which are characteristic of several species in the erect lightly calcified genus *Menipea*. No one seems to have observed them in live material; Harmer (1923, pp. 318—319) reports on internal avicularia in the preserved Australian and New Zealand collections.

The existence of polymorphs indicates colony coordination and the behaviour of these polymorphs may indicate yet a higher level of colony integration. Marcus (1926, figures 31, 32) demonstrated the periodicity in movement of vibracularia. The time of the forward sweep of these elongate "whips" (a movement once every 150 sec) is presumably keyed to an environmental stimulus against which this frequency would be effective. The nature of that stimulus has yet to be demonstrated. Electrophysiological evidence now confirms the existence of a colonial nervous system (Thorpe et al., 1975a,b).

Canu (1900, pp. 343—345, in his only discussion of evolutionary principles that I have found) considers that the occurrence of avicularia absolutely confirm four laws of evolution, which are: (1) initial relative simplicity; (2) leading to multiplication of homologous parts (metamerization); (3) followed by differentiation of these parts; and then (4) the integration (cephalization) of the morphologic entities into a higher organization, as indicated by a fixed number, a symmetrical arrangement, integrative behaviour, and coalescing of discrete functions. Canu writes (p. 345): "At the beginning of the Cretaceous, the great majority of the cheilostomes have vicarious avicularia: their number continually diminishes until the present in which adventitious avicularia are the most numerous and occur almost exclusively in modern oceans." Canu explicitly uses this evolutionary pattern as a guide for his classification (p. 346).

Canu's historical summary contradicts the deductive prediction of Schopf (1973, p. 256) that vicarious avicularia would be a more difficult caste to evolve than adventitious forms. The modern distribution is that 85% of both the tropical and arctic faunas have adventitious avicularia; and 26% of the tropical faunas versus 6% of the arctic fauna have vicarious avicularia (for that portion of the fauna which has avicularia). The historical data suggest that vicarious avicularia are the easier caste to evolve and that the modern distributional pattern is more the result of historical consequences than of a continuously existing "steady-state" equilibrium. With regard to defensive polymorphs, cheilostome bryozoans appear to be increasingly *K*-selected as time proceeds.

Chambers within the bryozoan skeleton which have dimensions approxi-

mately the same size as these defensive polymorphs (but which lack any gener-
ally-agreed-upon modern analogue) are abundant in the Trepostomata, Cryptostomata and Cyclostomata (under a variety of names). Because of the ubiquity
of *defence* as a function of colonies of all types, it seems only reasonable to
begin with the idea that that function, in whatever way fits best, should be
ascribed to the second most common type of polymorph occurring in all of
these other diverse colonial groups. (The most common type of polymorph is
certainly the normal feeding zooids.) A defensive function might especially
apply to the "mesopores" of Trepostomata and "exilazooecia" in Cerama-
poroids.

Polymorphism: Spines

Silén's (1942b) "Testimony of the Spines", concluded that spines were
merely another type of polymorphic individual since they are budded, and
since they have a prominent constriction at the place of their connection with
the parental zooid. Moreover, spines and avicularia regularly replace in each
other in various species complexes [e.g., "*Membranipora unicornis*", see
Nordgaard (1906, p. 11); and in *Menipea erecta*, see Robertson (1905, p.
257)].

Spines differ in abundance (1 to 20 per zooid), position (limited to the orifice or found all around the frontal), orientation (upward, or arched inward
over the frontal), length and shape (rod-like, bifurcate, trifurcate, etc., flattened), mobility, and probably in other features as well. At least since Harmer's
work around 1900, it has generally been assumed that spines are yet another
answer of bryozoans to the problems of providing adequate protection. In only
one species has this been experimentally examined, as discussed later (p. 179).

Voigt (1939) reviewed the process of the fusion of spines in their contribution to the formation of a frontal shield, and then he provided a brief summary
of spine types in different lineages. Among the patterns cited were: the reduction of spines in *Membranipora* s.l., with the remaining spine becoming an enormous bowed sabre ("Krummsäbels") in *Amphiblestrum;* the development of
prongs on spines; the occurrence of a scutum or shield in the erect genus *Scrupocellaria* with a "Prinzip der Specializierung der Spinae" that "the lower spine
of the inner side will be formed into a lid- or shield-shaped organ. . ."; the rule
that "in especially open places, spines often become enormously enlarged"
(and he cites *Cornucopina*). Each of these trends was interpreted as an example of orthogenesis, and Voigt's views on this topic are presented later (p. 188).

Polymorphism: Brood chambers

Polymorphic individuals identified as reproductive in function exist in all
bryozoan orders (Brood, 1970, p. 62) but are least known in the Cryptostomata (Tavener-Smith, 1966; Stratton, 1975), except the Ctenostomata. The
polymorphs follow one of two pathways in that there is, in general, either one
brood chamber per zooid for many zooids in the colony (as in the Cheilostomata), or one (or a very few) brood chambers per colony, often spread over

several zooid surfaces (as in many Cyclostomata). These two contrasting patterns in abundance of brood chambers would seem to mean different things in terms of reproductive potential. In both the case of many brood chambers per colony and few brood chambers per colony, the maternal contribution to future offspring is identical since a colony is a single genotype. However, when many brood chambers occur (as in cheilostomes), the possibility for genetic variability is higher since each egg is individually fertilized rather than one fertilized egg giving rise to many embryos through polyembryony (as in cyclostomes). I know of no data that even suggest which method may yield the larger number of larvae per generation, but this would be very interesting to know. At least in double-walled cyclostomes there appear to be no structural reasons why individual brood chambers could not evolve, but even in the double-walled genus *Meliceritites* (which has avicularia!) only the colonial brood chamber is found.

From the point of view of natural selection it would appear risky to put all of one's reproductive efforts into a single or a few brood chambers, with only one insemination, as occurs in cyclostomes. This seems to be at the K end of the spectrum. Closer to the r end of the spectrum would be distributing one's reproductive effort at many places over a colony, and at different times of the year, as the colony enlarges, as in the Cheilostomata (and possibly also the Cryptostomata).

Among Cheilostomata some half dozen different types of brood chambers have been identified (Małecki, 1964; Levinsen, 1909; see Hyman, 1959, p. 335). In a few tropical species of *Thalamoporella* a flexible-walled two-valved type permits the development of 2—6 eggs per brood chamber (see Hyman, 1959, pp. 337, 339) but this is very exceptional. The vast majority of other brood chambers are of fixed size and contain only one embryo at a time. Even in the highly unusual occurrence of multiple brood chambers in British Columbia specimens of *Schizoporella unicornis*, Powell (1970, p. 1852) found that "never more than a single embryo is incubated by a "hyper-ovicellate" individual at any given time". In only one other species (the deep-water *Schizoporella dunkeri*) have multiple ovicells per zooid been observed (Hincks, 1880a, p. 240). Reasons for this condition are not known.

Different types of brood chambers may make different demands on colony resources. Sometimes brood chambers are specifically associated with defensive polymorphs, and these may therefore be taken to involve the highest degree of coordination in their formation (and be closest to the K end of the spectrum). It would be of great interest to know, for a dozen "representative" genera along a presumed r to K gradient, the length of time the larvae spend in the brood chamber, the percentage of the time a brood chamber is occupied, and the total contribution of each brood chamber to total colony production. If one could deductively assign a "scale of complexity" to different types of brood chambers, a fairly extensive faunal list (keyed to literature figures) is available for tropical and arctic faunas (Schopf, 1973). A comparison of brood chamber types along a latitude gradient might instruct us about the degree of specialization of the different types.

Polymorphism: Subcolonies

A fourth type of polymorphism exists at the colony level of organization in the subdivision of a larger colony into subcolonies, as indicated above in the summary of morphology of *Diplosolen*. Among genera of Trepostomata, Cystoporata, Cryptostomata and Cyclostomata (in both double-walled and single-walled forms), continuous gradients occur in zooid size, packing and total shape over the colony surface. The centres of these gradients are at the raised prominences of enlarged zooid chambers called monticules (Anstey et al., 1976). Growth hormones (perhaps similar to plant auxins?) are inferred to have been transported away from monticular centres with the result that subcolonies maintain a spatial and temporal continuity. In some cases, the monticules are clearly raised mounds with radiating ridges and grooves. In this instance they may have also functioned as centres for excurrent flow of water away from the colony surface, as occurs in similarly shaped mounds of modern cheilostomes (Banta et al., 1974).

Colony development and colony shape

The notions of r versus K selection may be especially useful to apply in considering colony development (Kaufmann, 1970, 1971a, 1973; Schopf, 1971). Most colonies begin their growth with a single ancestrula. However, twice the growth potential is achieved in *Membranipora* with its "Twin" ancestrula. After settlement, most species bud only one to three zooids from one ancestrula (e.g., *Hippopodina feegeensis* forms a tetrad; Powell, 1968). However, in others the ancestrula is immediately surrounded by five (Mawatari, 1946) or six zooids (Stach, 1938a), and in at least two of the genera of free-living Bryozoa (*Cupuladria* and *Selenaria*), the single ancestrula is immediately surrounded by eight other zooids (Waters, 1925a). Thus, the third generation of budding occurs with the potential for very rapid growth (and also maintaining the balance of the colony, as pointed out by Silén, 1942a, p. 20; and Powell, 1966, p. 30).

As formulated by Kaufmann (1970, 1971a, 1973), the further development of bryozoan colonies can be grouped into three general types: linear, sheet, and erect growth (Figs. 8 and 9). (1) Linear growth is proportional directly to T, where T is time. (2) Most encrusting colonies, however, grow like a sheet and bud zooids radially from an initial zooid. This results in the colony growth that is proportional to πR^2, where R is the radius, or more simply, T^2, since R is proportional to T. Observed patterns of growth in encrusting species are discussed in an excellent and highly original fashion by Schaaf (1974). Among 26 possible patterns of growth, the bryozoans he considered chiefly utilized only three of these, which are among the most highly ordered or the various possibilities. (3) Finally, the number of zooids in erect colonies periodically doubles over the whole colony; this growth is proportional to $2^{(1/x)T}$, where x is the number of zooids between bifurcations of branches. Stebbing (1971) documents the log growth of *Flustra foliacea*.

Each method of adding zooids has particular advantages and disadvantages for producing larvae. As presented by Kaufmann, an encrusting growth enables

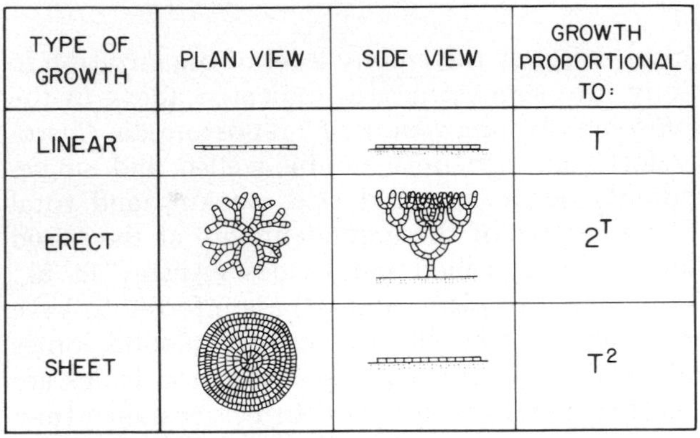

TYPE OF GROWTH	PLAN VIEW	SIDE VIEW	GROWTH PROPORTIONAL TO:
LINEAR			T
ERECT			2^T
SHEET			T^2

Fig. 8. Diagrammatic representation of three major types of growth of ectoprocts, after the formulation of Kaufmann (1970).

maximum production of zooids for a small amount of time. Erect growth adds zooids proportional to 2^T and would produce the greatest number of zooids given a long time. Since the number of zooids is related to reproductive capacity, the ability to contribute offspring to the next generation can be strongly affected by colony growth form. The maintenance of space is also increased by the ability of colonies to regenerate year after year (Harmer, 1891; Levinsen, 1902, p. 27; Zirpolo, 1921; Cummings, 1975).

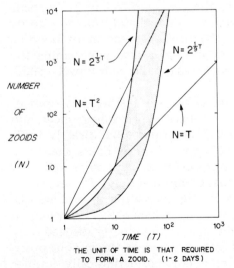

THE UNIT OF TIME IS THAT REQUIRED
TO FORM A ZOOID.   (1-2 DAYS)

Fig. 9. Addition of zooids (N) as a function of time (T). For periods of time up to 2—3 weeks, linear growth ($N = T$) and sheet growth ($N = T^2$) produces the largest number of zooids. After 4—6 weeks, erect growth ($N = 2^{(1/3)T}$) and ($N = 2^{(1/9)T}$) offers the greatest potential for producing zooids. The superscripts 1/3 and 1/9 respectively refer to the distance of 3 and 9 zooids between bifurcations; these are common situations in nature. The representation is after the formulation of Kaufmann (1970).

The usual interpretation of growth forms [of which at least eighteen types have been defined (Schopf, 1969b, appendix)] is in terms of the physical stresses of the environment. Carefully documented work has shown, for example, that both within species and between species of cyclostomes, the degree of calcification and extent of branching is strongly correlated with light intensity and water motion (Harmelin, 1968, 1973). However, the different rates of growth and budding along an environmental gradient must also result in reproductive consequences for the colony, and these have yet to be examined. Further studies of population structure (Håkansson, 1975) will be of great value.

Behavioural patterns

Defence
With all the previous discussion about defensive specializations through calcification, avicularia, and spines, we might well ask where are the intended recipients of these aggressive responses. All of the nudibranchs that eat Bryozoa are too large to be affected by avicularia. However, as documented by Kaufmann (1971b) for the cheilostome *Bugula*, amphipods build tubes of detritus on the colony and these tubes smother zooids. Avicularia of *Bugula* easily catch these crustaceans and inhibit their movement. To an as yet unknown extent, this retards the rate of tube formation, and thus lengthens the life of the bryozoan colony. A similar inhibitory effect by avicularia was observed for *Bugula* accosted by pycnogonids which eat detritus trapped in the colony (Wyer and King, 1973). In a study of patterns of seasonal settlement, Mawatari and Kobayashi (1954, p. 4) show that sedentary amphipods, with "their large muddy type nests. . .", have two seasonal swarms — in June and December. It may well be that the failure to observe a major and significant recipient of avicularian action in other species of Bryozoa is due to the fact that colonies were not observed in nature at the right time of the year. For encrusting species, there is not a single instance in which it is known what it is that the avicularia could be defending against!

Defence of space by spines has been proposed and documented by Stebbing (1973b) in the case of *Electra pilosa*. He wrote, ". . . when *E. pilosa* and *Alcyonidium polyoum* grew together and touched, spines were often restricted to that part of the colony that was about to be overgrown by *A. polyoum*." In several instances, when a colony of *E. pilosa* formed long spines adjacent to a threatening colony, long spines would also form at the other end of the *E. pilosa* colony. As Stebbing states, "It appears that here is an example of interzoidal communication of a 'message' of functional significance for the colony as a whole." Presumably spines of other species function in a similar manner (Stebbing, 1973a). An additional defensive mechanism of *E. pilosa* is simply to grow faster than the opposition, and thereby avoid death.

Large, gelatinous non-calcified colonies of both fresh-water and marine bryozoans emit poisonous toxins. In exposure to freshly cut material, gill tissue of fresh-water fish and salamander larvae become blistered and are destroyed, and these predators are suffocated (Collins et al., 1966, and references cited therein). The toxin is located in cells of the body tissue and not in the coelomic fluid, and thus would be released only when the bryozoans were bitten (Meacham and Woolcott, 1968).

Marine forms of the ctenostome genus *Alcyonidium* cause a severe contact dermatitis in fishermen cleaning it from their nets in the North Sea. This ailment has come to be known as the "Dogger Bank Itch" (literature reviewed by Ryland, 1967). Whether or not this toxin is effective on fish is not known. A defensive secretion was suggested by Harmer (1930, p. 102) for polymorphs of the cyclostome *Diplosolen*. Defensive secretions may also occur in many lineages of living cheilostomes which have glands that open to the exterior at the mandible of avicularia (Waters, 1892; 1925b, p. 539). Since avicularia cannot feed, I am not able to think of another plausible function for these glands. Lutaud (1969, and references cited therein) has identified similarly shaped vestibular glands in zooids with polypides, and there they contain bacteria which are species-specific, but whose function is also unknown. It may be the case that chemical defences of various sorts are common and widespread in the bryozoans in general, but their discovery, if present, awaits application of (i.a.) chromatographic techniques to a wide spectrum of species.

Other biotic associations

A very limited but highly interesting literature suggests that arthropod relations with bryozoans are significant for both groups, indicating in some cases a co-evolutionary development. Kaufmann (1971b) discussed the *Bugula*-amphipod story reviewed above. In addition, mimetic coloration is attributed to caprellid amphipods associated with *Bugula neritina* off southern California (Keith, 1971) and with *Bugula* sp. off northern France (Caullery, 1926). Caullery believes the coloration is protective for the caprellid against their predators, porcellane crabs, some of which are themselves patterned like the bryozoan! Neither of these species of *Bugula* appear to have avicularia, however, and the biological interaction between these bryozoan—arthropod species pairs needs to be examined. Four species-specific relationships between copepods and encrusting highly calcified cheilostomes (with avicularia) have been reported (Medioni and Soyer, 1966, 1967) but no interactions regarding the avicularia in the bryozoan hosts have been discussed. Species-specific chironomid midge—bryozoan relationships are known for fresh-water forms (Ertlova, 1974).

Several field observations and laboratory experiments show that some species of bryozoans are uniformly associated with specific substrates (Ryland, 1962, 1967, 1970; Soule and Soule, 1969; Eggleston, 1972b). These substrates can be shells, stones, tubes and spines of polychaetes, chitons, algae, crabs, hydroids, specific bivalve genera, and probably others. The bryozoan—hermit crab association is well documented for the Tertiary (review by Pouyet, 1973, p. 128) and is reported to have existed for more than 150 m.y. (since the Jurassic: Palmer and Hancock, 1973). No systematic analysis has yet been made of the taxonomic or morphologic patterns of adaptation in these cases of association. Presumably the specification of a substrate (at the K end of the spectrum) is balanced by assuring less competition for space and therefore a breeding scheme in balance with the host. Reproduction of the ctenostome *Triticella* is synchronized with the time of yearly moulting of its decapod host (Eggleston, 1971).

Species of *Membranipora* that are almost limited to a single species of algae (*Laminaria* in some parts of the world, *Macrocystis* in others) must perforce have several generations per year since the substrate may exist for only a period of weeks (Yoshioka, 1973). And yet species of the very similar genus *Electra* may settle on seventeen species of algae at Woods Hole (Rogick and Croasdale, 1949; see also Ryland and Stebbing, 1971). By having a whole variety of substrates *Electra* achieves the insurance needed so that, for example, there may only be two generations per year in *Electra angulata* (Mawatari, 1953, p. 9). There has not yet been a systematic summary of generations per year versus nature of substrate, and number of offspring produced, for various bryozoan species. In the annual eel grass *Posidonia oceanica* of the Aegean Sea, only a few of the encrusting species produce the theoretically expected small, circular colonies which reproduce at an early age; reasons for this are not known. The proportion of annual versus perennial species as a function of environmental variability is also just beginning to be sorted out (Eggleston, 1972a; Abbott, 1975).

Bryozoan substrate preferences change as a function of depth. In material from the continental shelf off New England, nearly 50% of the species whose substrate could be determined were attached to hydroids in collections from 0 to 50 m, but less than 10% of the species utilized hydroids at 200—250 m (Schopf, 1969b, figure 4). Rocks and shells increased over this same depth interval to providing from about 25% of the substrates to nearly 75%. Yet rocks and shells become increasingly rare as depth increases, as the sediments become finer. Thus it appears that bryozoans are making increased use of rock and shell as adequate substrates become scarcer. Concomitant with this is the evolution of species that can attach to foraminifera and sand grains by means of tendrils, and taxa with this modification become increasingly important in the deep sea (Schopf, 1969a). See *Euginoma* figured by Schopf (1976).

The notion of a bryozoan—algal consortium has been strongly supported for some Palaeozoic species, most especially in the famous "cork-screw" genus *Archimedes* (Condra and Elias, 1944, p. 35), but for other genera as well (Condra and Elias, 1945; Elias and Condra, 1957; Rigby, 1957). For modern forms, Cuffey (1970, p. 44) has stated: "Single-celled algae (zooxanthellae) live commensally in the soft tissues of a few modern marine bryozoans", but he has no observations of this, and the literature he cites is equivocal. Indeed, I can find no evidence to support the view that any modern bryozoan has a symbiotic relationship with single-celled algae, such as is known to occur in colonial corals, and in some foraminifera. Golden or yellow plant cells have been reported in brown bodies or in the gut wall (Brandt, 1883; MacMunn, 1887; Oltmanns, 1923; Zirpolo, 1923). However, the cells in brown bodies may simply represent undigested food. And Zirpolo's views (which are the most explicit in their attribution of zooxanthellae to cells of the gut wall) are evaluated by Buchner (1930, pp. 145—146) as: "However we must confess that that which concerns *Zoobotryon* and also *Bicellaria*, we are not yet able to have been persuaded of the correctness of the observations on the basis of sections through these objects." It would be a major discovery to show conclusively that zooxanthellae have a commensal relationship with any living bryozoan. This would certainly be at the *K* end of the adaptational spectrum.

	A. ZOOID VERTICAL WALLS (1 2 3 4 5)	B. INTERZOOIDAL CONNECTION (1 2 3 4 5)	C. EXTRAZOOIDAL PARTS (1 2 3 4)	D. ASTOGENY (1 2 3 4 5)	AVERAGE INTEGRATIVE PROPORTION
RECENT CYCLOSTOMES					
SINGLE-WALLED — stomatoporids					0.46
crisiids					0.46
some idmoneids					0.46
Diplosolen					0.46
? *Heteropora pacifica*					0.46
DOUBLE-WALLED — other heteroporids					0.61
lichenoporids					0.61
some crisinids					0.68
some hornerids					0.68
PALEOZOIC STENOLAEMATES					
SINGLE-WALLED CYCLOSTOMES stomatoporids					0.36
crownoporids					0.36
UNASSIGNED — corynotrypids (exterior-walled)					0.41
Pseudohornera (double-walled)					0.62
Diploclema (single-walled)					0.65
DOUBLE-WALLED ORDERS — some FENESTRATIDS					0.69
typical TREPOSTOMES					0.74
some bifoliate CRYPTOSTOMES					0.75
CYSTOPORATIDS some fistuliporids					0.80
some ceramoporids					0.84

Fig. 10. Summary of integrative states of stenolaemate Bryozoa. States of each character arranged as in series A—D. States observed in one or more taxa in each group are indicated by solid rectangles; those based only on inference or speculation are indicated by hollow rectangles. The average proportion of complete integration for the four characters is given in column to right. From Boardman and Cheetham (1973, figure 13c).

Levels of colonial integration

Boardman and Cheetham (1973) classified various morphologic features and patterns according to the extent to which they contribute to colonial integration. Chief emphasis was placed on the nature of vertical walls, interzoidal connections, extrazoidal parts, astogeny, and morphologic polymorphism. Their data are summarized in Figs. 10, 11, and 12. Variability decreases in some (but not all) characters as the level of colonial integration increases (Cheetham, 1975).

Among Stenolaemata, note that the summed average integrative proportion rises from the Palaeozoic to the Recent in cyclostomes, and that double-walled taxa have a higher level of integration than single-walled taxa (Fig. 10). Among cheilostomes, the maximum level of integration rises from the Jurassic to the Present (Fig. 12). The level of integration (which also is roughly indicative of morphologic complexity) closely follows the rise in taxonomic diversity. These patterns can also be interpreted as being examples of the increasing effect of K selection through geological time, and are consistent with the model of speciation described above (p. 162).

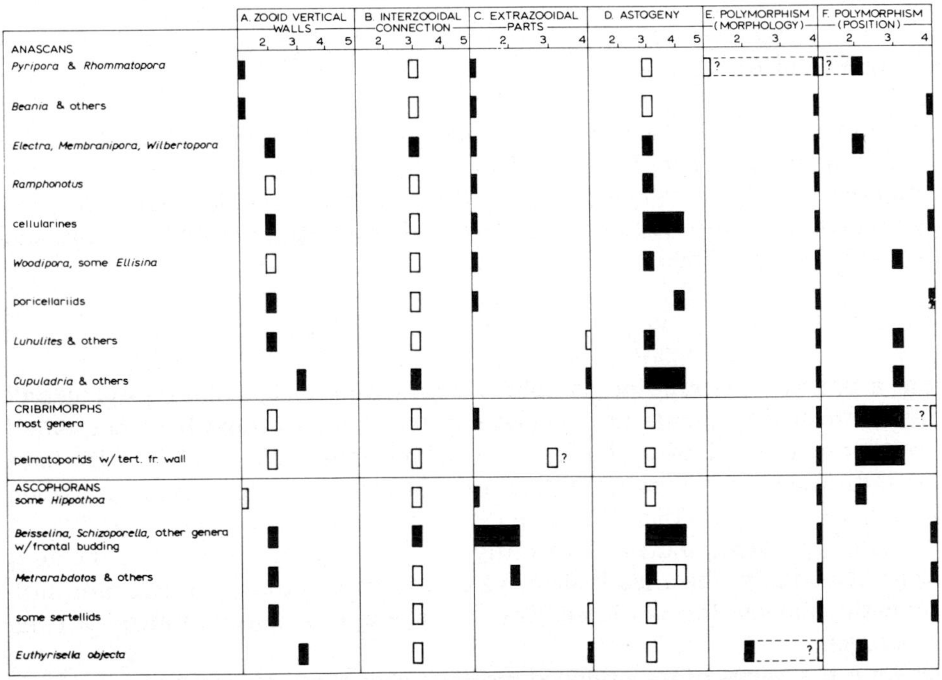

Fig. 11. Summary of integrative states in groups of cheilostome genera. States of each character arranged as in series A—F. States observed in one or more taxa in each group are indicated by solid rectangles; those based only on inference are indicated by hollow rectangles, with queries marking most speculative inferences. From Boardman and Cheetham (1973, figure 20).

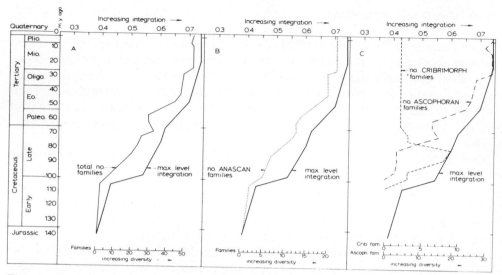

Fig. 12. Evolutionary relationship between level of integration and taxonomic diversity in cheilostomes. Maximum level of integration is compared with: A, total number of cheilostome families; B, number of anascan families; and C, numbers of cribrimorph and ascophoran families. From Boardman and Cheetham (1973, figure 26).

Themes of Evolution

In this section I will examine several general evolutionary themes from the point of view of data from Bryozoa. These themes are: (1) ontogeny recapitulates phylogeny; (2) convergence; (3) orthogenesis and Dollo's Law; (4) size and symmetry; (5) rates of speciation; (6) "living fossils"; (7) survivorship; and (8) diversity through time.

Ontogeny recapitulates phylogeny

From a practical standpoint, the phrase "ontogeny recapitulates phylogeny" has been a recurring theme in bryozoan systematics. Authors have compared the first formed zooid of a cheilostome colony (the "ancestrula") with later formed individuals in colony growth. From Harmer (1923, p. 321) to Levinsen (1909, pp. 18—19), to Voigt (1939, p. 90, 1959), to Silén (1942b, p. 27), to Powell (1967, p. 213), and most recently to Dzik (1975, p. 402), the theme is repeated: "Generally, the ancestrula has a considerably more primitive morphology than the zooecia formed later. This is a specific recapitulation of phylogeny in astogeny".

These are the views of experienced workers over three quarters of a century, and although the conclusion of recapitulation is in need of palaeontological evidence (Cheetham, 1969, p. 229), we need to ask why it has retained its appeal. There seem to be two main answers. First, as Cheetham (1969, p. 229) notes, "In a broad way, the general progression of structural complexity in both ontogeny and astogeny of many cheilostome species repeats the strati-graphic sequence of zooecial types". And second, the ancestrula approaches a generalist (or *r*-selected) zooid morphology from which it is possible to derive almost any addition type of morphologic complexity.

These and similar views of others focus on an historical explanation for why the ancestrula looks as it does. However, two other non-historical explanations may be equally valid. It may be more nearly true that strong stabilizing selec-tion renewed each generation, has channeled the morphology of that first individual into a very limited number of responses. For example, in the absence of a calcified frontal surface, the occurrence of spines is a common defensive mechanism. Moreover, even this limited *K*-selected adaptation is missing from the ancestrula of a species typically inhabiting the most unstable of all marine environments — the uppermost metre (*Cryptosula pallasiana*, see Waters, 1926, p. 428). Secondly, with regard to the limited number of spines (and indeed the small range in morphology in general), how complex (and varied) can a single zooid become? Much of the simplicity may be purely an artifact of the limi-tations of simple starting materials. Until cheilostome evolutionary pathways are worked out, and comparison is *then* made of ancestrular form, it will be necessary to reserve judgement on the phylogenetic significance of the "taxa".

Evidence of the "biogenic law" occurs in some erect colonies which branch. In a species of *Bugula*, Marcus illustrated that the first few individuals of a new branch lack any defensive or reproductive polymorphs and indeed closely resemble the ancestrula of the colony as a whole (1938, pl. 15, figs. 30, 40, pl. 16, fig. 39; simplified diagram in Boardman et al., 1970, p. 299). The occur-rence of this phenomenon in several groups of Bryozoa, and also in corals and plants, was supported by Lang (1921, p. xxii), who has also cited other instances of "periodicity" in cheilostome evolution (Lang, 1919b, pp. 219—220).

Strong support for the "biogenic law" has appeared in writings on the Cyclo-stomata (many papers reviewed by Cummings, 1910). However, this theme is not often referred to and appears not to be highly regarded by modern wor-kers on cyclostomes.

When bryozoan larvae settle and metamorphose, the larval posterior becomes the budding surface of the newly formed individual. Jebram (1973a, p. 568) states that "the polarity change during the metamorphosis of the gymnolaema-tous larvae is a clear ontogenetic indicator of the change of growth direction during the phylogenetic development from the Phylactolaemata to the Gymno-laemata". This supposition forms one of the bases on which Jebram has recon-structed the phylogeny of the Bryozoa.

Convergence

Convergence has been referred to many times in the Bryozoa, as is fitting in a group with forms illustrating variations on a limited number of morphologic themes. In practice, the question of course is whether resemblance is due to common ancestry, or to similar environmental conditions, or to similar physical constraints, as was discussed previously with regard to the ancestrula of cheilo-stomes. The matter is well stated by Waters (1921, p. 407): "Put shortly, are there a number of Bryozoa from different families with quite different zooecial characters which have taken on the same way of growth and subbasal charac-ters, or have related forms with similar growth gradually assumed more diver-gent characters?" In theory, the way to recognize convergence is by examining in detail a complex structure and then asking what the probability is that con-vergence occurred or, alternatively, whether or not genetic relationships are likely to be the underlying cause for the similarity. Although some claim this is impossible to assess, others try, based on their experience and judgement.

Cheetham (1969, p. 229) cites evidence in cheilostomes that "evolutionary trends in polymorphism may thus be either progressive or retrogressive in dif-ferent lineages". Harmelin (1975) cites "regressive forms" in considering ten evolutionary trends in cyclostomes. Stratigraphic and biogeographic evidence must often be used.

Tubular forms are probably the most homeomorphic among all Bryozoa. To illustrate this point. Hinds (1975, p. 901) noted that: "Of the twelve *Idmonea* species described by Canu and Bassler (1920) from Lower Tertiary sediments many were either found to belong to genera other than *Idmonea*, or not generi-cally identifiable on the basis of existing literature and external appearance of the type specimens". In the Cretaceous *Spiropora*, Voigt (1968, p. 51) studied "a zorial growth form which arose at several times in different phylogenetic lineages". Indeed, fron "one species", Voigt and Flor (1970) are able to dis-tinguish a dozen species belonging to four genera, as shown by differences in the brood chambers.

In search for less homeomorphic characters in tubular Bryozoa, Hinds (1975, p. 908) stressed the pattern of budding, "inasmuch as the growth mode is inter-preted to be a conservative character. . .". However, Boardman (1968) and Boardman and McKinney (1976) described a budding pattern yielding "vir-tually identical" zooids which are square (!) in cross section for six genera (three Ordovician, another Silurian, the fifth Devonian, and the last Late Palaeozoic) in six families. This example is readily recognized because of the unusual pattern. One wonders if less bizarre examples of homeomorphy in budding pattern (see Illies, 1973, 1975) would simply tend to be overlooked.

Homeomorphy was also considered to have been rampant in the multitude of Cretaceous cribrimorphs. Lang (1919b, pp. 217, 227) wrote: "The whole tendency of this paper is to show that parallelism of evolution occurs in the characters of the Pelmatoporinae. . . . The Pelmatoporinae form a natural group which contains earlier and simpler forms that may be seen to pass, as a rule through continuous gradations, to species that appear later in time and are more specialized in structure. Fundamentally, the specialization consists in lay-ing down more Calcium Carbonate along definite tracks. This results in paral-lelism in the evolution of each lineage, tending to produce homeomorphic forms, and suggesting a predisposition in the parental stock to evolve along determined lines; but the distinguishing marks of each lineage are to be seen in the concentration of effort on certain points in the building of a secondary skeleton as well as in the method and details of architecture." Many other examples of convergence could be cited.

The fact of the existence of convergence is demonstrated. Not a great deal would seem to be gained at this point by searching for more examples. On the other hand, it would be of considerable interest to have a probability distribution of its expected occurrence among animals of the same grade of complexity as bryozoans, and to know how that compares with groups of similar taxonomic diversity. Characters can only be recognized as convergent if they are sufficiently complex to have been previously recognized as divergent. For that pool of complex characters, what is the percentage of species existent at any given time that displays convergence (or parallelism) during, say, a one million year span? My guess would be that virtually all species display convergence in some character of taxonomic interest by this broad criterion. Using characters of "major" taxonomic interest, my guess would be that for living bryozoans that convergence could be demonstrated in perhaps 95% of the cyclostomes, decreasing to perhaps 60% in the structurally more complex cheilostomes, although that figure could well be higher. As major taxonomic reports become completed for different geographic regions, it would be of interest to have this information.

Orthogenesis and Dollo's Law

Without question, the most published proponent of orthogenesis among bryozoan workers, and perhaps among palaeontologists in general, was W.D. Lang. In papers that appeared over the first quarter of this century, Lang contributed original work on corals (1923b), reviewed data on ammonites (1919a), oysters and butterflies (1923a), drew parallels with Freudian psychology (1925), and contributed a substantial body of data on the cribimorph and other Bryozoa (especially, 1921 and 1922). Lang tried to separate that part of genetic change which results in morphologic change from the total genetic change: *"The morphic expression of the difference between two forms is by no means proportional to the fundamental or genetic difference"* (italics his; 1921, p. iv). Lang considered that the metabolism of an individual colony was outside the control of natural selection and he talked about "the piling up of calcareous matter, so that further evolution is inconceivable. . ." (1921, p. vii); or, "Moreover, as evolution proceeds, the shell becomes absurdly over-adequate for protection (a Hippurite-shell, for instance, comes readily to mind) and, unless the tendency ever to secrete more calcium carbonate is checked, becomes a burden too heavy for the organism's bionomy, and ultimately causes its extinction" (1921, p. viii). This view of excess secretion was thought not to be a response to the environment since not all forms in any given environment proceeded in the same way; rather the secretion was attributed to a "compulsion from within" which "introduces an element of inevitability in evolution, and makes it possible to speak of the evolutionary aim of a lineage, without thereby implying anything of what is usually understood by teleology". "Such predetermined evolution is very like what has been termed Orthogenesis or Programme-Evolution" (1921, p. xviii). Thus, "Whenever in the evolution of chitinous lineages this equilibrium is lost, and calcium carbonate secretion is begun, not only does the process prove irreversible, but it increases in intensity, until, in the final terms of the lineages, all the energies of the organism are given over to calcium carbonate secretion and the Polyzoan becomes buried under the products of its own activities. Lineages which secrete calcium carbonate are, therefore, doomed lineages" (1921, p. xxv; see also Lang, 1916, p. 76; 1919b, pp. 194—196).

And what is the role of natural selection in this since Lang was, after all, a professed Darwinian? "But the evolution of Cretaceous cheilostomes suggests that, in so far as Natural Selection is present, its action is rather a negative one — insuring that the superfluous calcareous matter should be laid down where it is least in the organism's way, than that the structure it builds should be directly useful" (1921, p. xxv).

Voigt (1939) found abundant evidence in trends in spine morphology to support an extreme form of Dollo's Law of irreversibility of development. A correct statement of Dollo's Law is that "complex structures cannot be precisely re-evolved" (see Gould, 1970, for a review of how Dollo's Law has been misinterpreted). What Voigt claimed was biological determinism: The orthogenetic view is that evolution was programmed so as to prevent any irreversibility. In discovering the formation of a large shield-like spine in *Scrupocellaria*, he raises the question as to why it should only develop on one side, and not be symmetrical. And he answers his question with the opinion that once a difference between the inner and outer side develops, that the difference is enhanced, "a trend which has been designated orthogenesis since the time of Haack" (Voigt, 1939, p. 104). And he goes on (p. 105) to say in his eloquent summary that "The examples described in this article of differentiation of spines proceed so clearly and so unequivocally in the way of an orthogenetic ("orthogenetischen") development — which has already been shown by Lang for the Cretaceous cribrimorphs — that the examples cited in this paper in all probability can be placed with the well known classical examples of the biological law of momentum ("Trägheitsgesetz") in the sense of Abel (1929)". As for this Law of Momentum, Simpson (1953, p. 282) notes that Abel was trying to apply physical notions (i.e., momentum, inertia) directly to biology. It is only another variant on the theme of orthogenesis.

In the modern view, Lang's general observation of increasing complexity in calcification, with subsequent extinction of many of these lineages, is confirmed (Larwood, 1962, 1969). But we would neither interpret this, nor the trends noted by Voigt (1939), as being due to a vital force gone beserk, for which no other evidence has been uncovered in this or any other group of organisms. Rather, each of these increasingly complex evolved forms, owing their evolution in the first place to a stable ecologic setting (Schopf, 1973), are seen as having lost their selective advantage as the environment changed. For the cribrimorphs in the latest Cretaceous, this change appears to parallel the considerable withdrawal of water from the continental platform, and reduction in shallow continental seas. In short, the evolutionary significance of the extraordinarily complexly calcified overarching spines was probably defence and this is clearly at the K end of the evolutionary spectrum. It is the extreme specialization at this end which is, on the average, most often removed by natural selection, owing to the lack of selective advantage as the environment shifts.

Rate of speciation

The model of speciation presented in the introduction emphasized timelessness. We nevertheless ask whether the (perceived) rate of speciation has in fact been independent of time. The rate of speciation in any given time plane of short duration, say 100—1000 generations, is believed to be most likely a function of the rate of occurrence of factors causing geographic isolation. As long as species can extend their ranges "without bound", i.e. over distances very large in comparison to dispersal, then the chance of a physical subdivision of the initial distance is, on the average, higher than if a species has a very limited distribution. Accordingly, geographic speciation should be enhanced by environmental conditions that permit species to extend their ranges and then be subdivided. This appears to have been the case in the Hawaiian Islands where (rare?) dispersal by logs and larvae, coupled with island isolation over nearly 1000 km has resulted in ecologic specialization, as studied in the 28 genera of the cheilostome family Smittinidae (Soule and Soule, 1973).

The probability of a species extending its range may be significantly different in ecologically distinct areas. For the tropics and the deep sea, the critical factors in encouraging the diversification of species appear to be the combination of: (1) a high probability of extending a species range because of uniformity of habitat; (2) a corresponding low probability of maintaining gene flow over the range of a species; (3) a consistency of ecologic conditions in food and substrate type which therefore permits a higher probability that specialization will occur; and therefore (4) that when geographic ranges are subdivided, new species will result. Geographic speciation would therefore seem to have a higher probability of occurring in the spatially expansive deep-sea and the shallow-water tropics compared with the spatially constricting temperate and high latitude shelf regions.

Temperate latitude shelves are characterized by strong temperature changes at their northern and southern boundaries especially on western sides of oceans (for example, between Cape Hatteras and Cape Cod). Thus gene flow may be especially important in maintaining species coherence in that region, and environmental variation may be particularly noticeable.

When new species are observed in temperate latitude deposits they are often found to coincide with changing benthic conditions. For this reason, Buge (1948, p. 77) wrote: "Species evolution is not so much the result of the factor of time by itself as it is the result of conditions which proceed from environmental change." And Voigt (1972, p. 45) added that, "Because of their dependence on facies, bryozoans are not continuously represented in sections, and it is practically impossible to follow them great distances vertically." I wonder if these observations are not partly a function of looking at faunas of temperate latitudes. In the tropics, continuous gradation through more stable environmental conditions may result in a different perception of phyletic evolution through time.

If saturation is achieved in species diversity, what is the time course? In a uniform (i.e. stable and predictable) environment, the saturation level would be higher and time to evolutionary saturation should be longer than in more variable environments (see Fig. 13). However, environmental disturbance is likely to occur more often in less uniform environments and therefore saturation may be less commonly achieved than occurs in uniform environments. Once saturation is achieved, then I believe that the evolutionary importance of allopatric speciation is reduced and that of phyletic speciation is increased. The innate difficulty of correctly determining the rate of phyletic speciation is considerable (Schopf et al., 1975). Biochemical data on the rate of change in proteins (i.e., in the genes that code for the proteins) of species in different ecologic regimes would aid us in determining if phyletic evolution is more or less rapid in morphologic generalist versus morphologic specialist species.

Empirical estimates of the rate of speciation through time based on palaeontological data are likely to be lower than values which would be derived if we knew the changes in the genetic composition of species. Sibling species of very similar morphology but divided by behavioural traits, for example, will not be recognized in gross morphology. Further, if species A has 10 potentially useful morphologic features, and species B has 100, and if the probability of a morphologic change occurring in each species is the same, say 10%, then we have one potential character to change in species A, but 10 potential characters in species B. Thus, we are more likely to see morphologic changes in species with a more complicated anatomy of fossilized hard parts.

The presence of continuous versus discontinuous variations in bryozoan morphology is of interest in considering rates of evolution in that historically one tends to rely upon discontinuous variation, or "punctuated equilibria" (Eldredge and Gould, 1972), to characterize distinct species, as the normal result of allopatric speciation. Phyletic evolution of different species through

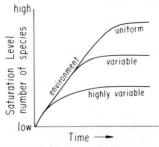

Fig. 13. Predicted relationship of saturation level, and time to saturation, for environments of different degrees of variability. As discussed in the text, the probability of speciation is considered to be higher in more uniform environments.

time, however, may be characterized by continuous variation. Gregory (1896, pp. 27—28) explicitly considered the question of continuous versus discontinuous variation in the Cyclostomata. He reports that species alive at any given time are characterized by discontinuous variation, but that "if we compare the forms of Bryozoa that lived in successive zones, the species are continuous. Variation in this case is therefore continuous in time and discontinuous in space." Data from a living cheilostome show previously unsuspected clinal variation in both gene frequencies and in morphology (Schopf, 1974; Schopf and Dutton, 1976). And Voigt (1959, 1975a) emphasizes the occurrence of non-polymorphic discontinuous variation within single colonies of various species. These observations suggest to me that even in ecologic time, and very likely over evolutionary time, phyletic speciation is very significant.

For a dozen major groups, a range in morphologic complexity by a factor of 5 (as measured by richness of morphologic terms) is correlated with a change in rate of evolution of about the same extent (Schopf et al., 1975, figure 1). This places an approximate upper boundary on rates of evolution that can be accounted for simply by changes in morphologic complexity. Among Bryozoa, rates of production of new genera per million years differs by a factor of 4 (in the morphologically complex cheilostomes versus the morphologically simple cyclostomes; Newell, 1952, figure 2). The argument which I wish to make is that there is some, perhaps very significant, amount, of evolutionary change which is obscured in forms of relatively simple, external morphology. It may even be the case that, at least for Bryozoa, rates of speciation are approximately the same, and that the differences which are observed are (to put it boldly) purely an artifact of changing morphologic complexity.

The net result of these considerations is that rates of speciation as determined from palaeontological sequences are likely to greatly underestimate "true rates" of speciation. It also follows that our closest approximation of "true" rates would come from the fastest rates observed in the fossil record.

We are now in a position to make some additional predictions regarding the rate of species proliferation through time. Specifically, diversity should vary as the ratio of stable to unstable environments changes. One indication of an increase in stable environments would be a reduction in the latitudinal temperature gradient. With ameliorating conditions, species would extend their ranges and be more prone to have geographic subdivisions within the range. Some support for this idea results from plotting the temperature through time versus number of new cheilostome families, as in Fig. 14. The times of greatest expansion of new families are in the Upper Cretaceous and Upper Eocene, both of which come at the end of a cycle of climatic amelioration. That this proliferation of species may be related to expanding ranges, is suggested by the statement of Voigt (1972, p. 50): "One must not forget that during the Upper Maestrichtien there was an undoubted migration of tropical forms toward the north". The most extreme climatic conditions are during the Palaeocene, Oligocene, and Recent. By this accounting, the present is a time of consolidation of the diversification of cheilostomes and we should not expect to see a major adaptive radiation until the climate ameliorates considerably. Note that

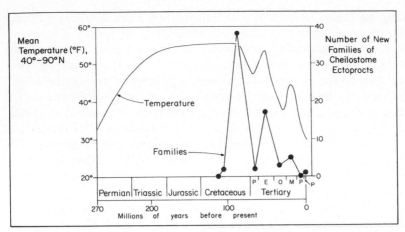

Fig. 14. Cheilostome diversity versus environmental temperature. Environmental data after Dorf (1960); families after Larwood et al. (1967).

not all evolutionary bursts have been of equal intensity or length. The initial radiation during the Cretaceous was most striking perhaps because it was longer or perhaps because there were more "niches" to be filled.

Living fossils

The possibility of significantly different rates of evolution have always fascinated the general public and professionals alike, but alas this quaint notion conjures up more in imagination than in reality. The bryozoans include, of course, their share of archetypical, long-persisting forms — indeed, *every* group has these types, and they all suffer a similar demise. As soon as new characters are used, their ranges are drastically shortened, they no longer appear so simple, and, in time, all that remains is the legend.

Ryland (1970, p. 111) does his best to retain the legend as fact and notes that some genera of the Cyclostomata "are among the longest persisting in the animal kingdom, for both *Stomatopora* and *Berenicea* first appeared in the Ordovician". *Stomatopora*, Brood (1972, p. 221) wrote, "is supposed to be the most primitive of the post-Paleozoic cyclostomes, and also one of the oldest. All modern representatives as well as all sufficiently described Mesozoic species are provided with interzooidal pores and pseudopores. These pores are not hitherto described from the Paleozoic species assigned to this genus. It is therefore possible that the present genus is not older than the Triassic or Lower Jurassic and that the Paleozoic representatives have to be placed in another genus." A similar fate has been prescribed for *Berenicea* which Brood (1972, p. 175; see *Diastopora*) also cites as ranging only from the Triassic to Recent, less than half its previous presumed persistence.

Borg's (1933, p. 382) candidates for living fossils among the Cyclostomata were the Recent Heteroporidae, a group which he and others presumed to include Palaeozoic species (p. 258). But again, as analysis of characters has progresssed, that family is now considered to include only post-Triassic forms (Brood, 1972, p. 413). In addition, as Boardman (1975, p. 600) writes: "[*Heteropora*] is presently identified by a robust growth habit and two sizes of zooecia. There is in fact little else to be seen externally in fossil forms. . . species with a wider range of differences in internal characters than can be illustrated here have been assigned to the genus. . . . Based on a wider experience with Paleozoic stenolaemates, however, the predicted distribution of wall structure suggests that several genera and more than one family are presently included in the genus *Heteropora*." Obviously *Heteropora* is no "living fossil".

Among the Cheilostomata, the genus *Neilla* is perhaps the leading candidate for "living fossil" status, having persisted from the Eocene to the present. This genus is perhaps the simplest cheilostome of its type, and the possibility of seeing morphologic change in such forms is now demonstrated to be markedly less than in forms with a more complex external morphology (Schopf et al., 1975).

Buge (1972, p. 56) makes the interesting observation that not only is the number of living species known from the Eocene very few, "if there are any", but that this number "diminishes in proportion to the increase in our knowledge and as the possibility increases for distinguishing as different species those which former authors saw only as one form with a great persistence." He then cites two examples, in one of which a single species was studied in great detail by modern methods and subdivided into a dozen species and subspecies as it became part of the subject of a monograph on its genus (*Metrarabdotos*) and a second example in which a species range was considerably shortened by subsequent study (*Onychonella angulosa*). Voigt (1975b) adds yet another case of a strongly restricted range: the Cretaceous cyclostome *Coronopora truncata* is decidedly different from the Recent *Domopora stellata* which therefore "is by no means a 'living fossil' ".

If most morphologic *and* genomic change is a function of the speciation event *per se*, then the degree of change is simply a function of the number of such events. If, on the other hand, morphological change is not proportional to the underlying genetic change, then all bets are off as to what palaeontologists OUGHT to find in terms of a model of speciation *through* time. The few data which explicitly compare genetic change vis-a-vis organismal change indicate that the two are *not* proportional to each other. Avise (1975, and references cited therein) summarizes data on genetic change versus morphological change through time for fish, and he finds the many branch points in minnow evolution have not resulted in any greater degree of genetic divergence in modern minnows than has occurred after only a few branch points in the evolution of modern sunfish. Wilson (1975, and references cited therein) finds very large genomic changes occurring in the absence of organismal (skeletal) change in frogs compared to mammals, and that humans and chimpanzees are genetically extremely similar, but morphologically are widely divergent.

To return to bryozoans, there is no reason to consider this group anything special with regard to the "living fossil" question. With all of these clearly documented cases of artifact — of both morphologic and genetic types — it is easy to be strongly sceptical about there being *any* bona fide candidates of "living fossil" canonization, though of pretenders there are many.

Survivorship

Interest in survivorship curves for major taxa has greatly accelerated since Van Valen (1973) asserted that "all groups for which data exist go extinct at a rate that is constant for a given group". Van Valen's Law, as it was labeled (Raup, 1975), has since been questioned concerning the methodology of survivorship curve formation (Raup, 1975), the bias introduced by using the standard time scale for recording lengths of time which taxa live (Sepkoski, 1975), and has been attacked and reinterpreted by various authors.

As Van Valen stated (1973), a survivorship curve "is a simple plot of the proportion of the original sample that survive for various intervals." Although simple in principle, the construction of a life table involving both living and extinct forms is complicated (Caughley, 1966).

As shown in Fig. 15, data for extinct families appear to be distributed in approximately a linear fashion for longevities of 5—165 m.y. The half-life for extinct families is about 85 m.y. In terms of macarthurs (Van Valen, 1973), one finds the half-life, inverts, and standardizes to 500 years (or in this instance, $(1/8.5) \times [10^7/(5 \times 10^2)] = 6 \times 10^{-6}$ macarthurs; or 6 μma).

The right-hand end of the survivorship curve beyond 165 m.y. is quite steep. As Raup (1975) states, this may be due to the tendency for taxonomists to sub-

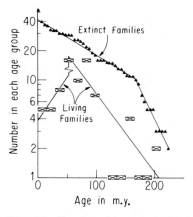

Fig. 15. Survivorship curves for extinct families (triangles), and for living families (crossed rectangles, summed for 15-m.y. periods); life table available from author. The next lowest point for extinct families is at 410 m.y. Note that the slope for living families (for longevity of 45 to 200 m.y.) is approximately twice as steep as for extinct families (for longevity of 5 to 165 m.y.). Data for extinct families include Cheilostomata (15, 9 of which are 0—5 m.y.), Cyclostomata (11), Cryptostomata (10), Trepostomata (8), Cystoporata (4), and Ctenostomata (3). Data for living families includes Cheilostomata (58), Cyclostomata (18), and Ctenostomata (1). Data on longevity from Larwood et al. (1967); with emendations to ranges from Lagaaij (1968), Brood (1974), and Cheetham (1972).

divide a group when it gets too large. Indeed, between the time that Van Valen (1973) made his graph for Bryozoa, and the present version, two long-lived living cyclostome families (one was said to begin in the Devonian and the other in the Ordovician) have been revised to begin in the Triassic, with the Palaeozoic groups occurring in new shorter-lived taxa. Accordingly if there is anything fundamental to the notion of a constant rate of extinction, then we must discount (for Bryozoa) all records of longevity greater than approximately 165 m.y.

Data for living families present other problems. The number in each group deviates strongly from a regular decline with increasing age (Fig. 15), and so either they (1) do not represent a stable age distribution, and/or (2) the sample of bryozoan families is too small and too biased to be useful in estimating the total number of families at the start of each interval (the l_x schedule). For purposes of discussion, I present the data as they exist, with a smoothed line of estimate of best fit (fitted by eye from a moving mean). Sufficient independent data on extinction in each age group would permit an independent calculation of mortality rate (as in Krebs, 1972, table 10), and this would overcome the problem of reversals in the schedule of number in each age group. However, the resulting mortality rates in this case are so widely varying and the data are so few that it seems even less profitable to present this estimate than simply to plot the data as they are.

The smoothed line for living families of longevity 45—200 m.y. represents a half life of 40 m.y. (or 12.5 μma). Most of this curve is attributed to the Cheilostomata. The smoothed curve is thus another way to suggest that cheilostomes have (or appeared to have) evolved much more rapidly than the non-cheilostome bryozoans. In the portion of the curve from 0 to 45 m.y., there are far "too few" families recognized if that portion of the curve is to be consistent with the rest of the curve. Bryozoologists may have been reluctant to recognize as distinct families groups which include no genera older than the Oligocene. Alternatively, the change in slope may be "real" to the extent that the Late Cretaceous—Early Tertiary rapid expansion of families of cheilostomes follows the development of distinct faunal provinces as continents drift apart and gene flow is reduced. Subsequent replacement of families in these endemic families may be at a much lower rate than is the initial rise to "saturation".

Species survivorship has been estimated by Stach (1938b, p. 80) and accepted by Buge (1972, p. 56) and Weisbord (1967, p. 15). The longevity of Tertiary Bryozoa is distributed as follows: Eocene 2—5% living species; Oligocene 7—15%; Miocene 20—30%; Pliocene 60—80%; and Pleistocene 70—100%. David (1964) uses essentially the reverse relationship and considers the number of species that are held in common with the Eocene. The ratio of cyclostomes to cheilostomes decreases through the Cretaceous and Tertiary, and this is sometimes used (e.g., Ghiurca, 1964) as an index to stratigraphic placement. The half life of a living cheilostome bryozoan species is 5—10 m.y. (dating from the Late Miocene, Early Pliocene). Pouyet (1971, pp. 194—195) discusses strict ecologic replacement of Miocene and Pliocene cellepores by living species of *Schizoporella*, thus indicating an ecological steady state.

Diversity through time

Müller (1958) compiled generic diversity for all Bryozoa (Fig. 16) and for the cyclostomes (Fig. 17A), cryptostomes and trepostomes (Fig. 17B) and cheilostomes (Fig. 17C), using the most complete coverage available even today (Bassler, 1953). A view of diversity through time of cheilostome families based on more recent information is also available (Fig. 18).

A "traditionalist" view of these patterns would be that the trepostomes and cryptostomes filled similar niches, but the (more complex) cryptostomes won out, only to be exterminated by a change in the environment (end of Permian); at that point the cyclostomes were permitted to expand to fill the same niches, and they in turn were then replaced by the competitively superior (and more complex) cheilostomes. For example, in explaining why the cyclostomes have

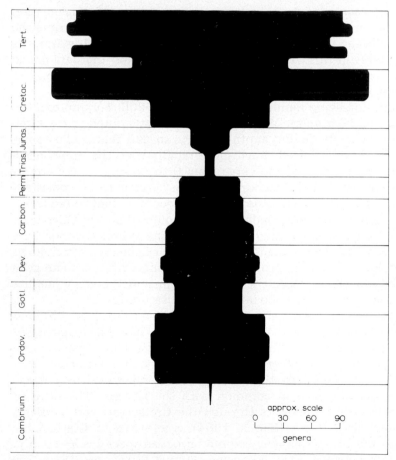

Fig. 16. Approximate number and time distribution of 900 genera of Bryozoa, including 273 Cyclostomata, 103 Trepostomata, 126 Cryptostomata, 387 Cheilostomata, and 11 Ctenostomata. From Müller (1958, p. 264).

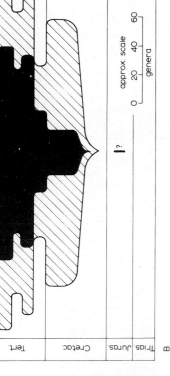

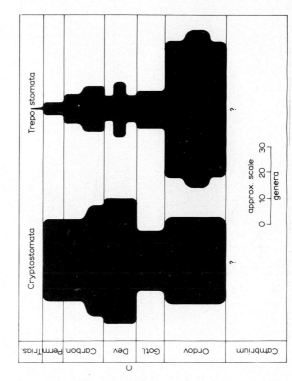

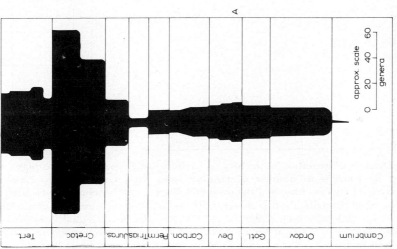

Fig. 17. A. Approximate number and time distribution of 273 genera of cyclostomes. From Müller (1958, p. 269). B. Approximate num-
ber and time distribution of 387 genera of Cheilostomata; 184 are Anasca (outer grey part), and 203 are Ascophora (inner black part).
From Müller (1958, p. 285). C. Approximate number and time distribution of 126 genera of Cryptostomata (left); and of 103 genera of
Trepostomata (right). From Müller (1958, p. 275).

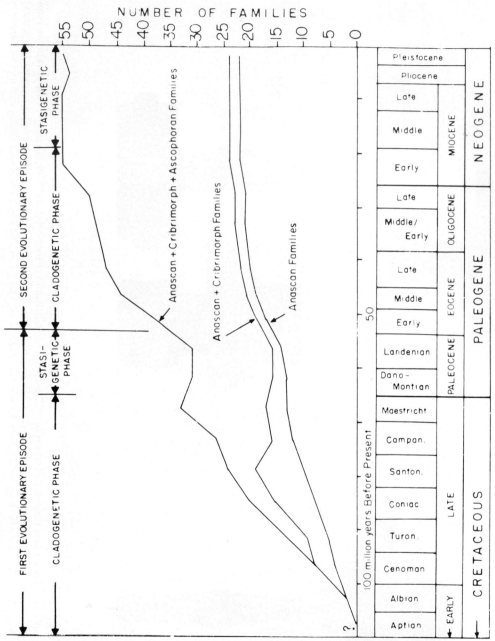

Fig. 18. Division of evolutionary history of cheilostome Bryozoa into two major episodes, as suggested by changes in number of families from Early Cretaceous to Pleistocene. Two roughly sigmoid curves, Albian—Paleocene and Eocene—Pleistocene, are each separated into a phase of rapid diversification (cladogenesis) and a following quiescent phase (stasigenesis). Because of the generally long ranges of families and the nearly equal intervals of time over which the census was taken, it seems inappropriate to calculate frequencies as time-frequencies. Changes in diversity at the family level, as shown here, may be an inadequate expression of evolutionary activity at lower categorical levels. The earlier evolutionary episode shows increase in all three major groups of cheilostomes — anascan, cribrimorph, and ascophoran. Most increase in the later episode has been in the ascophoran cheilostomes. From Cheetham (1971, figure 1).

the pattern they do, Newell (1952, p. 375) wrote, "Perhaps this group was inhibited by competition until an environmental niche was opened up by disappearance of competitors. Such competing forms may have been the cryptostomes and trepostomes, which became extinct at the close of the Paleozoic." Cheetham (1971) even goes so far as to label parts of the change in slope of curve of diversification of cheilostomes a "cladogenetic pulse" and a "stasigenetic pulse".

Each of the various ways in which a strictly deterministic view can be brought to bear on bryozoan evolution has, however, a stochastic counterpart. That is, given a probability distribution for rates of extinction and rates of speciation, clades can be stochastically generated. These clades (Raup et al., 1973) are shown in Fig. 19 and they are very similar to the shape of real-world clades. The cyclostomes, for example, are like Fig. 19, *C*, *7*; cryptostomes, Fig. 19 *D*, *3*; trepostomes, Fig. 19, *D*, *18*; and cheilostomes, Fig. 19, *C*, *20*; and Bryozoa as a whole in Fig. 19, *A*, *7*. In other words, we can view bryozoan evolution

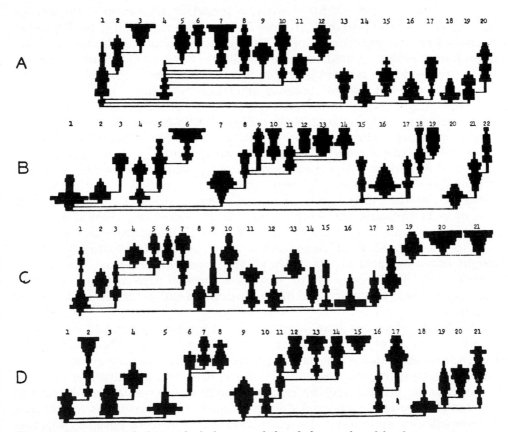

Fig. 19. Diversity variation and phylogeny of the clades produced by four computer runs. Each run used the same input constants, but a different random number sequence. From Raup et al. (1973, figure 5).

as merely another example of faunal turnover (and expansion and contraction) in a stochastic world where species are behaving in a statistical sense like so many gas molecules which enter and leave a room which was being alternately expanded and contracted in volume, at a fixed temperature.

General equilibrium values for bryozoan diversity can, I think, be assumed because of the strong relationships with available substrate area (Fig. 1), and with the latitudinal gradient. The fact that these same relationships have been found in major taxon after major taxon shows that faunal diversity is strongly controlled. World faunal diversity for all taxa seems to be a first-order function of (1) number of adaptive zones which are colonized, and (2) number of faunal provinces. Since both of these change through geological time, the total diversity will also change. But all changes are within a given equilibrium for each adaptive zone.

Summary

In the initial section of this chapter, a model of speciation is emphasized which stresses the importance of gene flow (= dispersal ability, for bryozoa). Patterns of genetic and morphologic differentiation parallel each other in Bryozoa and are inversely related to dispersal ability. Speciation is held to be a stochastic process which is more probable in situations where species can extend their ranges without bound, because gene flow is reduced and because the larger the range the higher the probability that some part of the range will become geographically isolated. Such ecologic settings occur in the tropics and in the deep-sea where bryozoan specialization is in fact most pronounced.

The bulk of the chapter is concerned with morphologic and biogeographic patterns of evolution. With regard to morphologic features, a gradient in degree of specialization often exists. Morphologies can be arranged so that patterns of r versus K selection appear to occur in the degree of calcification, the development of polymorphism (of several kinds), in reproductive potential of different colony patterns of growth, in levels of colonial integration, and in adaptations to feeding. Each of these trends along an adaptive pathway can be thought of as an evolutionary "vector" going, in a general way, from r to K selected types. Evidence exists of sequences through time of r to K morphologies in defensive polymorphism of Cheilostomata and in the levels of colonial integration.

Data from bryozoans have been applied to many arguments concerning general themes of evolution. In each of these (for example, the biogenic law, orthogenesis, and rates of speciation), a very deterministic, historical explanation has been sought in the previous literature. However, the nature of the overall evolutionary process and the resulting products can just as well be viewed from the perspective of a general stochastic model of speciation no matter how deterministic the cause a specific adaptation may appear to have. By this model, specialization (and speciation) continues to the extent that environmental stimuli are repetitive, and thus able to be "keyed upon". Once saturation is reached, replacement would be at a uniform rate. It appears inevitable that, given this speciation model, any given character complex, or lineage, will, on the average, appear through time to go from the r to the K end of the spectrum. The r and K selections are therefore viewed as a consequence of this stochastic model of speciation.

In a similar fashion, even though each gas molecule of a given volume has its own deterministic history, there is a pattern to the properties of all of the molecules taken together which can be described independently of knowledge of the individual molecules. Explanations for trends in morphologic complexity, in survivorship or in diversity through time can be sought for both at the deterministic, historical level, or at the stochastic, equilibrium level. At any given time, some proportion of the local populations in existence is going to yield new species while others go extinct; and some proportion of the morphologic features will be continued with greater specialization keying to specific environmental stimuli, and others will be terminated; and some genera and higher taxa will multiply while others will cease to exist. Many of the patterns and trends of evolution among the bryozoa can be viewed from both deterministic and stochastic perspectives, and it has been the larger purpose of this chapter to begin to indicate that both of these levels of explanation can be applied.

Acknowledgements

Much of my experience with bryozoans has been obtained during summers of the past ten years at the Marine Biological Laboratory, Woods Hole. I have benefited from discussions with Karl W. Kaufmann, whose work on r and K selection among the Bryozoa provides the basis of a major theme of this present study. I thank David M. Raup for extensive discussion on survivorship curves. I thank Lynn H. Throckmorton and Ralph G. Johnson for reviewing the text. The paper is dedicated to Lars G. Silén, Ehrhard Voigt and Eveline du Bois-Reymond Marcus, each of whom has continuously published interesting work on the Bryozoa for more than forty years.

References

Abel, O., 1929. Das biologische Trägheitsgesetz. Palaeontol. Z., 11: 7—17.
Abbott, M.B., 1975. Relationship of temperature to patterns of sexual reproduction in some recent encrusting Cheilostomata. In: Bryozoa 1974. Doc. Lab. Géol. Fac. Sci. Lyon, H.S., 3(1): 37—50.
Anstey, R.L., Pachut, J.F. and Prezbindowski, D.R., 1976. Morphogenetic gradients in Paleozoic bryozoan colonies. Paleobiology, 2: 131—146.
Avise, J.C., 1976. Genetic differentiation during speciation. In: F.J. Ayala (Editor), Molecular Evolution. Sinauer Associates, Sunderland, Mass., pp. 106—122.
Banta, W.C., 1973. Evolution of avicularia in cheilostome Bryozoa. In: R.S. Boardman, A.H. Cheetham and W.A. Oliver (Editors), Animal Colonies. Dowden, Hutchinson and Ross, Stroudsburg, Penn., pp. 295—303.
Banta, W.C., 1975. Origin and evolution of cheilostome Bryozoa. In: Bryozoa 1974. Doc. Lab. Géol. Fac. Sci. Lyon, H.S., 3(2): 565—582.
Banta, W.C., McKinney, F.K. and Zimmer, R.L., 1974. Bryozoan monticules: Excurrent water outlets? Science, 185: 783—784.

Bassler, R.S., 1953. Bryozoa. In: R.C. Moore (Editor), Treatise on Invertebrate Paleontology, Part G. University of Kansas Press, Lawrence, Kansas, 253 pp.

Blake, D.B., 1975. The order Cryptostomata resurrected. In: Bryozoa 1974. Doc. Lab. Géol. Fac. Sci. Lyon, H.S., 3(1): 211—224.

Boardman, R.S., 1968. Colony development and convergent evolution of budding pattern in "Rhombotrypid" Bryozoa. Atti Soc. Ital. Sci. Nat. Mus. Civ. Stor. Nat. Milano, 108: 179—184.

Boardman, R.S., 1975. Taxonomic characters for phylogenetic classifications of cyclostome Bryozoa. In: Bryozoa 1974. Doc. Lab. Géol. Fac. Sci. Lyon, H.S., 3(2): 595—606.

Boardman, R.S. and Cheetham, A.H., 1973. Degrees of colony dominance in stenolaemate and gymnolaemate Bryozoa. In: R.S. Boardman, A.H. Cheetham and W.A. Oliver (Editors), Animal Colonies, Dowden, Hutchinson and Ross, Stroudsburg, Penn., pp. 121—220.

Boardman, R.S. and McKinney, F.K., 1976. Skeletal architecture and preserved organs of four-sided zooids in convergent genera of Paleozoic Trepostomata (Bryozoa). J. Paleontol., 50: 25—78.

Boardman, R.S., Cheetham, A.H. and Cook, P.L., 1970. Intracolony variation and the genus concept in Bryozoa. Proc. North Am. Paleontol. Conv., Part C, pp. 294—320.

Boardman, R.S., Cheetham, A.H. and Oliver, W.A. (Editors), 1973. Animal Colonies. Dowden, Hutchinson and Ross, Stroudsburg, Penn., 603 pp.

Borg, F., 1933. A revision of the recent Heteroporidae (Bryozoa). Zool. Bidr. Upps., 14: 253—394.

Borg, F., 1965. A comparative and phyletic study on fossil and recent Bryozoa of the suborders Cyclostomata and Trepostomata. Ark. Zool., Ser. 2, 17: 1—91.

Brandt, K., 1883. Über die morphologische und physiologische Bedeutung des Chlorophylls bei Tieren. II. Artikel: Mittheilungen aus der Zoologischen Station zu Neapel, 4: 191—302.

Brood, K., 1970. On two species of Saffordotaxis (Bryozoa) from the Silurian of Gotland, Sweden, Stockh. Contrib. Geol., 21: 57—68.

Brood, K., 1972. Cyclostomatous Bryozoa from the Upper Cretaceous and Danian in Scandinavia. Stockh. Contrib. Geol., 26: 1—464.

Brood, K., 1974. Bryozoa from the Ludlovian of Bjärsjölagård in Skåne. Geol. Fören. Stockh. Förh., 96: 381—388.

Brood, K., 1975. Cyclostomatous Bryozoa from the Silurian of Gotland. Stockh. Contrib. Geol., 28: 45—119.

Buchner, P., 1930. Tiere und Pflanze in Symbiose. Borntraeger, Berlin, 900 pp.

Buge, E., 1948. Les Bryozoaires du Savignéen (Helvétien) de Touraine. Mém. Mus. Natl. Hist. Nat., N.S., 27: 63—94.

Buge, E., 1972. Remarques sur les méthodes d'utilisation stratigraphique des Bryozoaires postpaléozoïques. Mém. Bur. Rech. Géol. Min. Fr., 77: 55—58.

Canu, F., 1900. Révision des Bryozoaires du Crétacé figurés par d'Orbigny, Part 2 — Cheilostomata. Bull. Soc. Géol. Fr., Sér. 3, 28: 334—463.

Canu, F. and Bassler, R.S., 1920. North American Early Tertiary Bryozoa. U.S. Natl. Mus., Bull. 106: 879 pp. (2 vols.).

Caughley, G., 1966. Mortality patterns in mammals. Ecology, 47: 906—918.

Caullery, M., 1926. Aspect mimétique de Caprella acanthifera Leach sur les Bugula. Bull., Biol. Fr. Belg., 60: 126—133.

Cheetham, A.H., 1968. Morphology and systematics of the bryozoan genus Metrarabdotos. Smithson. Miscell. Collect., 153: 1—121.

Cheetham, A.H., 1969. In: R.S. Boardman and A.H. Cheetham, Skeletal Growth, Intracolony Variation, and Evolution in Bryozoa: A Review. J. Paleontol., 43: 205—243.

Cheetham, A.H., 1971. Functional morphology and biofacies distribution of cheilostome Bryozoa in the Danian Stage (Paleocene) of southern Scandinavia. Smithson. Contrib. Paleobiol., No. 6: 1—87.

Cheetham, A.H., 1972. Cheilostome Bryozoa of Late Eocene age from Eua, Tonga. U.S. Geol. Surv., Prof. Pap., 640-E: E1—E26.

Cheetham, A.H., 1975. Taxonomic significance of autozooid size and shape in some early multiserial cheilostomes from the Gulf Coast of the U.S.A. In: Bryozoa 1974. Doc. Lab. Géol. Fac. Sci. Lyon, H.S., 3(2): 547—564.

Collins, E.J., Tenney, W.R. and Woolcott, W.S., 1966. Histological effects of the poison of Lophopodella carteri (Hyatt) on the gills of Carassius auratus (Linnaeus) and larval Ambystoma opacum (Gravenhorst). Va. J. Sci., July, 1966: 155—163.

Condra, G.E. and Elias, M.K., 1944. Study and revision of Archimedes (Hall). Geol. Soc. Am. Spec. Pap., 53: 243 pp.

Condra, G.E. and Elias, M.K., 1945. Bicorbula, a new Permian Bryozoan, probably a bryozoan—algal consortium. J. Paleontol., 19: 116—125.

Cuffey, R.J., 1970. Bryozoan—environment interrelationships — An overview of bryozoan paleoecology and ecology. Penn. State Univ., Earth Min. Sci., 39: 41—45.

Cummings, E.R., 1910. Paleontology and the Recapitulation Theory. Proc. Indiana Acad. Sci., 1909: 305—340.

Cummings, S.G., 1975. Zooid regression in Schizoporella unicornis floridana. (Bryozoa, Cheilostomata). Chesapeake Sci., 16: 93—103.

David, L., 1964. Méthode d'utilisation des Bryozoaires pour la stratigraphie du Néogène. Inst. "Lucas Mallada", C.S.I.C. (España), Curs. Conf., 9: 147—151.

Deevey Jr., E.S., 1947. Life tables for natural populations of animals. Q. Rev. Biol., 22: 283—314.

Dorf, E., 1960. Climatic changes of the past and present. Am. Sci., 48: 341—364.

Dzik, J., 1975. The origin and early phylogeny of the cheilostomatous Bryozoa. Acta Palaeontol. Polon., 20: 395—423.

Eggleston, D., 1971. Synchronization between moulting in Calocaris macandreae (Decapoda) and reproduction in its epibiont Tricella koreni (Polyzoa, Ectoprocta). J. Mar. Biol. Assoc. U.K., 51: 409—410.

Eggleston, D., 1972a. Patterns of reproduction in the marine Ectoprocta of the Isle of Man. J. Nat. Hist., 6: 31—38.

Eggleston, D., 1972b. Factors influencing the distribution of sub-littoral ectoprocts off the south of the Isle of Man (Irish Sea). J. Nat. Hist., 6: 247—260.

Eldredge, N. and Gould, S.J., 1972. Punctuated equilibria: An alternative to phyletic gradualism. In: T.J.M. Schopf (Editor), Models in Paleobiology, Freeman, Cooper, San Francisco, Calif., pp. 82—115.

Elias, M.K. and Condra, G.E., 1957. Fenestella from the Permian of West Texas. Geol. Soc. Am., Mem., 70: 158 pp.

Ertlova, E., 1974. Some remarks on midges (Diptera, Chironomidae) from Bryozoans. Biol. (Slovenská akademia vied), Bratislava, 29: 869—876.

Ghiurca, V., 1964. Contributii la Cunoasterea Fauneii de Briozoare din Transilvania (v). Brizoarele tortoniene de la Lopadea Veche (Raionul Aiud). Stud. Univ. Babes-Bolyai, Ser. Geol.-Geogr., 1: 45—50.

Gould, S.J., 1970. Dollo on Dollo's Law: Irreversibility and the status of evolutionary laws. J. Hist. Biol., 3: 189—212.

Gregory, J.W., 1896. Catalogue of the Fossil Bryozoa in the Department of Geology British Museum (Natural History). The Jurassic Bryozoa. Longmans, London, 239 pp.

Håkansson, E., 1973. Mode of growth on the Cupuladriidae (Bryozoa, Cheilostomata). In: G.P. Larwood (Editor), Living and Fossil Bryozoa. Academic Press, London, pp. 287—298.

Håkansson, E., 1975. Population structure of colonial organisms: A palaeoecological study of some free-living Cretaceous bryozoans. In: Bryozoa 1974. Doc. Lab. Géol. Fac. Sci. Lyon, H.S., 3(2): 385—399.

Harmelin, J.-G., 1968. Contribution a l'étude des Bryozoaires Cyclostomes de Méditerranée: les Crisia des Côtes de Provence. Bull. Mus. Natl. Hist. Nat., Sér. 2, 40: 413—437.

Harmelin, J.-G., 1973. Morphological variations and ecology of the recent cyclostome bryozoan "Idmonea" atlantica from the Mediterranean. In: G.P. Larwood (Editor), Living and Fossil Bryozoa. Academic Press, London, pp. 95—106.

Harmelin, J.-G., 1975. Evolutionary trends within three Tubuliporina families (Bryozoa, Cyclostomata). In: Bryozoa 1974. Doc. Lab. Géol. Fac. Sci. Lyon, H.S., 3(2): 607—616.

Harmer, S.F., 1891. On the regeneration of lost parts in Polyzoa. Rep. 60th Meet. Br. Assoc. Adv. Sci., Leeds, pp. 862—863.

Harmer, S.F., 1923. On cellularine and other Polyzoa. J. Linn. Soc., 35: 293—361.

Harmer, S.F., 1930. Presidential Address. Proc. Linn. Soc. Lond., 141st Sess., pp. 68—118.

Hayward, P.J., 1975. Observations on the bryozoan epiphytes of Posidonia oceanica from the Island of Chios (Aegean Sea). In: Bryozoa 1974. Doc. Lab. Géol. Fac. Sci. Lyon, H.S., 3(2): 347—356.

Hincks, T., 1880a. A History of the British Marine Polyzoa. John Van Voorst, London, Vol. I (text), 601 pp.; Vol. II (plates), 83.

Hincks, T., 1880b. Contributions towards a general history of the marine Polyzoa. Ann. Mag. Nat. Hist., Ser. 5, 6: 69—93.

Hincks, T., 1881. Contributions towards a general history of the marine Polyzoa. VI. Polyzoa from Bass's Straits. Ann. Mag. Nat. Hist., Ser. 5, 8: 1—14.

Hincks, T., 1882. On certain remarkable modifications of the avicularium in a species of Polyzoon; and on the relation of the vibraculum to the avicularium. Ann. Mag. Nat. Hist., Ser. 5, 9: 20—25.

Hincks, T., 1890. Critical notes on the Polyzoa. Part II. Classification. Ann. Mag. Nat. Hist., Ser. 6, 9: 83—103.

Hinds, R.W., 1975. Growth mode and homeomorphism in cyclostome Bryozoa. J. Paleontol., 49: 875—910.

Hutchinson, G.E., 1968. When are species necessary? In: R.C. Lewontin (Editor), Population Biology and Evolution. Syracuse University Press, Syracuse, N.Y., pp. 177—186.

Hyman, L.H., 1959. The Invertebrates: Smaller Coelomate Groups, V. McGraw-Hill, New York, N.Y., 783 pp.

Illies, G., 1973. Different budding patterns in the genus Stomatopora (Bryozoa, Cyclostomata). In: G.P. Larwood (Editor), Living and Fossil Bryozoa. Academic Press, London, pp. 307—315.

Illies, G., 1975. On the genus Stomatoporina Balavoine, 1958 (Bryozoa, Cyclostomata). In: Bryozoa 1974. Doc. Lab. Géol. Fac. Sci. Lyon, H.S., 3(1): 51—58.

Jebram, D., 1973a. The importance of different growth directions in the Phylactolaemata and Gymnolaemata for reconstructing the phylogeny of the Bryozoa. In: G.P. Larwood (Editor), Living and Fossil Bryozoa. Academic Press, London, pp. 565—576.

Jebram, D., 1973b. Stolonen-Entwicklung und Systematik bei den Bryozoa Ctenostomata. Z. Zool. Syst. Evolutionsforsch., 11: 1—48.

Kaufmann, K.W., 1970. A model for predicting the influence of colony morphology on reproductive potential in the Phylum Ectoprocta. Biol. Bull., 139: 426.

Kaufmann, K.W., 1971a. The effect of colony morphology on the life history strategy of Bryozoans. Geol. Soc. Am., Abstr. Progr., 3(7): 618.

Kaufmann, K.W., 1971b. The form and functions of the avicularia of *Bugula* (Phylum Ectoprocta). Postilla, 151: 26 pp.
Kaufmann, K.W., 1973. The effect of colony morphology on life-history parameters of colonial animals. In: R.S. Boardman, A.H. Cheetham and W.A. Oliver (Editors), Animal Colonies. Dowden, Hutchinson and Ross, Stroudsburg, Penn., pp. 221—222.
Keith, D.E., 1971. Substrate selection in caprellid amphipods of Southern California, with emphasis on *Caprella californica* Stimpson and *Caprella equilibria* Say (Amphipoda). Pac. Sci., 25: 387—394.
King, C.E., 1964. Relative abundance of species and MacArthur's model. Ecology, 45: 716—727.
Krebs, C.J., 1972. Ecology. Harper and Row, New York, N.Y., 694 pp.
Lagaaij, R., 1968. First fossil finds of six genera of Bryozoa Cheilostomata. Atti Soc. Ital. Sci. Nat. Mus. Civic. Stor. Nat. Milano, 108: 345—360.
Lang, W.D., 1916. Calcium carbonate and evolution in Polyzoa. Geol. Mag., Ser. 6, 3: 73—77.
Lang, W.D., 1919a. The evolution of ammonites. Proc. Geol. Assoc., 30: 49—65.
Lang, W.D., 1919b. The Pelmatoporinae, an essay on the evolution of a group of Cretaceous Polyzoa. Philos. Trans. R. Soc. Lond., Ser. B, 209: 191—228.
Lang, W.D., 1921. Catalogue of the fossil Brvozoa (Polyzoa) in the Department of Geology, British Museum (Natural History). The Cretaceous Bryozoa (Polyzoa). Vol. III. The Cribrimorphs — Part I. Longmans, London, 269 pp.
Lang, W.D., 1922. Catalogue of the fossil Bryozoa (Polyzoa) in the Department of Geology, British Museum (Natural History). The Cretaceous Bryozoa (Polyzoa). Vol. IV. The Cribrimorphs — Part II. Longmans, London, 404 pp.
Lang, W.D., 1923a. Evolution: A resultant. Proc. Geol. Assoc., 34: 7—20.
Lang, W.D., 1923b. Trends in British Carboniferous corals. Proc. Geol. Assoc., 34: 120—136.
Lang, W.D., 1925. Persistence in fossils. Proc. Geol. Assoc., 36: 227—239.
Larwood, G.P., 1962. The morphology and systematics of some Cretaceous cribrimorph Polyzoa (Pelmatoporinae). Bull. Br. Mus. (Nat. Hist.), Geol., 6: 1—285.
Larwood, G.P., 1969. Frontal calcification and its function in some Cretaceous and recent cribrimorph and other cheilostome Bryozoa. Bull. Br. Mus. (Nat. Hist.), Zool., 18: 171—182.
Larwood, G.P., Medd, A.W., Owen, D.E. and Tavener-Smith, R., 1967. Bryozoa. In: W.B. Harland et al. (Editors), The Fossil Record. Geological Society of London, pp. 379—395.
Levinsen, G.M.R., 1902. Studies on Bryozoa. Vidensk. Medd. Nat. Fören., pp. 1—31.
Levinsen, G.M.R., 1909. Morphological and Systematic Studies on the Cheilostomatous Bryozoa. Nationale Forfatteres Forlag, Copenhagen, 431 pp.
Levinsen, G.M.R., 1912. Studies on the Cyclostomata Operculata. Mém. Acad. R. Sci. Lettr. Dan., Copenhague, Ser. 7, 10: 1—52.
Lutaud, G., 1969. La nature des corps funiculaires des Cellularines, Bryozoaires Chilostomes. Arch. Zool. Expér. Gén., 110: 5—30.
MacArthur, R.H., 1957. On the relative abundance of bird species. Proc. Nat. Acad. Sci. (U.S.), 43: 293—295.
MacArthur, R.H., 1972. Geographical Ecology: Patterns in the Distribution of Species. Harper and Row, New York, N.Y., 269 pp.
MacArthur, R.H. and Wilson, E.O., 1967. The Theory of Island Biogeography. Princeton University Press, Princeton, N.J. 203 pp.
MacMunn, C.A., 1887. Chromatology of Sponges. Proc. Physiol. Soc. No. III (March 12, 1887): xi—xii. [See J. Physiol.].
Małecki, J., 1964. On two new genera of Bryozoa Cheilostomata from the Tortonian of Poland. Acta Palaeontol. Polon., 9: 499—512.
Marcus, E., 1921. Über die Verbreitung der Meeresbryozoen. Zool. Anz., 53: 205—221.
Marcus, E., 1926. Beobachtungen und Versuche an lebenden Meeresbryozoen. Zool. Jahrb., 52: 1—102.
Marcus, E., 1938. Bryozoarios Marinhos Brasileiros II. Univ. São Paulo, Fac. Filos. Ciênc. Letr., Bol. No. 4, Zool., No. 2: 1—196.
Mawatari, S., 1946. On the metamorphosis of *Smittina collifera* Robertson. Studies on the fouling Bryozoa, No. 4. Misc. Rep. Res. Inst. Nat. Resour., No. 10: 31—35.
Mawatari, S., 1953. On *Electra angulata* Levinsen, one of the fouling bryozoans in Japan. Misc. Rep. Res. Inst. Nat. Resour., No. 32: 5—11.
Mawatari, S., 1968. A pedunculate ctenostome from the Antarctic Region. Proc. Jap. Soc. Syst. Zool., No. 4: 42—45.
Mawatari, S., 1973. Studies on Japanese Anascan Bryozoa, 1. Inovicellata. Bull. Nat. Sci. Mus., 16: 409—428.
Mawatari, S., 1974. Studies on Japanese Anascan Bryozoa, 3. Division Malacostega (1). Bull. Natl. Sci. Mus., 17: 17—55.
Mawatari, S. and Kobayashi, S., 1954. Seasonal settlement of animal fouling organisms in Ago Bay, middle part of Japan II. Misc. Rep. Res. Inst. Nat. Resour., No. 36: 1—8.
McKinney, F.K., 1975. Autozooecial budding patterns in dendroid stenolaemate bryozoans. In: Bryozoa 1974. Doc. Lab. Géol. Fac. Sci. Lyon, H.S., 3(1): 65—76.
Meacham Jr., R.H. and Woolcott, W.S., 1968. Studies of the coelomic fluid and isotonic homogenates of the freshwater bryozoan *Lophopodella carteri* (Hyatt) on fish tissues. Va. J. Sci., 19: 143—146.
Medioni, A. and Soyer, J., 1966. *Laophonte ? drachi* n. sp. Copépode Harpacticoïde associé au Bryozo-

aire *Schismopora armata* (Hincks, 1860). Vie Milieu Sér. A, Biol. Mar., 27: 1053—1063.

Medioni, A. and Soyer, J., 1967. Copépodes Harpacticoïdes de Banyuls-sur-Mer. 6. Nouvelles formes associées à des Bryozoaires. Vie Milieu, Sér. A, Biol. Mar., 28: 317—343.

Miller, M.C., 1965. Grazing carnivore: some sea-slugs feeding on sedentary invertebrates. Poiaieria, 3: 1—11 [Conchology section, Auckland Institute and Museum].

Monniot, F. and Buge, E., 1971. Les spicules d'Ascidies fossiles et actuelles. Ann. Paléontol., Invertébrés, 57: 93—105.

Moyano, G., H.I., 1975. El polimorfismo de los Bryozoa Antarticos como un indice de estabilidad ambiental. Gayana. Inst. Biol., Zool., No. 33: 42 pp.

Müller, A.H., 1958. Lehrbuch der Palaozoologie. Band II. Invertebraten, Teil 1. Protozoa—Mollusca, 1. Gustav Fischer Verlag, Jena, 566 pp.

Newell, N.D., 1952. Periodicity in invertebrate evolution. J. Paleontol., 26: 371—385.

Nielsen, C., 1971. Entoproct life-cycles and the entoproct/ectoproct relationship. Ophelia, 9: 209—341.

Nordgaard, O., 1906. Bryozoa from the 2nd Fram Expedition 1898—1902. Report of the Second Norwegian Arctic Expedition in the "Fram" 1898—1902, No. 8: 44 pp.

Nye, O.B., 1968. Aspects of microstructure in post-Paleozoic Cyclostomata. Atti Soc. Ital. Sci. Nat. Mus. Civ. Stor. Nat. Milano, 108: 111—114.

Nye, O.B., 1976. Generic revision and skeletal morphology of some cerioporid cyclostomes (Bryozoa). Bull. Am. Paleontol., 69(291): 1—222.

Oltmans, F., 1923. Morphologie und Biologie der Algen. Fischer, Jena, second revised edition, 530 pp.

Osman, R.W., 1977. The establishment and development of a marine epifaunal community. Ecol. Monogr., in press.

Palmer, T.J. and Hancock, C.D., 1973. Symbiotic relationships between ectoprocts and gastropods, and ectoprocts and hermit crabs in the French Jurassic. Palaeontology, 16: 563—566.

Pianka, E.R., 1970. On *r*- and *K*-selection. Am. Nat., 104: 592—597.

Pouyet, S., 1971. *Schizoporella violacea* (Canu et Bassler, 1930) (Bryozoa, Cheilostomata): Variations et croissance zoariale. Geobios, 4: 185—197.

Pouyet, S., 1973. Révision systématique des Cellépores (Bryozoa, Cheilostomata) et des espèces fossiles européennes. Analyse de quelques populations à Cellépores dans le Néogène du Bassin rhodanien. Doc. Lab. Géol. Fac. Sci. Lyon, No. 55: 266 pp.

Powell, N.A., 1966. Colony formation in *Selenaria nitida* Maplestone. Rec. Aust. Mus., 27: 27—32.

Powell, N.A., 1967. Polyzoa (Bryozoa) — Ascophora — From North New Zealand. Discovery Rep., 34: 199—394.

Powell, N.A., 1968. Early astogeny in *Hippopodina feegeensis* (Busk) (Polyzoa — Ascophora). Nat. Mus. Can. Bull., No. 223, Contrib. Zool., IV: 1—4.

Powell, N.A., 1970. *Schizoporella unicornis* — An alien bryozoan introduced into the Strait of Georgia. J. Fish. Res. Board Can., 27: 1847—1853.

Raup, D.M., 1975. Taxonomic survivorship curves and Van Valen's Law. Paleobiology, 1: 82—96.

Raup, D.M., Gould, S.J., Schopf, T.J.M. and Simberloff, D.S., 1973. Stochastic models of phylogeny and the evolution of diversity. J. Geol., 81: 525—542.

Reguant, S., 1959. Algunas consideraciones sobre las ideas actuales acerca de la filogenia de los Briozoos ectoproctos. Publ. Inst. Biol. Apl., 30: 87—103.

Rigby, J.K., 1957. Relationships between *Acanthocladia guadalupensis* and *Solenopora texana* and the bryozoan—algal consortium hypothesis. J. Paleontol., 31: 603—606.

Robertson, A., 1905. Non-incrusting chilostomatous Bryozoa of the west coast of North America. Univ. Calif. Publ., Zool., 2: 235—322.

Rogick, M.D. and Croasdale, H., 1949. Studies on marine Bryozoa III. Woods Hole region Bryozoa associated with algae. Biol. Bull., 96: 32—69.

Ross, J.P., 1964. Morphology and phylogeny of early Ectoprocta (Bryozoa). Geol. Soc. Am. Bull., 75: 927—948.

Ryland, J.S., 1962. The association between Polyzoa and algal substrata. J. Anim. Ecol., 31: 331—338.

Ryland, J.S., 1967. Polyzoa. Oceanogr. Mar. Biol. Annu. Rev., 5: 343—369.

Ryland, J.S., 1970. Bryozoans. Hutchinson University Library, London, 175 pp.

Ryland, J.S. and Stebbing, A.R.D., 1971. Settlement and orientated growth in epiphytic and epizoic bryozoans. In: D.J. Crisp (Editor), Fourth European Marine Biology Symposium. Cambridge University Press, London, pp. 105—123.

Schaaf, A., 1974. Les modalités de la croissance et ses altérations chez quelques Bryozoaires cheilostomes. Application à l'espèce *Steginoporella rhodanica* Buge and David, 1967. Doc. Lab. Géol. Fac. Sci. Lyon, No. 60: 1—81.

Schopf, T.J.M., 1969a. Geographic and depth distribution of the Phylum Ectoprocta from 200 to 6,000 meters. Proc. Am. Philos. Soc., 113: 464—474.

Schopf, T.J.M., 1969b. Paleoecology of ectoprocts (bryozoans). J. Paleontol., 43: 234—244.

Schopf, T.J.M., 1971. An approach to understanding evolutionary relationships in the Phylum Ectoprocta. In: A.L. Meyerson and C.S. Zois (Editors), Papers in Marine Science. The Link Lecture Ser., pp. 1—10.

Schopf, T.J.M., 1972. Varieties of paleobiologic experience. In: T.J.M. Schopf (Editor), Models of Paleobiology. Freeman, Cooper, San Francisco, Calif., pp. 8—25.

Schopf, T.J.M., 1973. Ergonomics of polymorphism: Its relation to the colony as the unit of natural selection in species of the Phylum Ectoprocta. In: R.S. Boardman, A.H. Cheetham and W.A. Oliver (Editors), Animal Colonies. Dowden, Hutchinson and Ross, Stroudsburg, Penn., pp. 247—294.

Schopf, T.J.M., 1974. Survey of genetic differentiation in a coastal zone invertebrate: the ectoproct *Schizoporella errata*. Biol. Bull., 146: 78—87.

Schopf, T.J.M., 1976. Environmental versus genetic causes of morphologic variability in bryozoan colonies from the deep sea. Paleobiology, 2: 156—165.

Schopf, T.J.M., 1977a. Population genetics of ectoprocts. In: R.A. Wollacott and R.S. Zimmer (Editors), Biology of Bryozoa. Academic Press, London, in press.

Schopf, T.J.M., 1977b. The role of biogeographic provinces in regulating marine faunal diversity thru geologic time. In: J. Gray and A.J. Boucot (Editors), Historical Biogeography: Plate Tectonics and the Changing Environment. 37th Ann. Biol. Colloq. Ore. State Univ., in press.

Schopf, T.J.M. and Dutton, A.R., 1976. Parallel clines in morphologic and genetic differentiation in a coastal zone marine invertebrate: The ectoproct *Schizoporella errata*. In preparation.

Schopf, T.J.M., Raup, D.M., Gould, S.J. and Simberloff, D.S., 1975. Genomic versus morphologic rates of evolution: Influence of morphologic complexity. Paleobiology, 1: 63—70.

Sepkoski, J.J., 1975. Stratigraphic biases in the analysis of taxonomic survivorship. Paleobiology 1: 343—355.

Silén, L., 1938. Zur Kenntnis des Polymorphismus der Bryozoen. Die Avicularien der Cheilostomata Anasca. Zool. Bidr. Upps., 17: 149—366.

Silén, L., 1942a. On spiral growth of the zoaria of certain Bryozoa. Ark. Zool., 34A: 1—22.

Silén, L., 1942b. Origin and development of the cheilo-ctenostomatous stem of Bryozoa. Zool. Bid. Upps., 22: 1—59.

Silén, L., 1944. The anatomy of *Labiostomella gisleni* Silen (Bryozoa, Protocheilostomata). K. Sven. Vetenskapsakad. Handl., 21: 111 pp.

Silén, L. and Harmelin, J.-G., 1974. Observations on living Diastoporidae (Bryozoa, Cyclostomata), with special regard to polymorphism. Acta Zool., 55: 81—96.

Simpson, G.G., 1953. The Major Features of Evolution. Columbia University Press, New York, N.Y., 434 pp.

Soule, D.F. and Soule, J.D., 1973. Morphology and speciation of Hawaiian and Eastern Pacific Smittinidae (Bryozoa, Ectoprocta). Bull. Am. Mus. Nat. Hist., 152: 365—440.

Soule, J.D. and Soule, D.F., 1969. Systematics and biogeography of burrowing bryozoans. Am. Zool., 9: 791—802.

Spjeldnaes, N., 1963. Climatically induced faunal migrations: Examples from the littoral fauna of the Late Pleistocene of Norway. In: A.E.M. Nairn (Editor), Problems in Palaeoclimatology. Interscience, London, pp. 353—357.

Stach, L.W., 1938a. Colony-formation in *Smittina papillifera* (MacGillivray, 1869) (Bryozoa). Proc. Zool. Soc. Lond., Ser. B, 108: 401—415.

Stach, L.W., 1938b. The application of the Bryozoa in Cainozoic stratigraphy. Rep. 23rd Meet. Aust. N.Z. Assoc. Adv. Sci., pp. 80—83.

Stanley, S.M., 1973. An explanation for Cope's Rule. Evolution, 27: 1—26.

Stebbing, A.R.D., 1971. Growth of *Flustra foliacea* (bryozoan). Mar. Biol., 9: 267—272.

Stebbing, A.R.D., 1973a. Competition for space between the epiphytes of *Fucus serratus* L. J. Mar. Biol. Assoc. U.K., 53: 247—261.

Stebbing, A.R.D., 1973b. Observations on colony overgrowth and spatial competition. In: G.P. Larwood (Editor), Living and Fossil Bryozoa. Academic Press, London, pp. 173—183.

Stratton, J.F., 1975. Ovicells in *Fenestella* from the Speed Member, North Vernon Limestone (Eifelian, Middle Devonian) in southern Indiana, U.S.A. In: Bryozoa 1974. Doc. Lab. Géol. Fac. Sci. Lyon, H.S., 3(1): 169—177.

Tavener-Smith, R., 1966. Ovicells in fenestrate cryptosomes of Visean age. J. Paleontol., 40: 190—198.

Tavener-Smith, R., 1969. Skeletal structure and growth in the Fenestellidae (Bryozoa). Palaeontology, 12: 281—309.

Tavener-Smith, R., 1975. The phylogenetic affinities of fenestelloid bryozoans. Palaeontology, 18: 1—17.

Tavener-Smith, R. and Williams, A., 1972. The secretion and structure of the skeleton of living and fossil Bryozoa. Philos. Trans. R. Soc. Lond., Ser. B, 264: 97—159.

Terborgh, J., 1973. On the notion of favorableness in plant ecology. Am. Nat., 107: 481—501.

Thorpe, J.P., Shelton, G.A.B. and Laverack, M.S., 1975a. Colonial nervous control of lophophore retraction in cheilostome Bryozoa. Science, 189: 60—61.

Thorpe, J.P., Shelton, G.A.B. and Laverack, M.S., 1975b. Electrophysiology and co-ordinated behavioural responses in the colonial bryozoan *Membranipora membranacea* (L). J. Exper. Biol., 62: 389—404.

Toriumi, M., 1956. Taxonomical study on fresh-water Bryozoa XVII. General consideration: Interspecific relation of described species and phylogenic consideration. Sci. Rep. Tôhoku Univ., Ser. 4, Biol., 22: 57—88.

Utgaard, J., 1973. Mode of colony growth, autozooids and polymorphism in the bryozoan Order Cystoporata. In: R.S. Boardman, A.H. Cheetham and W.A. Oliver (Editors), Animal Colonies. Dowden, Hutchinson and Ross, Stroudsburg, Penn., pp. 317—360.

Van Valen, L., 1973. A new evolutionary law. Evol. Theory, 1: 1—30.

Voigt, E., 1939. Über die Dornenspezialisation bei cheilostomen Bryozoen und die Nichtumkehrbarkeit der Entwicklung. Palaeontol. Z., 21: 87—107.

Voigt, E., 1959. Sur les différents stades de l'astogénèse de certains Bryozoaires cheilostomes. Bull. Soc. Géol. Fr., 7e Sér. 1: 697—704.

Voigt, E., 1968. Homoeomorphy in cyclostomatous Bryozoa as demonstrated in *Spiropora*. Atti Soc. Ital. Sci. Nat. Mus. Civ. Stor. Nat. Milano, 108: 43—53.

Voigt, E., 1972. Les méthodes d'utilisation stratigraphiques des Bryozoaires du Crétacé supérieur. Mem. Bur. Rech. Géol. Min., Fr., No. 77: 45—53.

Voigt, E. 1975a. Heteromorphy in Cretaceous Bryozoa. In: Bryozoa 1974. Doc. Lab. Géol. Fac. Sci. Lyon, H.S., 3(1): 77—96.

Voigt, E., 1975b. Ist *Domopora stellata* (Goldfuss, 1826) (Bryoz. Cyclost.) = *Coronopora truncata* (Fleming, 1828) ein lebendes Fossil aus der Oberkreide? Zool. Anz., 195: 186—200.

Voigt, E. and Flor, F.D., 1970. Homöomorphien bei fossilen cyclostomen Bryozoen, dargestellt am Beispiel der Gattung *Spiropora* Lamouroux 1821. Mitt. Geol.-Paläontol. Inst. Univ. Hamburg, 39: 7—96.

Waters, A.W., 1892. Observations on the gland-like bodies in the Bryozoa. J. Linn. Soc., 24: 272—278.

Waters, A.W., 1921. Observations upon the relationships of the (Bryozoa) Selenariadae, Conescharellinidae, etc., fossil and recent. J. Linn. Soc. Lond., 34: 399—426.

Waters, A.W., 1925a. Ancestrulae of cheilostomatous Bryozoa, Part II. Ann. Mag. Nat. Hist., Ser. 9, 15: 341—352.

Waters, A.W., 1925b. Ancestrulae of cheilostomatous Bryozoa. Part III. *Schizoporella*, etc. Ann. Mag. Nat. Hist., Ser. 9, 16: 529—545.

Waters, A.W., 1926. Ancestrulae and frontal of cheilostomatous Bryozoa. Part IV. Ann. Mag. Nat. Hist., Ser. 9, 17: 425—439.

Weisbord, N.E., 1967. Some Late Cenozoic Bryozoa from Cabo Blanco, Venezuela. Bull. Am. Paleontol., 53(237): 1—247.

Wilson, A.C., 1976. Gene regulation in evolution. In: F.J. Ayala (Editor), Molecular Evolution. Sinauer Associates, Sunderland, Mass., pp. 225—234.

Wyer, D.W. and King, P.E., 1973. Relationships between some British littoral and sublittoral bryozoans and pycnogonids. In: G.P. Larwood (Editor), Living and Fossil Bryozoa. Academic Press, London, pp. 199—207.

Yoshioka, P.M., 1973. The population dynamics and ecology of the encrusting ectoproct *Membranipora serrilamella*. Contrib. Scripps Inst. Oceanogr., 43: 1814—1815.

Zirpolo, G., 1921. Ricerche sulla rigenerazione del *Zoobotryon pellucidum* Ehrbg. Boll. Soc. Nat. Napoli, 33: 98—101. (Ser. 2, Vol. 13).

Zirpolo, G., 1923. Nuovo caso di simbiosi fra Zooxantelle e *Bicellaria ciliata* briozoo del Golfo di Napoli. Rendiconti 14th assemblea ordinaria e del Convegno dell'Unione Zoologica Italiana in Genova, 8—11 Oct., 1923.

CHAPTER 7

TRENDS, RATES, AND PATTERNS OF EVOLUTION IN THE BIVALVIA

STEVEN M. STANLEY

Introduction

This chapter represents an attempt to gain insight into large-scale evolution through the study of a class of animals that is well represented both in the modern world and in the fossil record. Little reference will be made to the experience that I have valued most in preparing it: hundreds of hours spent digging, gathering, and observing clams and their relatives to learn how they live and how their shell form reflects their life habits (Stanley, 1970). The analysis of evolution presented here rests on this kind of natural history and also on a new methodology by which certain techniques of demography are adapted to the study of fossil taxa. In fact, this methodology has proved as useful in uncovering patterns of evolution as in interpreting them. Basically, it provides for the estimation of rates of multiplication and termination of lineages. I have come to believe that these rates are the primary determinants of major trends and patterns of evolution. It has been possible to treat phylogeny only briefly. For this reason I have emphasized the Cambro-Ordovician emergence of the class, which sets the evolutionary stage for later events and which has been the subject of important new research.

Bivalve Phylogeny

Students of the Bivalvia face a familiar palaeontological problem: few transitions between subclasses or orders are clearly documented by the known fossil record. As might be expected, this problem takes us back to the Early Palaeozoic, when the initial divergence of major subtaxa took place. It is compounded by new evidence that at least one genus of bivalves seems to have existed in the Early Cambrian. The ranges of other unequivocal genera are restricted to post-Cambrian strata.

Before discussing the nature of early divergence, it is apt to note that a variety of morphological characters represent key variables in the definition of higher bivalve taxa. Foremost among these are hinge dentition (or lack thereof), pattern of shell ornamentation, morphology of the ligament, shell microstructure, gross shell form, degree of mantle fusion, presence or absence of an adult byssus (array of threads for attachment), and anatomy of the ctenidia and stomach. The last four features are non-skeletal, but the presence in life of siphons formed by mantle fusion is usually betrayed by a posterior pallial sinus, and the presence in life of an adult byssus can usually be inferred from various aspects of shell morphology (Stanley, 1972). Byssally attached taxa living

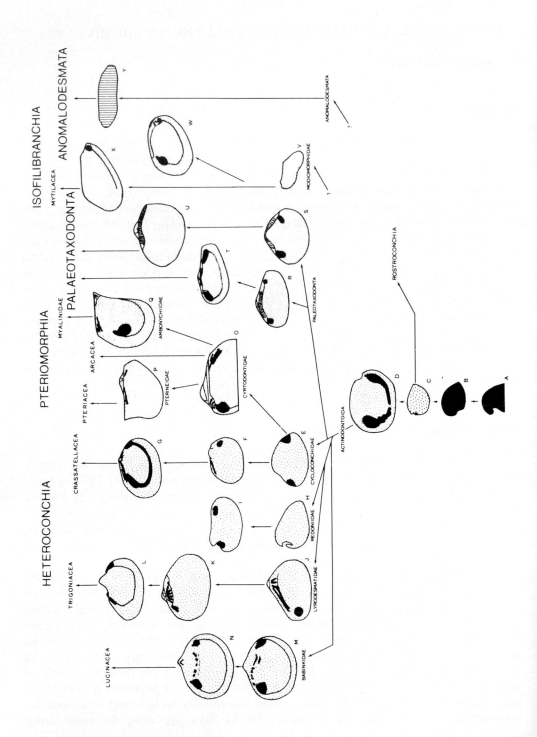

partly buried in soft substrata will be termed *endobyssate*, and those living on the surface of the substrata will be termed *epibyssate* (Stanley, 1972).

While many questions remain, recent studies have shed much light on the early adaptive divergence of the Bivalvia (Newell, 1965, 1969; Stanley, 1968, 1972; Pojeta, 1971; Pojeta et al., 1972; Pojeta et al., 1973; Pojeta and Runnegar, 1974; Runnegar and Pojeta, 1974; Pojeta, 1975). Modifying the scheme of Newell (1965; 1969) somewhat, Pojeta (1971, 1975) has recognized five subclasses of the Bivalvia, all of which originated in Cambro-Ordovician times. These are displayed in Fig. 1 and I shall introduce them briefly here.

Heteroconchia

Among the most distinctive features of the Heteroconchia are crossed-lamellar shell structure and heterodont dentition. This subclass includes the majority of post-Palaeozoic burrowing clams. A few atypical representatives are worthy of mention. The Lucinacea are aberrant heteroconchs that use an elongated foot to produce an inhalent mucus tube (Allen, 1958). They seem never to have given rise to another superfamily. Another specialized group, the Trigoniacea, expanded dramatically in the Mesozoic through the evolution of somewhat unusual adaptations for efficient burrowing (Stanley, in preparation). Because of their unusual morphology, they too became an evolutionary cul-de-sac, after relatively unspecialized early representatives gave rise to the freshwater Unionacea. Newell and Boyd (1975) report the apparent origin of the latter in the Triassic. The rudists (Hippuritacea), which originated in the Jurassic, represent another divergent group of heteroconchs. Their bizarre form, which evolved with the assumption of a habit of attachment by cementation, led to formation of extensive bioherms (Kauffman and Sohl, 1974). Near the end of the Cretaceous, the rudists declined and became extinct without issue.

Fig. 1. Phylogenetic scheme postulated by Pojeta (1975) for Cambro-Ordovician Bivalvia and their immediate precursors. All figures semi-diagrammatic. A. *Latouchella* (Tommotian, Lower Cambrian); B, *Anabarella* (Tommotian, Lower Cambrian); C, *Heraultia* (Georgian, Lower Cambrian); D, *Fordilla troyensis* (upper Lower Cambrian); E, *Actinodonta secunda* (upper Arenigian); F, *Actinodonta naranjoana* (Llandeilian); G, *Cycloconcha* (Cincinnatian); H, *Redonia prisca* (lower Arenigian); I, *Redonia bohemica* (Llanvirnian); J, *Lyrodesma armoricana* (upper Arenigian); K, *Lyrodesma conradi* (Edenian); L, *Lyrodesma poststriatum* (Maysvillian); M, *Babinka oelandensis* (upper Arenigian); N, *Babinka prima* (Llanvirnian); O, *Cyrtodonta* cf. *C. huronensis* (Wildernessian); P, *Palaeopteria*, (Barneveldian); Q, *Ambonychia radiata* (Cincinnatian); R, *Ctenodonta* (upper Canadian); S, *Deceptrix?* *oehlerti* (upper Arenigian); T, *Ctenodonta nasuta* (Mohawkian); U, *Deceptrix* aff. *D. hartsvillensis* (Barneveldian); V, *Modiolopsis davyi* (upper Arenigian); W, *Modiolodon oviformis* (Barneveldian); X, *Modiolopsis modiolaris* (Maysvillian); Y, *Rhytimya mickelboroughi* (Maysvillian). Additional information provided by Pojeta (1975).

Pteriomorphia

The Pteriomorphia are mostly anisomyarian and monomyarian bivalves (the arcoids being the chief exceptions). They include most modern species of epifaunal bivalves. The earliest representatives were small, circular burrowing cyrtodontids (Fig. 1, O) (Pojeta, 1971; Stanley, 1972). Pojeta reasonably concluded that these may have arisen from cycloconchid heteroconchs (Fig. 1, E and O). At least, the two groups must have had similar ancestors. All free-burrowing pteriomorphs had disappeared by mid-Palaeozoic times, but their semi-infaunal and epifaunal descendents have been legion. Most important from our Recent vantage point are the Anomiacea (jingle shells), Limacea (file shells) and especially the Pteriacea (wing shells, pen shells, and pearl "oysters"), Pectinacea (scallops), Ostreacea (oysters), and Arcacea (arc shells). Although the general adaptive radiation of pteriomorphs has been into epifaunal adaptive zones, some post-Palaeozoic arcaceans have reverted to the primitive burrowing habit (Stanley, 1972).

Palaeotaxodonta

The Palaeotaxodonta include forms traditionally called "protobranchs", with reference to their gill morphology. Most have taxodont dentition. The major order Nuculoida contains small deposit-feeding taxa that, since early in the Palaeozoic, have tended to occupy muddy substrata rich in organic matter.

Isofilibranchia

The Isofilibranchia, a relatively small group of elongate, anisomyarian taxa, include modern marine mussels of the superfamily Mytilacea and their apparent ancestors, the Palaeozoic Modiomorphacea. All known isofilibranchs have morphologies reflecting byssal attachment.

Anomalodesmata

The Anomalodesmata are phylogenetically coherent but difficult to diagnose. Nacro-prismatic shell structure is a fundamental feature of the group, and nearly all representatives have had an infaunal or semi-faunal habit. Runnegar (1974) provided a useful account of the composition and evolution of this subclass, which radiated into a broad adaptive zone in the Late Palaeozoic but is less conspicuous today than the Heteroconchia.

The Anomalodesmata and Isofilibranchia may be closely related (Carter, 1976). Many Palaeozoic anomalodesmatids have the endobyssate shape of *Modiolus*-like mytilids (Fig. 1, X) (Stanley, 1972). Also, many members of these groups share with a few other taxa a distinctive micro-ornamentation consisting of small, periostracal spikes or granules of calcium carbonate (Carter, 1976). Among the other similarly endowed taxa are the Gastrochaenacea, which bore into hard substrata, and the Myidae, which contain the living steamer clam. If this assemblage indeed becomes recognized as a coherent phylogenetic group, Carter suggests calling it the Granulata. Some of the potential members have previously been placed in the Heteroconchia, and it may yet turn out that some should remain there. It seems likely, however, that most of the earlier assignments were incorrect, reflecting over-reliance on dentition and shell structure as key features in classification.

Methods of Phyletic Reconstruction

The suggestion that the Granulata represent a natural unit does not come just from the study of micro-ornamentation, which *a priori* deserves no more weight than does the study of the micro-architecture of the shell. It comes also from the study of stratigraphic relationships (Carter, 1976) and similarities in gross shell form and mode of life (Stanley, 1972; Runnegar, 1974; Carter, 1976). All of the groups in question contain a large percentage of genera with

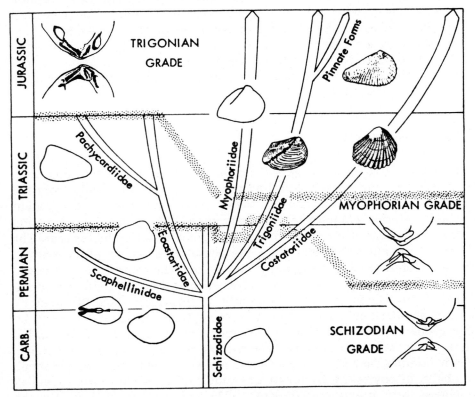

Fig. 2. Inferred phylogeny of the Trigoniacea reflecting the view that shell ornamentation is a more conservative feature than detailed hinge morphology. Grades of hinge dentition were attained independently by more than one family. Derivation of the Pachycardiidae from the Eoastartidae is conjectural. (From Newell and Boyd, 1975.)

elongate shells. These are variously adapted to endobyssate, deep-burrowing and boring habits.

The somewhat arbitrary use of key characters that has previously typified taxonomic designations and phylogenetic reconstructions within the Bivalvia simply has not worked. As one example, so-called heterodont dentition has apparently evolved polyphyletically. Similarly, crossed-lamellar shell structure has evolved polyphyletically from the primitive prismato-nacreous condition (Taylor, 1973). In undertaking large-scale phylogenetic reconstruction within the Bivalvia, the most fruitful procedure now seems to be to use a variety of characters, letting the emphasis of certain kinds of characters be dictated by the evidence of morphological intergradation and by the constraints imposed by the stratigraphic occurrence. Newell and Boyd (1975), for example, found that hinge morphology, long used as a key feature in the classification of the Trigoniacea, is less conservative in evolution than shell ornamentation. The two advanced grades of dentition previously used to define families turn out to be grades of evolution that were attained polyphyletically (Fig. 2).

Recent progress in reconstructing phylogenies has come not through focus on the characters of living taxa, but through heavy emphasis on fossil morphology and occurrençe, with subordinate use of Recent data. The historical neglect of Early Palaeozoic bivalves has ended in the past five years, as demonstrated by the emergence of the classification and partial phylogeny depicted in Fig. 1. This scheme is neither totally new nor without uncertainties, but represents a major advance that has been achieved largely through evaluation of Cambro-Ordovician faunas.

The Cambro-Ordovician Divergence

Pojeta (1971, 1975) has recognized *Fordilla* (Fig. 1, D) as the only valid Cambrian genus of currently known bivalves, yet it is reported only from the Lower Cambrian. *Fordilla* is assigned to the Bivalvia on the basis of its apparent anterior and posterior adductor muscle scars, moniliform pallial line, and anterior pedal muscle scar at the dorsal margin of the anterior adductor scar. The muscle scars extending ventrally from the adductor scars are problematical. *Fordilla* lacks lateral hinge teeth, and it may be that the ventral muscles served in the place of such teeth to keep the valves aligned. Pojeta (1975) has recognized only 5 bivalve species from the earliest Ordovician stage (Tremadoc), which lasted about 10 m.y. The succeeding Arenig Stage, which lasted about 20 m.y., has yielded several genera of the Heteroconchia belonging to four families (Fig. 1, E, H, J, M). It is possible that the various subclasses of the Bivalvia arose from *Fordilla* via early heteroconchs, some of which also have had small, circular shells. In any event, by about the middle of the Middle Ordovician, all five subclasses were in existence (Pojeta, 1971). This is about 35 m.y. after the mid-Tremadoc, which gives us an estimate of the minimum time for divergence of the subclasses.

Some might question whether *Fordilla* is, in fact, a true bivalve or even represents a group ancestral to the Bivalvia. I find it difficult to believe that the many bivalve-like features of this genus represent convergence with the Bivalvia. Furthermore, potential steps in the Cambrian origin of the class from univalved Mollusca are coming to light. Pojeta et al. (1972) have recognized a new class of primitive pseudobivalved molluscs, the Rostroconchia, which originated in the Cambrian. One of the prominent features of this class is a univalved larval shell, which contrasts to the double-valved larval shell of the Bivalvia. Noting similarities of morphology, Runnegar and Pojeta (1974) suggest derivation of *Fordilla* (hence the Bivalvia) from the oldest known (Lower Cambrian) rostroconch genus *Heraultia* (Fig. 3).

From anatomical studies, zoologists have long assumed that the Bivalvia were primitively infaunal (e.g., Yonge, 1953). Basic features that seem to have originated in association with burrowing habits include a laterally compressed shape, a sack-like foot, and co-adapted features like the ligament and adductor muscles that today form a unique burrowing mechanism. The fossil record concurs: taxa with shapes adapted for infaunal life predominate in faunas of the Early and Middle Ordovician (Pojeta, 1971). The five known species of the Early Ordovician that Pojeta (1975) considers possibly to be valid Bivalvia are all burrowing palaeotaxodonts. This is interesting, in that members of this protobranchiate group have traditionally been viewed as the most primitive Bivalvia (Yonge, 1939). On the other hand, Pojeta (1971) regards them as

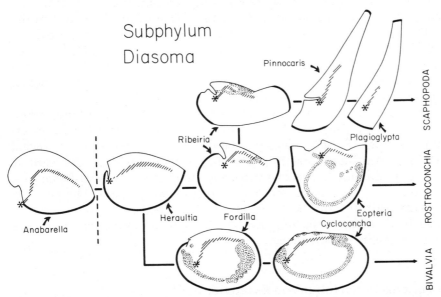

Fig. 3. Postulated early phylogeny of the subphylum Diasoma. The three classes labelled at the right were all apparently infaunal or semi-infaunal. *Heraultia*, the first rostroconch, is thought to have been derived from a primitive exogastrically coiled mollusc like *Anabarella*. Thick lines show extent of shell apertures. Stippled areas represent muscle insertions. Hashed areas show probable position of gut with mouth indicated by an asterisk. (From Runnegar and Pojeta, 1974.)

having arisen secondarily from early heteroconch burrowers. In any event, several different types of valid heteroconchs are known from the Arenig, as already discussed, and so are a few isofilibranchs having endobyssate shapes. Thus, the idea of Yonge (1962) that the bivalve byssus arose as a post-larval organ of attachment and was later passed on to adult stages by neoteny is supported by the fossil record. Virtually all living taxa of burrowing bivalves employ a post-larval byssus. Retention of the byssus into the adult stage was an important aspect of the initial adaptive radiation of the class, providing a pathway to epifaunal life. Though I know of no epifaunal bivalve taxon from the Lower Ordovician, epifaunal Pteriomorphia are quite conspicuous in Upper Ordovician faunas.

Another line of evidence suggesting infaunal ancestry for the Bivalvia is that the prosogyre shape and blunt anterior that aid typical modern bivalves in burrowing (Stanley, 1975c) are present in *Fordilla*. The laterally compressed shape and sack-like foot of bivalves are not readily employed for epifaunal life except at small body sizes (Stanley, 1972). Despite all of these arguments, the suggestion that the very earliest bivalves were epifaunal (Valentine and Gertman, 1972) is not ruled out. Two lines of evidence suggest that it is unlikely, however. One is that it would be a remarkable occurrence of preadaptation for the morphology of *Fordilla*, which includes features known to be adaptations for burrowing in younger clams, to have arisen for some unknown non-burrowing function. Another is that the phylogenetic scheme emerging from the study of Cambrian molluscs indicates a close relationship among the Rostroconchia, Bivalvia, and Scaphopoda. Runnegar and Pojeta (1974) unite these taxa in the subphylum Diasoma (Fig. 3). Conocardiid rostroconchs, earliest Ordovician bivalves, and all known scaphopods display features reflecting semi-infaunal or infaunal habits. It therefore appears likely that these habits originated with the emergence of the subphylum. Only the ancestral class, the Rostroconchia, could then have had an epifaunal ancestry.

Why the radiation of the Bivalvia was delayed until the Early Ordovician, assuming that the Early Cambrian *Fordilla* was indeed a bivalve, remains uncertain. It may be that *Fordilla* lacked adaptations for efficient burrowing (Stanley, 1975c). Another possibility (Stanley, 1975a) is that the earliest bivalves were only marginally successful because they lacked a post-larval byssus, an organ that is unique to the Bivalvia among living molluscs. The possible absence of hinge teeth in *Fordilla* could also have represented something of a barrier. Hinge teeth seem to be required in clams to prevent shearing of the valves in the movements of burrowing.

Trends and Patterns of Evolution and Extinction

Nearly all prominent adaptations that have appeared in Bivalvia since the Early Ordovician have arisen polyphyletically. The multiple origins of many morphological features are clear expressions of the inherent limitations of the bivalve body plan.

Mantle fusion

In the evolution of burrowing clams, fusion of segments of the mantle margins has served two important functions. First, it has produced siphons (Yonge, 1962), and second, it has provided a tight seal for the mantle cavity during the hydraulic process of burrowing, thus improving the efficiency of burrowing (Trueman, 1966). To a considerable degree, mantle fusion has accounted for the post-Palaeozoic radiation of burrowing bivalves (Stanley, 1968). The number of families of free-burrowing, non-siphonate, suspension-feeding clams is about the same today as it was in the Devonian (Fig. 4). In contrast, the number of siphonate families has risen from perhaps zero to about forty. Estimates of diversity by Boss (1971) indicate that 85—90% of living species of suspension-feeding marine clams are siphonate.

Deposit-feeding, nuculanid protobranchs have apparently possessed siphons since the Devonian, but their siphons are unique, primitive structures that have not been passed on to suspension-feeding clams. The aberrant Ordovician genus *Lyrodesma*, which is placed in a family by itself, exhibits a very shallow pallial sinus, but apparently died out without issue. The Permian trigoniacean genus *Scaphellina*, also placed in a family by itself (Newell and Ciriacks, 1962), displays a distinct pallial sinus and siphonal gape, but it too seems to have become extinct without leaving descendents. Both of the above genera had filibranch gills. It may be that the eulamellibranch gill has been a prerequisite

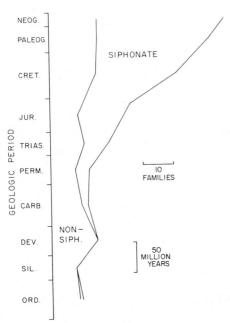

Fig. 4. Diversity of families of non-siphonate and siphonate free-burrowing clams. Minute symbiotic leptonacean families are excluded, as are the Lucinidae, which have just one siphon, and the Nuculacea. (Data from Moore, 1969.)

for the successful large-scale deployment of siphons (Stanley, 1968).

The first real pulse of diversification of siphonate forms came with the emergence of advanced anomalodesmatids in the Carboniferous (Stanley, 1968, 1972; Runnegar, 1974). The second, larger pulse, was the Mesozoic—Cenozoic radiation of the Heterodonta, which was not in full swing until the Cretaceous. The Lucinidae, which employ a single (exhalent) posterior siphon, display an unusual evolutionary pattern. They arose in the Silurian, but remained obscure until late in the Mesozoic, when they began a radiation that, according to Boss (1971), has yielded about 200 living species. Many living species are confined to seagrass environments (Allen, 1958; Stanley, 1970; Jackson, 1973), and it would seem no accident that the modern radiation of the lucinids coincided with the evolutionary emergence of rooted seagrass (Brasier, 1975). Jackson (1973) has described the ability of lucinids to feed upon coarse particulate seagrass debris, but I would place greater evolutionary emphasis on the dense root and rhizome system of the grass in providing lucinids with physical stability. The fragile nature of the unique inhalent mucus tube constructed by the lucinoid foot may make life on barren, shallow sea floors unusually difficult.

Suspension-feeding clams of the Early and Middle Palaeozoic were clearly sluggish creatures that could not successfully have colonized habits character-

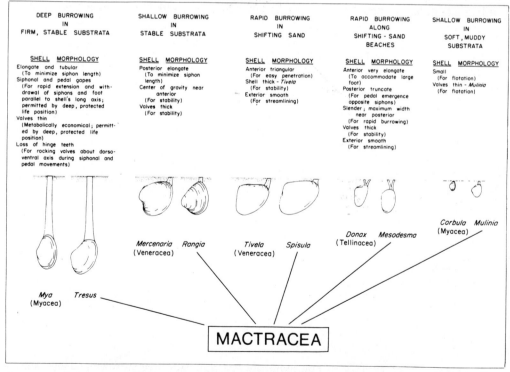

Fig. 5. Adaptive divergence of the Mactracea and convergence in form and habit with unrelated burrowing taxa. (From Stanley, 1972.)

ized by shifting sands (tidal bars and sandy beaches, for example). They must also have had limited success in intertidal settings, where deep-burrowing tends to reduce environmental stress. The radiation of siphonate taxa has clearly expanded the adaptive zone of the Bivalvia. It will also be argued subsequently that siphonate forms have probably succeeded preferentially in habitats previously well colonized by non-siphonate infaunal bivalves, through greater resistance to modern predators.

The phylogeny of siphon-feeders is replete with examples of adaptive convergence. Some of these are illustrated in Fig. 5.

Byssal attachment

Convergence and parallelism have also been prevalent in the evolution of epifaunal habits. As depicted in Fig. 6, certain members of the Arcoida (Pteriomorphia) evolved endobyssate habits by retention of the post-larval byssus of free-living ancestors. From some of these, epifaunal forms arose. Neotenous retention of the byssus and large-scale trends towards epifaunal existence are found in other taxa as well. The independent appearance of epibyssate species in several families of the Pterioida (Stanley, 1972) was the major source of increase in relative epifaunal diversity during the Palaeozoic (Fig. 7). In many instances, bivalve species retaining endobyssate habits have persisted along with their epifaunal descendents.

Neotenous retention of the byssus has sometimes occurred suddenly by the evolution of reproductive maturation in the post-larval stage. In other instances it has occurred by gradual or progressive persistence of byssal attachment into the adult stage. The minute Erycinacea (Leptonacea) have almost certainly arisen by the first process, probably polyphyletically (Stanley, 1972). Ockelman (1964) reports that one living species that has been assigned to this group is actually a neotenous venerid!

A particularly interesting aspect of arcoid evolution (Fig. 6) has been the polyphyletic reversion from epifaunal to burrowing habits. Comparable evolutionary reversals are seen in other taxa as well. Paradoxically, here too neoteny seems to have been the evolutionary mechanism (Stanley, 1972). The post-larval stage of a bivalve normally retains both an active foot and a byssus. Regardless of which dominates in the adult, the other remains "stored" in the juvenile for potential transfer to the adult stage. In fact, groups like the Carditidae display a spectrum of adult modes of life, ranging from free-burrowing to endobyssate attachment combined with a capacity for burrowing, to epibyssate attachment with no capacity for burrowing. Evolution can shift habits in either direction within this spectrum, though if the burrowing function of the foot is lost even in the juvenile stage, as in the Mytilidae, a taxon may be confined to the end of the spectrum characterized by byssate adults. Adults of such taxa can still penetrate soft substrata by pulling the shell towards objects of attachment and fixing byssal threads at successively deeper levels.

There is, in fact, a strong random element in the evolution of the Bivalvia. This random element is seen particularly on small taxonomic scales, as in the

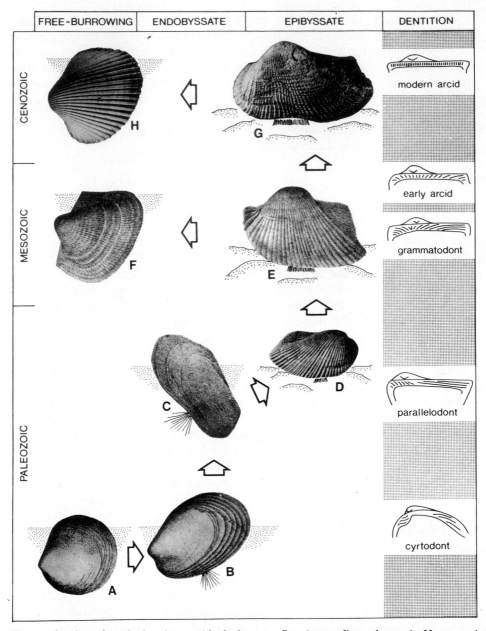

Fig. 6. Grades of evolution in arcoid phylogeny. Specimens figured are: A, *Vanuxemia subrotunda* Ulrich, Upper Ordovician, ×1.1 (from Ulrich, 1897); B, *Cyrtodonta subovata* Ulrich, Upper Ordovician, ×6.3 (from Ulrich, 1897); C, *Parallelodon (Cosmetodon) marshallensis* (Winchell), Mississippian, ×1.1 (from Driscoll, 1965); D, *Parallelodon striatus* (Schlotheim), Permian, ×1.8 (from Logan, 1967); E, *Grammatodon carinatus* (Sowerby), Lower Cretaceous, ×1.1 (from Woods, 1899); F, *Cucullaea glabra* Parkinson, Lower Cretaceous, ×5.4 (from Woods, 1899); G, *Arca imbricata* Bruguiere, Recent, ×1.1 (from Stanley, 1970); H, *Anadara ovalis* (Bruguiere), Recent, ×1.1 (from Stanley, 1970). Grades of evolution in dentition from Newell (1954). (From Stanley, 1970.)

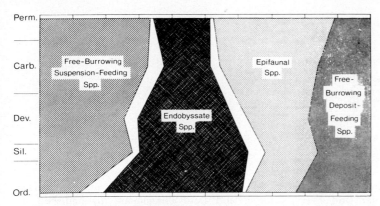

Fig. 7. Changes in the life-habit spectrum of the Bivalvia during the Palaeozoic. The plot is of the mean percentage of species in each life-habit group for faunas of each period randomly chosen from the literature. Blank species indicate mean percentages of species whose life habits could not be assigned to either neighbouring category with a high degree of certainty. For details see Stanley (1972).

appearance of new species and genera. Local conditions and chance factors seem to govern the direction in which evolution proceeds at a given place and time, as from the endobyssate condition to the epibyssate condition, or vice-versa. Superimposed on this fine-scale randomness are long-term directional features. Among the most salient of these have been the general decline of endobyssate taxa (Fig. 7) and the previously discussed ascendancy of siphonate infauna (Fig. 4). These characteristics of bivalve evolution will be analyzed in the final section of this chapter.

Mass extinction

On the whole, bivalves and gastropods have been unusually resistant to mass extinction. This point is illustrated especially well by the relatively weak effect on these taxa of the most severe Phanerozoic marine faunal crisis, that of the Permo-Triassic (Nakazawa and Runnegar, 1973). Observed diversities of genera and families for intervals of the Permian and Triassic are illustrated in Fig. 8A. Fig. 8B summarizes estimates by Nakazawa and Runnegar of actual diversity, based on knowledge that certain taxa had to exist at various times because they are known from earlier and later intervals or because at least one taxon is required to provide a transition between ancestral and descendent taxa separated by stratigraphic gaps. Note that there is very little impact of the faunal crisis at the family or superfamily level. Comparison of the two plots for genera reveals that the Early Triassic fossil record is abnormally poor. The upward adjustment here to fifty genera may perhaps be considered minimal, but the crisis clearly led to a modest decline in generic diversity. Nakazawa and Runnegar show that epifaunal pteriomorphs may have suffered disproportionately. Burrowing anomalodesmatids also declined notably, however. The seemingly weaker decline of other groups may be largely a matter of preservational bias.

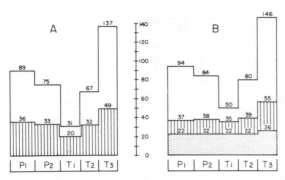

Fig. 8. Permo-Triassic diversities of genera, families, and superfamilies of marine bivalves. Early and Middle Permian and Early, Middle, and Late Triassic are distinguished. A. Observed occurrences of genera (solid line) and families (ruled area). B. Estimated numbers of genera (solid line), families (ruled), and superfamilies (stippled). (From Nakazawa and Runnegar, 1973.)

The effect of the Permo-Triassic crisis is revealed in other ways as well. Nakazawa and Runnegar note that there was a sharp decline in the number of new genera appearing in the Late Permian. It seems possible that a reduced rate of speciation was more important in reducing diversity than was an increase in rate of extinction. Zoogeographic effects are also apparent. Late Palaeozoic genera and species display a marked provinciality. The earliest Triassic faunas, however, are strikingly cosmopolitan. *Claraia*, which appears suddenly at the base of the Triassic throughout the world, and *Bakevellia*, *Promyalina*, and *Unionites* are particularly conspicuous. Nakazawa and Runnegar also show, however, the impact of the Permo-Triassic crisis was far heavier on groups other than bivalves and gastropods. Latest Late Permian faunas of Japan, and faunas of the same age in North America (Nicol, 1965; Ciriacks, 1963), differ markedly from early Late Permian faunas in containing few forms not belonging to these two molluscan classes. In contrast to what happened within groups like the articulate brachiopods, fusulinids, corals, ammonoids, and crinoids, enough bivalve species survived the crisis so that there was almost no decline in suprageneric diversity (Fig. 8). Although we still lack an explanation, eurytopy may have played a role in the preferential survival of bivalves and gastropods (Beurlen, 1956; Ciriacks, 1963; Fischer, 1964; Nakazawa and Runnegar, 1973). Another trait contributing to the stability of these taxa, their characteristically low rates of evolutionary turnover, will be analyzed below.

Rates of Evolution and Extinction

It is my general view that trends and patterns of large-scale evolution reflect differential rates of speciation and extinction. It will be necessary to investigate the measurement of such rates before analyzing the factors that govern them. Although Mayr (1963) has suggested that the species is the basic unit of macroevolution, study of the fossil record at the species level has generally been avoided because of the many apparent problems. I have proposed certain methods for avoiding many of these problems in the study of Late Cenozoic taxa (Stanley, 1975b, 1976) and will extend these methods in the following paragraphs and apply them to the Bivalvia.

The classic comparison by Simpson (1944, 1953) of generic longevity within the Mammalia and Bivalvia has been updated by Van Valen (1973) without altering the conclusion: an average genus of bivalves lasts about ten times as long as an average genus of mammals. Because rate of extinction is roughly the reciprocal of mean duration, the inference has been that rate of extinction is much higher for mammals than for bivalves. A basic problem with this kind of comparison is that a genus of living bivalves contains at least ten species (Moore, 1969; Boss, 1971), whereas a genus of living mammals contains only about three (Walker, 1968). This means that we are not dealing with comparable entities, because a genus dies out only when all component species happen to have become extinct. The species is the basic unit of extinction, and ideally it is species longevity that we should attempt to study. I will pursue the comparison of bivalves and mammals here, as done in the past (Stanley, 1973, 1975b, 1976), because it seems especially illuminating. Calculation of rates of evolution and extinction for one group in isolation raises fewer significant questions.

Rates at the species level

There is a simple technique for estimating the mean duration for species within a taxon (Stanley, 1975b). The procedure is to determine the Cenozoic interval in which average fossil faunas contain 50% extant species. This will approximate half of the turnover time for species, or half of an average species duration, regardless of the frequency distribution of durations. This technique shows that an average mammal species lasts more than 1 m.y., while an average bivalve species lasts something like 7 m.y. (precise data for the Bivalvia are not yet available).

A more refined technique permits the construction of an entire histogram of species durations for a higher taxon. This is obtained indirectly from fossil data by way of a survivorship curve, which depicts the progressive extinction of a typical set of species arrayed so as to originate simultaneously. The curve is, in effect, a reverse cumulative curve of species durations. It has been shown (Stanley, 1976) that a survivorship curve can be produced simply by doubling the time scale of an accurate curve depicting the percentage of living species found in Cenozoic faunas of various ages. The technique is summarized in Fig. 9, where it is applied to the Plio-Pleistocene mammal faunas of Europe. The beginning of the Würm has been used as an end point instead of the Recent in order to avoid the effects of the famous Würm—Recent mass extinction. The histogram (Fig. 9E) derived from the survivorship curve displays a mean duration of about 1.2 m.y., a mode of about 0.7 m.y., and very few species lasting less than 0.3—0.4 m.y. The raw data, which are taken from a summary by Kurtén (1968), support the conclusion that there are very few short-lived species. For example, of almost 200 species considered to have existed in the

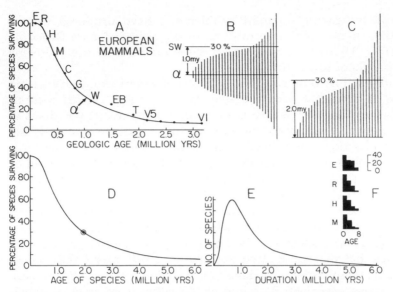

Fig. 9. Construction of a histogram for species durations of Late Cenozoic mammals. A is an extinction curve for the Plio-Pleistocene mammal species of Europe. Geological time extends backwards toward the right. On the left, the curve terminates at the beginning of the Würm. Each point represents' the percentage of all species occurring in a given stage that survived into the Würm. Points are placed at the midpoints of stages. An average stage contains more than 90 species. Percentages are derived from Kurtén (1968). Dates for stages are from Berggren and Van Couvering (1974). Points within the Villafranchian are equally spaced, in the absence of an established chronology. Stage abbreviations: E = Eem; R = Riss; H = Holstein; M = Mindel; C = Cromerian; G = Günz; W = Waalian; EB = Eburonian; T = Tegelen (Tiglian); $V5$ and $V1$ = last and first phases of the Villafranchian. At any time, some species will be newly formed and some will be near the ends of their stratigraphic ranges. It is assumed that an average species at any time is in mid-range (this assumption is justified below). Then B and C show how a survivorship curve for species (D) is formed by doubling the time-scale of A. B is a hypothetical set of average ranges (vertical bars) of species existing at time α, approximately 1 m.y. before the start of the Würm (SW). A survivorship curve depicts decline starting with a "cohort" of brand new species. Therefore to obtain a point for a survivorship curve from information provided by the α point we must realign the ranges of B so that all species originate simultaneously, as shown in C. Twice as long is then required for the decline to 30%. The point derived from the α point of A is therefore plotted at 2 m.y. in D, and, by extrapolation, the entire survivorship curve is plotted by doubling the time scale of A. A histogram of species durations (E) is derived directly from D. The vertical scale is arbitrary. The assumption that at any time an average species is in mid-range amounts to the assumption that rates of speciation and extinction are more-or-less constant during the interval of A, or that a stable age distribution is maintained. Major deviations from these conditions would be required for the technique to be invalidated, and as shown in F, age-frequency distributions for species entering the four final stages of A are generally similar. Ages in F are measured by number of stages of existence; species older than eight stages not plotted. The assumption that at any time an average species is in mid-range is also justified in part by the smoothness of the empirical curve (A). (From Stanley, 1976.)

Middle Pleistocene, only 3 are known from only one Pleistocene stage, and it is inconceivable that this small number could result from incompleteness of the fossil record. In the first place, incomplete preservation not only eliminates short-lived taxa from the record, but also yields artificially short ranges for others (Sepkoski, 1975). In the second place, the stratigraphic data are, in this case, excellent. They come from thousands of collecting sites, and in fact about 85% of living European mammal species have a Pleistocene record.

The analysis depicted in Fig. 9 strongly favours the Punctuational Model of evolution, which holds that most evolution occurs during or immediately after speciation (Ruzhentsev, 1964; Nevesskaya, 1967; Ovcharenko, 1969; Eldredge, 1971; Eldredge and Gould, 1972; Stanley, 1975b), as opposed to Phyletic Gradualism, the idea that most evolution occurs by phyletic change within well established species. The reason is that during the Plio-Pleistocene, much large-scale evolution was occurring within the Mammalia. For example, distinctive new genera of rodents, carnivores, and artiodactyls were arising. The left flank of Fig. 9E shows too few species of short duration for phyletic evolution to have produced these genera. Furthermore, longevities of species displayed in the histogram reflect not only extinction by phyletic transition, but also extinction by the termination of lineages. If we could plot a histogram representing only durations determined by phyletic extinction, there would be even fewer short-lived species.

Clearly phyletic evolution has been very slow in Late Cenozoic mammals, and distinctive new genera must have formed by divergent speciation events. In many instances two or more branching events may have been required, but this is no problem. Speciation events can come in rapid succession. This is not to say that most evolution necessarily occurs during the evolution of reproductive isolation, but it must occur rapidly in the emergence of certain species. Once established, most mammal species last for at least a few hundred thousand years. In other words, environmental conditions that are sufficiently propitious to allow a new species to blossom tend not to deteriorate to a critical degree over such a period. Clearly, some small isolated populations may attain reproductive isolation with little divergence and survive for only a brief "moment" of geological time. Assuming that such entities, which are technically species, arise with some frequency, a second mode should be present in Fig 9E adjacent to the ordinate and so slender as to be almost invisible. Such "species" are unimportant to the present analysis, however, because they never really became established. The successful establishment of a species seems to be a "boom or bust" phenomenon.

The technique depicted in Fig. 9 has not yet been applied to bivalves, but it is reasonable to assume that a representative histogram of species durations for bivalves will resemble Fig. 9E in general shape. Simply doubling the age (before the Würm) of the "50% surviving" point of Fig. 9A gives 1.3 m.y. as an estimate of mean duration for mammals, which is close to the mean of 1.2 m.y. calculated from the entire histogram. Thus, the simpler "50% method" of estimating the mean appears to be reliable. As mentioned earlier, my current estimate for bivalves using this method is about 7 m.y., or five times as long as the mean for mammals. This is less than the disparity in generic longevities, as would be expected from the larger number of species per genus in the Bivalvia.

Acceptance of the idea that most evolution occurs in association with the multiplication of species is fully compatible with ... the recognition of phyletic evolution. In fact, phyletic evolution is well documented for a number of fossil lineages (review by Stanley, 1976). This documentation, however, supports the Punctuational Model because it invariably reveals slow phyletic transition. Species of invertebrates typically last several million years before undergoing phyletic extinction, and phyletic durations of mammal species (Maglio, 1973; Gingerich, 1974) are compatible with the distribution displayed in Fig. 9E.

From data representing longevity of species, we can calculate the mean rate of extinction, E. Previously E was estimated as the reciprocal of mean species duration (Stanley, 1975b). The accuracy of this estimate will depend on the distribution of species durations. To obtain an estimate of E that is more generally accurate we must take into account the distribution of species longevities. Let us assume that this distribution is stochastically constant during an interval of time. This assumption has been shown to be reasonable for European mammals existing during the interval of Fig. 9A, and again this group will serve as our model. Then during the interval in question E will be more-or-less constant. Every species of the interval may be viewed as having an individual "rate of extinction" equal to the inverse of its duration. For example, a species that lasts 5 m.y. can be viewed as having an extinction rate of 0.2 m.y.$^{-1}$. Then the total rate of extinction E at any time t will equal the average of the individual "rates" of the N_t species existing at that time. If at time t there are η_i species of duration x_i, then:

$$E = \sum_i \eta_i x_i^{-1} / N_t \tag{1}$$

The instantaneous distribution of species durations will not be the same as the distribution for the entire interval (Fig. 9E). In the instantaneous distribution, long-ranging species will be better represented than short-ranging species. In fact, representation will be proportional to duration. Thus, if there are n_i species of duration x_i in Fig. 9E, the number of species of this duration in the instantaneous distribution will be:

$$\eta_i = c x_i n_i$$

where c is a constant, the value of which will be determined by the species diversity at time t. The instantaneous distribution derived from Fig. 9E is shown in Fig. 10, and the extinction rate, calculated from eq. 1, is 0.66 m.y.$^{-1}$.

Though data comparable with those employed in Fig. 9E are not yet available for bivalves, it is reasonable to use Fig. 9E as a model distribution for other taxa, including the bivalves. For mammals:

$$E = (1/D)\, 0.80$$

Given a D for bivalves of approximately 7 m.y., E for this group will be about 0.11 m.y.$^{-1}$.

The estimate of E for mammals or bivalves can be combined with an estimate for another parameter, R, which is the rate of net increase in the number of species early in adaptive radiation, to produce an approximate figure for rate

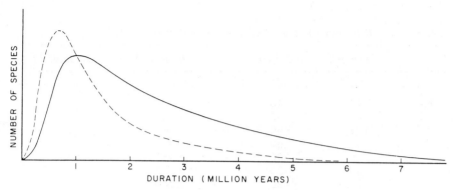

Fig. 10. Conversion of the histogram of Fig. 9E (dashed line), to a histogram representing species durations for Plio-Pleistocene mammals at a single instant in time (solid line).

of speciation early in adaptive radiation. (I restrict the term speciation here to the splitting off of new species; phyletic "speciation" is not counted.) We can estimate R as follows. If we focus upon the early stage of adaptive radiation, in which the net increase is crudely geometrical (exponential), we can apply the equation used to represent such an increase within populations of organisms:

$$N = N_0 \exp(Rt) \tag{2}$$

where N_0 is the original number of species; N is the final number; exp denotes e, the base of natural logarithms, raised to the power indicated by the term in parentheses; and t is geological time since origination. Assuming monophyly ($N_0 = 1$), N, t, and R are shown for various radiating families of bivalves and mammals in Table I, which is an expanded and amended version of a table published previously (Stanley, 1975b). The value of R is consistently much higher for mammals. Even if monophyly is not the rule, only a minor error is likely to be introduced by its assumption, because N_0 enters into the calculation only as ln N_0. Furthermore, it is reasonable to assume that the degree of polyphyly is approximately the same for mammals and bivalves, so that the relative values of R would not be altered appreciably if N_0 were known.

   Just as the rate of population growth equals birth rate minus death rate, the value of R here will equal the rate of splitting of lineages, which will be termed the speciation rate (S), minus the rate of termination of lineages (E'). Transposing:

$$S = R + E'$$

The chief source of inaccuracy here is that E, which we have calculated, represents both E' and pseudo-extinction. Inasmuch as we are especially concerned with the relative values for bivalves and mammals, we can again simply assume that the incidence of pseudo-extinction is the same for the two classes. Because our parameters become exponents in calculations, however, distortions from the use of E in place of E' will not be proportional for the two groups. Even so, the direct use of E gives maximum values of S, and the difference is so great

TABLE I

Estimates of t, N, R and $\bar{R}$ for radiating families of bivalves and mammals [1]

Families	t (m.y.)	N (species)	R (m.y.$^{-1}$)	$\bar{R}$ (m.y.$^{-1}$)
Bivalvia				
Mesodesmatidae	47	40	0.078	
Cardiliidae	31	5	0.052	
Tellinidae	122	350	0.048	
Semelidae	47	60	0.087	
Veneridae	122	500	0.051	0.063
Petricolidae	47	30	0.072	
Myidae	61	20	0.049	
Teredinidae	61	66	0.069	
Lyonsiidae	47	20	0.064	
Mammalia				
Bovidae	31	115	0.15	
Cervidae	19	53	0.21	
Muridae	19	844	0.35	
Cercopithecidae	19	60	0.22	0.22
Cebidae	19	37	0.19	
Cricetidae	35	714	0.19	

[1] Values of t from Moore (1969) and Romer (1966). Values of N from Boss (1971) and Walker (1968).

between the two classes (Table II) that there is no chance that the actual values could be even close to one another. For example, even if the incidence of pseudo-extinction in mammals were so high that for them $E' = 0.3\,E$, and there were no pseudo-extinction for bivalves, speciation rate would be more than twice as high for mammals. One major reason for the difference in speciation rates for the two groups must be the greater dispersal powers of most marine

TABLE II

Estimates of S and $\hat{N}$ after 20 m.y. of radiation for average families of the Bivalvia and Mammalia using various incidences of pseudo-extinction

	E'/E	E'	S	$\hat{N}$
Bivalvia	1.00	0.11	0.17	10
($R = 0.063$, $E = 0.11$)	0.80	0.088	0.15	8
	0.50	0.055	0.12	7
	0.30	0.033	0.10	5
Mammalia	1.00	0.67	0.89	328
($R = 0.22$, $E = 0.67$)	0.80	0.54	0.76	280
	0.50	0.34	0.56	206
	0.30	0.20	0.42	155

taxa and correspondingly reduced chances for geographical isolation (Day, 1963).

Obviously the exponential type of analysis is not perfectly accurate. I have introduced it out of the conviction that because adaptive radiation is a splitting phenomenon, it is appropriately analyzed only by using parameters and equations depicting geometrical increase. Furthermore, the techniques advocated here circumvent many of the problems normally encountered when using species as units in measuring rates of macro-evolution: (1) In the estimation of E using the "50% extant" method, only adequate statistical samples of ancient faunas are needed, as opposed to data representing nearly complete preservation. (2) In the estimation of R, the only fossil evidence needed is the geological time of origin, which we have, to a good approximation; numbers of fossil species, which are poorly known, need not be tabulated. (3) The number of living species (N), though not always known with great accuracy, need only be estimated because it enters into the calculation of R as ln N.

Extrapolation to macro-evolution

The differences between E, R, and S for mammals and bivalves have great significance. A tangible way of portraying the significance of R is to compute from it a doubling time (t_2), which is analogous to half-life in negative-exponential radioactive decay. This can be obtained by setting $N/N_0 = 2$ in eq. 2:

$t_2 = \ln 2/R$

Doubling times for radiating families of mammals average 3.15 m.y., whereas for bivalve families they average about 11 m.y. Consequently while in 20 m.y. an average new monophyletic mammal family would produce a standing diversity of about eighty species, an average bivalve family would produce only three or four species.

For various reasons, some of which are analyzed by Schopf et al. (1975), we cannot directly compare rates of appearance of genera or families from class to class or from phylum to phylum. We have no way of showing that groups given equal taxonomic rank are equivalent in any biological sense. On the other hand, rates of appearance of genera and families are so much higher for the Mammalia than for the Bivalvia (Figs. 12, 13) that it is hard to believe that the average percentage of morphological change per unit time could be the same for the two groups. Schopf et al. (1975) correctly called attention to the idea that the degree of morphological complexity within a given kind of organism may influence the rate of evolution that we perceive. There is, however, much evidence opposing their suggestion that rates of evolution may actually be equal in taxa like bivalves and mammals. First of all, the analysis summarized in Fig. 9 and explained more fully by Stanley (1976) confirms the inference from other fossil evidence (Stanley, 1975b) that most evolution occurs in association with speciation. Clearly, most evolution also occurs in adaptive radiation. Given the above demonstration that in adaptive radiation mammals speciate at much higher rates than bivalves, rates of large-scale evolution must be higher for mammals. (On the basis of earlier arguments, we can de-emphasize phyletic evolution within well-established species as being a minor factor in the overall change that occurs in adaptive radiation.)

To be more precise, let us assume that the degree of evolutionary divergence in speciation, when measured for example as a percentage of genetic, morpho-logical, or ecological change, is a variable having the same frequency distribu-tion for bivalves and mammals. Then the amount of large-scale evolution in a given interval of time early in adaptive radiation will tend to be proportional to the number of speciation events. The equation suggested previously (Stanley, 1975b) for estimating the total number of speciation events ($\hat{N}$), or new lineages formed, early in adaptive radiation is incorrect. I am indebted to Owen M. Phillips for his help in deriving the valid equation, as follows. If we look back in time from N living species in a taxon currently in the midst of adaptive radiation, the number of lineages at some previous time T will be given by:

$$N_T = N \exp(S - E')t = N \exp(Rt)$$

where t is negative, equalling $-T$, and E' is the rate of termination of lineages. The total number of lineages terminated in the next time interval is E' times the number of lineages present:

$$E'N \exp(Rt)$$

The total number terminated since diversification began is the summation over all past time intervals:

$$E'N \int_{-\infty}^{0} \exp(Rt)\, \mathrm{d}t = \frac{E'N}{R}$$

The total number of lineages that have existed ($\hat{N}$) is the total number termi-nated plus the number (N) still in existence, so that:

$$\hat{N} = \frac{E'N}{R} + N = \frac{SN}{R}$$

The only highly uncertain value in the calculation of $\hat{N}$ will be that of E'. We have estimated E, which includes pseudo-extinction, instead of E'. As it turns out, the values of R, S, and E are so different for bivalves and mammals that the incidence of pseudo-extinction cannot determine the general outcome of the comparison. Table II displays the effects on $\hat{N}$ of lowering E by 0, 20, 50, and 70% to represent possible incidences of pseudo-extinction. Even if for mammals E' equalled $0.3E$ and there were no pseudo-extinction in bivalve evo-lution (an absurd contrast), there would be 155 speciation events in the first 20 m.y. of radiation of a typical mammal family, as compared with about 10 speciation events in the same interval for a typical bivalve family! This compar-ison has significance even in a framework of phyletic gradualism, for even if most evolutionary change were to occur within well-established species, there would only be a small fraction of lineages that would produce dramatic change of the kind that leads to new genera and families. All else being equal, the number of markedly divergent lineages would tend to be proportional to the total number of lineages ($\hat{N}$).

The arguments of the preceding paragraphs are supported by direct evidence from the fossil record. Rates of appearance of genera and families are much higher for the Mammalia (Figs. 11, 12). In fact, rates of appearance of bivalve genera are comparable with rates of appearance of mammal families. A way of avoiding the problem that the taxa of different phyla are difficult to compare biologically is to consider how rapidly each class has filled its adaptive zone. Again, the difference is striking. As Fig. 11B shows, mammals reached their peak number of families during the Oligocene, only 30—40 m.y. after their major adaptive radiation began. Furthermore, very little of their modern adaptive zone (with the exception of the human niche) had not been invaded by that time. In other words, the basic adaptive limits of the mammalian body plan had been approached. Since then, most change has been a matter of refinement.

In contrast, bivalve families, to say nothing of genera, have continued to proliferate to the present day, over 400 m.y. after the start of radiation early in the Ordovician (Fig. 12). Whether or not bivalve families are in any way comparable with mammal families, this proliferation reflects the fact that many basic modes of life of modern bivalves have taken long spans of geological time to arise (Fig. 13). For example, the evolution of both free life on the substratum and epifaunal attachment by cementation required more than 200 m.y. The origin of deep-burrowing habits (except perhaps in the aberrant Lucinidae) required something like 150 m.y. Both the ability to burrow rapidly by modern standards and the ability to secrete shell ornamentation of the sort that aids burrowing required even longer. It seems to have taken nearly 100 m.y. simply

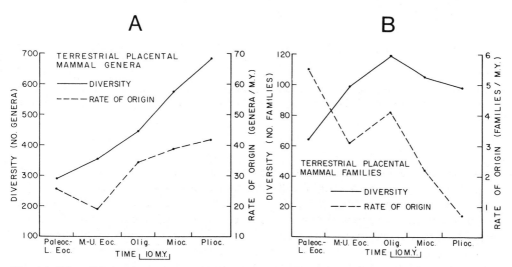

Fig. 11. Diversities and rates of origin of recognized fossil mammal genera (A) and families (B) during the Cenozoic. Data from Romer (1966). (From Stanley, 1973.)

A

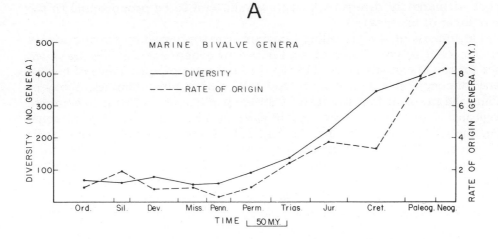

B

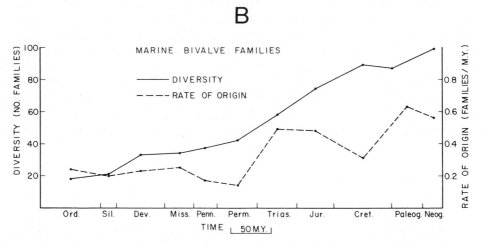

Fig. 12. Diversities and rates of origin of recognized fossil bivalve genera (A) and families (B) during the Phanerozoic. Data from the *Treatise on Invertebrate Paleontology*. (From Stanley, 1973.)

for the first large burrowing clam (maximum dimension about 10 cm) to arise. The honour may belong to the Middle Silurian genus *Megalomoidea*.

There is nothing to indicate that the evolution of these or other slow-to-appear adaptations required the crossing of greater adaptive thresholds than were transgressed by mammals, for example, in the invasion of air or water. The fact that all of the adaptations of the Bivalvia have arisen repeatedly and in distantly related subtaxa indicates that they are readily accessible in the context of the group's basic body plan. Competition from brachiopods is an

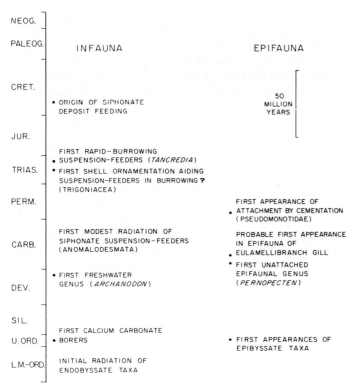

Fig. 13. Highlights in the adaptive radiation of the Bivalvia, with emphasis on the timing of first appearances.

unlikely delaying mechanism in Palaeozoic times because, among other reasons, the major taxa arising in the Mesozoic took as long to evolve the adaptations as the taxa that evolved them in the Palaeozoic. Furthermore, the adaptive zone of infaunal bivalves overlaps very little with that of brachiopods, yet evolution was no more rapid here than in the epifaunal realm. What we see in bivalve phylogeny is a gradual unfolding of basic adaptations that has not ceased to the present day. Its tempo makes the Palaeocene-to-Oligocene adaptive radiation of modern mammals seem instantaneous by comparison.

The exponential nature of diversification accounts for the pinched tail at the base of typical balloon diagrams (Stanley, 1975b). For groups like the bivalves, the stratigraphic range of this tail can be quite long. Here we may have an explanation for the long interval commonly observed between the first recorded appearance and the dramatic expansion of many higher taxa of bivalves. The Veneracea and Cardiacea, for example, which today contain hundreds of species, arose in the Late Triassic, but did not become really conspicuous until the Late Cretaceous. The same explanation may apply to the gradual emergence of advanced Bivalvia from *Fordilla*.

Evolutionary stability

Obviously an abrupt change in diversity may result from a change in speciation rate, extinction rate, or both. As noted above, Nakazawa and Runnegar
(1973) suggested that the reduced rate of proliferation of new genera (and, we
may infer, species) was the cause of the Permo-Triassic decline in bivalve diversity. Whether this has generally been true for episodes of extinction within the
Bivalvia or other taxa remains to be determined. In this light, the exponential
analysis undertaken earlier permits the evaluation of another important difference between bivalves and mammals. This is a difference in evolutionary stability. The first consideration here is that the values of E and S tend to be positively correlated among groups of organisms. Several factors may be at the root
of this relationship. One is that in groups characterized by intense interspecific
competition and high rate of extinction, the rate of speciation should be elevated by the frequent evacuation of niches. This concept is implicit, for
example, in the work of Van Valen (1973) and is incorporated into the simulations of cladogenesis by Raup et al. (1973). Another factor may be that traits
of organisms tending to promote high rates of extinction may happen to be
traits that also promote high rates of speciation. Poor dispersal ability, which
tends to produce high endemicity, is probably one such trait (Jackson, 1973).
Yet another factor may be that groups happening to have inherently high rates
of extinction die' out rapidly unless they also happen to have high rates of
speciation. In any event, using crude unpublished data, I find that the correlation between S and E exists. It is upheld by the data for mammals and bivalves
(Table II). The important point here is that taxa such as mammals, in which
typical values of S and E are relatively high, must inherently be more suscepti-
ble to adaptive radiation and more vulnerable to mass extinction than are taxa
with low values of S and E. Fig. 14 illustrates the mathematical explanation for
this phenomenon. A particular value of R can result either from a high of both
S and E or from a low value of each. Perturbation of S and/or E by a given
percentage has a greater effect, however, if the normal values of S and E are
large rather than small.

The value of R will tend to fluctuate widely about the norm in groups with
high values of S and E. In some taxa, brief pulses of extinction may be offset
by immediate increases in speciation rate, as just discussed, but adverse conditions that persist must often act as a double-edged sword, accelerating extinction and suppressing speciation by preventing peripheral isolates from blossoming into new species. Instability from high values of S and E is what we seem to
see in taxa like ammonites (Arkell, 1957) and trilobites (Palmer, 1965), as well
as in groups of terrestrial tetrapods, where near catastrophic extinction and
rapid re-diversification are the norm (Bakker, this volume). The bivalves, characterized by low values of S and E, have a much more stable evolutionary
record, as already discussed with respect to their resistance to mass extinction.
Certainly factors like eurytopy and large population size may have contributed
to this stability, but low values of S and E must also have played a role. Predictably, the aberrant rudists, which underwent unusually rapid diversification and

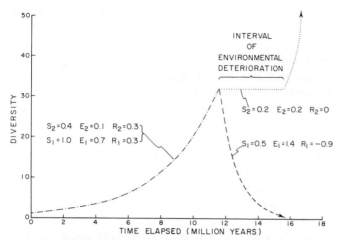

Fig. 14. Illustration of the effects of values of S and E on evolutionary stability. Values of S, E, and R are in units of m.y.$^{-1}$. Hypothetical taxa 1 and 2 initially radiate with equal values of R, but this net rate of increase results from low values of S and E in Taxon 1 and high values in Taxon 2. The environment suddenly deteriorates, reducing S for both taxa to 50% and doubling E for both taxa. For Taxon 1, this yields a high negative value of R, leading to rapid extinction (the exponential treatment disallows attainment of zero diversity, but chance factors are likely to cause extinction as zero is approached). For Taxon 2, R declines only to zero during the 2 m.y. interval of environmental deterioration; at the end of this interval, diversification resumes at the original rate ($R = 0.3$).

generic turnover (Stanley, 1973), also suffered a dramatic final mass extinction.

While stability is attained with moderate diversity, early in adaptive radiation the low R of the Bivalvia is disadvantageous, leaving incipient higher taxa at low diversities for long periods of time. (Recall that after 20 m.y. monophyletic, expanding clade of the sort that may give rise to a new family will typically contain only three or four species.) During this initial period of typically slow diversification, bivalve clades are vulnerable to extinction from chance fluctuations in R or episodes of increased environmental stress. Here we may have another factor in the disappearance without issue of seemingly advanced but monogeneric families like the siphonate Ordovician Lyrodesmatidae and Permian Scaphellinidae. In other words, chance must play a particularly important role in the establishment of higher bivalve taxa.

Biological Agents in Macro-Evolution

Species selection

From the great longevity of bivalve species cited in the previous section we can infer that long-term phyletic evolution plays a minor part in the guidance of large-scale trends. Most evolutionary change occurs during or immediately after speciation events, yet there is a great deal of randomness in speciation

(Mayr, 1963). Chance factors play a major part in determining the timing of speciation, the habitat in which it occurs (the kinds of selection pressures that guide it), and the particular sub-population that forms the nucleus for a new species. Species are in a sense experiments, in that while each will be well adapted to the local conditions under which it has formed, many will subsequently fare poorly in a broader environmental context. These will tend to become extinct more rapidly than average and will therefore have a smaller chance of leaving daughter species. On the average, they will also spawn fewer daughter species per unit time because of the lower incidence of adaptive success of isolated demes having their characteristics. These failures represent poor performance in a process called *species selection* that must be primarily responsible for the course of large-scale evolution (Stanley, 1975b).

Species selection is analogous to natural selection, but operates at the level of the species, via differential speciation and extinction, rather than at the level of the individual, via differential reproduction and mortality. It is not a form of group selection of the sort advocated by Wynne-Edwards (1962) and rejected by Williams (1966), among others. This is because it does not assert that new alleles or allelic combinations become fixed because they are of value to species. Certainly most become established because they are of value to individuals. It is only their fate in macro-evolution that is determined largely by species selection.

The fabric of bivalve evolution discussed earlier is easily viewed in this framework. The iterative, haphazard appearance of various features and the common reversal of life-habit transitions seem to reflect the random aspect of speciation. Net, long-term trends, such as the decline of endobyssate taxa and the ascendancy of siphonate taxa, have been superimposed on such fine-scale events by the process of species selection. Some speciation events have led to evolutionary dead-ends, but others have introduced traits that have become important raw material for species selection.

MacArthur (1972) recognized that the basic agents that limit population sizes within species are also the agents of extinction. These are predation, competition, physical environmental factors, and chance factors. They also act as the agents of species selection, not only in causing extinction but also in permitting speciation to occur at different rates in different species. The role of biological interactions in macro-evolution has been emphasized by Bock (1972) and Stanley (1973). One simple line of evidence suggests that they are especially important in guiding long-term species selection. This is that most changes in the physical environment are temporary and localized. On a scale of tens or hundreds of millions of years, few changes other than those involving world climate are likely to be directional. Climate is undoubtedly of greater importance in determining large-scale evolutionary trends for terrestrial vertebrates than for most marine invertebrates. This is true in part because species of the former tend to be relatively confined geographically.

Rate of evolution is another matter. Continental movements, for example, must significantly influence rates of speciation, and hence rates of large-scale change, for all taxa. The analogy here with population biology is specifically with Fisher's Fundamental Theorem of Natural Selection, which states that rate of evolution by natural selection is proportional to genetic variability. The rate of macro-evolution should vary with variability at the species level, or with the rate of speciation (Stanley, 1975b). Though stated in a slightly different way, this idea has been incorporated into the preceding discussion of rates of evolution.

As long-term directional agents in species selection, changes in biological interactions must be particularly important for all taxa. These interactions are complex, but from them has come much obvious, long-term directional change. The exact kinds of interaction causing this change are not always obvious. It is on the two most important kinds, predation and competition, that the remainder of this chapter will focus.

Predation

In contrasting the evolution of bivalves with the evolution of mammals, I have emphasized the relatively great impact of predation on Recent bivalve faunas (Stanley, 1973).

Predation is often extremely heavy on juveniles because their small size leaves them vulnerable to attack or direct consumption (Thorson, 1966; Muus, 1973). Even adults of many species suffer heavy losses. In a gross sense, therefore, species of bivalves, like those of many other benthic invertebrates, tend to be opportunistic or fugitive species. Many have the potential to invade a habitat rapidly by virtue of an enormous broadcasting of planktonic larvae, and yet predation and other forms of disturbance after settlement commonly decimate a cohort.

The effects of predation and other agents of mortality on bivalve distribution and abundance are well illustrated by the Mytilidae, or sea mussels, some of which are endobyssate and others of which are epibyssate (Stanley, 1970, 1972). Along the Atlantic Coast of the United States, north of Cape Hatteras, two mytilid species abound in the intertidal zone. The epibyssate species *Mytilus edulis* Linné forms dense mats on many rocky surfaces, also occurring on sand flats attached in clumps to coarse debris. The endobyssate species *Modiolus demissus* (Dillwyn) flourishes especially in high intertidal marshes, where it normally lives mostly buried in peaty substrata. Both species commonly form local populations containing millions of individuals. Both species also occur subtidally, but generally in reduced density. The lower limit of most intertidal *Mytilus* populations is determined by intense subtidal predation (Dexter, 1947; Paine, 1966). Dense populations of *Mytilus* that do occur subtidally consist mostly of large individuals that have grown to a critical size that renders them largely invulnerable to predation (Dayton, 1971). Such populations are relatively rare. Those species of mytilids that characteristically occur in subtidal settings also form sparser populations. The only possible exception known to me is *Modiolus modiolus*, which grows to enormous body size. In the tropics, where predation is especially intense, all subtidal species seem to occur only sparsely. Many species are difficult to locate within their reported geographic ranges, even with the use of SCUBA equipment.

The modern pattern of occurrence of the Mytilidae is in sharp contrast to that reconstructed for the Modiomorphidae (Fig. 1) and certain other endobyssate taxa of the Palaeozoic, which even our incomplete fossil record shows to have been abundant in subtidal settings. In fact, a broader generalization is possible. Soft sea floors of the modern ocean appear quite barren of solitary epifauna and semi-infauna compared with reconstructions of comparable Palaeozoic habitats (Ziegler et al., 1968; Bretsky, 1969; Bowen et al., 1974; Thayer, 1974; Babin and Racheboeuf, 1975). In contrast, there has been a considerable increase in the diversity of shelled infauna since the Early Palaeozoic (Stanley, 1968). Burrowing clams (Fig. 4) form the most significant proportion of this increase, but echinoids and gastropods, which also require special adaptations for infaunal life, have greatly expanded their infaunal adaptive zones in post-Palaeozoic times. The gradual invasion of infaunal adaptive zones can be explained by the inherent difficulty of infaunal existence, which requires special adaptations like siphons and petaloid tube feet. Why has there been a contrasting decline of exposed benthos? Intensifying predation pressure has probably been a major agent of species selection against such creatures. Among these are endobyssate bivalves, which are usually partly exposed in life. Since the Ordovician, these forms have undergone a decrease in diversity relative to other life-habit groups (Fig. 7). They are, in fact, quite uncommon today. Perhaps the most conspicuous group, the pinnids, generally live with their viscera below the sediment—water interface and have a retractable mantle. Most endobyssate taxa that survive are confined to stable habitats, such as seagrass meadows. It seems plausible that such habitats are refugia, where reduced mortality from physical disturbance largely offsets the effects of modern predators. If advanced predation has had an evolutionary impact on endobyssate taxa, it should also have had an impact on epifauna. In fact, there is only one diverse epifaunal group occupying soft substrata today. This is the scallops which, most significantly, can swim at the approach of a predator.

It seems that nearly every epifaunal species of bivalves that is abundant in modern seas displays at least one of the following features that serve to reduce predation (Stanley, 1974): (1) general confinement to the intertidal zone, where predation is reduced (*Mytilus*); (2) growth to unusually large adult size (uncommon: *Placuna*); (3) secretion of a very thick shell (spondylids, chamids, ostreids); (4) secretion of heavy spines along the commissure (spondylids, chamids); (5) adoption of habits that are not truly epifaunal, but entail nestling (cryptic occurrence within crevices or beneath rocks or corals) or partial boring (arcids, *Hiatella*, carditids, limids, pectinids, *Botula* (mytilid), isognomonids); (6) attachment above the sea floor on alcyonarians (*Pteria*); (7) possession of the ability to swim away from predators (Pectinacea).

Evidence from other taxa strengthens the argument. I have attributed the well-known decline of articulate brachiopods since the Late Mesozoic to the appearance and Cretaceous adaptive radiation of very important groups of modern predators, including teleost fish, carnivorous snails, and crabs (Stanley, 1974). Fish have undergone marked adaptive advancement since the Ordovician, when they lacked jaws, and the modern teleosts represent the culmination of this modernization. Being small and thin-shelled relative to bivalves, articulate brachiopods are especially vulnerable to predation. Most living species occupy cryptic habitats beneath rocks and corals. Diversity is also higher in the temperate regions than in the tropics, where predation is particularly intense. Preliminary experiments kindly conducted by Mr. Brian Keller at my request showed that a few Jamaican forms living beneath corals suffered heavy predation (probably by fish) within a day or two of being exposed by being turned upside-down. The decline of stalked crinoids and their restriction to relatively deep water today probably also reflects the intensification of benthic predation. Crinoids, having an unprotected ectoderm, would seem to be especially vulnerable to predation. Significantly, the alcyonarians, often viewed as the ecological successors of crinoids, are virtually inedible. It is striking that the comatulids are the only crinoids that today abound in shallow water where predation is most intense. Like scallops, they can swim away from predators.

Even without evidence, we would expect to find in the history of the marine ecosystem an upward spiral in efficiency of capture by predators and escape by prey. We have direct evidence of such a spiral within the Cenozoic Mammalia, both for locomotory adaptations and for intelligence. Relative brain size has increased in both carnivorous and herbivorous mammals, and yet predators have necessarily held an edge throughout (Jerison, 1973).

Competition

Density of a population may be limited by predation or some other form of disturbance, or by intraspecific or interspecific competition. Between these two extremes there exists a spectrum of intermediate possibilities. The preceding paragraphs reflect the view (Stanley, 1973) that predation is the dominant biological factor for most living species of bivalves and probably has played a similar role in the past, though in circumstances of less sophisticated predation and predator avoidance. The rudists represent a likely exception. Like modern corals, rudists probably competed for limited space and, in fact, Kauffman and Sohl (1974) believe that, during their heyday, they tended to exclude corals from bioherms.

Various lines of evidence attest to a more prominent role for competition in the evolution of the Mammalia (Stanley, 1973). The idea of monopolization of adaptive zones, which implies competitive exclusion, pervades the literature on mammalian evolution. Van Valen (1971) concluded that the adaptive zones of major orders of mammals tend to be quite distinct. The major exception, the large zonal overlap between perissodactyls and artiodactyls, seems to have resulted in displacement of the former by the latter (Simpson, 1944, 1953). Simpson also contended that the Carnivora displaced the Creodonta and Van Valen and Sloan (1966) concluded that competition from condylarths, primates, and rodents led to the extinction of the multituberculates. It is generally believed that Mesozoic reptiles, especially dinosaurs, confined Mesozoic Mammalia to niches requiring small body size. The sudden radiation of the Mammalia following the Late Cretaceous extinction was indeed remarkable. It is also generally accepted that geographical areas barren of competitors are required for the adaptive radiation of terrestrial quadrupeds. Large-scale convergence in the Mammalia is generally restricted to taxa occupying separate land masses (Kurtén, 1969). The marsupial radiation of Australia and the marsupial—placental radiation of South America are prime examples.

The displacement of some South American marsupial species by North American placentals following the Late Pliocene uplift of the Isthmus of Panama also offers evidence of competition, although some extinctions may have resulted from intensified predation. The most striking extinctions were among carnivorous groups (Patterson and Pascual, 1968), which is perhaps predictable because carnivores tend to suffer less predation than herbivores and are commonly food-limited. Patterson and Pascual (1968) noted that a great deal of faunal mixing was actually tolerated in the north—south interchange of non-carnivores and attributed this mixing to a minor degree of overlap between many niches and to the successful narrowing of many niches where overlap did occur.

It seems that the mixing of diverse mammalian faunas is generally more successful than sympatric adaptive radiation into occupied adaptive zones. In other words, mammalian adaptive zones tend to become saturated, but saturation is especially effective against the diversification of primitive taxa that would otherwise be capable of radiating. One reason would seem to be that expansion into vacant niches can only be saltational to a limited degree. A major evolutionary transition must normally proceed via a certain number of "stepping-stone" niches, and if every possible path is blocked by occupancy of key niches, the major transition is unlikely to take place. Furthermore, a fully formed species entering a region is more likely to force one or more pre-existing species into a narrower niche than is a small, isolated population that even in the absence of competition would face poor odds of blossoming into a new species.

Most mammals, like most birds and certain invertebrate taxa, suffer less intense predation than is typically observed in the Bivalvia. Even when food is not limiting for mammals, territoriality often guarantees the individual having adequate food at times of scarcity (Klopfer, 1962). One factor here is the lower incidence of predation, but another is the possession of advanced systems of sensory perception, behaviour, and mobility, which are absent from lowly bivalves and brachiopods. Much of the evidence of competitive influence in mammals takes the form of direct observation of territorial behaviour, aggressive interaction, and abutment of geographical ranges (see review by Stanley, 1973).

The fossil record of the Bivalvia differs quite strikingly from that of the Mammalia in ways that seem to reflect relatively weaker competition, especially among burrowing clams. One striking feature is the Cenozoic success of numerous bivalve taxa that are adaptively primitive. The Glycymeridae, which Yonge (1953) suggested serve as a good model of the primitive body plan of suspension-feeding bivalves (i.e., a body plan that would have typified Cambro-

Ordovician taxa), actually arose in the Cretaceous (Nicol, 1950). During the Cenozoic, this family has expanded quite successfully throughout the world. Similarly, the two families of the Lucinacea that Allen (1958) considered most primitive arose in the Mesozoic, whereas the apparently advanced Lucinidae are actually the ancestral group, having arisen as early as the Silurian. Within the Arcoida, the Anadarinae have undergone even more extensive Cenozoic diversification than the Glycimeridae. Like the latter, they have filibranch gills, lack siphons, and are sluggish, shallow burrowers (Stanley, 1970). They are, in all apparent aspects, adaptively primitive, yet they have not been excluded by the more advanced siphonate eulamellibranch taxa that are not only their contemporaries but have diversified simultaneously. An equivalent phenomenon in mammalian evolution would be a Late Cenozoic appearance and radiation without geographical isolation of taxa equivalent in adaptive level to the multituberculates, condylarths, or creodonts. The preposterous nature of this proposition underscores the fundamental difference between macro-evolution in the Bivalvia and Mammalia. What few primitive survivors there are in the Mammalia have persisted in special circumstances. Typically they occur in geographical refugia or possess extraordinarily broad niches that make extinction unlikely. Keast (1972, p. 207), for example, has written of the Australian platypus and echidna:

"Both monotremes are relatively free of competition in their respective feeding zones. . . Monotremes very versatile in terms of habitat, the platypus inhabiting both alpine streams at an altitude of 5,000 feet, and sluggish subtropical Queensland coastal rivers. Echidnas occur in all classes of country from rainforest to stony areas in the central Australian desert, one of the widest habitat tolerances of any mammal."

The ability of opossums to survive competition with advanced mammals is attributed to their high reproductive rate and "to a truly incredible ability to subsist on almost any conceivable food source" (Carlquist, 1965, p. 136).

Ecological relationships of living bivalves offer direct evidence that interspecific competition is relatively weak. Certainly instances of competition exist among locally dense populations of living bivalves, but the densities required are the exception rather than the rule, especially for suspension-feeding species occupying soft substrata. Even on densely colonized rocky shores, competition seems to have a weak evolutionary role. For example, along the Pacific Coast of North America, *Mytilus edulis* and *M. californianus* sometimes co-occur on rocky intertidal surfaces, where predation is relatively weak. *M. californianus* tends to overgrow and exclude *M. edulis*, but even here total local exclusion is rare and physical disturbance and predation prevent it from being permanent (Harger, 1972). Furthermore, the two species contain enormous numbers of individuals and their populations do not always happen to co-occur, even where their geographical ranges overlap. Some deposit feeders render muddy substrata inhospitable to certain kinds of suspension feeders (Rhoads and Young, 1970), but this is a restricted form of competition. Some dense populations of suspension feeders effectively strain larvae from the water and kill them (review by Woodin, 1976) and some intertidal suspension feeders exhibit reduced growth rates when densely aggregated (Hancock, 1971), but the densities required are relatively high. Another factor is that, like many other taxa having poor sensory perception, limited mobility, and primitive behaviour patterns, bivalves are incapable of aggressive interaction.

My argument is perhaps best summed up in the context of biogeography. Predation and physical disturbance tend to restrict the distribution of competitive interactions in space and time with respect to the occurrence of an entire deme or species of bivalves. Mammals disperse as adults, so that their distribution patterns tend to be more continuous. Even where patchiness is induced by discontinuities in habitat, mobility of adults leads to occupancy of appropriate habitats that are barren or sparsely populated. Dispersion is not passive, as is most dispersion in bivalves, but is guided by conscious behaviour. Populations tend to spread as continuous entities. Species swarm into vacant areas, disperse, and stake out individual territories. Thus, mammal communities can saturate entire continents to the point of suppressing the adaptive radiation of groups that have the evolutionary potential to occupy the same adaptive zone.

In contrast, bivalves are in general opportunistic or fugitive species, which, by definition, are not adapted for competitive interactions. Within some dense patches, competition for space or food may reduce growth rates or impede recruitment, but barren or sparsely populated areas remain. The high incidence of predation and other forms of disturbance within the Bivalvia produces heterogeneous patterns of distribution. Passive larval dispersal also tends to produce disjunct and fluctuating spatial patterns. Local populations not only fluctuate in density, but even appear and disappear, depending on where larvae happen to settle and survive. The sedentary habits of adults tend to prevent dispersal and the partitioning of space into territories. Distribution within local patches therefore remains heterogeneous, and barren areas persist between patches. Population sizes also tend to be orders of magnitude larger than those of mammals. Seldom will any kind of competitive interaction affect all or even most of the scattered demes of a species. The disjunct populations of certain species on opposite sides of the Atlantic (Scheltema, 1968) form a dramatic example. Consequently, I envision populations as having the ability to enter and leave well-populated, geographical regions without entering into competitive interactions that are of any great evolutionary consequence. I know of no example of abutment of the geographical ranges of two bivalve species not attributable to a major physiographical barrier. Sympatry of similar congeneric species is widespread, as opposed to the condition in mammals (Hutchinson, 1959; Kohn and Orians, 1962). Also in contrast to the condition for mammals, adaptive zones between major orders and classes of bivalves overlap considerably, and convergent evolution, as between the Mactracea and other taxa (Fig. 6), occurs extensively without geographical separation.

Elastic and inelastic taxa

The difference between mammals and bivalves in intensity of competition seems to be reflected in a difference in general pattern of adaptive radiation.

As noted above, nearly all of the basic adaptive types of modern Mammalia had arisen by the Oligocene, 30—40 m.y. after the group began to radiate extensively. With the exception of human phylogeny, macro-evolution has since been largely a process of refinement of adaptation within what seems to have been a generally saturated and only slightly expanding adaptive zone. The number of families has actually declined since that time (Fig. 11B). Bivalves display no such pattern. Both the number of bivalve families and the rate of origin of families have continued to increase into the Cenozoic (Fig. 12B), but the pattern is not simply one of continued increase in number of taxa. Adaptive breakthroughs of all types have occurred again and again, in one group after another, with no obvious suppression by the presence of similarly adapted groups already in existence.

It seems reasonable to postulate an evolutionary relationship between the importance of competition in limiting population size and the characteristic rate of re-diversification following mass extinction. A group in which competitive relationships are important has in effect a built-in system of re-diversification; the elimination of species also eliminates a key factor that previously suppressed the formation of new species. Such a group has a kind of *elastic* quality, in that it should tend automatically to rebound rapidly through the diversification of surviving taxa as soon as environmental conditions return to normal. We do, in fact, see an elastic quality, not only in the evolution of mammals, as when they diversified after the mass extinction of Mesozoic reptiles, but in the iterative evolution of the trilobites (Palmer, 1965) and ammonites (Arkell, 1957). It is perhaps no accident that cephalopods and arthropods are quite vertebrate-like in their sensory perception, mobility, and terrestrial behaviour (Stanley, 1973), which function in competitive interactions.

We might expect the evolution of bivalves to exhibit a contrasting *inelastic* quality, because the extinction of bivalves should leave predators to continue to limit the populations of surviving species and to suppress speciation. The problem with investigating this effect by studying mass extinctions is that predatory groups commonly suffer as well. At the end of the Palaeozoic, for example, ammonoids, which probably preyed upon bivalves to some degree, nearly disappeared, and pavement-toothed sharks, which certainly preyed upon bivalves, declined markedly (Romer, 1966). Still, the elastic—inelastic distinction may apply to more limited extinctions.

Another aspect of elasticity is seen at the other end of adaptive radiations. Not only have the Mammalia evolved rapidly, but, as we have seen, their diversification has been rather abruptly braked by the abutment of adaptive zones of subtaxa within the potential range of adaptation of the class. Subsequently, new taxa have been added primarily through speciation following extinction or through the compression of adaptive zones of existing taxa. The two aspects of elastic taxa are embodied in the steady-state view of diversity adopted by some students of mammalian evolution (Webb, 1969; Van Valen, 1973). According to this view, diversity tends to be maintained at a level approaching saturation of the habitat, and any niches vacated by extinction are rapidly filled by the expansion of niches or the evolution of new species. There is, of

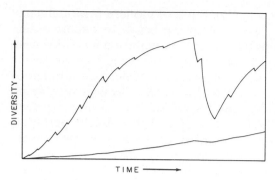

Fig. 15. Diagrammatic view of patterns of diversification for mammals (upper curve) and bivalves (lower curve). The mammal curve is crudely sigmoid, but diversification is interrupted by sudden declines in diversity, in part reflecting the instability contributed by high rates of speciation and extinction. The bivalve curve displays a much lower net rate of increase and also greater evolutionary stability. The overall shape of the bivalve curve must also be sigmoid, but when the large-scale inflection point may be reached is uncertain (there is no evidence that it has yet been reached).

course, no such thing as complete saturation of a habitat, only a decline to near zero of the probability that a new species will arise. Counteracting the general tendency for the saturation level to be approached is the evolutionary instability of the Mammalia arising from the high rates of speciation and extinction (Fig. 14).

Because of the great overlap tolerated among adaptive zones of bivalve subtaxa, there seems to be no clear-cut limit to diversification within the basic adaptive zone of the class. What reduction there has been in the rate of appearance of taxa above the family level seems to have resulted from a decreasing flexibility resulting from increasing specialization (Stanley, 1973). Fig. 15 depicts this "unrestrained" quality and the slow and steady nature of diversification that reflects low values of R, S, and E. Where diversification of living families will end, barring wholesale environmental deterioration, is difficult to imagine.

A remarkable phenomenon, perhaps related to intensity of competition, is revealed when we compare values of R for mammals calculated at the family level (Table I) with values calculated at the genus level (Table III). The latter are more variable because they represent smaller entities, but are also typically several times higher. The difference in mean value is enormous in the light of the fact that the effect of R is exponential. It must reflect a basic pattern of phylogeny in which there is a distinct hierarchy of clades. Our schemes of classification attempt to represent this pattern, though they do so imperfectly. The point is that gaps in the overall pattern exist on several scales. The higher the taxon, the more scales of gaps will be present and the smaller R will be. The phylogeny of a family, for example, includes clusters of species that are the products of bursts of evolution and that we tend to recognize as genera (Fig. 16A).

Species diversity for living genera of the Bivalvia is less well documented. A compilation currently in progress however (Stanley, 1976), is revealing a sim-

TABLE III

Estimates of R for living genera of the Bovidae that have arisen since the Miocene and contain four or more living species [1]

Genus	t (m.y.)	N (species)	R (m.y.$^{-1}$)	$\bar{R}$ (m.y.$^{-1}$)
Tragelephas	1.3	6	1.4	
Bos	1	7	1.9	
Cephalophus	1	10	2.3	
Kobus	1	6	1.8	1.4
Oryx	1.3	4	1.1	
Damaliscus	1.3	6	1.4	
Capra	3	5	0.5	
Ovis	4	7	0.5	

[1] Data from Walker (1968) and Romer (1966).

pler phylogenetic pattern than that inferred for mammals. Values of R for radiating bivalve genera are not greatly higher than values for families (Table I). The implication is that genera of bivalves are less discrete than genera of mammals. The pattern of phylogeny within bivalve families is apparently more homogeneous, perhaps looking more like Fig. 16B than like Fig. 16A.

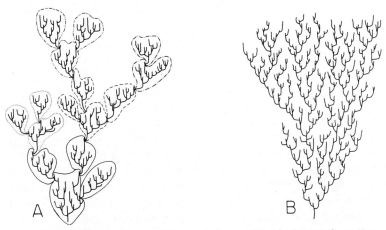

Fig. 16. Schematic representation of alternative phylogenetic patterns. The horizontal scale represents morphological condition and the vertical scale, time. Most evolutionary change is shown as being associated with speciation events (the Punctuational Model). A depicts the pattern inferred for mammals. Lineages are clumped into distinct clades, which are recognized as genera (circumscribed by patterned lines). Genera are clustered into families, each indicated by a distinctive line pattern. (Morphological differences between genera and families would be more distinct if plotted in n dimensions.) Radiating genera have higher values of R than families, and families have higher values of R than the entire clade. B depicts a pattern that is perhaps approached within the Bivalvia. Clades are indistinctly clumped, making taxonomic boundaries more arbitrary. Sets of species that become recognized as genera will generally display values of R that are not appreciably higher than the value for the entire phylogeny, which might represent a familial clade.

The somewhat confused state of generic taxonomy in the Bivalvia may well reflect a relatively high degree of polyphyly in the origin of bivalve genera. The simple morphology of bivalves relative to that of mammals might contribute to such a pattern. Convergence so precise as to go undetected is less likely for taxa of complex morphology, so that polyphyly is less likely. Even so, polyphyly is not likely to alter average values of R very greatly because the depression of R for certain clades by artificial loss of species will tend to be balanced by the elevation of R for clades to which these species are artificially added. The lack of clustering of species within familial phylogenies of the Bivalvia cannot be explained by polyphyly. Even distinctive and apparently monophyletic genera, like *Pandora*, exhibit low values of R.

Perhaps degree of competitive interaction is again a key factor. There may be a tendency for a distinctive new genus of mammals to speciate rapidly, but only to the point at which it occupies all readily available niches in accessible geographical areas. (The frequent abutment of geographical ranges of congeneric mammal species was noted earlier.) Then only divergent new species, namely ones representing new genera or leading to them, will be likely to arise. New genera thus formed will often expand rapidly for a brief time (some, of course, will not expand at all). Bursts of evolution would also be expected to follow major extinctions in highly competitive and territorial animals like mammals. Furthermore, because congeneric species of mammals tend to be more restricted geographically than are congeneric species of bivalves, they would be expected to be extirpated more frequently by single events of environmental deterioration.

The weaker cladistic pattern of bivalve phylogeny is perhaps predictable because competition is weaker. Diversification of a new genus is not likely to be stifled quickly because adaptive zones of genera overlap extensively. If a new bivalve genus survives beyond the stage at which it contains only one or two localized species, it is likely to continue to diversify slowly for millions of years, perhaps spreading over an enormous geographical area. Its rate of diversification may be viewed as dwindling if species of the clade which it forms become recognized as new genera. This is a kind of diffuse pseudo-extinction. Rate of diversification of a cosmopolitan genus may also dwindle if the component species, or their peripheral isolates, do not fare well in time against changing environmental conditions, but geographical and ecological inhomogeneities should often prevent such a decline from being sudden.

Certainly the contrasting patterns of phylogeny in the Bivalvia and Mammalia will bear further investigation, but their discovery offers evidence of the value of exponential analysis in the study of the diversification of species.

Summary

The earliest Bivalvia were burrowing clams related to other infaunal and semi-infaunal classes of Cambro-Ordovician molluscs. Neotenous retention of the post-larval byssus led to adult endobyssate and, finally, epibyssate habits.

The oldest epifaunal bivalves known are from the Middle Ordovician.

The fossil record shows that most evolutionary change must occur during or shortly after speciation. Species are therefore the basic units of macro-evolution. Differential rates of origin and extinction of species are the source of large-scale trends and patterns in phylogeny. These differential rates represent species selection, which is analogous to natural selection but operates on species rather than individuals. Rates of speciation and extinction can be studied by techniques adapted from demography.

Large-scale evolution in the Bivalvia has been very slow, reflecting a demonstrably slow rate of speciation within the class. Many basic adaptations have arisen polyphyletically, but only after hundreds of millions of years of evolution. On a small scale, a distinct randomness is observed, in the form of reversals in the direction of evolution. This reflects random aspects of the process of speciation. Superimposed on this randomness are long-term trends. Endobyssate taxa formed the most common life-habit group of the Ordovician Period, but subsequently declined, while free-burrowing and epifaunal taxa radiated. Mantle fusion and the formation of siphons played a major role in the post-Palaeozoic expansion of burrowing taxa, permitting more rapid and deeper burrowing. In general, the Bivalvia display a remarkable degree of evolutionary stability. In part, this reflects eurytopy, but in part it reflects low rates of branching and extinction of lineages. It can be shown mathematically that taxa having higher turnover rates are more likely to undergo dramatic fluctuations in diversity. Dramatic episodes of diversification and extinction in the evolution of terrestrial tetrapods, ammonites, and trilobites accord with this relationship.

Factors that limit population size are the agents of species selection. Predation has been an important factor of this type for the Bivalvia. The evolution of modern predators has probably been responsible for the decline of endobyssate taxa and the survival of only those epifaunal taxa with mechanisms for avoiding predation. Heavy predation and other forms of disturbance, primitive sensory perception and behavioural patterns, relative immobility of adult animals, and vagaries of larval transport tend to prevent competition from affecting higher bivalve species uniformly and persistently enough to play a major role in large-scale evolution. Adaptive zones of major taxa overlap considerably, and limits of diversification are not easy to envisage. There has been no abatement in the rate of appearance of families since the Ordovician. Genera are not as distinct entities as in the Mammalia, where they seem to represent monophyletic bursts of speciation into unoccupied adaptive zones. In the Bivalvia net rates of diversification are nearly the same at the genus and family levels, indicating that clades within families are intergradational and indistinct. Weak competition within the Bivalvia is a possible source of this pattern.

References

Allen, J.A., 1958. On the basic form and adaptations to habitat in the Lucinacea (Eulamellibranchia). Philos. Trans. Soc. Lond., 241B: 421—484.

Arkell, W.J., 1957. Introduction to Mesozoic Ammonoidea. In: R.C. Moore (Editor), Treatise on Inverte-
    brate Paleontology, Part L. Mollusca 4. Geological Society of America and University of Kansas Press,
    Lawrence, Kansas, pp. 81—129.
Babin, C. and Racheboeuf, P.R., 1975. Réflexions sur le benthos Dévonien du Massif Armoricain replacé
    dans le cadre de l'Europe occidentale. Extr. de Geobios, 8: 241—257.
Berggren, W.A. and Van Couvering, J.A., 1974. The Late Neogene. Developments in Palaeontology and
    Stratigraphy. 2. Elsevier, Amsterdam, 216 pp.
Beurlen, K., 1956. Der Faunenschnitt an der Perm—Triasgrenze. Z. Dtsch. Geol. Ges., 108: 88—99.
Bock, W.J., 1972. Species interaction and macroevolution. Evol. Biol., 5: 1—24.
Boss, K.J. 1971. Critical estimate of the number of Recent Mollusca. Occas. Pap. Mollusks, 3: 81—135.
Bowen, Z.P., Rhoads, D.C. and McAlester, A.L., 1974. Marine benthic communities in the Upper Devo-
    nian of New York. Lethaia, 7: 93—120.
Brasier, M.D., 1975. An outline history of seagrass communities. Palaeontology, 18: 681—702.
Bretsky, P.W., 1969. Central Appalachian Late Ordovician communities. Geol. Soc. Am. Bull., 80: 193—
    212.
Carlquist, S., 1965. Island Life. The Natural History Press, Garden City, N. Y., 451 pp.
Carter, J.G., 1977. Ecology and evolution of the Gastrochaenacea (Mollusca, Bivalvia) with notes on the
    status of the Myoida. Yale Univ. Peabody Mus. Bull. (in press).
Ciriacks, K.W., 1963. Permian and Eotriassic bivalves of the Middle Rockies. Am. Mus. Nat. Hist. Bull.,
    125: 1—100.
Day, J.H., 1963. The complexity of the biotic environment. Syst. Assoc. Publ., No 5: 31—49.
Dayton, P.K., 1971. Competition, disturbance, and community organization: the provision and sub-
    sequent utilization of space in a rocky intertidal community. Ecol. Monogr., 41: 351—389.
Dexter, R.W., 1947. The marine communities of a tidal inlet at Cape Ann, Massachusetts: A study in bio-
    ecology. Ecol. Monogr., 17: 261—294.
Eldredge, N., 1971. The allopatric model and phylogeny in Paleozoic invertebrates. Evolution, 25: 156—
    167.
Eldredge, N. and Gould, S.J., 1972. Punctuated equilibria: an alternative to phyletic gradualism. In:
    T.S.M. Schopf (Editor), Models in Paleobiology. Freeman, San Francisco, Calif., 82—115.
Fischer, A.G., 1964. Brackish oceans as the cause of the Permo—Triassic marine faunal crisis. In: A.E.M.
    Nairn (Editor), Problems in Palaeoclimatology. Interscience, London, pp. 566—574.
Gingerich, P.D. 1974. Stratigraphic record of Early Eocene *Hyopsodus* and the geometry of mammalian
    phylogeny. Nature, 248: 107—109.
Hancock, D.A., 1971. The role of predators and parasites in a fishery for the mollusc *Cardium edule* L.
    In: P.J. den Boer and G.R. Gradwell (Editors), Dynamics of Populations. Centre for Agricultural Pub-
    lishing and Documentation, Wageningen, pp. 419—439.
Harger, J.R., 1972. Competitive co-existence: maintenance of interacting associations of the sea mussels
    Mytilus edulis and *Mytilus californianus*. Veliger, 14: 387—410.
Hutchinson, G.E. 1959. Homage to Santa Rosalia or why are there so many kinds of animals? Am. Nat.,
    93: 145—159.
Jackson, J.B.C., 1973. The ecology of molluscs of *Thalassia* communities, Jamaica, West Indies. I. Distri-
    bution, environmental physiology, and ecology of common shallow-water species. Bull. Mar. Sci.,
    23: 313—350.
Jerison, H.J. 1973. Evolution of the Brain and Intelligence. Academic Press, New York, N.Y., 482 pp.
Kauffman, E.G. and Sohl, N.F., 1974. Structure and evolution of Antillean Cretaceous rudist frameworks.
    Verh. Naturf. Ges. Basel, 84: 399—467.
Keast, A., 1972. Australian mammals: zoogeography and evolution. In: A. Keast, F.C. Erk, and B. Glass
    (Editors), Evolution, Mammals, and Southern Continents. State University of New York Press,
    Albany, N.Y., pp. 195—246.
Klopfer, P.H. 1962. Behavioral Aspects of Ecology. Prentice-Hall, Englewood Cliffs, N. J., 166 pp.
Kohn, A. and Orians, G.H. 1962. Ecological data in the classification of closely related species. Syst.
    Zool., 11: 119—126.
Kurtén, B., 1968. Pleistocene Mammals of Europe. Aldine Publishing Co., Chicago, Ill., 317 pp.
Kurtén, B., 1969. Continental drift and evolution. Sci. Am., 220: 54—64.
MacArthur, R.H., 1972. Geographical Ecology. Harper and Row, New York, N.Y., 269 pp.
Maglio, V.J., 1973. Origin and evolution of the Elephantidae. Trans. Am. Philos. Soc., 63(3): 1—149.
Mayr, E., 1963. Animal Species and Evolution. Harvard University Press, Cambridge, Mass., 796 pp.
Moore, R.C. (Editor), 1969. Treatise on Invertebrate Paleontology. Part N. Mollusca 6. Bivalvia. Geo-
    logical Society of America and University of Kansas Press, Lawrence, Kansas, 1224 pp.
Muus, K., 1973. Settling, growth and mortality of young bivalves in Øresund. Ophelia, 12: 79—116.
Nakazawa, K. and Runnegar, B., 1973. The Permian—Triassic boundary: a crisis for bivalves? Alta. Soc.
    of Pet. Geol. Mem., 2: 608—621.
Nevesskaya, L.A., 1967. Problems of species differentiation in light of paleontological data. Paleontol. J.,
    1967: 1—17.
Newell, N.D., 1965. Classification of the Bivalvia. Am. Mus. Nat. Hist. Novit., No. 2206: 25 pp.
Newell, N.D., 1969. Classification of Bivalvia. In: R.C. Moore (Editor), Treatise on Invertebrate Paleon-
    tology, Part N. Mollusca 6. Geological Society of America and University of Kansas Press, Lawrence,
    Kansas, pp. 205—224.

Newell, N.D. and Boyd, D.W., 1975. Parallel evolution in early trigoniacean bivalves. Am. Mus. Nat. Hist. Bull., 154: 55—162.

Newell, N.D. and Ciriacks, K.W., 1962. A new bivalve from the Permian of the western United States. Am. Mus. Novit., No. 221: 4 pp.

Nicol, D., 1950. Origin of the pelecypod family Glycymeridae. J. Paleontol., 24: 89—98.

Nicol, D., 1965. An ecological analysis of four Permian molluscan faunas. Nautilus, 78: 86—95.

Ockelman, K.W., 1964. *Turtonia minuta* (Fabricius), a neotenous veneracean bivalve. Ophelia, 1: 121—146.

Ovcharenko, V.N., 1969. Transitional forms and species differentiation of brachiopods. Paleontol. J., 1969: 67—73.

Paine, R.T., 1966. Food web complexity and species diversity. Am. Nat., 100: 65—76.

Palmer, A.R., 1965. Biomere — a new kind of stratigraphic unit. J. Paleontol., 39: 149—153.

Patterson, B. and Pascual, R., 1968. The fossil mammal fauna of South America. Q. Rev. Biol., 43: 409—451.

Pojeta, J., 1971. Review of Ordovician pelecypods. U. S. Geol. Surv. Prof. Pap., 695: 46 pp.

Pojeta, J., 1975. *Fordilla troyensis* Barrande and early pelecypod phylogeny. Bull. Am. Paleontol., 67: 363—384.

Pojeta, J. and Runnegar, B., 1974. *Fordilla troyensis* and the early history of the pelecypod mollusks. Am. Sci., 62: 706—711.

Pojeta, J., Runnegar, B., Morris, N.B. and Newell, N.D., 1972. Rostroconchia: a new class of bivalved mollusks. Science, 177: 264—267.

Pojeta, J., Runnegar, B. and Kříž, J., 1973. *Fordilla troyensis* Barrande: the oldest known pelecypod. Science, 180: 866—868.

Raup, D.M., Gould, S.J., Schopf, T.J.M. and Simberloff, D.S., 1973. Stochastic models of phylogeny and the evolution of diversity. J. Geol., 81: 525—542.

Rhoads, D.C. and Young, D.K., 1970. The influence of deposit-feeding organisms on sediment stability and community·trophic structure. J. Mar. Res., 28: 150—178.

Romer, A.S., 1966. Vertebrate Paleontology. University of Chicago Press, Chicago, Ill., 486 pp.

Runnegar, B., 1974. Evolutionary history of the bivalve subclass Anomalodesmata. J. Paleontol., 48: 904—940.

Runnegar, B. and Pojeta, J., 1974. Molluscan phylogeny: the paleontological viewpoint. Science, 186: 311—317.

Ruzhentsev, V. Ye., 1964. The problem of transition in paleontology. Int. Geol. Rev., 6: 2204—2213.

Scheltema, R.S. 1968. Dispersal of larvae by equatorial ocean currents and its importance to the zoogeography of shoal-water tropical species. Nature, 217: 1159—1162.

Schopf, T.J.M., Raup, D.M., Gould, S.J. and Simberloff, D.S., 1975. Genomic versus morphologic rates of evolution: influence of morphologic complexity. Paleobiology, 1: 63—70.

Sepkoski, J.J., 1975. Stratigraphic biases in the analysis of taxonomic survivorship. Paleobiology, 1: 343—355.

Simpson, G.G., 1944. Tempo and Mode in Evolution. Columbia University Press, New York, N.Y., 237 pp.

Simpson, G.G., 1953. The Major Features of Evolution. Columbia University Press, New York, N.Y., 434 pp.

Stanley, S.M., 1968. Post-Paleozoic adaptive radiation of infaunal bivalve molluscs — a consequence of mantle fusion and siphon formation. J. Paleontol., 42: 214—229.

Stanley, S.M., 1970. Relation of shell form to life habits in the Bivalvia. Geol. Soc. Am. Mem., 125: 296 pp.

Stanley, S.M., 1972. Functional morphology and evolution of byssally attached bivalve mollusks. J. Paleontol., 46: 165—212.

Stanley, S.M., 1973. Effects of competition on rates of evolution, with special reference to bivalve mollusks and mammals. Syst. Zool., 22: 486—506.

Stanley, S.M., 1974. What has happened to the articulate brachiopods? Geol. Soc. Am. Abstr. Progr., 6: 966—967.

Stanley, S.M., 1975a. Adaptive themes in the evolution of the Bivalvia (Mollusca). Ann. Rev. Earth Planet. Sci., 3: 361—385.

Stanley, S.M., 1975b. A theory of evolution above the species level. Proc. Natl. Acad. Sci. (U.S.A.), 72: 646—650.

Stanley, S.M., 1975c. Why clams have the shape they have: an experimental analysis of burrowing. Paleobiology, 1: 48—58.

Stanley, S.M., 1976. Stability of species in geologic time. Science, in press.

Taylor, J.D., 1973. The structural evolution of the bivalve shell. Palaeontology, 16: 519—534.

Thayer, C.W., 1974. Marine paleoecology in the Upper Devonian of New York. Lethaia, 7: 121—155.

Thorson, G., 1966. Some factors influencing the recruitment and establishment of marine benthic communities. Neth. J. Sea Res., 3: 267—293.

Trueman, E.R., 1966. Bivalve mollusks: fluid dynamics of burrowing. Science 152: 523—525.

Valentine, J.W. and Gertman, R.L., 1972. The primitive ecospace of the Pelecypoda. Geol. Soc. Am., Abstr. Progr., 4: 696.

Van Valen, L., 1971. Adaptive zones and the orders of mammals. Evolution, 25: 420—428.

Van Valen, L., 1973. A new evolutionary law. Evol. Theory, 1: 1—30.

Van Valen, L. and Sloan, R.E., 1966. The extinction of the multituberculates. Syst. Zool., 15: 261—278.

Walker, E.P., 1968. Mammals of the World. The Johns Hopkins Press, Baltimore, Md., 2nd ed., 1500 pp.

Webb, S.D., 1969. Extinction—origination equilibria in late Cenozoic land mammals of North America. Evolution, 23: 688—702.

Williams, G.C., 1966. Adaptation and Natural Selection. Princeton University Press, Princeton, N.J., 307 pp.

Woodin, S.A., 1976. Adult—larval interactions in dense infaunal assemblages: Patterns of abundance. J. Mar. Res., 34, in press.

Wynne-Edwards, V.C., 1962. Animal Dispersion in Relation to Social Behavior. Hafner, New York, N.Y., 653 pp.

Yonge, C.M., 1939. The protobranchiate Mollusca: a functional interpretation of their structure and evolution. Philos. Trans. R. Soc. Lond., 237B: 335—374.

Yonge, C.M., 1953. The monomyarian condition in the Lamellibranchia. Trans. R. Soc. Edinb., 62: 443—478.

Yonge, C.M., 1962. On the primitive significance of the byssus in the Bivalvia and its effects in evolution. J. Mar. Biol. Assoc. U.K., 42: 112—125.

Ziegler, A.M., Cocks, L.R.M. and Bambach, R.K., 1968. The composition and structure of Lower Silurian marine communities. Lethaia, 1: 1—27.

CHAPTER 8

AMMONITE EVOLUTION

W.J. KENNEDY

Introduction

Ammonites, that is to say members of the predominantly Jurassic and Cretaceous cephalopod orders Phylloceratida, Lytoceratida and Ammonitida, are the group of fossil invertebrates which, more than any other, have attracted the attention of observers since antiquity (Nelson, 1968). During the nineteenth and early twentieth centuries, their widespread occurrence and excellent preservation, together with the fact that the complete development of shell form from larval stages to maturity could be readily studied led to their use in support of a whole range of evolutionary laws and principles from iterative evolution to Haeckel's Biogenetic Law. The conclusions of Alpheus Hyatt (1866, 1889, 1894, 1903) and S.S. Buckman (1887—1907) based upon the study of ammonite development and presumed evolution, had a major effect on contemporary palaeontological thought, and many of the concepts introduced by them are, albeit erroneous, still widely quoted. In this essay, I have attempted firstly to outline the major features of ammonoid evolution, to place the group in context; thereafter I discuss various evolutionary patterns displayed by the group, including examples of lineage studies, convergence and like. Available data on evolutionary rates — expressed in terms of faunal turnover and species longevity — are reviewed, with a final section on extinction.

Ammonite Biology

Ammonites were a diverse, variable and successful group, exploiting a range of niches within marine ecosystems. Nothing is known of their external soft tissues, and details of their internal organs are only sketchily known and equivocal. As a large and diverse group, generalisation about habits is difficult, although relevant to understanding controlling mechanisms and possible selective pressures acting on them. These topics are beyond the scope of this chapter, but the reader is referred to a recent review on the subject (Kennedy and Cobban, 1976) as a companion to the present essay.

The Major Features of Ammonoid Evolution

Ammonoids as a whole first appeared in the early Devonian (Figs. 1, 3). The obvious origin for the group is within the coiled nautiloids, and this view was most positively stated by L.F. Spath (1933, 1936), and still finds some support

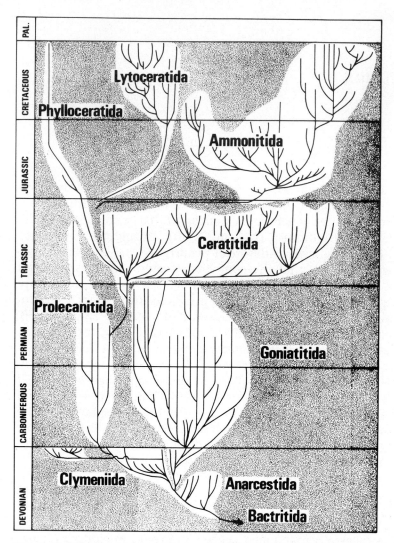

Fig. 1. Outline phylogeny of the ammonoids (modified after Teichert, 1967, and reproduced with the permission of the author and the University of Kansas Press).

(e.g. Donovan, 1964; Arkell et al., 1957; Teichert et al., 1964). There are, however, fundamental differences in ontogenetic development, the nature of protoconch etc. between nautiloids and ammonoids, and these led Schindewolf (1932, 1933, 1934, 1939) to propose an origin in the bactritids, an inconspicuous group of small orthoconic to cyrtoconic shells with marginal siphuncles, orthochoanitic septal necks, sutures with gentle lateral flexures but a distinctive and narrow ventral lobe, and a bulbous protoconch. Because of their rather different combination of features to that seen in other cephalopods the taxonomic position of the bactritids has long been disputed (Teichert, 1967 reviews the problem); the current view is to regard them either as a suborder of ammo-

Fig. 2. Basic ammonoid morphotypes. 1—3. Goniatitina: *Tornoceras simplex* from the Upper Devonian of Bundesheim, Eifel. *Tornoceras* is believed to be representative of the rootstock from which the Goniatitida evolved. 4—6. Clymeniina: *Clymenia laevigata* from the Upper Devonian Clymenienkalk of Schubelhammer, Bavaria. 7,8. *Goniatites crenistria* from the Lower Carboniferous Limestone of the U.K. A typical goniatite with angular, goniatitic sutures. 9. Prolecanitina: *Prolecanites compressus* from the Lower Carboniferous Limestone, Scarlett, Isle of Man. 10, 11. *Ceratites nodosus*, a typical ceratite from the Middle Triassic Muschelkalk of Württemberg, Germany.

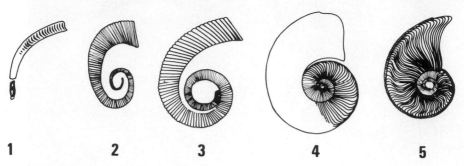

1   2   3   4   5

Fig. 3. Ammonoid origins: the bactritid to anarcestid transition during Early Devonian time. *1, Cyrtobactrites; 2, Anetoceras; 3, Erbenoceras; 4, Teicherticeras; 5, Mimagoniatites.* (Modified after Teichert, 1967, and reproduced by permission of the author and the University of Kansas Press.)

noids (Erben, 1964) or a group of ordinal status between orthocerids and ammonoids (Teichert et al., 1964).

Final resolution of the problem of ammonoid origins came from the work of Erben (1960, 1962, 1964, 1965, 1966) who discovered a series of intermediates between cyrtoconic bactritids and the earliest true ammonoids in the Early Devonian Hunsrück Shale of Western Germany (Fig. 3). Although precise stratigraphic control is complicated by the deformed nature of the sequences studied, this progression appears valid, with a sequence from *Bactrites* to *Lobobactrites*, *Cyrtobactrites*, *Anetoceras*, *Teichertoceras*, and *Mimagoniatites* to the anarcestid ammonoids. According to Erben the progression follows a series of evolutionary trends: (1) progressive coiling from loose gyroconic to involute; (2) progressive increase in depth of lobes in the suture line; (3) increasing sinuosity of growth lines and a change in their overall direction from rursiradiate to prorsiradiate; and (4) changes in whorl shape from rounded, to compressed, to depressed. There also appears to be a progressive increase in size.

The earliest ammonoids are classed in the suborder Anarcestina (Figs. 1, 2) represented by some tens of genera only, and ranging from Early to Late Devonian in age. These were rather small animals, their shells usually ornamented only by growth striae, compressed and involute, with gently sinuous sutures, although some develop stronger ornament and exhibit multiplication of sutural elements. The group gave rise to three other groups, descent being traceable on criteria of sutural development (see Kullmann and Wiedmann's 1970 summary of the significance of sutures in ammonoid phylogeny) and other features. The first of these groups are the short-lived (Late Devonian only) clymeniids, a remarkable group of around thirty genera, characterized by coiled shells and dorsal siphuncle. Donovan (1964) regarded the clymeniids as a group of nautiloids; their sutures suggest, however, an origin in the Anarcestina. The Goniatitina (150—200 genera) are in contrast a long-ranging group whose origins lie in the long ranging genus *Tornoceras*. They range from the mid-Devonian through the Late Palaeozoic with a peak in diversity during the Late Carboniferous and Early Permian; a single genus *Pseudogastrioceras* survives into the Early Triassic. As a group, goniatites are characterized by a globose to discoidal shell form, weak ornament or none, and high degrees of involution, although compressed, evolute and coarsely ribbed forms are known.

The classic goniatitic suture line (Fig. 4) is characterized by having most or all of the lobes and saddles entire, the lobes being angular and the saddles rounded, whilst the ventral lobe is bifid. In ontogeny, the first adventitious lobes appear prior to the umbilical lobe. There is a tendency for an overall increase in the number of elements as a result of the addition of adventitious lobes by subdivision of the lateral saddle. Ceratitic (with rounded unbroken

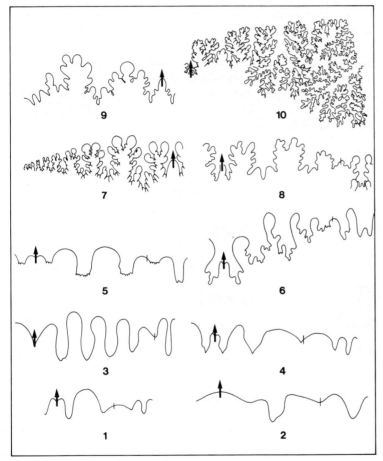

Fig. 4. The main ammonoid suture types. *1, Manticoceras* (Devonian), a goniatite; *2, Kosmo-clymenia* (Upper Devonian), a clymeniid; *3, Prolecanites* (Lower Carboniferous), a prolecani-tid; *4, Goniatites* (Lower Carboniferous), a goniatite; *5, Xenodiscus* (Upper Permian), an early ceratite; *6, Monophyllites* (Middle Triassic), an early phylloceratid; *7, Holcophylloceras* (Lower Cretaceous), a late phylloceratid; *8, Derolytoceras* (Lower Jurassic), a lytoceratid; *9, Caloceras* (Lower Jurassic), an ammonitid; *10, Pseudhelicoceras* (Lower Cretaceous), an ancyloceratid. (After Kullman and Wiedmann, 1970, and Arkell et al., 1957, with permission.)

saddles and denticulate lobes) and ammonitic sutures (with denticulate or frilled lobes and saddles) appear independently in several goniatite lineages, notably the Perrinitidae (Early to Middle Permian), Cyclobaceae (Late Carboniferous to Late Permian) and Thalassoceratidae (Late Carboniferous to Middle Permian).

In striking contrast to the goniatites, and the key to the origin of the Mesozoic ammonoids are the Prolecanitida (Late Devonian/Early Carboniferous to Late Triassic, about fifty genera). Early members of the group are evolute, discoidal forms with slender whorls, generally lacking in ornament; some later forms are oxyconic (*Prodromites*) involute and carinate (*Sageceras*) or may bear ventrolateral nodes (*Artinskia*). The sutures are highly distinctive (Fig. 4), goniatitic, consisting of narrow, lanceolate lobes and tear-shaped saddles in early forms, with subdivision of lobes to give a ceratitic form in some later taxa (e.g., Sageceratidae).

Whereas in goniatites, increase in sutural elements was by subdivision of the lateral saddle to give *adventitious* lobes, increase in the prolecanitids is by the development of *auxiliary* elements from the umbilical lobe, between the lateral lobe and the umbilicus (Fig. 5). In addition, some subdivision of the ventral saddle may occur (Teichert, 1967, p. 189).

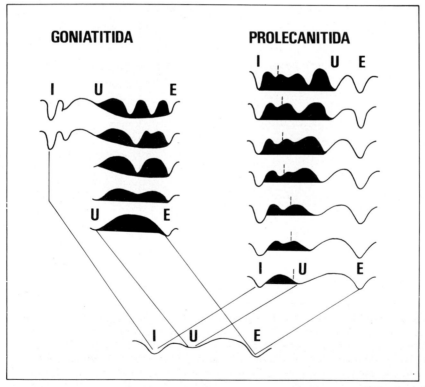

Fig. 5. Ontogenetic development of sutures in Goniatitida and Prolecanitida (after Schindewolf, 1954a).

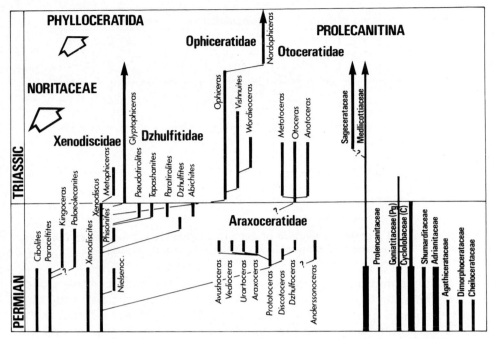

Fig. 6. Ammonoid extinctions and origins at the Palaeozoic/Mesozoic boundary [after Wiedmann (1973c, *Biol. Rev.*, 48), reproduced by permission of the author and editors].

From the prolecanitids there is a simple linear descent to the Middle to Late
Permian Family Xenodiscidae, closely similar to prolecanitids but with cera-
titic sutures, and thence to the Ophiceratidae, the root stock of the major
Triassic radiation of the ammonoids.

Only three ammonoid groups survived the Permian to Triassic phase of
extinctions (Fig. 6) and, of these, the goniatites survived as only a single genus
to die out in the Early Triassic. The prolecanitids survived in the genus *Episage-
ceras* which gave rise to a handful of compressed forms with ceratitic sutures,
the Sageceratidae, which persisted until the Late Triassic (Carnian). The
otoceratids, in contrast, are the rootstock for a major radiation during Triassic
time; as Teichert (1967, p. 191) notes, the ceratites comprise something like
a quarter of all known ammonoid genera but occur in about a tenth of the total
time span of the group.

The bulk of Triassic ammonoids are usually referred to the Ceratitida, with
only a handful of genera placed in an order Phylloceratida. The ceratites are
typified by possession of ceratitic sutures, and the majority show an advance
over Palaeozoic ammonoids in that the primary suture has gained an extra element, becoming quadrilobate (Fig. 7). If Palaeozoic ammonoids show little
morphological diversity, the Triassic ceratites are spectacular in their radiation
with a host of different forms. Involute and evolute taxa evolved, ribbed, carinate and tuberculate; oxycone genera like *Pinacoceras* show the acme of ammonoid sutural complexity by the proliferation of auxiliary elements. There is

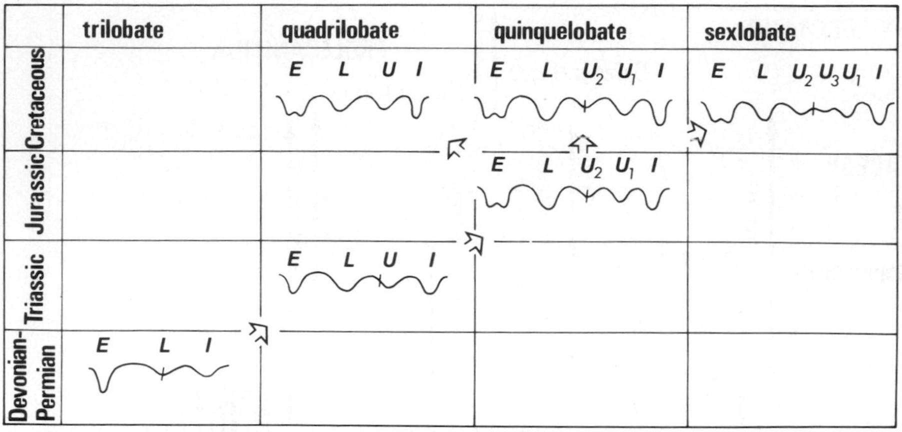

	trilobate	quadrilobate	quinquelobate	sexlobate
Jurassic|Cretaceous		$E \quad L \quad U \quad I$	$E \quad L \quad U_2 \; U_1 \quad I$	$E \quad L \quad U_2 \, U_3 U_1 \; I$
			$E \quad L \quad U_2 \; U_1 \; I$	
Triassic		$E \quad L \quad U \quad I$		
Devonian-Permian	$E \quad L \quad I$			

Fig. 7. Progressive and regressive evolution of ammonoid primary sutures [after Wiedmann (1969, *Biol. Rev.*, 44), reproduced by permission of the author and editors].

widespread homeomorphy at various levels between the main ceratite groups, and during the Late Triassic loosely coiled taxa appear for the first time since the Early Devonian genesis of the ammonoids (Fig. 8). These, referred to the family Choristoceratidae (see Wiedmann, 1973a for a full discussion and illustration), may be planispiral with loosely coiled body chambers (*Choristoceras*), turrilicone (*Cochloceras*) or straight (*Rhabdoceras*). Although a short-lived radiation, these heteromorphs are the earliest manifestation of one of the dominant evolutionary patterns — towards uncoiling — seen in later Cretaceous ammonites.

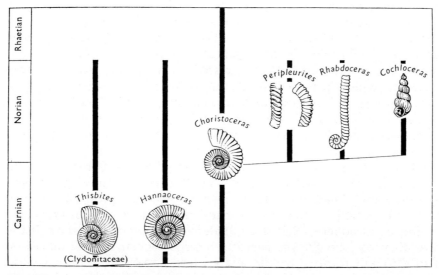

Fig. 8. The phylogeny of the Triassic heteromorphs [after Wiedmann (1969, *Biol. Rev.*, 44), reproduced by permission of the author and editors].

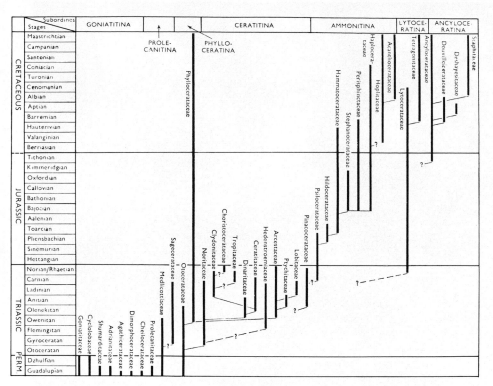

Fig. 9. The phylogeny of the Mesozoic ammonoids [modified after Wiedmann (1973c, *Biol. Rev.* 48), reproduced by permission of the author and editors].

The decline of the ceratites is as marked as their early radiation. The peak of diversity is seen in the 150 or so Carnian genera; this falls to less than a hundred in the succeeding Norian stage, and in the Rhaetian the number is down to single figures.

In contrast, the earliest *ammonites*, as considered here, have a long and inconspicuous history. These earliest forms are the Phylloceratida (Figs. 9, 10) which first appear in the Early Triassic and range to the Late Cretaceous; they are the rootstock of all post-Triassic ammonoids, about 30 genera being referred to the group.

The typical phylloceratid (Fig. 10) is thin-shelled, discoidal, rather involute, with gently inflated flanks and rounded venter. Ornament generally consists of fine striae, or merely growth lines; a few develop strong, fold-like ribs or flares, and constrictions on internal moulds. The suture is highly distinctive, primarily quadrilobate in Triassic forms and quinquilobate in Jurassic and Cretaceous representatives. The internal suture is perhaps the most diagnostic feature, having a simple, unfrilled, *lituid* internal lobe (Wiedmann, 1968a; Kullman and Wiedmann, 1970, pp. 11—14), whilst at a more superficial level, the phylloid, leaf-shaped folioles (saddle endings) are distinctive. Triassic phylloceratids are referred to two families, the Ussuritidae, with typically monophyllic saddles,

Fig. 10. The basic Early Jurassic ammonite stocks. 1, 2. *Lytoceras fimbriatum*, a Pliens-bachian lytoceratid from the Lower Lias of Lyme Regis, Dorset (×0.60). 3, 4. *Phylloceras heterophyllum*, a Hettangian phylloceratid from the Upper Lias of Whitby, Yorkshire (×0.60). 5—7. *Psiloceras erugatum*, a representative of the earliest Jurassic ammonitids from the Lower Lias (Hettangian) of Whitby, Yorks., and 8—10, *Caloceras intermedium*, its descendant from the Lower Lias (Hettangian) of Island Magee, near Belfast, Northern Ireland.

and the Discophyllitidae which show di- and triphyllic terminations. These in turn gave rise to the Phylloceratidae which show further sutural complexity, including tetraphyllic forms and those with spatulate as well as phylloid saddle terminations. These latter are morphologically conservative, and range to the Late Cretaceous.

The origins of the remaining Jurassic and Cretaceous ammonites has been a matter of dispute for many years. In the *Treatise* (Arkell et al., 1957) Arkell admitted that no phylogenetic classification was available and that the origins of many Jurassic groups was unknown; in consequence he followed a scheme in which iterative evolution was given a prominent position with repeated off-shoots from two conservative rootstocks, the phylloceratids and lytoceratids (Arkell et al., 1957, fig. 150), resulting in a 'horizontal' classification, placing the bulk of the ornamented 'trachyostracan' genera into a polyphyletic order Ammonitida.

Since this time, largely as a result of detailed ontogenetic studies by Otto Schindewolf and his school (Schindewolf, 1961—1968; Wiedmann, 1962a—c, 1963, 1965, 1966, 1968a, 1969, 1970, 1973a—c; Kullmann and Wiedmann, 1970), a much clearer picture of ammonite evolution has become available, with classification reflecting phylogeny.

There is thus a complex pattern of mosaic evolution during the latest Triassic (Wiedmann, 1973c; Fig. 11): the phylloceratids continued with little modification and developed a quinquilobate primary suture, whilst a second group, the Lytoceratida, arose perhaps as early as the Norian [*Trachyphyllites* is a lytoceratid according to Wiedmann (1970, pl. 5, fig. 6 and text fig. 8a), although its age is disputed (Wiedmann, 1973c, p. 171, footnote)], probably from the monophyllic *Leiophyllites* (Anisian), although there is a morphological and temporal gap between that genus and the first lytoceratid *Trachyphyllites*.

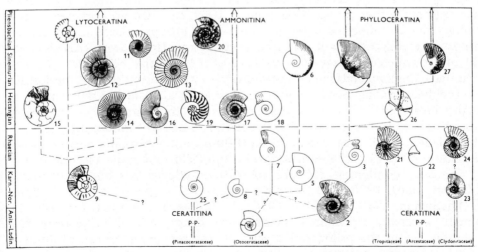

Fig. 11. Microevolution at the Jurassic/Triassic boundary [modified after Wiedmann (1973c, *Biol. Rev.*, 48), reproduced by permission of the author and editors].

Trachyphyllites displays many of the characteristic features of the lytoceratids, being evolute with a rounded whorl section and fine ornament, interrupted by periodic flares and associated constrictions. Late members of the group (Fig. 10) typically develop fine crinkled ribs, a quinquilobate primary suture with a *septal lobe* — a bifid extension of the internal lobe which climbs and is attached to the face of the preceding septum. Lobes and saddles are subdivided into moss-like terminations, including the internal lobe, producing a very different suture to that seen in phylloceratids (Fig. 4). Like phylloceratids, however, the basic lytoceratid stock (approximately 30 genera) is long ranging and conservative in superficial external shell morphology, as is reflected in the long time range of many genera. The group is, however, highly progressive in two respects.

First, in the Barremian they give rise to the Tetragonitaceae, (approximately twelve genera) which possess a sexlobate primary suture and are in this respect the most advanced ammonites (Fig. 7), although their external shell morphology is essentially lytoceratid with weak ornament, flares and constrictions only. Secondly, during the latest Jurassic (Tithonian) and earliest Cretaceous (Berriasian/Ryazanian) the group gave rise to the loosely coiled crioceratitids with quadrilobate primary sutures (Wiedmann, 1969, 1973b; Kullmann and Wiedmann, 1970) which in turn are the origin of a major monophyletic group, the Anycloceratina (over 150 genera), which include the Turrilitaceae, Scaphitaceae and Anyclocerataceae of Arkell et al. (1957), previously regarded as separate offshoots of the lytoceratid rootstock. Within this major group, variation in coiling is fully explored, with helicoid, gyrocone, straight and even more bizarre developments. In some groups (e.g., the baculitids) a loose, uncoiled morphology is retained throughout the history of the group. In the majority of heteromorph families, however, there is a tendency towards progressive re-coiling. This trend was demonstrated independently in crioceratitids and ancycloceratids by Wiedmann (1962c) and Casey (1960), respectively, and is now known in many other groups (Wiedmann, 1969, fig. 16). So complete is this re-coiling, that a number of heteromorph-derived groups were formerly classed with normally coiled forms in the thus polyphyletic Hoplitaceae (Arkell et al., 1957) or Douvilleicerataceae (Casey, 1961). These "false hoplitids" (Douvilleiceratidae, Cheloniceratidae, Parahoplitidae, Acanthoplitidae, Astiericeratidae, Trochleiceratidae, Mathoceratidae, Deshayesitidae) are better classed in the Anycloceratina, whilst, as discussed later, the Jurassic heteromorphs (Family Spiroceratidae) are regarded as derivatives of true Ammonitida.

The origins of the true Ammonitida (800 genera approximately) lie in the Late Triassic (Fig. 11). The earliest group, the Psilocerataceae, are dominantly Early Jurassic, combining features of both phylloceratids and lytoceratids (Wiedmann, 1970, 1973c). Their origin lies in the Carnian—Rhaetian *Phyllytoceras* (Wiedmann, 1970, pp. 1009—1010, pl. 10, figs. 5, 6 and text figs. 16, 17). Although possessing a primarily quadrilobate suture, it is in other respects a psiloceratid. From the psiloceratids arise all the remaining groups of Jurassic and Cretaceous ammonites; there is no support for an independent origin of the eoderoceratids in the lytoceratid rootstock, for they show sutural confor-

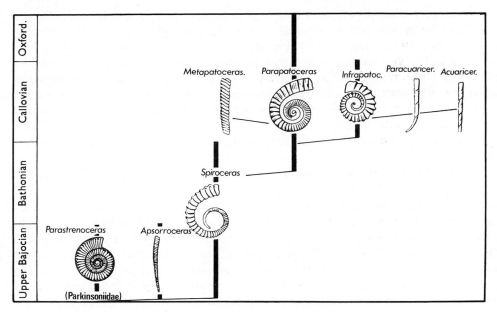

Fig. 12. The phylogeny of Jurassic heteromorphs [after Wiedmann (1969, *Biol. Rev.*, 44), reproduced by permission of the author and publishers].

mity to the psiloceratid pattern (Schindewolf, 1962; Kullman and Wiedmann, 1970, p. 15), as do the dactylioceratids. Relationships of the Middle Jurassic ammonites still require minor clarification, but a monophyletic psiloceratid origin has been demonstrated. The Middle Jurassic heteromorph family Spiro-ceratidae (Fig. 12) can be derived as a monophyletic unit from *Strenoceras*, a loosely coiled member of the Parkinsoniidae (Schindewolf, 1961—1968; Westermann, 1956).

The supposed Pliensbachian heteromorph *Arcuceras* Potonié, 1929, is a pyritised crinoid stem (Donovan and Hölder, 1958). If the false hoplitids were to be removed, there would be few problems in deciphering the broad phylogenetic relationships of Cretaceous Ammonitina (Fig. 9). The Desmocerataceae (and hence the 'true' hoplitids and the acanthoceratids) derive from the haploceratids (Wiedmann, 1966) rather than the Phylloceratina, as suggested by Wright (1952, 1955), Arkell et al. (1957), and Casey (1957, 1961); there is no transition from the lituid internal lobe of phylloceratids to the denticulate internal lobe of the Ammonitida.

The Origins of Higher Categories

It appears to be generally true that higher categories arise and diversify rapidly, that ancestral forms are commonly rather small, generalized, and often rare. The group demonstrates, in other words, many of the universal aspects of origin of higher categories so clearly summarised by Simpson (1953).

As examples one can cite the origin of ammonoids as a whole in the bactritids of the Early Devonian, although here accompanied by a well documented transition at generic level (p. 254, Fig. 3). The origin of the whole of the goniatitids lies in the generalized anarcestid *Manticoceras*. The aberrant Clymeniina with dorsal siphuncles are of cryptic origin, appearing and diversifying suddenly during the Fammenian (Late Devonian: see House, 1961). The Lytoceratida are separated from their presumed ancestor *Trachyphyllites* by a temporal and morphological hiatus, whilst the protean lytoceratid *Phyllytoceras* is known from rare tiny juveniles only. The origins of the Anycloceratina are sudden but unknown.

At lower taxonomic levels, similar criteria apply. Indeed, the diagnoses of many families and superfamilies given by Arkell et al. (1957) are concerned more with origins than morphological criteria, as they rightly should be, and an inferred monophyletic origin in a single genus is the general pattern. The origin of the whole of the superfamily Scaphitaceae thus lies in the diminutive and rather rare *Eoscaphites*, whilst the Douvilliceratacea and Deshayesitaceae appear to have monophyletic origins in the genera *Paraspiticeras* and *Hemihoplites*, respectively (Wiedmann, 1966). The Baculitidae arise from the solitary genus *Lechites*. Amongst normally coiled forms, the family Acanthoceratidae, in excess of thirty genera, have their origins in species of *Stoliczkaia*, the Tetragonitaceae in the diminutive *Eogaudryceras*, the Echioceratidae in *Epophioceras*, and so on.

Evolutionary Patterns at Family Level

From the preceding paragraphs one can generalize on the monophyletic origin of families from single genera. Beyond this, common patterns frequently emerge, and although members of given families may be linked by special morphological features, such as the tabulate venters of kosmoceratids or corded keels of amaltheids, the passage across adaptive barriers by pioneer genera leads to repetitive gross morphology and homeomorphy. In the case of a few selected families I have attempted to summarize the data in Table I. To generalize fur-

TABLE I

Comparative morphotypes between superfamilies

Morphotype	Psilocerataceae	Eoderocerataceae	Acanthocerataceae
Platycone	*Psiloceras*	*Gemmellaroceras*	—
Oxycone	*Oxynoticeras*	*Pseudamatheus*	*Manuaniceras*
Capricorn	*Saxoceras*	*Androgynoceras*	*Brancoceras*
Coronate	*Pseudotropites*	*Coeloceras*	*Fagesia*
Serpenticone	*Laquaeoceras*	*Protechioceras*	*Parabrancoceras*
Ribbed/carinate	*Paracaloceras*	*Paltechioceras*	*Peroniceras*
Dwarf	*Cymbites*	*Pimelites*	*Falloticeras*

265

Fig. 13. The principal liparoceratid morphotypes. 1, 2. A capricorn, represented by the microconch *Androgynoceras maculatum* from the Lower Lias (Pliensbachian) of Robin Hood's Bay, Yorkshire. 3, 4. A sphaerocone, represented by the macroconch *Liparoceras cheltiense* from the Lower Lias (Pliensbachian) of Leckhampton, Gloucestershire. 5. A vari-costate, with capricorn inner and massive outer whorls, but in fact also a macroconch, represented by *Androgynoceras* aff. *intracapricornus* from the Lower Lias (Pliensbachian) of the Yorkshire Coast.

ther, one detects repeatedly the radiation of families into oxycone, cadicone, serpenticone and platycone taxa, with repeated dwarf, hypernodose and giant forms also arising. A consequence of this is the frequent linking of homeomorphs into polyphyletic groups, in ignorance of their actual descent.

General detailed studies on the evolution of particular families are available and have been held to provide interesting insights into the pattern of this type of radiation. Perhaps the best-known example is the family Liparoceratidae. This highly polymorphic group is restricted to the Early Jurassic (chiefly Pliensbachian) and is of world-wide distribution. Within the family, three broad morphotypes can be recognized (Fig. 13).

(1) Sphaerocones — that is to say involute, globular shells with a small, or sometimes occluded umbilicus, typified by the genera *Liparoceras* (including *Becheiceras, Parinodiceras, Vicinodiceras*).

(2) Capricorns — more evolute, round-whorled forms ornamented by strong blunt ribs, resembling a goat's horn, typified by the genera *Beaniceras, Oistoceras* and *Aegoceras.*

(3) "Hybrid" or varicostate forms with capricorn inner whorls, and feebly ornamented *Liparoceras*-like outer whorls. These are typified by certain *Liparoceras* species such as *L. naptonense, Androgynoceras* such as *A. subcontractum,* and *Oistoceras* such as *O. allaeotypum.*

The interest of the group lies in the varicostates. Initially, Hyatt (1889), in fortunate ignorance of stratigraphic relations, claimed that the three types were a clear example of recapitulation (Fig. 14). The sphaerocones were thus descended from the capricorns via the hybrids; the latter preserving irrefutable indication of ancestry in their inner whorls. Fifty years later, Spath (1938) produced a substantial monograph of the Liparoceratidae based on detailed collections from the Lias of the Dorset Coast in southern England. Spath's dia-

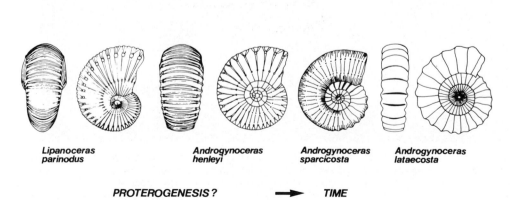

Fig. 14. Liparoceratid evolution. Hyatt's palingenesis or Spath's proterogenesis?

gram (a modified version of which is shown in Fig. 14) summarizes his view
that the series in fact demonstrated proterogenesis — that is to say, the capricorns evolved from the sphaerocones via the varicostates, and the inner whorls
of the latter foreshadowed the adult form of their descendants. Spath could
not, however, show definitely that varicostates in fact preceded capricorns,
for the critical parts of the section are poorly fossiliferous on the coast, most
varicostates studied by him coming from the English Midlands and lacking sufficient stratigraphic control to be placed in sequence. In spite of this, the
liparoceratids are cited as a classic example of proterogenesis.

Twenty-five years later, Callomon (1963) brought together the results of
detailed collecting of all forms in stratigraphic sequence in the knowledge that
sexual dimorphism was widely if not universally present within the ammonoids.
It is now clear that sphaerocones reach a large size and have simple apertures
when adult — they are macroconch females — and capricorns are mature at a
quarter of this size and develop rostra — they are microconch males. Both

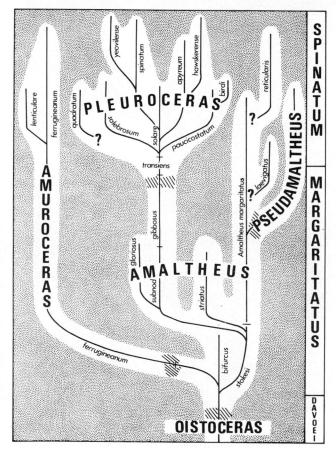

Fig. 15. The evolution of the Jurassic ammonite family Amaltheidae (redrawn after Howarth, 1958) (reproduced by permission of the author and the Palaeontographical Society).

Fig. 16. The basic amaltheid morphotypes. 1—3. *Oistoceras* aff. *angulatum*, the microconch liparoceratid which is believed to give rise to the amaltheids. Specimen is from the Lower Lias of Lincoln. 4—6. *Amaltheus bifurcus*, the earliest amaltheid. Note the progression in development of the chevron ornament from *Oistoceras* via this species to the *Amaltheus*

groups are said to range throughout the same interval (although this is not my own experience!), whilst the varicostates are also macroconchs, and thus the females of other microconch capricorns.

A second family whose evolution is widely cited is the Amaltheidae (Howarth, 1958). This group is derived from the liparoceratids, its origin lying in the planulate capricorn *Oistoceras* (Figs. 15, 16) which gave rise to the oxycone *Amaltheus stokesi* via *Amaltheus bifurcus*. This last species has compressed and evolute outer whorls and relatively evolute inner whorls of circular cross-section, the evolutionary sequence in the group being thus recorded in the individual development of the intermediate species.

From *Amaltheus stokesi* arise, in one direction, the dwarf, presumably neotonous *Amauroceras* and the oxyconic *Amaltheus margaritatus* which in turn gives rise to the extreme strigate oxycone *Pseudamaltheus*. In another direction arose a series of relatively inflated, strongly ribbed species (e.g., *A. subnodosus*, *A. gibbosus*), which in turn lead to the ribbed planulate *Pleuroceras*, itself the origin of a minor radiation of eight closely related species.

Howarth's work is of great precision, with careful stratigraphic control, and its importance is not only to precisely illustrate the diversification within a family, but to demonstrate that planulates could evolve into oxycones, and these in turn could lead to planulates — refuting Dollo's law of irreversibility of evolution as it is usually (and mistakenly — see Gould, 1970) represented.

Given the possibility of sexual dimorphism in the group we must surely re-examine this elegant picture, however, *Amaltheus stokesi* and *margaritatus* could be regarded as macroconch females, whilst the ancestral *A. bifurcus* is a micromorph, if not a microconch, as are *A. wertheri, striatus* and *gibbosus* amongst others, Equally, the *Pleuroceras* species are mature at dissimilar sizes such as to again suggest the possibility of dimorphism. It should be noted, however, that it is not possible to link all of the amaltheids into dimorphic pairs.

The somewhat later, mid-Jurassic family Cardioceratidae display a similar range of variation in morphology to the amaltheids, but have been cited as examples rather of evolutionary parallelism. Thus a broad sequence is held to lead from ancestral *Macrocephalites* to progressively more compressed genera, *Quenstedtoceras* → *Cardioceras* → to the oxycone *Amoeboceras* and 'dwarf' *Nannocardioceras* (Arkell and Moy-Thomas, 1940). It has, however, been envis-

---

margaritatus illustrated as 7—8. Specimen is from the Middle Lias Down Cliff Sands of Eypesmouth, Dorset (×1.6). 7, 8. *Amaltheus margaritatus*, a juvenile from the English Middle Lias. 9, 10. *Amaltheus subnodosus*, a specimen of the tuberculate amaltheids which are the earliest members of the line giving rise to *Pleuroceras*. Specimen is from the Middle Lias of Yeovil, Somerset. 11—13. *Pleuroceras transiens*, the earliest *Pleuroceras*, showing intermediate characters towards *Amaltheus*. From the Middle Lias of the Hebrides (?). 14—16. *Pleuroceras solare* showing an intermediate form tending towards the more characteristic square whorls of *Pleuroceras*, more fully developed in 17—19. *Pleuroceras hawskerense* from the Middle Lias of Whitby, Yorkshire. 20, 21. *Amauroceras ferrugineanum*, a typical member of the dwarf offshoot of the *stokesi-margaritatus* lineage. From the Middle Lias of Hawsker Bottoms, Yorkshire. 22. *Pseudamaltheus engleharti*, a juvenile specimen of the giant strigate offshoot of *Amaltheus* s.s. From the Middle Lias Marlstone Rock Bed of South Petherton, Somerset.

aged as a series of lineages crossing morphological grades in parallel, with genera as polyphyletic cross-sections of a plexus and its side branches. Rather, the scheme must be reappraised taking dimorphism into account and applying rational species concepts, for Arkell's parallel lines are, at any one stratigraphic level, adjacent sections of a morphological continuum.

The family Acanthoceratidae, a major group of normally coiled, strongly ribbed and tuberculate Cenomanian and Turonian genera provides a further example of this evolutionary pattern. The origin of this group lies in a single Late Albian genus, *Stoliczkaia*. Nuclei of this genus, which has feebly ribbed, generally non-tuberculate outer whorls, are miniatures of a range of Early Cenomanian genera, including *Mantelliceras, Sharpeiceras, Acompsoceras* and *Graysonites*, to which it gave rise through proterogenesis, whilst *Cottreauites* and *Paracalycoceras* appear as gradational forms, and a dwarf genus *Neosayno-ceras* is cryptogenic. The principal early Middle Cenomanian genera, *Calyco-ceras* and *Acanthoceras*, arise in quite different ways — *Acanthoceras* from *Acompsoceras*, nuclei of which are miniature versions of their descendant, whilst *Calycoceras* is of cryptic origin, perhaps in *Mantelliceras*. Both genera gave rise to a series of neotenous offshoots: *Acanthoceras* to *Protacanthoceras*, and *Calycoceras* (*Gentoniceras*) to subgenera such as *C.* (*Newboldiceras*) and *C.* (*Lotzeites*). There is also simple progression, as from *Acanthoceras* to *Euom-phaloceras* and *Kanabiceras*. Equally, progressive size increase may lead from *Protacanthoceras* to *Thomelites*, and, with a series of intermediates, to *Metoi-coceras* and *Jeanrogericeras*. Progression, size increase, and neoteny are thus recurring, and sometimes alternating themes, with the repeated appearance of additional genera of cryptic origins.

Evolution at the Species Level

Detailed studies of lineages have been a favourite exercise for ammonite workers, and have generally supported ideas of phyletic gradualism. Classic examples include the kosmoceratids and lyelliceratids documented by Brinkmann (1929, 1937), Howarth's amaltheids (Howarth, 1958), the gastroplitids (Reeside and Cobban, 1960), *Scaphites* and *Baculites* of the U.S. Interior (Cobban, 1951a, b et seq.) amongst many others. A selected set of such examples is discussed further.

Baculites and Scaphites from the U.S. Western Interior

Late Cretaceous faunas of the U.S. Western Interior Seaway are amongst the best documented in the world, and provide the most satisfactory examples of lineage studies, this time in heteromorphs.

The *Baculites* of the region (Fig. 17; Cobban, 1951a, 1952, 1958, 1962a,b, 1973; Scott and Cobban, 1965; Cobban and Scott, 1972) have their origins in a small, smooth species, identified as *B. yokoyamai* Tokunaga and Shimizu (otherwise best known from the Coniacian of Japan). This first appears in

Fig. 17. Typical representatives of the Upper Cretaceous baculitid radiation in the U.S. Western Interior. 1, 2. *Baculites* cf. *yokoyamai*, the earliest small, feebly ornamented species, from the Turonian Bridge Creek Limestone of Pueblo, Colorado. 3, 4. *Baculites codyensis*, a Santonian species developing flank ribbing typical of endemic species. 5, 6. *Baculites asper*, showing a similar basic type of ribbing to *B. codyensis* although differing sufficiently for specific separation. 7. *Baculites eliasi*, the last member of the lineage, from the Maas-

272

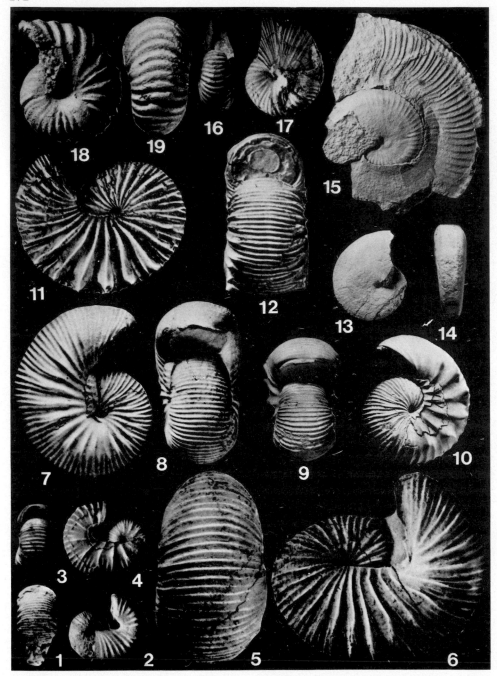

Fig. 18. The U.S. Western Interior endemic *Scaphites*. 1, 2. *Scaphites delicatulus*. The Late Cenomanian ancestor of the radiation. 3, 4. *Scaphites larvaeformis*, its Early Turonian descendent. 5, 6. *Scaphites mariasensis*, one of the larger forms which evolved from the tiny Cenomanian—Turonian rootstock. 7, 8. *Clioscaphites*, represented here by *montanensis*, are a group which develop trifid lobes, and tend to recoil, as in: 9, 10. *Clioscaphites saxitonianus*, a recoiled Santonian offshoot of the main *Clioscaphites* lineage. 11, 12. *Clioscaphites platygastrus*, a tabulate-ventered Middle Santonian species, ancestral to *Haresiceras*. 13—15.

Early Turonian sediments over an area extending from the north end of the Black Hills uplift to eastern Colorado and central Kansas. By mid-Turonian times the species is restricted to a narrow belt from Utah to central Kansas (Kennedy and Cobban, 1976a, fig. 10), after which it disappears briefly to reappear in the Late Turonian. The species is small and smooth with an elliptical or subelliptical whorl section and moderately simple suture; the only detectable evolutionary change seen during the Turonian is progressive size increase. During the Early Coniacian *B. yokoyamai* gives rise to *B. mariasensis*, which has a narrower venter, faint ventral ribbing, and nearly smooth flanks. From this species arises a lineage of ribbed species, slowly increasing in size, ornament and sutural complexity, as well as showing changes in whorl section. The lineage culminated in the Late Campanian species *Baculites rugosus*, which reaches diameters as large as 110 mm. Later forms tend to show a progressive *decrease* in size and simplification of sutures, the last of the twenty or so species in the lineage being *Baculites eliasi*, an Early Maastrichtian form.

This pattern is in marked contrast to that shown by the contemporaneous more widely distributed Gulf Coast and eastern seaboard species which appear to have evolved at a much slower rate, with species tending to be rather small and retaining a simple suture until the Late Campanian when there is rapid size increase. This latter group penetrated only occasionally into the Interior until the Late Campanian; but with the disappearance of the last endemic *Baculites*, they come to dominate Interior faunas.

The endemic scaphitids of the Interior (Fig. 18, 19; Reeside, 1927a—c; Cobban, 1951b, 1964, 1969, Cobban and Scott, 1964) have their origins in the Late Cenomanian, the earliest species being a small, widely occurring form *S. delicatulus*. From this arise a succession of Turonian species showing variable and sometimes restricted geographical distributions within the seaway (although some range as far as western Greenland). Evolutionary changes involve changes in coiling, ornament, section and sutures, together with progressive size increase, the latter culminating in the Early Santonian *Scaphites depressus* (Fig. 18) which is tightly coiled, with the most complex sutures. By mid-Santonian times, these heteromorphs had re-coiled and the sutures had developed trifid lobes such that they have been separated into a genus *Clioscaphites* restricted to the U.S. Interior and western Greenland. This in turn gave rise to two distinctive genera during the Santonian interval, *Desmoscaphites* with constrictions on the inner whorls, and compressed later whorls, and compressed involute *Haresiceras* with tabulate venters (Fig. 18). Within all these lineages, phyletic gradualism appears as the predominant mechanism of change as Cobban (1951b, 1964) has carefully described. *Desmoscaphites* is believed to originate from *Clioscaphites vermiformis*, and *Haresiceras* from the contempo-

---

Haresiceras spp., the recoiled, platycone end members of the *Clioscaphites—Haresiceras* lineage. 16—19. *Desmoscaphites erdmanni*, the first *Desmoscaphites* which evolved from *Clioscaphites* in the Late Santonian. (18, 19 are figured ×2.4 to show the diagnostic constricted inner whorls.

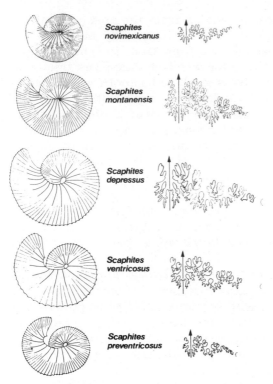

Fig. 19. Size change, recoiling and evolution of sutures in Upper Cretaceous scaphitids from the U.S. Western Interior (modifed after Cobban, 1951b, and reproduced with the author's permission).

raneous "var." *toolensis*, presumably as a result of temporary geographical isolation of populations of each.

A more detailed study which takes into account the universal sexual dimorphism in *Scaphites* is Cobban's work on the Late Santonian and Campanian species *S. leei* and *S. hippocrepis*. Collections from seventeen hundred feet of shales allowed division of both *leei* (older) and *hippocrepis* (younger) into three successive subspecies, all of which show dimorphism. The oldest form *S. leei* I has a coarsely ribbed body chamber and ventrolateral tubercles on both body chamber and phragmocone. *S. leei* II lacks nodes on the phragmocone. *S. leei* III is more densely ribbed than II, and females have fewer umbilical tubercles. *S. hippocrepis* I is in turn more densely ribbed and has weaker ventrolateral tubercles which tend to elongate. Increasing rib density and strengthening tubercles mark *S. hippocrepis* II; the final member of the lineage is very densely ribbed and developed a row of incipient lateral tubercles. This evolutionary sequence takes place in an interval of approximately four million years and appears gradational, although, as with the *Baculites* discussed, there are thicknesses of sediment between the levels of collections. Sharp changes, if they did occur, would have been minor, however.

Brinkmann's Kosmoceras

Perhaps the best known, and certainly the most widely cited example of gradual, progressive evolutionary change in ammonites, is Brinkmann's (1929) work on the Callovian genus *Kosmoceras*. Brinkmann collected over 3000 crushed specimens from the cyclic shale, bituminous shale, and shell bed sequences of the Lower Oxford Clay at Peterborough, England. Figs. 20 and 21 shows Brinkmann's essential conclusions, some of which have met with subsequent support, in particular from the work of Callomon (1955). There are thus progressive changes in size, in the ratio of outer ribs to tubercles, and in the length of lappets.

Brinkmann's four lineages were subdivided into a series of species and, without entering into sterile debate on species concepts, appear to illustrate both gradualism and parallel evolution. Our views on the latter must of course, be modified. Brinkmann's lineages consist of two macroconch and three microconch series, referred originally to *Zugokosmoceras* [M], *Kosmoceras* [M],

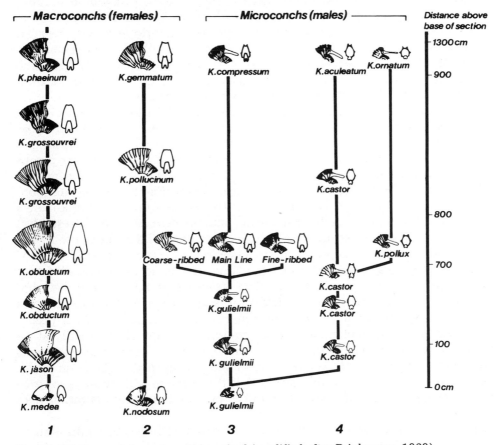

Fig. 20. Brinkmann's kosmoceratids revised (modified after Brinkmann, 1929).

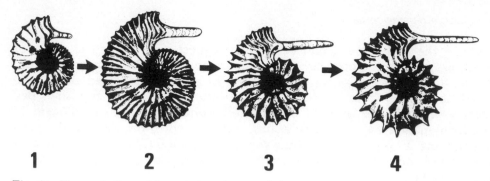

1   2   3   4

Fig. 21. The evolution of lappets in microconch kosmoceratids (modifed after Brinkmann, 1929).

Guieliemiceras (his "*Anakosmoceras*") [m] with two short lived offshoots, and *Spinikosmoceras* [m].

Not least amongst the difficulties with this study is (as we can now see it) the apparent sympatric speciation of males. More relevant is Callomon's (1963) conclusion that Brinkmann's "method of work" (Brinkmann, 1929, p. 27) was open to criticism. Not only do Brinkmann's statistics not do justice to the great variability of forms in fact found, but also Brinkmann collected his specimens, noted their horizon, referred them to *genus and species*, measured them and then discarded them (Callomon, 1963, p. 36). In other words his statistics could do nothing else but support his species determinations.

Let us give the final word to Callomon (1963). . . "Most of the shells in fact belong only to two of Brinkmann's subgenera, *Zugokosmoceras* and *Guieliemiceras*; the impression in the field that the rarer shells of other types belong to separate species is hard to resist. However, in cases where material was plentiful, I formed the opposite hypothesis (to Brinkmann's), that all the macroconchs and microconchs in one bed each form a single highly variable and nongaussian population, which is equally hard to refute. Certainly Brinkmann's samples were statistically quite inadequate, and in no case does he establish branching of lineages in terms of actual bi- or polymodal distributions. To my mind the most telling single piece of evidence in favour of the genetic unity of the whole group lies in the onset of bundling of secondary ribs in *all* forms simultaneously at the base of the Athleta Zone (level 1094 cm at Peterborough). It is too striking to be coincidental, and if we abandon Brinkmann's phylogenetic scheme the hypothesis of dimorphism does, as he himself points out, provide a simple explanation of the rigorous division of the family at all levels into macro and microconchs, and the remarkable parallel development occurring among them, both of which are undoubtedly real". The main evolutionary change in this classic lineage study is thus sudden with subsequent evolutionary changes being of a minor nature — punctuated equilibria rather than phyletic gradualism.

In terms of size there are, however, overall progressive changes (Brinkmann 1929, table 9, p. 103, table 63, p. 133; Callomon, 1955, figs. 3, 4), also in laplet length. This latter change (Fig. 21) takes us back to previous suggestions that evolution of this feature may be sexually selected, whilst the changes in ornament and size presumably had functional, perhaps mechanical significance. The enormous length of lappets in the *ornatum*-type of *Spinikosmoceras* (Fig. 21) recall Huxley's (1938) comment that "with increasing intensity of selection between males, epigamic characters that are hypertelic or even useless to the species are developed, and with polygamy, such characters may be evolved to the limits of mechanical possibility and even be disadvantageous to their possessors in all respects save mating". As in the Irish Elk *Megaceros*, so *Kosmoceras*?

Neogastroplites

Albian ammonite faunas of the northern parts of the Western Interior region of the U.S.A. are dominated by the hoplitid ammonite *Neogastroplites*, an essentially endemic taxon (Kennedy and Cobban, 1976a, p. 60) which comprises in excess of 95% of most collections of ammonites from the Mowry Shale and its equivalents of the area, being accompanied only by the pseudo-ceratite *Metengonoceras*. This group colonized a broad and generally rather shallow gulf (see Gill and Cobban, 1966, for map) of reduced salinity (Eicher, 1965), which extended from the Arctic to northern Wyoming, and provides what appears to be a clear example of phyletic gradualism. The sequence was the subject of a careful analysis by Reeside and Cobban (1960). These authors collected five large samples (approximately 300 to 800 individuals) from individual concretions, on the basis of which a remarkable range of intraspecific variability was documented. On the basis of stratigraphic relations and supplementary data from smaller collections from other localities, five successive species were recognized: *Neogastroplites haasi* (oldest); *N. cornutus*; *N. mulleri*; *N. americanus*; and *N. maclearni* (youngest).

Several general points of interest arise: (1) the overall range in gross morphology (e.g., compressed and smooth to depressed and spinose) is similar in all species, (2) the proportion of variants fluctuates drastically from level to level; (3) interspecific differences are subtle, involving details of ornament and form, and are overwhelmed by intra-specific variation.

This work includes a wealth of numerical data which should be subjected to analysis to reveal the overall evolutionary trends (if any) other than shifts in population structure, for these are at present most difficult to determine. The suite appears, however, to demonstrate gradualism, but as with so many other examples, the sampling is insufficient. In their table 2, Reeside and Cobban show substantial barren gaps in their sequence so that, once again, evolution may have been either step-wise or gradual.

Other, less completely documented examples of species transitions are numerous. Ziegler (1957, 1959) demonstrated size increase and decrease in umbilical diameter in male *Creniceras* from the Kimmeridgian (Late Jurassic) of southern Germany, the two being said to show no correlation. In fact, his succession shows an apparent jump in size increase. A further case of punctuated equilib-

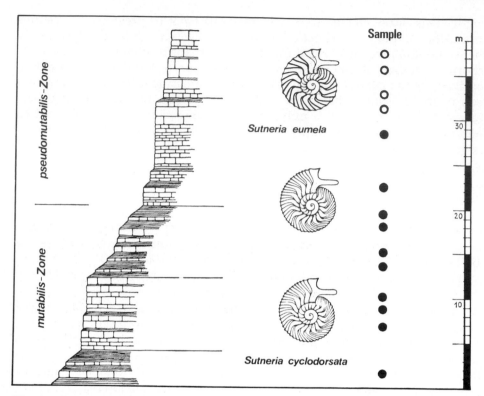

Fig. 22. The evolution of the Jurassic perisphinctid *Sutneria* [modified after Ziegler (1959, *Evolution*, 13), and reproduced by permission of the author and editors].

rium — saltation has been suggested in the *Sutneria cyclodorsata—S. eumela* sequence, again in the South German Kimmeridgian, the study being based on microconch males (Fig. 22). *S. cyclodorsata* is said to be rare, small, with rounded whorls and fine, triplicate and quadriplicate, rursiradiate (apically flexed) ribs; *S. eumela* shows coarse, duplicate ribbing. Other ammonites show no such hiatus in morphology at the same level, whilst there is said to be no sedimentological evidence for either condensation or non-sequence. Ziegler (1959, p. 233) also cites two further apparently sudden speciation events in *Sutneria*, with the earlier *Sutneria galator* to *platynota* transformation close to the Oxfordian—Kimmeridgian boundary. There is the possibility that sedimentary condensation may be a control, but the Middle Kimmeridgian *S. subeumela* to *S. eumela* transformation appears real.

Ziegler's (1958) *Glochiceras* study is more difficult to appraise, again involving males only, but demonstrates one of the common remarks made by workers describing what they believe to be horizons preserving explosive multiple speciation events — "owing to the great variability of morphological characters it is now very difficult to distinguish the various species". The range of variation seen in Cobban and Reeside's gastroplitids (p. 277) is probably typical of many ammonite groups, and the bulk of supposed co-occurrence and presumed sympatric speciation described in the literature represent in my view no more than the splitting of continuous variation series (or pairs of dimorphic series) into parallel lineages, as with the example of Westermann's (1966) revision of the Bajocian (mid-Jurassic) ammonite *Sonninia (Euhoploceras) adicra*. Sixty-four "species" described from England by S.S. Buckman and nine other European "species" all fall within the variation range of the macroconch form, whilst five other "species" are synonyms of the microconch (*S. subdecorata*).

The apparently high contemporaneous species diversity of genera and the consequent problem of mechanisms of evolutionary origins suggested by some workers are all an artefact of taxonomic splitting rather than a picture of phygeny, e.g., Spath (1923—43) dealing with Albian forms, Casey's (1960 onwards) deshayesitids, douvilleiceratids and otohoplitids, Arkell's (1935—1948) myriad contemporaneous cardioceratids and perisphinctids, Collignon's (1971) baculitids, and Van Hoepen's (1931—1966c) mortoniceratids. This has led to curious problems for ammonite workers, not the least of which are situations such as "species" of the contemporaneous Late Albian "genera" *Lepthoplites, Pleurohoplites* and *Callihoplites* giving rise to "species" of *Schloenbachia*, whereas but a single species plexus is involved.

Such problems extend to generic level. Thomel (1972, fig. 3) shows a series of more or less contemporaneous radiations amongst Cenomanian acanthoceratids in which new "genera" and "subgenera" are in some cases based on what I believe to be no more than variants of individual species.

Sutural Evolution

Viewed broadly, Mesozoic ammonoids are characterized by progressive developments of the suture, involving changes in the number of lobes in the primary suture and in the denticulation of the adult suture (Kullman and Wiedmann, 1970).

Considering the first point, the broad temporal changes are summarized in Fig. 17. Quadrilobate sutures are dominant in the Triassic ceratites, although a quinquilobate primary suture is seen in some, for instance the Arcestidae. The phylloceratids have a quadrilobate primary suture during the Triassic, whilst all Jurassic and Cretaceous forms have quinquilobate primary sutures. This is equally true of the Ammonitida and the bulk of the Lytoceratida (e.g., the Lytocerataceae), a sexlobate primary suture appearing only in the Tetragonitaceae. There are, however, reversions to the general trend; the anyloceratids, derived from quinquilobate lytoceratid ancestors, have a quadrilobate primary suture, as do their normally coiled descendants the 'false hoplitids' (Wiedmann, 1966). The second general trend towards frilling is widespread, both in the overall change from Triassic to Jurassic—Cretaceous forms, and within individual groups.

Detailed ontogenetic studies demonstrate allometric increase in sutural complexity (Newell, 1949), presumably as a mechanical response. Changes in sutural complexity demonstrated in detailed lineage studies (pp. 270—274)

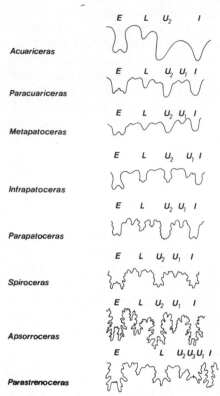

Fig. 23. The phylogeny of sutural simplification in Jurassic heteromorphs [modified after Wiedmann (1969, *Biol. Rev.*, 44), and reproduced by permission of the author and publishers].

commonly accompany and may be a result of size changes in symmetry and the configuration of subdivisions (as in *Clioscaphites*, p. 274), where other factors are clearly relevant.

Equally, simplification of sutures occurs, as is demonstrated by the various Mesozoic pseudoceratitic groups. Many forms with these secondarily entire lobes and saddles are cryptogenic (e.g., Flickiidae, Engonoceratidae), but progressive reduction has also been demonstrated — as in the Jurassic heteromorphs (Fig. 23).

Divergence, Convergence and Parallelism

Divergence

The examples of evolutionary patterns at high and low taxonomic levels already cited illustrate very clearly the widespread documentation of patterns of evolutionary divergence. On a broad view, the ceratites and ammonites are two major radiations, the products of one filling the ecological vacuum left by the demise of the other. At family and superfamily level, similar radiations leading to diverse morphotypes (with consequent homoeomorphy) is the rule, as with the superfamilies of Jurassic ammonites or, say, the families Kosmocerati-dae and Cardioceratidae. No further comment is necessary.

Convergence

Limited to a tubular external shell with finite limits on variation in coiling, ornament and sutures imposed by functional constraints, it is scarcely surprising that evolutionary convergence — homoeomorphy — is widespread in ammonites. Very complete reviews and documentation of the phenomenon are given by Schindewolf (1938, 1940), Haas (1942), Reyment (1955) and Kennedy and Cobban (1976a). It may extend to the whole shell (Fig. 24) or to specific features, such as the suture line. Synchronous and heterochronous homoeomorphy are widespread both between taxonomically distant groups, and between stratigraphically separated members of lineages, as with some of Cobban's baculitids. It is perhaps the repeated recurrence of similar morphotypes which, more than any other feature, should confirm the adaptive significance of shell shape and ornament to the doubting.

A brief glance through the pages of volume L of the *Treatise* (Arkell et al., 1957) is perhaps the best way to appreciate the scale of convergence in the group, whilst it should be added that some features of the classification adopted in the *Treatise* are in fact the result of the grouping of polyphyletic homoeomorphs (Wiedmann, 1963, 1966, 1968a, 1970). The stratigraphic and phylogenetic consequences of failure to recognize homoeomorphy in practice are discussed more fully elsewhere (Kennedy and Cobban, 1976a).

Parallelism

The literature of ammonite evolution contains many references to parallelism. Unfortunately, the most quoted examples appear to be spurious. The parallel changes in Brinkmann's kosmoceratids are for instance, parallelism between sexual dimorphs! The supposed parallelism in the Cardioceratidae, described by Arkell and noted previously (p. 269), in part represent this, and in part represent the failure to appreciate intraspecific variation with consequent over-splitting at successive horizons. It is hardly surprising that the compressed and hypernodose "species" at either end of successive populations evolve in parallel. One thus becomes very suspicious of Brinkmann's famous *Leymeriella* (1937) and the Wrights' *Hyphoplites* (1949).

There are, however, some more convincing examples, notably Wiedmann's crioceratitids and Casey's ancyloceratids (Fig. 25). Even here, however, the situation would be a happier one if the nature of the sexual dimorphism (if any) in these groups were known, and could be eliminated (if necessary) from the picture.

Fig. 24. Homeomorphy and evolutionary convergence. 1, 2. *Brancoceras helcion*, a Lower Cretaceous (Albian) brancoceratid from Escragnolles. 3, 4. *Androgynoceras artigyrus*, a Lower Jurassic (Pliensbachian) eoderoceratid from the Lower Lias of Stroud, Gloucs. 5. *Trachyceras (Protrachyceras) meeki*, a ceratite from the Middle Triassic of Cottonwood Canyon, Nevada. 6. *Discoscaphites conradi*, a re-coiled scaphitid heteromorph from the

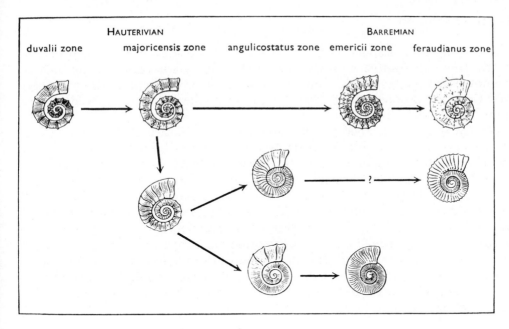

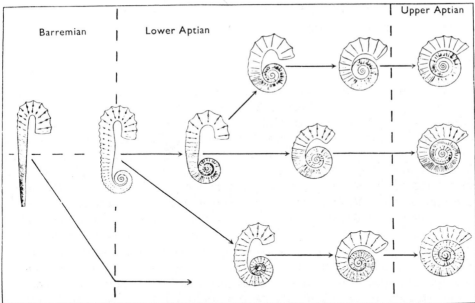

Fig. 25. Parallel evolution in Cretaceous heteromorphs [modified after Wiedmann (1969, *Biol. Rev.*, 44), and reproduced by permission of the author and publishers, and after Casey, with permission from the author and the Palaeontographical Society).

Upper Cretaceous (Maastrichtian) Fox Hills Sandstone of South Dakota. 7. *Discotropites sandlingensis*, a ceratite from the Late Triassic Hosselkus Limestone of Shasta County, California. 8. *Oxytropidoceras (Manuaniceras) hubbardi*, Lower Cretaceous (Albian) mojsiso-vicziid.

Iterative Evolution

The complex classification of Jurassic and Cretaceous ammonites proposed in the *Treatise*, together with the admitted polyphyletic origin of the bulk of decorated ammonites, stems from the acceptance of the hypothesis of iterative evolution first proposed by Salfeld (1913) and expounded by Frebold (1922), Spath, Arkell and others.

Briefly stated, this theory held that the conservative Phylloceratida and Lytoceratida gave rise to a series of decorated groups throughout the Jurassic and Cretaceous: the spurious Arcuceratidae, the Spiroceratidae, Derolytoceratidae, Dactylioceratidae, Protetragonitidae, Bochianitidae, Scaphitidae, Macroscaphitidae and Ptychoceratidae from the Lytoceratida; the Haploceratidae (?) Juraphyllitidae and Desmoceratidae from the Phylloceratida. As such, the group has become a key example of this type of evolutionary pattern (e.g., Raup and Stanley, 1971, fig. 10—26). As with the other "classic" evolutionary case histories drawn from the ammonites — Brinkmann's *Kosmoceras*, the cardioceratids, liparoceratids and the like — this story can no longer stand. Fig. 9 shows the current view of Jurassic and Cretaceous phylogeny; in terms of iterative evolution, nothing remains.

Palingenesis and Proterogenesis

Palingenesis

In 1828, Von Baer pointed out that animals resemble each other much more in their early developmental stages than when adult, and this is still as true for ammonites as it is for mammals. Unfortunately Von Baer's views were developed beyond the evidence supporting them, leading to Haeckel's Biogenetic Law — that ontogeny recapitulates phylogeny (1866). If true, the ammonites of all groups were a prime testing ground for this law — but find an individual, dissect it, and its ancestry is revealed. For half a century this view prevailed, and the names of Hyatt, Buckman, Perrin Smith, Trueman and others delineate a school which elaborated the theory with disastrous consequences. Perhaps the greatest criticism of many of the palingenic sequences proposed was their lack of support in terms of stratigraphic control; the theory was used to determine evolutionary sequences rather than evolutionary sequences being used to support theory. Donovan (1973) has recently produced a critique of this phase of ammonite studies; it has essentially been abandoned by workers in the group for half a century.

Proterogenesis

In 1901, Pavlow observed that the juveniles of certain Jurassic ammonites showed features of ornament and whorl shape which resembled the adult features of their descendants. This phenomenon, generally referred to as protero-

Fig. 26. Examples of neoteny. 1—4. *Mojsisoviczia*, and 5—7, *Falloticeras*, its neotenous derivative. Note how the inner whorls of the specimen shown in 2—4 are identical with the adult
Falloticeras. 8—11. An adult *Protacanthoceras*, and 12, 13, juvenile, and 14 adult *Acanthoceras* from which it evolved. Juvenile *Acanthoceras* and adult *Protacanthoceras* are superficially identical except the latter shows body chamber modifications as shown in 8 and 10.

genesis or caenogenesis, is widespread, and many instances have been documented by subsequent workers. It provides a mechanism for major steps in evolution, as is now well known. In general these changes appear to involve abrupt
neoteny, and new taxa are much smaller than their immediate ancestors.

I would suggest this type of origin for, for example, the following amongst
many:

(1) The Cenomanian acanthoceratid *Protacanthoceras* (adult size of early
species 15—30 mm) as a derivative of *Acanthoceras* (adults up to 500 mm)
(Fig. 26).

(2) The late Cretaceous gaudryceratid *Vertebrites* (adult at 45 mm) from
Gaudryceras of the *stefaninii* group (adult at approximately 200 mm).

(3) The Albian mojsisovicziid *Falloticeras* (adult at 20—30 mm) from *Mojsisoviczia* (adult at approximately 120 mm) (Fig. 26).

(4) The Albian—Cenomanian mortoniceratid *Euhystrichoceras* (commonly
adult at 15—20 mm) from Late Albian *Mortoniceras* (adult at 100 mm +).

It is equally tempting to regard neoteny as a mechanism by which to derive
the diminutive early representatives of various heteromorph groups — the diminutive *Eoscaphites* (30 mm) from the early whorls of *Hamites* of the *tenuis*
group (adults in excess of 130 mm maximum dimension), the early Baculitidae with their coiled larval shell and straight shaft from the Ptychoceratidae
with closely pressed parallel shafts, or the early turrilitids from the initial
whorls of some anisoceratid. The Cymbitinae; *Metacymbites* (both Lower
Jurassic); *Amauroceras* (Lower Jurassic); *Neosaynoceras* (Cenomanian); some
pseudoceratitic forms, e.g., *Trochleiceras* (Aptian—Albian); and the Flickiidae
(Albian—Cenomanian), all of cryptic origin, can equally be readily interpreted
as neotenous offshoots of normal stock.

The dangers of this approach are, however, obvious; not all diminutive taxa
can be so easily derived, for many of them are obviously microconchs, as with
equally small taxa such as *Creniceras, Oecotraustes, Horioceras, Diaphorites,
Glochiceras, Trilobiticeras, Bomburites* and many more, in which females are
of more normal dimensions.

Of the classic examples of proterogenesis, that of the liparoceratids has already
been dismissed (p. 299). It is equally unfortunate that the other prime example
of this pattern of evolution cited in the *Treatise* has also been demolished. This
involved the supposed transformation of the globose Early Callovian *Macrocephalites* into compressed, sharply ventered *Quenstedtoceras* via *Cadoceras*
and into compressed, tabulate ventered *Kosmoceras* via *Kepplerites* (Fig. 27)
(Pavlow, 1901; Schindewolf, 1936; Arkell et al., 1957, fig. 154).

Macrocephalites is, however, a tethyan form, and the cardioceratids and kosmoceratids essentially boreal, and detailed correlation between these two provinces is very difficult during most of the Jurassic. We now know that both
groups appear well *before* the *Macrocephalites* that were their supposed ancestors (Callomon, 1963, p. 33 *et seq.*, with references). The cardioceratids thus
originate in what is probably the Upper Bajocian, with early sphaeroceratid-like

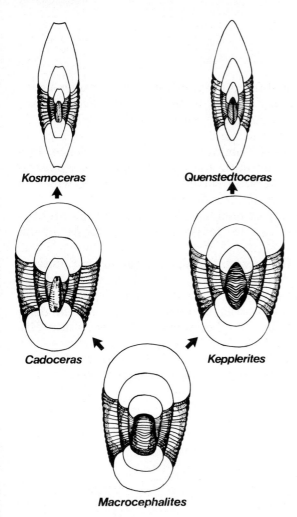

Fig. 27. Spurious proterogenesis in the supposed evolution of cardioceratids and kosmoceratids from macrocephalitids.

ancestors whose origins lie in the Stephanocerataceae, whilst the kosmoceratids originate in the Late Bathonian *variabile* Zone (Callomon, 1959) and early forms of this latter group closely resemble *Stephanoceras*. The sudden appearance of both groups in the Early Callovian of Western Europe is simply a reflection of their migration south into the area from northern regions. The admitted resemblance between the earliest European kosmoceratid, *Kepplerites keppleri* and *Macrocephalites (Dolikephalites)* is in fact a resemblance between a *macroconch* kosmoceratid and a *microconch* macrocephalid, and it is not seen in the other dimorphs.

Evolutionary Size Changes

Size increase

Cope's Rule — that there is a general trend in evolution towards larger size — has been discussed recently by Stanley (1973) in terms of the evolution of small forms to a larger, optimum size. Amongst the groups cited by Stanley in support of the rule were ammonites, and his data are reproduced as Fig. 28. This trend is in fact ubiquitous amongst ammonites, in my experience. In a general way, the largest Cretaceous ammonite, the giant *Parapuzosia seppenradiatus* is bigger than the largest Late Jurassic perisphinctid *Titanites*, which in turn exceeds the largest Triassic ammonoid *Pinacoceras*.

It is not, of course, true that the largest member of a group is necessarily the youngest, and this not being the case does *not* refute Cope's Rule, as Wiedmann (1969) suggested.

The most comprehensive, quantified account of size change in ammonites is given by Hallam (1975a) and is reproduced here as Table II. Apart from these quantified data, there are many other examples, including those given by Zeiss (1968) and Ziegler (1969) for various Jurassic ammonites, whilst size increase is also obvious in scaphitids and the baculitid heteromorph lineages studied by Cobban (p. 270), or in a general way from the diminutive *Eoscaphites* to the late Cretaceous *Trachyscaphites* and *Rhaeboceras*, or *Lechites* to the 2-m long Senonian *Baculites*. In other Cretaceous groups, progressive size increase in spe-

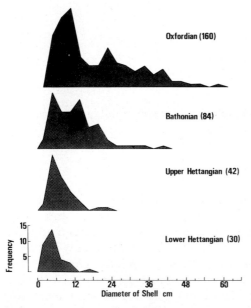

Fig. 28. Size increase in Jurassic ammonites [based on Stanley (1973, *Evolution*, 27), and reproduced with the permission of the author and publishers].

TABLE II

Size changes (after Hallam, 1975a)

Taxon	Age range	Number of zones	Increase in maximum size (mm)	milli-darwins
Arnioceras semicostatum	L. Sinemurian	4	55— 91	207
Aulacostephanus	L. Kimmeridgian	3	175—345	328
Caenisites—Asteroceras	Sinemurian	2	147—270	459
Cardioceras	L. Oxfordian	2	137—210	383
Coroniceras—Primarietites	L. Sinemurian	1	350—650	928
Creniceras dentatum	L. Kimmeridgian	3	20— 37	308
Garantiana—Parkinsonia	U. Bajocian	2	110—350	785
Glochiceras	U. Oxfordian—U. Kimmeridgian	10	39— 56	64
Grammoceras—Phlyseogrammoceras	U. Toarcian	2	114—194	425
Idoceras	U. Oxfordian—L. Kimmeridgian	5	154—221	143
Kosmoceras (Gulielmites)	M. Callovian	1	70—150	1071
Leioceras	Aalenian	1	60—100	833
Liparoceras	L. Pliensbachian	3	88—285	540
Metophioceras conybeari	L. Sinemurian	1/3	130—320	3690
Ochetoceras	U. Callovian—U. Kimmeridgian	15	79—134	57
Pectinatites	L. Tithonian	3	184—380	206
Pleuroceras	U. Pliensbachian	1	66—335	2538
Psiloceras	L. Hettangian	1	130—253	973
Taramelliceras	L. Oxfordian—L. Kimmeridgian	12	50—196	165

cific lineages can be recognized: in the Ancylocerataceae, Nostoceratidae, Anisoceratidae and Turrilitaceae amongst the heteromorphs, and in the Placenticeratidae, Pachydiscidae, Desmoceratidae, Lyelliceratidae, Acanthoceratidae, Collignoniceratidae, Coilopoceratidae and Sphenodiscidae, and perhaps most other normally coiled groups.

One consequence of Stanley's interpretation of Cope's Rule is, as Hallam (1975a) notes, that the organisms that increase in size more rapidly should become extinct more quickly, since extreme specialization would occur earlier and hence render them more vulnerable to extinction; also, given that food resources are likely to become scarcer, the number of individuals will be smaller, and hence larger species will once again be more liable to extinction. It is interesting to note that the available data on Jurassic ammonites (Hallam, 1975a), support this.

Size decrease

Whilst size increase is widespread amongst lineages, it is obvious that many new taxa must be much smaller than their ancestors. Progressive size decrease appears rare and is not as well documented as size increase. The only well documented examples I am aware of are amongst some of the Cretaceous scaphitids and *Baculites* described by Cobban (1951, pp. 8 and 9; Kennedy and Cobban, 1976a, p. 61), and here no quantification of data has been attempted. In contrast, abrupt size decrease appears a common phenomenon as a result, it would seem, of neoteny, and I would agree with Hallam (1975a, p. 496) that it may be one of the principal means by which new morphotypes arise (p. 286). I have already suggested neoteny followed by progressive size increase as the key, for instance, to the phylogeny of the Cretaceous Acanthocerataceae, and it is the obvious origin of some genuinely diminutive taxa, as discussed earlier (p. 286).

Rates of Evolution

Rates of evolution can be quantified in several ways, two of which are considered here: species longevity and taxonomic frequency.

Species longevity

Species longevity is of course dependent on the often rather subjective criteria used to differentiate between successive species, and also on the integration of biostratigraphical systems with radiometric ages. The only area in which this is currently possible is the U.S. Western Interior. Here, during Cretaceous times, there were deposited scores of thin volcanic ash bands (bentonites), intercalated with fossiliferous marine sediments. Radiometric dates on these ash bands have been published and summarised by Gill and Cobban (1966) and Obradovich and Cobban (1975). They can be applied to the *Baculites* and *Scaphites* lineages discussed previously (p. 270 ff.), where there is the added advantage that all the systematic work has been carried out by one worker (Cobban) giving a consistent species concept throughout. From this work (tabulated by Kennedy and Cobban, 1976a, table 4, p. 70) it appears that endemic baculitid species had life-span of between 500,000 and 700,000 years, whilst the endemic scaphitid species had life-spans of between 600,000 and 700,000 years. By integrating the Interior zonation with that of the Gulf Coast region of the United States, and by plotting the stray occurrences of immigrant species from elsewhere, more widely occurring species can be shown to have much longer life-spans — *Baculites undatus*, a Gulf Coast immigrant ranges through 3.3. m.y., whilst the scaphitid *Hoploscaphites constrictus* is said to span the whole of the Maastrichtian interval — nearly 4.0 m.y.

Records of very long ranges are usually based on taxa that have few obvious morphological features on which to subdivide lineages. As noted elsewhere (Kennedy and Cobban, 1976, p. 69 and 70), *Phylloceras thetys* has a recorded range of Valanginian to Cenomanian — 25 m.y.; "*Neophylloceras*" *ramosum*, Turonian to Coniacian — 12 m.y.; *Damesites sugata*, Coniacian to Lower Campanian — 8 m.y.; *Austiniceras austeni*, Cenomanian to Turonian — 8 m.y.; whilst *Phylloceras serum* spans the Late Tithonian to Barremian interval — an interval of over 20 m.y.

Average zonal durations also give a measure of species longevity; during the Triassic and Jurassic these average 1.2 m.y. and during the Cretaceous, nearer to 2 m.y. These are, however, rather broad zones; the limits of biostratigraphical accuracy in ammonites is lower — 0.2 to 0.9 m.y. in the U.S. Western Interior, Cenomanian to Maastrichtian, and on average 0.5 m.y. in the Early Jurassic of Western Europe. These minimum figures correspond quite well to Simpson's (1952) best estimate of the life-span of a species as between 0.5 and 2.75 m.y.

Although conclusive documentation is not fully available, there is an obvious suggestion that endemic taxa evolved at a higher rate (in terms of species longevity) than pandemic taxa.

Taxonomic frequency

Objective plots of taxonomic frequency through time suffer from problems similar to those involved in determining species longevity. The time-scale for the Triassic, Jurassic and Early Cretaceous is simply inadequate (indeed virtually non-existent!) for the determination of frequency rates, as is clear from discussions by the various authorities in the Geological Society of London's *Phanerozoic Time Scale* and subsequent supplements (e.g., Harland et al., 1964; Harland et al., 1971). In consequence, most published plots of data assume equal durations for stages, as for instance with Moore's compilation (in Arkell et al., 1957, fig. 159). In an attempt to improve on this, plots given here are based on the observations by Hallam (1975b) and Hallam and Sellwood (1976) on the average duration of Jurassic ammonite zones extended to the Triassic and Lower Cretaceous (e.g., stage duration is proportional to number of zones recognized).

This is obviously highly questionable, given the numerous suggestions that rates of evolution (of which species longevity and hence zonal duration are a function) vary in response to regressive and transgressive events, but is arguably a better guess than assuming that stages have equal duration.

A second problem is the monographic one; I have used the *Treatise* as a source of data and it is obvious that there is excessive taxonomic splitting in some groups at certain times and rather less at others. It is thus tempting to relate the Albian diversity peak to the monograph of the Gault ammonites by Spath (1923—1943; see also Wiedmann, 1969, 1973c); Donovan (1958) and Getty (1973) revised the Lower Jurassic family Echioceratidae in which the number of genera was reduced from nineteen to five (with three other genera regarded as *incertae sedis*). The same reduction in numbers is probably necessary to provide a real picture of the fragmented Perisphinctaceae, Hoplitidae, Stephanocerataceae and the like. It can only be a pious hope that the degree of fragmentation is relatively constant through time!

Sexual dimorphism equally confounds numerical analysis. In some groups, dimorphs are sufficiently similar to have been referred simply to different species or perhaps subgenera of the same genus. In others, the dimorphs are referred to different genera, giving anomalously high diversities in particular groups at different times. Many dimorphic pairs have been recognized, but this is not universal, whilst some workers still fail to accept the reality of dimorphism.

The data presented in Fig. 29 are, to say the least, open to criticism on a fair number of grounds! In spite of this the diagrams summarize very clearly the

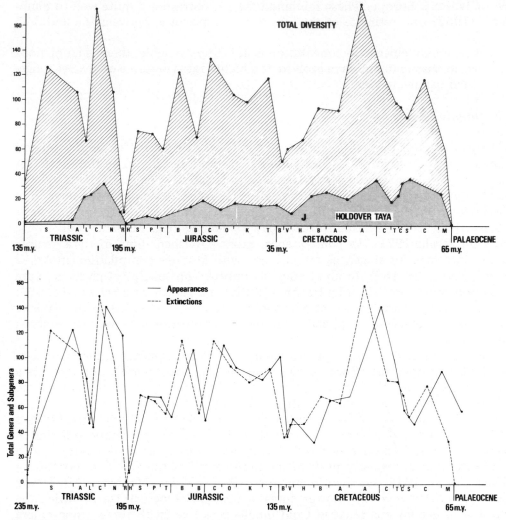

Fig. 29. Diversity changes in ammonoids during the Mesozoic.

overall course of changes in diversity of the Mesozoic ammonoids and the crises in their evolution, as are discussed further in the final section to come.

Racial Senility

During the last century and the early part of this, as still today, ammonite evolution has been cited as demonstrating the decline with age in a group, as in an individual. As many authors have pointed out, the analogy is unacceptable, for nothing in continuously reproducing populations does, or can possibly correspond to this ageing. Be it so, but the ammonites are the prime supposed example of racial senescence. The view stems mainly from Hyatt's work at the end of the last century when he pointed to the complexity of ornament and suture and the (subjectively assessed) bizarre coiling demonstrated by the group at the close of their history.

This hypothesis is clearly nonsense, although, as Wiedmann (1969, p. 563) notes, and experience supports, it is the heteromorphs which are still linked with notions of "aberrant shell form, degeneration, typolysis and phylogenetic extinction" by many non-specialists.

"There are furthermore aberrant forms which rapidly, one after another, show an ever stronger tendency to degenerate and produce biologically absurd structures which, if not directly lethal, have always been impartially understood as ridiculous for the basic concept of the ammonite form." (Translated from Dacqué, 1935, p. 32.)

"Just as the great ceratitoid group of ammonoids produced retrogressive as well as stationary and progressive forms during the Trias, so from one, or several, of the families just mentioned there arose decadent lines of descent. . . Thus in *Baculites* the whole organisation was affected by decadent influences, and it is therefore the most perfect impression of all-round retrogression among the ammonoids." (Swinnerton, 1950, p. 202.)

"It is of particular significance that the aberrant shell types with gastropod-type spirals and loosed whorls only appear in greater numbers in the crisis periods of ammonite development. . . The forms involved constitute short-lived peripheral lineages originating in evolutionary groups shortly before their extinction. It is therefore most likely by far that internal grounds are decisive for this extravagance of forms, which is, moreover, often found at the end of lineages undergoing extinction and commonly shows no sign of being adaptive. One may think of a senility of the stock or of a gradual decline of the strong determined form-control or of what one will; in any case no external factors are determinate. Against the interpretation of the peripheral lineages as results of adaptation is, in addition, their transience in comparison with the longevity of their parent stocks." (Translated from Schindewolf, 1936, pp. 74 f.)

"A survey of the development of the shelly cephalopods, above all the ammonoids, shows that in the last phase of their evolution, in the late Cretaceous, a hypertrophic transformation and dissolution of the shell type occurs (*Baculites, Scaphites, Crioceras*, etc.), and that the sculptural elements also to some extent show a hypertrophic dissolution (*Douvilleiceras, Inflaticeras*)." (Translated from Beurlen, 1937, p. 87.)

"Of extreme interest are the indications of degeneration of the cephalopods. Apart from the size excesses already mentioned they are manifest in a dissolution of the normal spiral coil, as well as in the imitation of ancient groups." (Translated from Erben, 1950, p. 120.)

"In the course of the typolysis (of the ammonites), which appears above all at the end of the last phase of vitality, the obedience to form embodied in the type again becomes weak. Numerous indications of decline and degeneration are to be seen. Particularly characteristic are regressive processes which to some extent "throw to the wind" that which was created by progressive evolution."

(Translated from Müller, 1955, pp. 16 f.)

"The reason for the extinction of the morphologically so varied Triassic ammonites on the boundary to the Jurassic must be sought in the internal organization of the animal since external factors of the environment do not come into consideration. . . The extinction is preceded by overspecialization and indications of senility of the most different kinds. The so stable and strenuously maintained basic form of a shell coiled in a plane spiral degenerates. . . Moreover, after the younger ammonites have experienced a phase of explosive development and a long period of gradually advancing specialization they also arrive at a phase of overspecialization and extravagance of form, quite analogous to the Triassic representatives. The nearer we approach the upper boundary of the

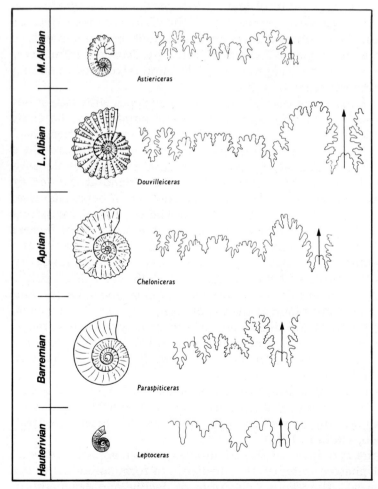

Fig. 30. The origin of the "false" hoplitid superfamily Douvilleicerataceae from heteromorph ancestors, and the re-appearance of uncoiled forms subsequently illustrates the evolutionary success of the heteromorphs.

Cretaceous and the final extinction of the ammonite stock, the more often do we encounter degenerate forms." (Translated from Schindewolf, 1950, pp. 168 ff.)

Simpson (1953, p. 292) dismissed these views briefly, whilst Wiedmann (1969) has exhaustively treated this subject. The facts give no support either to the view that heteromorphs are degenerate end-members of lineages, nor that they were unsuccessful and overspecialized. The three principal heteromorph groups, cochloceratids (Late Triassic), spiroceratids (mid-Jurassic) and ancyloceratids (principally Cretaceous) are all monophyletic offshoots, rather than groups of iterative end-forms.

Homeomorphy is as widespread in heteromorphs as in "normally" coiled forms, indicating (since shell form and coiling are highly functional features) that they represent strong positive selection rather than degeneracy.

Heteromorphy is not a terminal phenomenon in lineages, for as well as showing progressive uncoiling, there is progressive re-coiling as an equally widespread trend (Figs. 12, 30). This is as true at the point of origin of ammonoids in the bactritid—anarcestid sequence (p. 254, Fig. 3) as it is in ancyloceratids (Casey, 1960; Fig. 23), crioceratitids (Wiedmann, 1962c; Fig. 25) and scaphitids (Wiedmann, 1965; Fig. 19), or in the genesis of the 'false hoplitids'. With the sequence from heteromorphic *Leptoceras* and *Paraspiticeras* to the douvilleiceratids — which in turn uncoil — as in *Astiericeras* (Fig. 30) or the heteromorph *Hemihoplites* to normally coiled deshayesitids.

In terms of evolutionary "success" the heteromorphs also belie any view of racial senility; the earlier cochloceratids and spiroceratids are relatively minor heteromorphic radiations compared to the proliferation of ancyloceratid genera. Again, the Cretaceous heteromorphs are long ranging — the Baculitidae range from Albian to Maastrichtian and thus existed for more than 30 m.y., far longer than the majority of normally coiled families, whilst the ancyloceratids as a group range through close to 70 m.y., a span exceeded only by the phylloceratids, lytoceratids and haploceratids amongst normally coiled groups (Fig. 9). Generic longevity is also high, with *Scaphites*, for instance, ranging through over 30 m.y. and *Baculites* spanning close to 20 m.y. In terms of rates of evolution measured by species longevity, the heteromorphs again show a whole range from amongst the shortest known to medium length — 0.2 m.y. to over 3 m.y. (Kennedy and Cobban, 1976a, p. 69).

Finally, the youngest ammonites known, according to Hancock (1967) and Wiedmann (1969), include both normally coiled forms and several heteromorph groups: scaphitids, baculitids and diplomoceratids (Fig. 31). In no respect then does the history and pattern of heteromorph evolution mark them as degenerates, or the inhabitants of an evolutionary cul-de-sac.

Extinction

Extinction in ammonoids has been more significant at some times than others. There are peaks at: (1) the Mesozoic—Palaeozoic boundary (Figs. 6, 29)

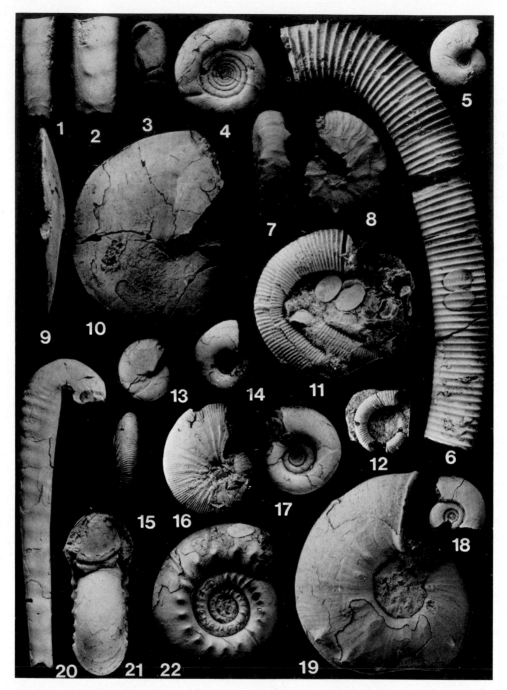

Fig. 31. The diversity of Late Cretaceous ammonites is illustrated by these representative genera from the Campanian to Lower Maastrichtian Valudayur group of southern India. 1, 2. *Baculites vagina.* 3, 4. *Vertebrites kayei.* 5. *Zelandites varuna.* 6. *Neohamites subcompressus.* 7, 8. *Indoscaphites cunliffei.* 9, 10. *Sphenodiscus siva.* 11, 12. *Glyptoxoceras indicum.* 13. *Desmophyllites diphylloides.* 14. *Pseudophyllites indra.* 15, 16. *Phylloceras surya.* 17. *Saghalinites cala.* 18. *Hauericeras remda.* 19. *Menuites menu.* 20. *Phylloptychoceras sipho.* 21, 22. *Brahmaites brahma.*

with a delay of at least one stage duration between maximum extinction and renewed radiation; (2) the Triassic—Jurassic boundary (Figs. 9, 28) with again a pause between extinction and renewal; (3) during the Early Cretaceous at the Berriasian—Valanginian boundary; and (4) final extinction during the Late Maastrichtian.

It is of course, no coincidence that these major and drastic extinctions correspond to the boundaries of eras, periods, or stages, for it was on just these changes that these units were first recognized. It is equally true that periods of high extinction rates in ammonoids correspond to those in other groups, both aquatic and terrestrial (Harland et al., 1967; Wiedmann, 1969, 1973c) and no special theory is therefore demanded to explain these events in the history of ammonoid evolution. The figures illustrate very clearly the detail of the first three crises in the evolution of the group, and the reader is referred to papers by Wiedmann (1967, 1968b, 1969, 1973c) for full documentation.

The final demise of the group is less readily plotted in this way (although shown in broad fashion in Fig. 9) because of problems of correlating the widely scattered occurrences of Maastrichtian ammonites. Hancock (1967) and Wiedmann (1969, 1973c) have reviewed the general pattern (Fig. 32). Some 22 families are present in the Cenomanian (at the beginning of the Upper Cretaceous), but only 11 in the Maastrichtian, although most of the decrease was pre-Santonian. Hancock notes 78 Campanian genera and about 34 in the Maastrichtian (it should be noted that these intervals are of rather different duration: Campanian ~ 11 m.y., Maastrichtian ~ 5 m.y.). This Maastrichtian figure is, however, no lower than say the Berriasian (basal Cretaceous) where only 35 genera spread amongst nine families are recorded. Of the 34 Maastrichtian genera, 23 do not range above the Lower Maastrichtian, although the remaining eleven are spread amongst five superfamilies.

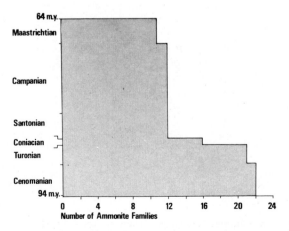

Fig. 32. The decline in ammonite diversity versus time during the Late Cretaceous (data from Hancock, 1967; and Obradovich and Cobban, 1975).

An alternative treatment of this data, in terms of decline of species vs. appearance of new characters (measured as the first appearance of families) gives a similar picture of final demise preceded by progressive decline. In my own experience, declining diversity is matched by a decrease in abundance, whilst Wiedmann (1973c, pp. 159—160, pl. 3, figs. 2—7) has suggested that dwarfing of faunas also occurs (though it is not clear to me if his figures of individuals showing environmental stunting, are based on juveniles, or genuinely diminutive species).

The proposed controls on extinction have been many and varied, including endogenous factors — racial senility, biological competition, food chain deficiency, orogeny, transgressions and regressions, climatic changes, salinity changes, oxygen deficiency, cosmic radiation changes, meteorite impacts, geomagnetic reversals and so on. Many of these can be readily dismissed, as in the excellent discussions by Wiedmann (1973c), Rhodes (1967), Lipps (1970), Valentine and Moores (1970) and Valentine (1971, 1973).

The complex pattern of early extinction phases and the progressive reduction of ammonite diversity and abundance during the Late Cretaceous appear to contradict any view of catastrophic events as causal factors. Rather, there is marked correspondence between radiation, extinction, and sea level changes, with regression and extinction matched, maximum regression (Late Permian,

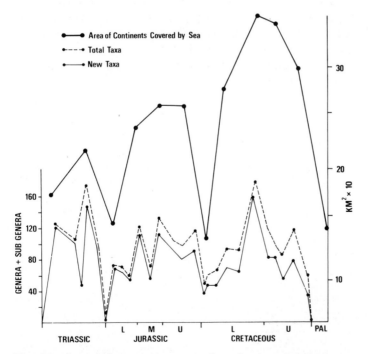

Fig. 33. Ammonite appearances and extinctions in relation to continental flooding during the Mesozoic [diversity data is from Fig. 28 herein; the seaboard curve from Hallam (1971), and is based on Termier's palaeogeographic maps].

Late Jurassic, Late Cretaceous) coinciding with maximum extinction (Fig. 33). In turn, adaptive radiations relate to the appearance of renewed marine transgression (Wiedmann, 1973c, pp. 185—189). This is held to be true not only on a broad scale, but also at the level of stage-by-stage changes in fauna, as shown in Wiedmann's detailed analysis of Late Cretaceous extinctions (1973c, fig. 9).

Summary

The Jurassic and Cretaceous ammonites are the products of the third and final major evolutionary radiation in the history of the Ammonoidea. As with the earlier radiations, the origins lay in what are in many respects conservative, and rather rare long-ranging groups.

The overall pattern of rise and decline of the ammonites, including their final extinction, can be linked to phases of radiation and extinction in other groups of organisms and, in turn, to periods of transgression and regression respectively.

Higher categories generally originate in individual genera or species groups, and from what are often rather generalized ancestors there are recurring patterns of evolutionary divergence as the families and superfamilies explored the limitations of the shell to which they were bound.

Evolutionary convergence, manifested as homeomorphy, is a further recurring phenomenon extending to the whole shell and to individual elements such as the septa, or pattern of coiling. Neoteny and progressive size increase are also widespread and recurrent evolutionary processes exhibited by the group.

Most of the 'classic' examples of ammonite evolution can be shown to be based on false premises, due to inadequate sampling, excessive taxonomic splitting, or failure to recognize sexual dimorphism, whilst revised views on the phylogenetic relationships within the group provide no evidence for the iterative evolution ammonites are purported to demonstrate.

Detailed lineage studies demonstrate both progressive and abrupt species transitions, but gaps in sampling, by virtue of the discontinuous nature of the fossil record, do not permit the demonstration of *complete* gradualism. If evolution was stepwise, then the steps were very small in some groups, however.

Patterns of evolutionary convergence, divergence and parallelism, together with the ubiquity of homeomorphy, confirm the adaptive and functional significance of most features of shell morphology. Two features suggest that other factors may have also been at work: the high intra-specific variability of some groups, and the striking differences between sexual dimorphs. The latter suggests that some aspects of ornament and structures such as lappets and rostra may have been sexually selected display features.

Acknowledgements

Support for much of the research incorporated into this contribution has come from many sources including The Lindemann Trust, Royal Society,

Natural Environment Research Council, British Association for the Advancement of Science, William Waldorf Astor Fund, and the Trustees of the Sir Henry Strakosh Bequest. For useful discussions and help in many ways, I thank Messrs. R.J. Cleevely, M.K. Howarth, J.M. Hancock, A. Hallam, E.G. Kauffman, H.C. Klinger, N.J. Morris, H.P. Powell, D. Phillips, R.A. Reyment, J. Sornay and J. Wiedmann. I am also grateful to the authors and organizations who have allowed me to reproduce illustrations and information from their publications as indicated in the text. Messrs. J.W. Valentine and M.R. Cooper kindly reviewed earlier drafts of the paper; their assistance is gratefully acknowledged.

References

Arkell, W.J., 1935—1948. A monograph of the ammonites of the English Corallian Beds. Palaeontogr. Soc. (Monogr.), 420 pp. (78 pls.).

Arkell, W.J., 1950. A classification of Jurassic ammonites. J. Paleontol., 24: 354—364.

Arkell, W.J. and Moy-Thomas, J.A., 1940. Palaeontology and the taxonomic problem. In: J. Huxley (Editor), The New Systematics. Oxford University Press, Oxford, pp. 395—410.

Arkell, W.J., Furnish, W.M., Kummel, B., Miller, A.K., Moore, R.C., Schindewolf, O.H., Sylvester-Bradley, P.C. and Wright, C.W., 1957. Treatise on Invertebrate Palaeontology. Part L. Mollusca, 4. Geological Society of America and University of Kansas Press, Lawrence, Kansas, 490 pp.

Beurlen, K., 1937. Die stammes geschichtliche Grundlagen der Abstammungslehre. Fischer, Jena, 264 pp.

Bramlette, M.N., 1965. Massive extinctions in biota at the end of Mesozoic time. Science, 148: 1696—1699.

Brinkmann, R., 1929. Statistischbiostratigraphische Untersuchungen an mitteljurassischen Ammoniten über Artbegriff und Stammensentwicklung. Abh. Ges. Wiss. Göttingen, Math.-Phys. Kl., N.F., 13: 1—249 (pls. 1—5).

Brinkmann, R., 1937. Biostratigraphie des Leymeriellestammes nebst Bemerkungen zur Paläogeographie des nord-westdeutschen Alb. Mitt. Geol. Staats inst. Hamb., 16: 1—18.

Buckman, S.S., 1887—1907. A monograph of the Ammonites of the Inferior Oolite Series. Palaeontogr. Soc. (Monogr.), 456 pp. (127 pls.).

Callomon, J.H., 1955. The ammonite succession in the Lower Oxford Clay and Kellaways beds at Kidlington, and the zones of the Callovian stage. Philos. Trans. R. Soc. Lond., 239 B: 215—264 (pls. 2—3).

Callomon, J.H., 1959. The ammonite zones of the Middle Jurassic Beds of East Greenland. Geol. Mag., 96: 505—513.

Callomon, J.H., 1963. Sexual dimorphism in Jurassic ammonites. Trans Leicester Litt. Philos. Soc., 57: 21—66.

Casey, R., 1957. The Cretaceous ammonite genus *Leymeriella*, with a systematic account of its British occurrences. Palaeontology, 1: 29—59 (pls. 7—10).

Casey, R., 1960. A monograph of the Ammonoidea of the Lower Greensand. Palaeontogr. Soc. (Monogr.), part I, 44 pp. (pls. 1—10).

Casey, R., 1961. A monograph of the Ammonoidea of the Lower Greensand. Palaeontogr. Soc. (Monogr.), part III, pp. 119—216 (pls. 26—35).

Cobban, W.A., 1951a. New species of *Baculites* from the Upper Cretaceous of Montana and South Dakota. J. Paleontol., 25: 817—821 (pl. 118).

Cobban, W.A., 1951b. Scaphitoid cephalopods of the Colorado group. Prof. Pap. U.S. Geol. Surv., 239: 1—42 (21 pls.).

Cobban, W.A., 1952. A new Upper Cretaceous Ammonite genus from Wyoming and Utah. J. Paleontol., 26: 758—760 (pl. 110).

Cobban, W.A., 1958. Two new species of *Baculites* from the Western Interior region. J. Paleontol., 32: 660—665 (pls. 9—11).

Cobban, W.A., 1962a. New *Baculites* from the Bearpaw Shale and equivalent rocks of the Western Interior. J. Paleontol., 36: 126—135 (pls. 25—28).

Cobban, W.A., 1962b. *Baculites* from the lower part of the Pierre Shale and equivalent rocks in the Western Interior. J. Paleontol., 36: 704—718 (pls. 105—108).

Cobban, W.A., 1964. The late Cretaceous cephalopod *Haresiceras* Reeside and its possible origin. Prof. Pap. U.S. Geol. Surv., 454-1: 1—21 (3 pls.).

Cobban, W.A., 1969. The Late Cretaceous ammonites *Scaphites leei* Reeside and *Scaphites hippocrepis* (De Kay) in the Western Interior of the United States. Prof. Pap. U.S. Geol. Surv., 619: 1—29 (5 pls.).

Cobban, W.A., 1973. The Late Cretaceous ammonite *Baculites undatus* Stephenson in Colorado and New Mexico. U.S. Geol. Surv. J. Res., 1: 459—465.

Cobban, W.A. and Scott, G., 1964. Multinodose scaphitid cephalopods from the Lower Part of the Pierre Shale and equivalent rocks in the Conterminous United States. Prof. Pap. U.S. Geol. Surv., 483-E: E1—E12 (4 pls.).

Cobban, W.A. and Scott, G., 1972. Stratigraphy and ammonite fauna of the Graneros Shale and Greenhorn Limestone near Pueblo, Colorado. Prof. Pap. U.S. Geol. Surv., 645: 1—108 (39 pls.).

Collignon, M., 1971. Atlas des fossiles caractéristiques de Madagascar. xvii (Maestrichtien). Service Géologique, Tananarive, 44 pp. (pls. 640—658).

Dacqué, E., 1935. Organische Morphologie und Paläontologie. Borntraeger, Berlin, 476 pp.

Donovan, D.T., 1958. The Lower Jurassic ammonite fauna from the fossil bed at Langeneckgrat, near Thun (median Pre-Alps). Schweiz. Paläontol. Abh., 74: 1—58 (pls. 1—17).

Donovan, D.T., 1964. Cephalopod phylogeny and classification. Biol. Rev., 39: 259—287.

Donovan, D.T., 1973. The influence of theoretical ideas on ammonite classification from Hyatt to Trueman. Paleontol. Contrib. Univ. Kansas, 62: 16 pp.

Donovan, D.T. and Hölder, H., 1958. On the existence of heteromorph ammonoids in the Lias. Neues Jahrb. Geol. Paläontol., Monatsh., 1958: 217—220.

Eicher, D.L., 1965. Foraminifera and biostratigraphy of the Graneros Shale. J. Paleontol., 39: 875—909.

Erben, H.K., 1950. Das stammesgeschichtliche Degenerieren und Aussterben. A.d. Heimat, 58: 116—123.

Erben, H.K., 1960. Primitive Ammonoidea aus dem Unterdevon Frankreichs und Deutschlands. Neues Jahrb. Geol. Paläontol., Abh., 110: 1—128 (6 pls.).

Erben, H.K., 1962. Uber böhmische und türkische Vertreter von Anetoceras (Ammon., Unterdevon). Paläontol. Z., 36: 14—27 (pls. 1—2).

Erben, H.K., 1964. Die Evolutionen der ältesten Ammonoidea (Lieferung I). Neues Jahrb. Geol. Paläontol. Abh., 120: 107—212 (pls. 7—10).

Erben, H.K., 1965. Die Evolutionen der ältesten Ammonoidea (Lieferung I). Neues Jahrb. Geol. Paläontol. Abh., 122: 275—312 (pls. 25—27).

Erben, H.K., 1966. Über den Ursprung der Ammonoidea. Biol. Rev., 41: 641—658.

Frebold, H., 1922. Phylogenie und Biostratigraphie der Amaltheen im mittleren Lias von Nordwestdeutschland. Jahrber. Niedersächs. Geol. Ver., 15: 1—26 (pls. 1—8).

Getty, T.A., 1973. A revision of the generic classification of the family Echioceratidae (Cephalopoda, Ammonoidea) (Lower Jurassic). Paleontol. Contrib. Univ. Kansas, 63: 32 pp. (5 pls.).

Gill, J.R. and Cobban, W.A., 1966. The Red Bird section of the Pierre Shale in Wyoming, with a section on a new echinoid from the Cretaceous Pierre Shale of eastern Wyoming by P.M. Kier. Prof. Pap. U.S. Geol. Surv., 393-A: 73 pp. (12 pls.).

Gould, S.J., 1970. Dollo on Dollo's law: Irreversibility and the status of evolutionary laws. J. Hist. Biol. 3: 189—212.

Haas, O., 1942. Recurrence of morphological types and evolutionary cycles in Mesozoic ammonites. J. Paleontol., 16: 643—650 (pls. 93—94).

Haeckel, E., 1866. Generelle Morphologie der Organismen. Tl. II. Allgemeine Entwicklungsgeschichte der Organismen. Reimer, Berlin, 462 pp.

Hallam, A., 1971. Re-evaluation of the palaeogeographic argument for an expanding Earth. Nature, 232: 180—182.

Hallam, A., 1975a. Evolutionary size increase and longevity in Jurassic bivalves and ammonites. Nature, 258: 493—496.

Hallam, A., 1975b. Jurassic Environments. Cambridge University Press, London, 280 pp.

Hallam, A. and Sellwood, B.W., 1976. Middle Mesozoic sedimentation in relation to tectonics in the British area. J. Geol., 84: 301—321.

Hancock, J.M., 1967. Some Cretaceous/Tertiary marine faunal changes. In: W.B. Harland, C.H. Holland, M.R. House, N.F. Hughes, A.B. Reynolds, M.H.S. Rudwick, G.E. Satterthwaite, L.B.H. Tarlo and E.C. Willey (Editors), The Fossil Record. Geological Society of London, London, pp. 91—104.

Harland, W.B., Smith, A.G. and Wilcox, B., 1964. The Phanerozoic Time Scale. Q. J. Geol. Soc. Lond., Suppl. to Vol. 120: 458 pp.

Harland, W.B., Holland, C.H., House, M.R., Hughes, N.F., Reynolds, A.B., Rudwick, M.J.S., Satterthwaite, G.E., Tarlo, L.B.H. and Willey, E.G. (Editors), 1967. The Fossil Record. Geological Society of London, London, 827 pp.

Harland, W.B., Francis, E.H. and Evans, P., 1971. The Phanerozoic time scale — a supplement. Spec. Publ. Geol. Soc. Lond., 5: 356 pp.

House, M.R., 1961. Acanthoclymenia, the supposed earliest Devonian clymenid, is a Manticoceras. Palaeontology, 3: 472—476 (pl. 75).

Howarth, M.K., 1958. A monograph of the ammonites of the Liassic Family Amaltheidae in Britain. Palaeontogr. Soc. (Monogr.), 53 pp. (10 pls.).

Huxley, J., 1938. The present standing of the theory of sexual selection. In: G.R. de Beer (Editor), Evolution. Oxford University Press, Oxford, pp. 11—42.

Hyatt, A., 1866. On the agreement between the different periods in the life of the individual shell and the collective life of the tetrabranchiate cephalopods. Proc. Boston Soc. Nat. Hist., 10: 302—303.

Hyatt, A., 1889. Genesis of the Arietitidae. Smithson. Contrib. Knowl., 26—673: 238·pp. (14 pls.).

Hyatt, A., 1894. Phylogeny of acquired characteristics. Proc. Am. Philos. Soc., 32: 349—647 (14 pls.).

Hyatt, A., 1903. Pseudoceratites of the Cretaceous. Monogr. U.S. Geol. Surv., 44: 351 pp. (47 pls.), (Posthumously published and edited by T.W. Stanton.)

Kennedy, W.J. and Cobban, W.A., 1976a. Aspects of ammonite biology, biostratigraphy and biogeography. Spec. Pap. Palaeontol., 17: 94 pp. (11 pls.).

Kennedy, W.J. and Cobban, W.A., 1976b. The role of ammonites in biostratigraphy. In: E.G. Kauffman and J.E. Hazel (Editors), Concepts and Methods in Biostratigraphy. Dowden, Hutchinson and Ross, Stroudsburg, Pa., in press.

Kullman, J. and Wiedmann, J., 1970. Significance of sutures in phylogeny of Ammonoidea. Paleontol. Contrib. Univ. Kansas, 47: 32 pp.

Lipps, J.H., 1970. Plankton evolution. Evolution, 24: 1—22.

Moore, R.C., 1950. Evolution of the Crinoidea in relation to major paleogeographic changes in Earth history. Rep. 18th Sess. Int. Geol. Congr., Great Britain, 1948, 12: 27—53.

Moore, R.C., 1952. Evolution rates among crinoids. J. Paleontol., 26: 338—352.

Moore, R.C., 1954. Evolution of late Paleozoic invertebrates in response to major oscillations of shallow seas. Bull. Mus. Comp. Zool., 112: 259—286.

Müller, A.H., 1955. Der Grobablauf der stammesgeschichtlichen Entwicklung. Fischer, Jena, 50 pp.

Nelson, C.M., 1968. Ammonites: Ammon's horns into Cephalopods. J. Soc. Bibliogr. Nat. Hist., 5: 1—18.

Newell, N.D., 1949. Phyletic size increase — an important trend illustrated by fossil invertebrates. Evolution, 3: 103—124.

Nicol, D., 1961. Biotic association and extinction. Syst. Zool., 10: 35—41.

Obradovich, J.D. and Cobban, W.A., 1975. A time scale for the Late Cretaceous of the Western Interior of North America. In: W.G.E. Caldwell (Editor), The Cretaceous System in the Western Interior of North America. Geol. Soc. Can., Spec. Publ., 13: 31—54.

Packard, A., 1972. Cephalopods and fish; the limits of convergence. Biol. Rev., 47: 241—307.

Pavlow, A.P., 1901. Le Crétacé Inférieur de la Russie et sa faune. Mém. Soc. Imp. Nat. Moscou, N.S., 21: 87 pp. (8 pls.).

Raup, D.M., 1966. Geometric analysis of shell coiling: general problems. J. Paleontol., 40: 1178—1190.

Raup, D.M., 1967. Geometric analysis of shell coiling: coiling in ammonoids. J. Paleontol., 41: 43—65.

Raup, D.M. and Stanley, S.M., 1971. Principles of Paleontology. Freeman, San Francisco, Calif., 388 pp.

Reeside, Jr., J.B., 1927a. Cephalopods from the lower part of the Cody Shale of the Oregon Basin, Wyoming. Prof. Pap. U.S. Geol. Surv., 150-A: A1—A19 (8 pls.).

Reeside Jr., J.B., 1927b. The cephalopods of the Eagle Sandstone and related formations in the Western Interior of the United States. Prof. Pap. U.S. Geol. Surv., 151: 1—87 (45 pls.).

Reeside Jr., J.B., 1927c. The Scaphites, an Upper Cretaceous ammonite group. Prof. Pap. U.S. Geol. Surv., 150-B: B21—B40 (pls. 9—11).

Reeside Jr., J.B. and Cobban, W.A., 1960. Studies of the Mowry Shale (Cretaceous) and contemporary formations in the United States and Canada. Prof. Pap. U.S. Geol. Surv., 335: 1—126 (58 pls.).

Reyment, R.A., 1955. Some examples of homeomorphy in Nigerian Cretaceous ammonites. Geol. För. Stockh. Förh., 77: 567—594.

Rhodes, F.H.T., 1967. Permo-Triassic extinction. In: W.B. Harland, C.H. Holland, M.R. House, N.F. Hughes, A.B. Reynolds, M.J.S. Rudwick, G.E. Satterthwaite, L.B.H. Tarlo and E.C. Willey (Editors), The Fossil Record. Geological Society of London, London, pp. 57—76.

Salfeld, H., 1913. Über Artbildung bei Ammoniten. Z. Dtsch. Geol. Ges., 65: 437—440.

Schindewolf, O.H., 1932. Zur Stammesgeschichte der Ammoneen. Paläontol. Z., 14: 164—181.

Schindewolf, O.H., 1933. Vergleichende Morphologie und Phylogenie der Anfangskammern tetrabrachiater Cephalopoden. Einer Studie über Herkunft, Stammesentwicklung und System der niederen Ammoneen. Abh. Preuss. Geol. Landesanst., N.F., 148: 1—115 (pls. 1—4).

Schindewolf, O.H., 1934. Zur Stammesgeschichte der Cephalopoden. Abh. Preuss. Geol. Landesanst., Jahrb., 55: 258—283 (pls. 19—22).

Schindewolf, O.H., 1936. Paläontologie, Entwicklungslehre und Genetik. Borntraeger, Berlin, 108 pp.

Schindewolf, O.H., 1938. Über parallele Reihentwicklung bei Ammoneen. Fortschr. Geol. Palaeontol., 12: 387—491.

Schindewolf, O.H., 1939. Über den Bau Karbonischer Goniatiten. Paläontol. Z., 21: 42—67 (4 pls.).

Schindewolf, O.H., 1940. Konvergenzen bei Korallen und Ammoneen. Fortschr. Geol. Palaeontol., 12: 387—491.

Schindewolf, O.H., 1961—1968. Studien zur Stammesgeschichte der Ammoniten. Parts 1—7. Abh. Math.-Naturwiss. Kl. Akad. Wiss. Mainz, Part 1 (1961), 1960: p. 1—109 (pls. 1—2); Part 2 (1962), 1962: 111—257 (pl. 3); Part 3 (1964), 1963: 259—406; Part 5 (1966), 1966: 509—640; Part 6 (1967a), 1966: 641—730; Part 7 (1968), 1968: 731—901.

Scott, G.R. and Cobban, W.A., 1965. Geologic and biostratigraphic map of the Pierre Shale between Jarre Creek and Loveland, Colorado, U.S. Geol. Surv. Misc. Field Inv., Map I-439.

Simpson, G.G., 1952. How many species? Evolution, 6: 342.

Simpson, G.G., 1953. The Major Features of Evolution. Columbia University Press, New York, N.Y., 434 pp.

Spath, L.F., 1933. The evolution of the Cephalopoda. Biol. Rev., 8: 418—462.

Spath, L.F., 1936. The phylogeny of the Cephalopoda. Paläontol. Z., 18: 156—181.

Spath, L.F., 1938. A Catalogue of the Ammonites of the Liassic Family Liparoceratidae. British Museum (Natural History), London 191 pp. (26 pls.).

Spath, L.F., 1923—1943. A monograph of the Ammonoidea of the Gault. Palaeontogr. Soc. (Monogr.), 787 pp. (72 pls.).

Stanley, S.M., 1973. An explanation for Cope's Rule. Evolution, 27: 1—26.

Strakhov, N.M., 1948. Principles of Historical Geology. II. Gosgeolizdat, Moscow, 396 pp (in Russian).

Swinnerton, H.H., 1950. Outlines of Palaeontology. Arnold, London, 3rd ed., 393 pp.

Teichert, C., 1967. Major features of cephalopod evolution. In: C. Teichert and E.L. Yochelson (Editors), Essays on Palaeontology and Stratigraphy. Univ. Kansas Press, Spec. Publ., 2: 162—201.

Teichert, C., Kummel, B., Sweet, W.C., Stenzel, H.B., Furnish, W.M., Glenister, B.F., Erben, H.K., Moore, R.C. and Nodine Zeller, D.E., 1964. Treatise on Invertebrate Palaeontology Part K. Mollusca (3). Geological Society of America and University of Kansas Press, Lawrence, Kansas, pp. K1—K519.

Thomel, G., 1972. Les Acanthoceratidae cénomaniens de Chaînes subalpines méridionales. Mém. Soc. Géol. Fr., N.S., 116: 204 pp. (88 pls.).

Umbgrove, J.F.H., 1942. The Pulse of the Earth. Nijhoff, The Hague, 358 pp.

Umbgrove, J.F.H., 1947. The Pulse of the Earth. Nijhoff, The Hague, 2nd ed., 358 pp.

Valentine, J.W., 1971. Plate tectonics and shallow marine diversity and endemism, an actualistic model. Syst. Zool., 20: 253—264.

Valentine, J.W., 1973. Evolutionary Palaeoecology of the Marine Biosphere. Prentice-Hall, Englewood Cliffs, N.J., 511 pp.

Valentine, J.W. and Moores, E.M., 1970. Plate-tectonic regulation of faunal diversity and sea level: a model. Nature, 228: 657—659.

Van Hoepen, E.C.N., 1931. Die Krytfauna van Soeloeland. 2. Voorlopige Beskrywing van enige Soeloelandse Ammoniete. I. *Lophoceras, Rhytidoceras, Drepanoceras* en *Deiradoceras*. Paleontol. Navors. Nas. Mus. Bloemfontein, 1: 39—54 (14 figs.)

Van Hoepen, E.C.N., 1941. Die gekielde Ammoniete van die Suid-Afrikaanse Gault. I. Diploceratidae, Cechenoceratidae en Drepanoceratidae. Paleontol. Navors. Nas. Mus. Bloemfontein, 1: 55—90 (figs. 1—55, pls. 8—10).

Van Hoepen, E.C.N., 1942. Die gekielde Ammoniete van die Suid-Afrikaanse Gault. II. Drepanoceratidae, Pervinquieridae, Arestoceratidae, Cainoceratidae. Paleontol. Navors. Nas. Mus. Bloemfontein, 1: 91—157 (figs. 56—173).

Van Hoepen, E.C.N., 1944. Die gekielde Ammoniete van die Suid-Afrikaanse Gault. III. Pervinquieridae en Brancoceratidae. Paleontol. Navors. Nas. Mus. Bloemfontein, 1: 159—198 (pls. 20—26).

Van Hoepen, E.C.N., 1946. Die gekielde Ammoniete van die Suid-Afrikaanse Gault. IV. Cechenoceratidae, Diploceratidae, Drepanoceratidae, Arestoceratidae, (and) V. Monophyletism or polyphyletism in connection with the ammonites of the South African Gault. Paleontol. Navors. Nas. Mus. Bloemfontein, 1: 199—271 (figs. 174—268).

Van Hoepen, E.C.N., 1951a. Die gekielde Ammoniete van die Suid-Afrikaanse Gault. VI. The so-called old mouth-edges of the ammonite shell. Paleontol. Navors. Nas. Mus. Bloemfontein, 1: 273—284 (figs. 269—287).

Van Hoepen, E.C.N., 1951b. Die gekielde Ammoniete van die Suid-Afrikaanse Gault. VII. Pervinquieridae, Arestoceratidae, Cainoceratidae. Paleontol. Navors. Nas. Mus. Bloemfontein, 1: 285—342.

Van Hoepen, E.C.N., 1951c. A remarkable desmoceratid from the South African Albian. Paleontol. Navors. Nas. Mus. Bloemfontein, 1: 345—349 (3 figs.).

Van Hoepen, E.C.N., 1955a. New and little-known ammonites from the Albian of Zululand. S. Afr. J. Sci., Cape Town, 51: 355—377 (figs. 1—31).

Van Hoepen, E.C.N., 1966c. The Peroniceratidae and allied forms of Zululand. Mem. Geol. Surv. Rep. S. Sci., Cape Town, 51: 377—382 (figs. 32—36).

Van Hoepen, E.C.N., 1966a. New and little known Zululand and Pondoland ammonites. Ann. Geol. Surv. Pretoria, 4: 158—172 (12 pls.).

Van Hoepen, E.C.N., 1966b. New ammonites from Zululand. Ann. Geol. Surv. Pretoria, 4: 183—186 (7 pls.).

Van Hoepen, E.C.N., 1966c. The Peroniceratidae and allied forms of Zululand. Mem. Geol. Surv. Rep. S. Afr., Pretoria, 55: 70 pp. (27 pls.).

Westermann, G.E.G., 1956. Phylogenie der Stephanocerataceae und Perisphinctaceae des Dogger. Neues Jahrb. Paläontol., Abh., 103: 233—279.

Westermann, G.E.G., 1966. Covariance and taxonomy of the Jurassic ammonite *Sonninia adicra* (Waagen). Neues Jahrb. Geol. Paläontol., Abh., 124: 289—312.

Wiedmann, J., 1962a. Ammoniten aus der Vascogotischen Kreide (Nordspanien). 1. Phylloceratida, Lytoceratida. Palaeontographica, 118A: 119—237 (pls. 8—14).

Wiedmann, J., 1962b. Die systematische Stellung von *Hypophylloceras* Salfeld. Neues Jahrb. Geol. Paläontol. Abh., 115: 243—262 (pl. 16).

Wiedmann, J., 1962c. Unterkreide-Ammoniten von Mallorca. 1, Lytoceratina, Aptychi. Abh. Math.-Naturwiss. Kl. Akad. Wiss. Mainz, 1962: 1—148 (10 pls.).

Wiedmann, J., 1963. Entwicklungsprinzipien der Kreide Ammoniten. Paläontol. Z., 37: 103—121.

Wiedmann, J., 1965. Origin, limits, and systematic position of *Scaphites*. Palaeontology, 8: 397—453 (pls. 53—60).

Wiedmann, J., 1966. Stammesgeschichte und System der posttriadischen Ammonoideen; Neues Jahrb. Geol. Paläontol., Abh., 125: 49—79; 127: 13—81 (pls. 3—6).

Wiedmann, J., 1967. Due Jura—Kreide-Grenze und Fragen stratigraphischer Nomenklatur. Neues Jahrb. Geol. Paläontol., Monatsh., 1967: 736—746.

Wiedmann, J., 1968a. Evolución y clasificación de los ammonites del Cretácico. Bol. Geol. Fac. Petrol. Univ. Ind. Santander, 24: 23—45 (2 pls.).

Wiedmann, J., 1968b. Das Problem stratigraphischer Grenzziehung und die Jura—Kreide-Grenze. Eclogae Geol. Helv., 61: 321—386.

Wiedmann, J., 1969. The heteromorphs and ammonoid extinction. Biol. Rev., 44: 563—602 (3 pls.).

Wiedmann, J., 1970. Über den Ursprung der Neoammonoideen. Das Problem einer Typogenese. Eclogae Geol. Helv., 63: 929—1020 (10 pls.).

Wiedmann, J., 1973a. Upper Triassic heteromorph ammonites. In: A. Hallam (Editor), Atlas of Palaeobiogeography. Elsevier, Amsterdam, pp. 235—249.

Wiedmann, J., 1973b. Ancyloceratina (Ammonoidea) at the Jurassic/Cretaceous boundary. In: A. Hallam (Editor), Atlas of Palaeobiogeography. Elsevier, Amsterdam, pp. 309—316.

Wiedmann, J., 1973c. Evolution or revolution of ammonoids at Mesozoic system boundaries? Biol. Rev., 48: 159—194.

Wiedmann, J. and Dieni, T., 1968. Die Kreide Sardiniens und ihre Cephalopoden. Paleontol. Ital., 34: 1—171 (pls. 1—18).

Woodford, A.O., 1965. Historical Geology. Freeman, San Francisco, Calif., 512 pp.

Wright, C.W., 1952. A classification of Cretaceous ammonites. J. Paleontol., 26: 213—222.

Wright, C.W., 1955. Notes on Cretaceous Ammonites. II. The phylogeny of the Desmocerataceae and the Hoplitaceae. Ann. Mag. Nat. Hist., (12)8: 561—575.

Wright, C.W. and Wright, E.V., 1949. The Cretaceous ammonite genera *Discohoplites* Spath and *Hypoplites* Spath. Q. J. Geol. Soc. Lond., 104: 477—497 (pls. 28—32).

Zeiss, A., 1968. Untersuchungen zur Paläontologie der Cephalopoden des Unter-Tithon der Südlichen Frankenalb. Abh. Bayer. Akad. Wiss., 132: 190 pp. (27 pls.).

Ziegler, B., 1957. *Creniceras dentatum* (Ammonitacea) im Mittel-Malm Südwestdeutschlands. Neues Jahrb. Geol. Paläontol., Monatsh., 1956: 553—575.

Ziegler, B., 1958. Monographie der Ammonitengattung *Glochiceras* im epikontinentalen Weissjura Mitteleuropas. Palaeontographica, A, 110: 93—164.

Ziegler, B., 1959. Evolution in Upper Jurassic ammonites. Evolution, 13: 229—235.

CHAPTER 9

TRILOBITES AND EVOLUTIONARY PATTERNS

NILES ELDREDGE

Introduction

In recent years we have seen a tremendous boom in our knowledge of the palaeobiology of the Class Trilobita. In many respects this welcome state of affairs merely reflects a general trend in the entire field of palaeontology — a movement towards a more complete integration of the diverse principles of biology with the enormous body of information we have already amassed about fossils. Trilobite studies have kept pace with this expansion of palaeontological interests: in addition to the hundreds of new genera described since the publication of the trilobite volume of the *Treatise* (Harrington et al., 1959), we are witnessing a resurrection of earlier interests in comparative anatomy, functional morphology, palaeoecology, biogeography, phylogenetic reconstruction, and, perhaps most auspiciously, an explicit attention to evolutionary and ecological theory. There is also a subtle shift in emphasis from a predominantly faunal approach, wherein an entire collection of trilobites from a given stratigraphic setting is described, to a phylogenetic approach in which students become familiar with the systematics of a monophyletic group throughout its entire geographical and stratigraphic range. The recent (Martinsson, 1975) publication of the proceedings of the 1973 NATO Congress (held in Oslo, Norway) serves as the single most important symbol of recent advances in the continuing study of the palaeobiology of trilobites.

There are two basic reasons for studying fossils in a biological framework. First, there is the simple, intrinsic reason: we just want to find out more about their biology. Studies of the anatomy, functional morphology, palaeoecology, biogeography, and phylogenetic relationships of trilobites are of intrinsic interest. All contain the essential properties of scientific hypotheses and their testing (see Nelson, 1970, for a discussion of the logical and scientific structure of comparative biology). These subjects stand by themselves as proper and legitimate areas of scientific focus.

Yet there does seem to be a further reason to study fossils from a biological point of view: to see how they fit into the general scheme of things — meaning our theories on the evolution of life in the context of the history of this planet. In other words, how does evolutionary and ecological theory relate to our understanding of fossils, and — even more to the point — how does our understanding of fossils relate back to the essentially neontological evolutionary and ecological theories we are all working with? There has been as much of an upsurge of concern about biological theory in recent years in palaeontology as there has been a flowering of studies in the diverse specific topics mentioned above, and trilobite students have been in the thick of things. The main thrust of this paper will be to review particularly the impact of evolutionary and ecological theory on trilobite studies, and to evaluate the problems and degree of success that students of trilobites have experienced and effected in both interpreting their data in the light of theory and shedding new light on that theory.

As stimulating as the application of ecological and evolutionary theory to palaeontological data has been and can be, surely our ultimate goal is to improve on these very theories. How are we to do this? In general, our approach is to analyze patterns of distribution of taxa (and their constituent morphological attributes) in space and time; this primary data gathering and first-order analytic approach should be initiated with some specific problem in mind (e.g., differential rates of morphological evolution or taxic production, modes of speciation, etc.). The resultant patterns (e.g., of relationships and distribution of taxa; distribution of plesiomorphic and apomorphic character states; etc.) are then further analyzed in terms of conflicting hypotheses of processes derived from ecological and evolutionary theory. We reject all but the least ineffective hypothesis of those that we test; essential to this entire process, of course, is that hypotheses have to be sufficiently precise and circumspect to allow valid testing in the context of the available information — no easy task. In our flush of enthusiasm with biological theory, a major failing to date has been the formulation of overly general hypotheses ("models") which *per se* are insufficiently precise and/or realistic (see Levins, 1968, for a discussion of the three aspects of "models") to allow us to perform tests which could theoretically result in the rejection of all or part of the model.

In any case, the main stumbling block which has effectively prevented substantive contributions from palaeontology to either evolutionary or ecological theory stems not so much from methodological failings, but from (1) the nature of the data itself, and (2) more importantly, a manifestly incorrect bias concerning the fundamental nature of the evolutionary process itself. We are all aware of the problems with data represented by fossils — their very nature precludes both the derivation and, in most instances, the practical application of such fundamental concepts as selection, mutation, recombination, and polymorphism (all within-population and essentially genetic concepts), as well as speciation itself. However, carefully delimited studies of judiciously chosen material *can* lead to studies of within and among population variation, as well as speciation itself (e.g., Eldredge and Gould, 1972; Gingerich, 1976).

We are left with a conceptual bias as the main reason why palaeontology has to date contributed almost nothing to evolutionary theory. There are two fundamentally different ways of looking at the evolutionary process, boiling down to a conflict between two alternate forms of the fundamental question in evolution. In one view, evolution is defined as change in gene content and frequency, and the central question is: how are genes and their phenotypic expression (including physiology and behaviour) modified in the evolutionary process? The other view sees evolution as the origin of taxa (populations and/or species) and the morphological change as a subordinate (yet still interesting) issue to be approached strictly within the context of speciation; in this view, the central question of evolution is: how do new species originate?

This is not merely a semantic distinction. Although elements of both views permeate the approach we all have to evolution (neo-darwinism, for example, is a nearly total admixture of the two approaches) and although the vocabulary of nearly all evolutionary biologists is the same ("genes", "alleles", "chromosomes", "polymorphism", "mutation", "populations", "species", "characters", "character states", etc.) none the less, the difference between the two approaches is a real one and leads to totally different conceptions of problems and choices of hypotheses to explain them.

A single example should suffice to clarify this point. Westoll (1949) and Schaeffer (1952) published analyses of the bradytelic lungfishes and coelacanths, respectively. Everyone would agree that both groups have not changed much since an initial Devonian episode of morphological diversity. Rate of morphological change has been slow. Under the view that evolutionary problems centre around explanations of morphological change (and, of course, relative lack of that change) *per se*, we have four or five competing hypotheses (involving mutation rates, homeostasis, centripetal selection, etc. — they are not all mutually exclusive; see Eldredge, 1974b, for a similar discussion pertaining to xiphosuran evolution) that are usually brought out to explain such phenomena. As a corollary, under this view, we note that since the Devonian there are relatively few taxa (species; genera) within these lineages when compared to other, related lineages (e.g., actinopterygians), effectively removing the objection of poor preservation. Under this view of evolution as a strictly morphological problem, there are so few taxa simply because the group has changed so slowly that there are relatively few morphological differences to which we can point and which we may use to delineate separate taxa.

But there is patently another way of looking at this problem. If we assume that populations and species are real entities (see Ghiselin, 1974; Eldredge, 1977) and not merely segments of lineages arbitrarily subdivided by sequential changes in character states, we can ask if diversity — actual numbers of species — is in reality rather low in these lineages. In other words, the problem may be stated the other way around, and we may hypothesize that the coelacanth and lungfish lineages have exhibited extraordinarily low rates of speciation in post-Devonian times. We might further assume, as a partial corollary, that low speciation rates imply low rates of cumulative morphological change. We are then at liberty to examine the corpus of evolutionary and ecological theory which might shed some light on differential rates of speciation — particularly, in this case, on factors which may dampen speciation rates. Thus, ecological strategies (eurytopy and stenotopy), geographical distributions, and competition (see Stanley, 1973, for an excellent discussion) are concepts which can be brought to bear on the problem which would most likely be overlooked if bradytelic rates were to be treated solely as a problem of low rates of morphological change.

Following application of such concepts — in the form of specific, testable hypotheses — to particular cases, we are then in a position to feed back positive information to improve and correct aspects of ecological and evolutionary theory. For at this point we can actually evaluate the degree of congruence that specific neontological models of process have with the actual results of the evolutionary process.

A major theme of this paper is that, to the present day, we palaeontologists have managed to contribute relatively little to explicit theories of the evolutionary process, despite a tradition of examining "major patterns of evolution". The main reason stems from a nearly total commitment to viewing evolution strictly as a problem of the transformation of morphology, despite the fact that it is the origin of taxa — species and populations — which is the only context in which evolution is actually known to act. An additional complicating factor has been the conspicuous failure (analyzed below for trilobites) in formulating reliable theories of phylogenetic relationship among our taxa. Lack of a coherent theory of relationships among the families of trilobites has greatly impeded progress in our understanding of the major patterns of their evolution.

The remainder of this paper will be devoted to an examination of what progress has been made in understanding modes of origin of taxa from studies of trilobites. We shall then examine some of the "major features" of trilobite evolution pursued under a strictly morphological point of view, and see how these and related problems may be translated into the terms of taxic evolution. Finally, we shall examine the current state of progress in understanding trilobite relationships, both within the class, and among the Arthropoda.

Speciation in Trilobites

There is every reason to believe that trilobites are essentially like any other component of the vagile benthos in terms of the evolutionary patterns they should be expected to demonstrate. As interest in micro-evolutionary processes increases in palaeontology, trilobites — as complicated organisms exhibiting morphological change perhaps on a more easily detectable level than most other common invertebrates (Schopf et al., 1975) — should figure prominently as palaeontology moves towards its deserved status as a truly evolutionary discipline. Indeed, I (Eldredge, 1971) made the rather bold assertion that all micro-evolutionary phenomena known or suspected to affect morphology among Recent organisms are potentially documentable in the fossil record, and their analysis could go far towards shedding further light on the very processes themselves.

Perhaps the best example of a careful analysis of species-level evolutionary phenomena from the older literature is Kaufmann's (1933) study of *Olenus*, resurrected from obscurity by Simpson, 1953, and discussed further by Henningsmoen (1964), Eldredge (1971), and Fortey (1974). In recent years, such studies have come back into vogue, and some of the more important issues in contemporary palaeontology have been discussed in conjunction with trilobite data. Thus, debates on primary controls of diversity and on the actual modes of origin of new species have both drawn heavily on studies of trilobites.

Nowhere is the dichotomy in evolutionary thinking (see introduction) more graphically seen in palaeontology than in the two basic and largely contradictory models of phyletic gradualism versus speciation ("punctuated equilibria" of Eldredge and Gould, 1972). Harper (1975) has claimed these views to be merely extremes of a spectrum of possibilities, missing the point that one (gradualism) stems wholly from a formal, morphological view of the nature of the evolutionary process, whereas the other (speciation) sees the central problem as the origin of new taxa *per se*. Thus, these two "models" are, strictly speaking, not even comparable. Trilobite data have figured prominently in the argument on this issue.

When working on Devonian phacopid trilobites (Eldredge, 1971, 1972, 1973; also Eldredge and Gould, 1972), I concluded that species and subspecies tend to remain relatively unchanged (at least in terms of inter-taxic *differentia*) throughout their stratigraphic ranges and I produced the following model for the evolution of *Phacops rana* subspp.: Species spread throughout the available habitats within a depositional basin as far as they can, limited by the physical and biotic constraints of the environment as it varies from place to place. When new habitats appear through transgression, the species will go with it. *P. rana rana* was limited to the marginal sea for about two million years because two other closely related subspecies were already occupying the available habitat space over the cratonal interior. On the other hand, regressions destroy habitats, and species become locally extinct. *Phacops rana crassituberculata* and *Phacops rana milleri*, the two subspecies dominating the epeiric sea of the cratonal interior, had nowhere to go when the large regression came near the end of Cazenovian times, and they disappeared completely; the Centerfield transgression subsequently restored suitable habitats over the cratonal interior, and *Phacops rana rana*, already living in the marginal sea to the east, simply came along with it, no longer competitively excluded by close relatives.

Origins of the new (sub)species seemed best explained by geographical differentiation and isolation; i.e., allopatric speciation. The best documented example (derivation of *Phacops rana rana* from *P. rana crassituberculata*) occurred in the near-shore environment of the marginal sea during the peak of a transgression, as though maximal extent of the epeiric sea was necessary before geographical isolation could be effected. But other than this, no other direct effects on speciation are attributable to transgressions or regressions in the evolution of the *Phacops rana* complex.

Further work with slightly older phacopids (Eldredge, 1973, 1974a) — especially the Emsian—Eifelian *Phacops cristata* — added the further point that during the maximum extent of a transgression, a species may diversify so that local populations living in the more environmentally stable sea (on a diurnal and perhaps seasonal basis, at least) may become rather more specialized and phenetically less variable than closely related populations living along the fringes of the sea in near-shore environments. This conclusion is in general agreement with the model proposed by Bretsky and Lorenz (1970). But even in this instance where some evolution is thought to have occurred (via simple geographic differentiation) in the cratonal interior, I (1974a) still conclude that new species tend to arise near-shore; *Phacops cristata* is the intermediate in a series of three species all referable to the same monophyletic taxon, and it is the variable near-shore populations of *Phacops cristata* which most closely resemble both the older species *P. logani*, and the younger species *P. iowensis*.

Thus the overriding conclusion from the data pertaining to within and

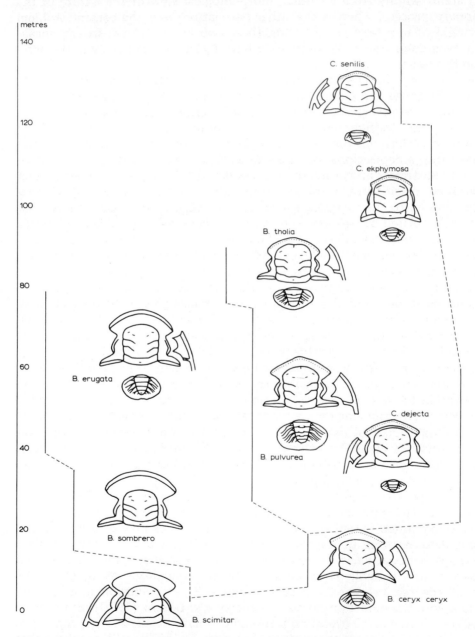

Fig. 1. Fortey's (1974, p. 16, fig. 2) reconstruction of the phylogeny of the Balnibarbiinae (Olenidae), interpreted by Fortey as indicative of allopatric speciation. See text.

among population variation of various phacopid taxa is that an extrapolation of the allopatric model of speciation, most cogently promulgated by Mayr (1942, 1963), is by far the most powerful explanatory hypothesis available. The alternative ("phyletic gradualism") was shown to be inconsistent with the phacopid data. Whether or not these findings can be extended to embrace the majority of metazoans, as contended or implied by myself and others (Eldredge, 1971, 1974a; Eldredge and Gould, 1972) — see Harper (1975) and Gingerich (1974, 1976) for contrasting views — further work with other trilobites (Fortey, 1974, on Ordovician olenids from Spitsbergen; Henry and Clarkson, 1975, on the cheirurid *Placoparia*) has tended to confirm the notion that trilobite species, at least, tend to "act" as discrete spatio-temporal entities. "If the duration of species with little morphological change is relatively long compared with that of their derivation from their ancestors, this implies that evolution in this case defines discrete morphological groups with a particular stratigraphic range — that is, that palaeontological species seem to have real meaning among these olenids rather than being arbitrary points on a continuous morphological spectrum ranging through time" (Fortey, 1974, p. 20; Fig. 1). However, Lespérance and Bertrand (1976) have concluded that the stratigraphic record of *Cryptolithus* in the Nicolet River Valley (St. Lawrence lowlands of Quebec, Canada) substantiates neither phyletic gradualism nor "punctuated equilibria".

Robison's (1975) recent convincing utilization of the concept of character displacement to explain species packing (density) comparatively in agnostoid faunas through time is an excellent example of the application of evolutionary theory to trilobite data. Robison found that in agnostoid faunas with three or more species, among those species comprising more than 10% of all the individuals (i.e., excluding rare species), there is developed a hierarchical arrangement of size (as measured by sagittal length of the pygidium) whereby each species along the size gradient is close to 1.28 times larger than the next larger species — approximating quite closely indeed the expected number proposed by Hutchinson (1959). This example of character displacement, as well as others from the fossil record including one dealing with phacopid trilobites, is reviewed and discussed further elsewhere (Eldredge, 1975).

Diversity Controls and Faunal Patterns of Speciation

Trilobite species have also figured prominently in what might be considered the next step in complexity of evolutionary patterns: changes in species diversity within monophyletic assemblages within a depositional basin. The data remain realistically firm — the stratigraphic ranges of discrete species — but the central item of interest is fluctuation in diversity, and in this case it is predominantly ecological theory that is brought to bear on the problem. This is a relatively new concern in palaeontology, and the history of thinking on Palmer's (1965a, 1965b) concept of the biomere (and attendant data!), based on Upper Cambrian trilobite species, tells us much about how palaeontologists in general approach problems of this nature.

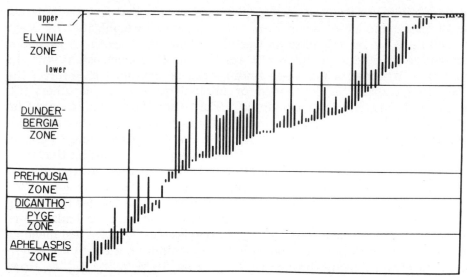

Fig. 2. Approximate stratigraphic ranges of ptychopariid trilobites of the Pterocephalid Biomere, Great Basin. (From Stitt, 1971, p. 178, text-fig. 1, after Palmer, 1965b.)

Palmer's original concept of a biomere — a biostratigraphic unit of stage magnitude bounded by abrupt and distinct "non-evolutionary" faunal breaks — was
based on a study of Upper Cambrian trilobites of the Great Basin of the western United States. First to be named was the Pterocephaliid Biomere (Fig. 2).
Most of the trilobites within that biomere, as in the subjacent *Crepicephalus*
Biomere and the superjacent Conaspid Biomere, belong to one or two families
— and diversity increases after the initial appearance of a few generalized forms.
Palmer (1965a, 1965b) thought the boundaries time-transgressive and not
marked by lithological changes. His basic interpretive model was that an extra-
cratonal stock of deep-water trilobites would invade the shelf and radiate, only
to be finally wiped out, presumably by a change (alleged to be a lowering) of
temperature. The cycle would then be repeated. Thus we have documentation,
on the species level, of relatively major evolutionary events (i.e., the radiation
of entire families) and extinction.
    Stitt (1971, 1975) has developed a more complex evolutionary model for
the trilobites of the various named biomeres, claiming that each typically represents an evolutionary pattern divisible into four separate, successive parts. The
first stage is rather brief, populated by relatively few species all of which are
rather generalized and variable, hence rather difficult to tell apart. Stitt (1975)
has likened this stage to Simpson's (1953) criteria for an adaptive radiation.
According to Stitt, this radiation is probably stimulated by "the opportunities
present on the cratonic shelf for these formerly oceanic trilobites" (1975, p.
383). Stage two is also a period of relatively low diversity, but the trilobites
are less variable and have longer stratigraphic ranges. Of stage two, Stitt (1975,
p. 384) says: "Stage two is what Simpson (1953, p. 229) calls the consolidation
or weeding out phase of adaptive radiation. Species of a few genera have

attained some sort of stability and equilibrium with the environment. These species become abundant. . . and other species, less well adapted, are slowly eliminated. The increase in diversity of cranidial morphologies evident in Stage two is an indication that the trilobite population has begun to establish distinct ecologic niches in the environment, and those forms best suited to particular niches are the ones that become abundant in the population".

Stage three has high diversity with taxa exhibiting long stratigraphic ranges and even less within-species variation. This is the stable, mature community. A fourth stage, consisting of relatively few species with short stratigraphic ranges, represents, according to Stitt (1975, p. 385), at least in so far as the Ptychaspidid Biomere of Oklahoma is concerned, a period of " 'evolutionary desperation' in which the established families of the biomere attempted to adjust to whatever environmental changes were causing the rapid extinction of the trilobite shelf community."

Stitt's (1971, 1975) explanatory models of the ecological and evolutionary dynamics accounting for biomere patterns are imaginative but cannot be taken as demonstrated until some more definitive, testable hypotheses are formulated and tested with the data. Indeed, Ashton and Rowell (1975), in a paper published just prior to the appearance of Stitt's second (1975) paper on evolutionary patterns in biomeres, tested an explicit contention in Stitt's basic characterization of his stages, viz., that within-taxon variation is markedly lower in stage three than in stage one. Measuring eight cranidial features of samples from both the Aphelaspis Zone (stage one of the Pterocephaliid Biomere) and the Dunderbergia Zone (stage three of the same biomere) at two different sections, they resoundingly rejected Stitt's contention of lowered variability within stage three as compared with stage one. In a stimulating discussion following the presentation of their results, Ashton and Rowell (1975, p. 169 ff.) discuss the applicability of the "niche-variation" model to the agreed-upon increase in diversity within biomeres. Bretsky and Lorenz (1970), in their paper arguing for a direct relationship between environmental stability, diversity, and narrowness of genetic and phenotypic variation, explained biomeres simply by postulating increasing environmental stability as time went on during the development of a biomere; when the environmental trend reversed, so did the pattern of high diversity of individual stenotopic species. Ashton and Rowell (1975) point out that an argument involving resource sharing (niche partitioning) which also demands a decrease in genetic variability cannot be supported by their results. Ashton and Rowell (1975) agree with the Bretsky and Lorenz (1970) model of biomeres to the extent of postulating increasing environmental stability as a general explanation for the observed increase in diversity (repeated in several biomeres) but they conclude that the concomitant assumption of decrease in within-species variability is not necessary, at the very least, and possibly altogether false (see Ayala et al., 1975, for additional arguments in support of this view).

Johnson (1974) utilized his model of "perched faunas" in a somewhat different application of evolutionary and ecological theory to explain biomere patterns. Johnson took sharp issue with Palmer's contention that biomere boundaries are time-transgressive and disputed Palmer's (1965a, 1965b) contention that there is no lithological evidence of environmental change associated with biomere boundaries. Johnson (1974) went on to cite some evidence from Lochman-Balk (1970) that the tops of biomeres are marked by the sudden onset of regressive conditions, and his model of biomeres conforms to his other examples in that paper: stenotopic, "perched" faunas are ripe victims for any sudden regression. Johnson's model for biomeres is in some ways the simplest, though of course it is not addressed to the underlying causes for the observed increase in diversity, the one characteristic of biomeres all authors seem willing to agree upon. Clearly the last word has not been written on the subject of biomeres and the evolutionary patterns they represent. These are precisely the right kinds of data, and the problem is of exactly the right order of magnitude for palaeontologists to come to grips with the evolution of taxa of higher categorical rank (families in this case) by analyzing evolutionary patterns at the species level.

Biogeography

The uproar in methodological theory of historical biogeography has seemed not to have affected palaeontologists to date. Thus there is no debate, involving trilobites or not, which argues the relative merits of the Croizat approach (e.g., Croizat et al., 1974) versus the more traditional approaches of Matthew (1915), Simpson (e.g., 1965) and others. Croizat et al. (1974) eschew such concepts as "centres of origin" and "migration" in favour of "tracks" and "generalized tracks" in attempts to reconstruct "ancestral biotas". Acceptance of historical changes in relative positions of plates has been widely heralded as vindication of Croizat, but palaeontologists have accepted plate tectonics with considerably more ease: we still simply map distributions of correlative biotas, but now we map them on correct contemporary, rather than anachronistic, geographies. The problem posed by the advent of plate tectonics to palaeontological biogeography has been rather different. First, some of us have used our data as "proof" of continental drift, instead of being content to accept drift on other grounds and then simply using other people's reconstructions of appropriate geographies for understanding the distributions of our fossils. There is an additional problem: we have been perhaps over enthusiastic in our rush to re-interpret all our data on trilobite distributions — being too quick to couch explanations of patterns as being due entirely to plate movements. Forgotten in the onslaught are some tried and true principles of intrinsic environmental factors affecting distribution. Perhaps the best paper yet written (Cook and Taylor, 1975) to make this point is based on trilobite data (Fig. 3). Cook and Taylor (1975) point out that a correspondence between certain Upper Cambrian faunas in China and western North America might be considered *prima facie* evidence for a close association between the two plates in Cambrian times. Cook and Taylor convincingly show, however, that precise reconstruction of the geographical relationships among the faunas within both China and western

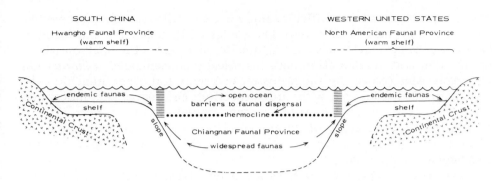

Fig. 3. Cook and Taylor's (1975, p. 561, fig. 4) "inferred habitat relationships between several Late Cambrian trilobite faunal provinces". See text.

North America (complete with regional stratigraphic and sedimentary petrological analysis) show that the two faunas which are so similar on the two landmasses represent very deep-water slope faunas, probably below the thermocline i.e., it is most likely that these faunas are open ocean and their occurrence on the two plates can at least as effectively be explained in terms of cosmopolitanism. This view is reinforced when the contemporary shelf trilobite faunas are compared and little or no correspondence is found between those of China and those of the western United States. The moral: facies, and the differing marine habitats they imply, are as important as ever before in the correct analysis of trilobite biogeography.

Schopf (1974) together with the companion paper by Simberloff (1974) has shown how a simple extrapolation of the species/area curve, together with other aspects of the MacArthur—Wilson brand of equilibrium biogeography, can be a powerful tool in analyzing diversity fluctuations. As Schopf (1974) points out, this approach follows logically from Valentine's (1973 and further references therein) modelling of the relationship between diversity and plate tectonics. To date, trilobite workers do not seem to have followed up this lead, but such an approach seems to be a most promising way of getting at the ultimate meaning of the scores of essentially descriptive biogeographical studies still being produced by students of trilobites (and all other fossils!).

Aspects of Evolutionary Adaptation in Trilobites

Trends, adaptive radiations, and studies of rates of morphological change — some of the more redoubtable "major features of evolution" — have been pursued nearly exclusively under the morphological view of the nature of the evolutionary process (see introduction), which is probably why such studies have only a vague conceptual nexus with the actual evolutionary process (speciation). Clearly all three topics ultimately can be phrased in terms of the taxic view of evolution.

In addition to this general problem, study of trends, adaptive radiations, and the like in trilobites has been difficult simply because trilobites are presumed to be extinct and we have such meagre knowledge of their feeding and locomotory systems, the two major anatomical/functional aspects of animals on which biologists interested in adaptation have historically concentrated. *Limulus* and certain crustaceans (especially the notostracan *Triops* and other branchiopods, and the marine Isopoda) can, with circumspection, be used as anatomical and behavioural (functional) analogues to a certain limited extent. In short, there is no lack of structural diversity of the dorsal exoskeleton among trilobites, but the functional significance underlying the different morphologies we see remains largely unexplained.

For these reasons there are perhaps fewer long-term "evolutionary trends" within lineages "documented" for trilobites than for many other groups. Many of these are ably summarized by Stubblefield (1960). In addition, Henningsmoen (1964) has discussed various aspects of trends using Upper Cambrian Olenidae as examples. But most studies of trilobite adaptation have focused on the functional morphology of single species and genera. Work has been sporadic. Following the monumental efforts of Rudolf Richter (especially the series "Von Bau und Leben der Trilobiten" written during the first quarter of this century) there followed a period of about forty years with little further progress, with the conspicuous exception of several papers by Whittington (e.g., 1941). Recently, more studies along these lines have appeared, most notably by Clarkson (e.g., 1966a–c, 1969, 1975; Clarkson and Henry, 1973). It is through studies such as Clarkson's on *Eophacops musheni* (1966c) and the odontopleurid *Leonaspis deflexa* (1969), largely devoted to orientation, thus behaviour, of the organism in life, that explanations of the adaptive reasons underlying morphological diversity in trilobites will ultimately emerge.

Rather than reciting a litany of adaptive trends, radiations, and the like from the literature, we shall instead focus on a single functional complex in two closely related trilobite groups, and discuss some of the phylogenetic patterns that have recently been described and analyzed by various workers. Campbell (in press) has discussed the functional significance of various sculpture patterns (spines and emarginations) along the anterior and lateral cephalic border and pygidial border of certain dalmanitacean trilobites. Generally, such trilobites are characterized by elaborate border morphology on either the cephalon or the pygidium; seldom are such structures encountered on both the cephalon and pygidium within the same species. Campbell considered various hypotheses as explanations for the (assumed) adaptive significance of such structures in the first place — hypotheses sufficiently flexible to account for alternate character states and especially evolutionary changes in such states. It is clear that the only unified explanation of all these structures (i.e., cephalic *and* pygidial) stems from considering the organisms when enrolled, i.e., when the pygidium is closely appressed to the ventral surface of the cephalon. Campbell further showed that, when enrolled, a system of slits around some or all of the periphery of the pygidial/cephalic juncture is created.

Campbell chose as most likely (least unlikely) the hypothesis that this slit system functioned to regulate the passage of water currents to and from the ventral (sternal) surface (with respiratory surfaces, be they on the epipodites as commonly assumed or the ventral membrane as suggested by Bergström, 1969; filter feeding is another possible function of inhalation of water currents in an enrolled state) and the external milieu. Sphaeromid isopods enroll, leaving a gap at the anterior portion of the "commissure", the limbs beat regularly, and water is taken in and expelled through the aperture. Campbell's explanation thus seems on very firm ground, and in the context of this model of the functional significance of the elaboration of cephalic and pygidial border morphology, we can re-examine a study of trends and phylogeny reconstruction (Lespérance and Bourque, 1971; Lespérance, 1975) of some (essentially) Appalachian Devonian Dalmanitacea.

Lespérance and Bourque (1971; later modified by Lespérance, 1975; see Fig. 4) published a phylogeny of the Synphoriinae (a newly erected subfamily of Dalmanitidae later raised to full familial status by Lespérance, 1975). The phylogeny was reconstructed in precisely the manner characterized as typical palaeontological procedure by Eldredge and Tattersall (1975), viz., compilation of all species comprising a genus, using morphological criteria and stratigraphic continuity or near continuity; then by extension, linking genera in ancestral—descendant sequence by morphological similarity and relative stratigraphic position. Of particular interest is the trend in crenulation morphology of the anterior and lateral cephalic border (Lespérance and Bourque, 1971), later modified into two trends by Lespérance (1975; see Fig. 5). (This modification resulted from a change in phylogenetic interpretation occasioned by the "downward extension" (i.e., stratigraphically) of one of the species; Lespérance, 1975.)

It is not entirely clear from either paper whether (1) the trend(s) in anterior border morphology were viewed initially as a morphocline, the analysis of which led in part at least to the elaboration of the phylogeny presented, or (2) the trend emerged separately after an analysis of relationships on other criteria. To be thoroughly convincing, a trend, i.e., an actual biostratigraphic character gradient, should be (1) a convincing morphocline, the polarity of which makes sense in and of itself, and (2) following an elaboration of a specific hypothesis of relationship, shown to be coincident, or nearly so, with the actual stratigraphic distribution of the taxa bearing the various character states. It is absolutely essential that, if the character suite in question is used to evaluate the phylogenetic relationships (and these are precisely the kinds of characters most useful for this purpose), that affinities be analyzed independently of stratigraphic occurrence. Otherwise we have a problem in circularity; many, if not most, "trends" in the palaeontological literature are selective renderings of the stratigraphic occurrences of character states which are not independently shown to have anything to do with an actual phylogeny. Again we see the pervasive influence of the morphological view of evolution, where morphological features are strung together without regard to the crucial intermediate step: analysis of the phylogenetic relationships among specific taxa.

Returning to the synphoriids: particularly in the case of the revised study (Lespérance, 1975), the morphoclines in anterior border morphology are in

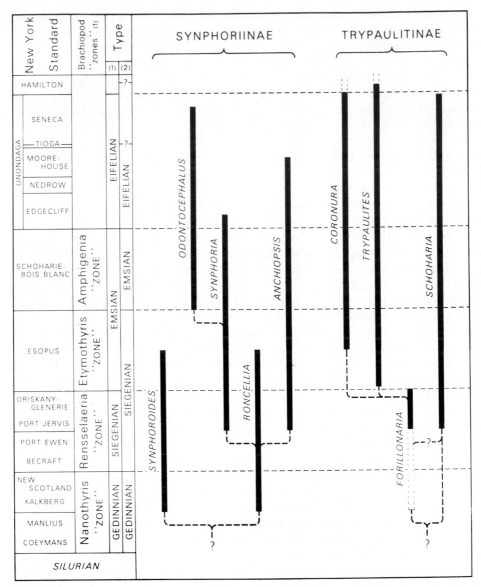

Fig. 4. Stratigraphic ranges and inferred phylogenetic relationships among North American Synphoriidae. (From Lespérance, 1975, p. 96, text-fig. 3.)

themselves plausible. The phylogeny also seems acceptable. Thus, as an hypo-thesis, the trends presented seem acceptable. In part at least they seem to imply progressively more sophisticated control and monitoring of the passage of water to the ventral surface of the enrolled trilobite. Accepting Lespérance's (1975) phylogeny and trend analysis at face value, there are several further interesting generalizations that we may reach. First we note that the "trends" are not entirely smoothly gradational: within the stratigraphic ranges of species

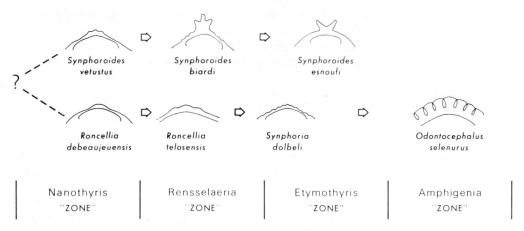

Fig. 5. Lespérance's (1975, p. 97, text-fig. 4) interpretation of two evolutionary trends in anterior cephalic border morphology among some Synphoriidae. See text.

and genera, there is little or no documentable change. Border morphology is basically stable within species, and even within genera. The trend appears in "step-wise" fashion among a series of genera whose stratigraphic ranges overlap. In this well-documented example, there is no room for phyletic gradualism.

Another point of evolutionary interest is that, within the subfamily Trypaulitinae, the anterior cephalic border remains unmodified (the primitive condition) throughout the entire stratigraphic range of the family, and that one genus from this subfamily (*Coronura*) independently developed a water-monitoring system by the addition of a series of spines along the periphery of the pygidium. Trends evidently involve only monophyletic subportions of monophyletic groups — another important *caveat* to those of us who would pick and choose among likely candidates whose stratigraphic positions conveniently fit a preconceived notion of a trend.

Morphology of the anterior cephalic border of the Synphoriidae is also instructive in at least one additional regard: there are, in the Devonian rocks of the Appalachian faunal province, several additional dalmanitacean genera with similar elaborate modifications of the cephalic border. Prior to the study of Lespérance and Bourque (1971), perhaps the best hypothesis was that *all* such dalmanitaceans of that province represented an adaptive radiation of rather large proportions. Lespérance and Bourque (1971), Lespérance (1975), and Campbell (1977) have provided criteria — in effect, shared, derived characters — which effectively segregate the Synphoriidae as the apomorphic sister group of the Dalmanitidae proper, to which the remainder of the Devonian Appalachian genera (including some with crenulations or spines on the cephalic border) evidently belong. This conflict in morphoclines (using one set of characters to reject a phylogenetic hypothesis based on another set) automatically results in the demonstration of parallelism. Accepting the analyses of the workers cited above, instead of one grand adaptive radiation in which the elaboration of cephalic and/or pygidial morphology played perhaps the most conspicuous role, we have two such radiations with similar trends developed in parallel. The interpretation of this situation in terms of micro-evolutionary processes is almost certain to be rather different from explanations based on the assumption that the entire assemblage is monophyletic.

Phylogenetic Relationships Among the Trilobites

Harrington (1959) has ably summarized the history of trilobite classification (the closest thing to a discussion of phylogeny in the *Treatise*) and the classification presented in his chapter is the best we have, despite its manifest problems (many of which were discussed by Harrington himself). But it is impossible to construct a single dendrogram of phylogenetic relationships among taxa of any categorical rank from the *Treatise* classification. In other words, there is no expression of relative affinity among subfamilies within families, of families within superfamilies, of superfamilies within orders, or among the orders of the Class. Whittington (1966) and Stubblefield (1960) have both noted that whereas many, if not most, trilobite families seem reasonably homogeneous (meaning more or less monophyletic — I would extend this generalization to the superfamilies *sensu Treatise*), no one has yet concocted a plausible hierarchical clustering of trilobites above this level. Nor is this merely a problem of historical lack of attention to the more precise formulation of methodology for reconstructing phylogenies as given, say, by Hennig (1950, 1966). Though this is indeed an aspect of the problem, it remains true that most genera of trilobites can easily be clustered with other genera on the grounds of morphological similarity (phenetics) and some such clusters can be related fairly easily to others; there are, to my subjective impression, more serious morphological gaps between the suborders and the orders (again, *sensu Treatise*) than is the case in most other animal "classes". Stubblefield (1960) has stressed the "cryptogenetic" origin of most trilobite families — though I do not share his view that a Simpsonian model of "quantum evolution" or, worse yet, some model of saltation need be invoked to explain trilobite evolution. In any case, these morphological gaps are going to prove difficult to surmount even with the sophisticated approach of "cladistics" which distinguishes between various kinds of morphological similarities. The problem with trilobite phylogeny, at least above the level of superfamilies, is simply that whatever similarities these taxa share with one another seem to be largely plesiomorphic for the class as a whole, and these in any case are masked by their differences, nearly all of which appear to be autapomorphies. The lichads, odontopleurids, trinucleids, and harpids are perhaps the most striking examples of this — all harpids look pretty much alike, and none of them resemble any non-harpid trilobite to any great extent at all except in "general" features which stamp them as trilobites in the first place!

Bergström (1973) has recently presented a new, even revolutionary theory of trilobite relationships (Fig. 6). This attempt is of course to be welcomed, and it would be grossly unfair to dismiss it out of hand without the extensive dis-

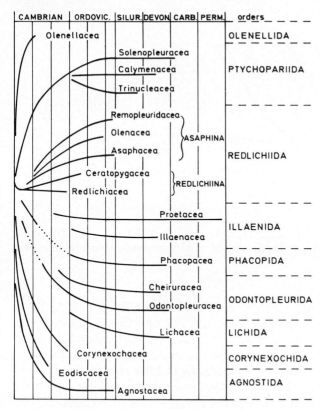

Fig. 6. Reconstitution of major groups of trilobites and their phylogenetic relationships as given by Bergström. (From Bergström, 1973, p. 38, fig. 12.)

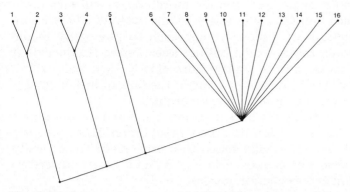

Fig. 7. Cladogram summary of relationships among major groups of trilobites, as given by Hahn and Hahn (1975, p. 448, fig. 1). The figure has been redrawn; spelling conforms to the original usage of Hahn and Hahn (1975). The Ptychopariina (group 6) are considered directly ancestral to groups 7—16 in their original diagram. 1, Agnostina; 2, Eodiscina; 3, Olenellida; 4, Redlichiida; 5, Corynexochida; 6, Ptychopariina; 7, Proetina; 8, Illaenida; 9, Asaphida; 10, Lichida; 11, Odontopleurida; 12, Calymenida; 13, Cheirurida; 14, Phacopida; 15, Trinucleida; 16, Harpida.

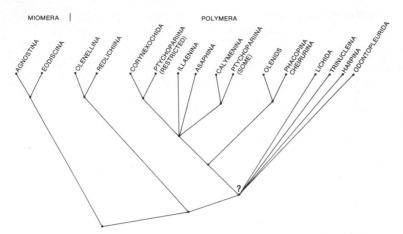

Fig. 8. Scheme of relationships among major groups of trilobites as suggested and discussed in this paper.

cussion it merits but which is not suitable here. Hahn and Hahn (1975) have critically appraised Bergström's phylogeny, though the one they propose as an alternative (redrawn as a cladogram in Fig. 7) reveals very little about the relationships among orders. I have not followed Bergström's phylogeny in the skeletal phylogeny (Fig. 8) for two reasons: Bergström (1973) relies primarily on enrollment patterns, states of which appear in some cases not wholly separable from others, and which are not defended against possible rampant convergence. Moreover, no attempt is made to defend the arrangement in terms of synapomorphies. On the other hand, the *Treatise* represents a consensus and historical compromise which, for all its defects, is obviously more completely phenetic (i.e., encompasses many more aspects of trilobite morphology) and such resultant classifications are much more likely to reflect actual evolutionary relationships. While I oppose adoption of a phenetic approach to phylogeny reconstruction, there is little doubt that the relationships as expressed (to the extent that they are!) in the *Treatise* mirror reality more closely than Bergström's attempt. Moreover, *some* of the diagnoses of some of the taxa of higher categorical rank in the *Treatise* contain manifest synapomorphies.

Thus the major groups of trilobites accepted here are based on the orders given in the *Treatise*, with the exception that the five suborders of the patently polyphyletic Ptychopariida are informally raised to ordinal rank. The diversity of these groups through time is summarized in Fig. 9 (total generic diversity) and by graphs of origination and extinction rates of families (Fig. 10).

I shall not attempt to justify all suggested relationships postulated in Fig. 8. However, discussion of selected points will illustrate some recent progress in trilobite comparative biology, highlighting both the nature of the morphological data utilized and the logic underlying its use. The two major sources of inference on primitive vs. derived states within a group of taxa remain (1) ontogeny, and (2) commonality; i.e., the distribution of a particular character state

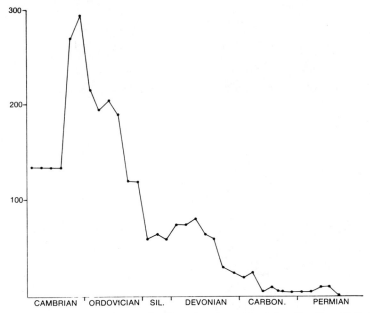

Fig. 9. Plot of total generic diversity, by series, for the Class Trilobita. Tremadocian is arbitrarily Ordovician. Data compiled from the *Treatise* (Harrington et al., 1959), thus horribly out of date. Data supplied by S.J. Gould, tabulated by a computer program written by J.J. Sepkoski.

both within the group under study, and especially among so-called "out groups" assumed to be related to the group under specific study in some (frequently only vaguely specified) manner. A third criterion, relative stratigraphic position, is of less readily perceived value and in any case is irrelevant to most of the problems discussed here.

Studies of trilobite ontogeny historically have had an important bearing on the derivation of notions of relationship. For good examples, see Robison (1967) on Corynexochida (see later), Temple (1956) on cheirurids and phacopids, and Hughes et al. (1975) on trinucleids. Another recent example is Jell's (1975) persuasive argument that the eodiscoids are the plesiomorphic sister group (my language, not Jell's!) of the agnostoids, and that together these taxa (as the Miomera of Jaekel, 1909) are the sister group of other trilobites (i.e., Polymera). In so arguing, Jell (1975, p. 13 ff.) viewed eodiscoid morphology and development in two separate ways: as a derived group *vis à vis* Polymera using comparative ontogenetic data, and as a primitive group *vis à vis* agnostoids, using comparative data of adult morphology. The latter argument is simply structured: eodiscoids have more recognizably trilobite-like characteristics than do agnostoids, and certain morphological trends within eodiscoids seem to approach agnostoid anatomy. Hence agnostoids are derived with respect to eodiscoids. The success of this argument hinges on the necessary demonstration that agnostoids and eodiscoids (Miomera) are in fact related in

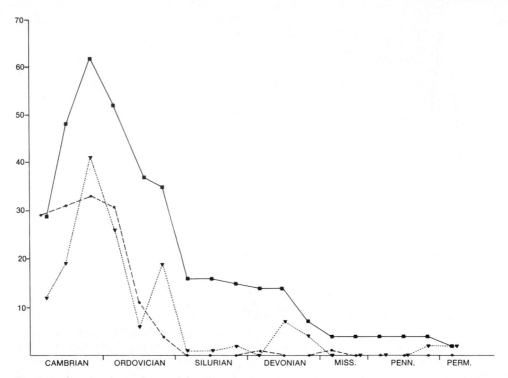

Fig. 10. Graph of numbers of families first appearing (●) and becoming extinct (▼) through time, plus total number of families present (■) during each interval. Data compiled from the *Treatise* (Harrington et al., 1959) and not quite as obsolete as in Fig. 9.

some manner to polymerid trilobites, and are not, e.g., simply to be regarded as a separate group of Palaeozoic arthropods of uncertain affinity. Jell cites six features of eodiscoids (he specifically stipulates "early" eodiscoids) which he claims are "features. . . shared by the early meraspides of many Cambrian polymerid trilobites" (Jell, 1975, p. 14). The situation is a bit tricky, as features shared among taxa in early ontogenetic stages are generally considered primitive. At the very least the similarities demonstrate a shared, derived mode of development of Miomera and Polymera, hence the conclusion is firmly established that, among known Arthropoda, they are sister groups. Technically, the data do not specify whether the Miomera arose from the Polymera by neoteny, or the Polymera from the Miomera by "acceleration", but the more economical hypothesis is surely the former, as Jell himself postulates.

Much of the discussion on trilobite phylogeny and classification has dwelt on the search for the "right" characters which will reveal the natural order of trilobites. Thus, facial sutures as well as axial cephalic and pygidial features are repeatedly raised as key features on which classifications are to be based. There are, of course, no such magical characters, but clearly some features show a distribution of character states which vary considerably among taxa (at a given hierarchical rank) and very little within groups. *This* is the reason why facial suture patterns have always been so appealing and why they retain their promise, among a host of other features, as useful characters for elucidating within-class relationships at a relatively high hierarchical level. So, the problem has been not so much the search for the magic characters as it has been the mishandling of the morphological data. The best illustration involves the central problem of trilobite relationships as is manifest in the *Treatise* and nearly all earlier work on trilobite phylogeny and classification: the aforementioned Ptychopariida. This group largely, but not wholly, corresponds to Beecher's (1897) Opisthoparia. The *Treatise* diagnosis of the order Ptychopariida contains not a single character state that can be taken as a synapomorphy among even two of the five suborders. Any of the many features cited in the diagnosis illustrate the main point, but let us focus on the premier example — the opisthoparian condition of the facial sutures.

Some genera within the vast array of Ptychopariida are gonatoparian or even proparian, but there is no doubt that the group is essentially opisthoparian, to the extent that it is essentially defined on that criterion. There are, of course, other orders recognized in the *Treatise* which are also wholly or largely opisthoparian; from time to time, these have all been included in the Opisthoparia (see Harrington, 1959, p. 145 ff. for a review) but have been split off in the *Treatise* together with some prior classifications because of their possession of distinct and obvious autapomorphies (i.e., synapomorphies allying all constituent taxa). Thus the Corynexochida are adjudged a discrete group, with distinctive pygidial features (especially relative size) and rostral plate morphology. Likewise, the Lichida and Odontopleurida are both opisthoparian, but are autapomorphic in so many respects that there is no problem in recognizing them as separate, distinct taxa — in fact the problem, as mentioned above, is the reverse — how are we to relate them to other trilobites?

On the other hand, the Redlichiina, olenellid-like trilobites with functional, dorsal opisthoparian sutures lacking in the Olenellina, have nevertheless frequently been linked to the Olenellina rather than the Opisthoparia or Ptychopariida. Although both the Redlichiina and Olenellina are on the whole rather primitive, the possibility definitely exists that at least some of the similarities they share are true synapomorphies. The redlichiid problem is put into more coherent perspective when we ask the following question: What general form of facial suture (dorsally expressed) is primitive for trilobites? The vast majority of trilobites are opisthoparian and this is surely reason enough to conclude that this condition is primitive! (Devotees of stratigraphic arguments can of course point to the additional fact that gonatoparian and proparian polymerid trilobites are virtually non-existent in the Cambrian.) If opisthoparian sutures are plesiomorphic for the Polymera, it follows that this character (as defined — variant versions of the opisthoparian condition are still potentially useful) is totally valueless in defining groups and postulating relationships among groups in which the condition is present. Failure to grasp this seemingly undeniable point has contributed much additional confusion to the already difficult problem of higher categorical phylogeny and classification of trilobites.

Returning briefly to the Redclichiida (Redlichiina + Olenellina), we can now ask: Is the opisthoparian condition primitive for all Polymera, or is olenellid morphology ("perrostral" suture) actually the primitive condition? If the latter is correct, then the Redlichiida are paraphyletic and the Redlichiina must be removed from the olenellids and associated with some other group (assuming the opisthoparian condition arose but once — certainly the most economical and reasonable view). However, in the actual diagnosis of the Olenellina in the *Treatise* (p. 0191), the facial sutures are said to be ankylosed — a conclusion with which I concur — and we can thus view olenellids as possessing an autapomorphous condition of the facial suture. Thus the way is finally cleared for concluding that the opisthoparian condition is plesiomorphic for Polymera (see also Stubbefield, 1960, p. 152). Opisthoparian sutures simply cannot be used as evidence of phylogenetic affinity. Future attempts at higher-level phylogenies within the Polymera simply must take this into consideration.

Robison (1967) provided convincing evidence that the Corynexochida are neotenously derived from the ptychopariids (*sensu stricto*), hence their representation as sister groups in Fig. 8. In this figure, the ptychopariids are considered to be a smaller, more restricted group of precisely unspecified composition, than in *Treatise* usage; Temple (1956) and Fortey (1974) have noted some suspicious similarities between olenids (classically considered Ptychopariina) and "phacopids" (herein understood as Cheirurina + Phacopina *sensu Treatise*, but not including Calymenina), hence the suggested relationships of Fig. 8. Likewise, the Calymenina (calymenids and homalonotids) seem to be related to certain earlier "ptychopariids" other than olenids — perhaps *Euloma* and its relatives. Not only will the Ptychopariida have to be split up, but the core plesiomorphic suborder Ptychopariina itself will have to be split up with segments allied with their appropriate, relatively more apomorphic, sister taxa.

As a final comment concerning the tentative phylogeny of Fig. 8, the asaphids and illaenids (including proetids; Fortey and Owens, 1975, have recently established the new order Proetida which explicitly excludes illaenids; following the *Treatise* here is not intended to reflect on the merits of their conclusions, but merely means their paper was received after this figure was drafted!) are of uncertain affinity within the central group of "ptychopariids." One misleading feature of the diagram is that the lichads, trinucleids, harpids and odontopleurids appear to form a sister group (of uncertain internal organization) to the other major cluster — the real point is that the actual position of each of these four groups *vis à vis* the central group is absolutely unknown at this point. Any one — or all four — may eventually be accurately incorporated into the central cluster, or their positions outside the cluster may ultimately become specifiable. I am not trying to say that the four form a separate group which is the sister of the central group. At the very least, the situation depicted in Fig. 9 speaks for itself in terms of the chaotic and rudimentary state of our understanding of trilobite relationships.

Relationships with Other Arthropods

The first question we must ask, in searching for trilobite relatives, is: Are trilobites themselves a monophyletic group? Secondly, we must determine their specific sister group, then successive sister groups, ultimately arriving at a general theory of relationships among Arthropoda. Discussion of the latter question is well beyond the bounds of the present paper, but we can and should enquire whether the Arthropoda as a whole can reasonably be regarded as a monophyletic taxon.

If trilobites are monophyletic, we should be able to specify, minimally, one (and hopefully more) feature which we can regard as synapomorphic, uniting the group. Immediately we are confronted by two fundamental, albeit dissimilar, problems. First, the crucial anatomical features (including morphogenesis) that have proved to be so important in phylogenetic reconstruction among Recent arthropods — not only limbs, but internal organs and development and segmentation of the head — are simply not there, for the most part, in the trilobite fossil record. The second problem is, if anything, even worse: to the extent that we understand anything at all about the basic anatomical organization of trilobites, in most features the group as a whole appears to be plesiomorphic with respect to such other arthropod groups as Crustacea and Chelicerata. And plesiomorphic groups are notoriously difficult (1) to prove to be monophyletic themselves and (2) to show to which group they are most closely related. Thus, as far as we know, trilobites possessed one pair of (?preoral) uniramous appendages (antennae) and three pairs of post-oral biramous appendages (Cisne, 1974, 1975; Whittington, 1975) on the cephalon, a pair of biramous appendages with each thoracic segment, and several pairs of biramous pygidial appendages. Significantly, the only serial differentiation among the biramous limbs appears to be size: there is as yet no evidence whatever that any structural differentiation of limbs was ever developed among trilobites. Now, this overall plan of limb organization is not known in any other arthropod group and is therefore useful in diagnosing the Trilobita. *But* we have no assurance that this structural plan in general, or any specific aspect of it, is synapomorphic for the group. For instance, we simply do not know whether the trilobite biramous limb is derived, or, instead, primitive with respect to the chelicerate limb (e.g., Størmer, 1959) or the crustacean limb (Hessler and Newman, 1975; Cisne, 1974) or both. Certainly, lack of serial differentiation can be adjudged primitive, and thus not a shared, derived feature uniting trilobites. Much the same could be said for what little (and conflicting) information we have concerning head segmentation in trilobites.

Returning to the features of the dorsal exoskeleton, one can of course speak of distinct body trilobation, as well as the characteristic subdivision of the body into cephalon, thorax, and pygidium. Trilobation, of course, is not unique to trilobites (most merostomes are distinctly trilobed). And, in terms of the three subdivisions of the body, the same argument pertains *vis à vis* trilobite organization as advanced earlier regarding the ventral appendages. Though seemingly unique in its details among arthropods, cephalic organization in fact seems primitive and thus probably not a good shared derived feature uniting the trilobites.

This hand-wringing aside, let us assume, as seems safe, that trilobites are indeed monophyletic. What then is their sister group? Until recently, the prevailing view has been Størmer's (e.g., 1944, 1959) that chelicerates and trilobitomorphs (= trilobites proper plus Trilobitoidea) are more closely related to each other than either is to any other known group. More recently, Cisne (1974) has compared the unusually well preserved details of limb morphology and musculature of *Triarthrus eatoni* with that of primitive crustaceans, and has concluded that trilobites are primitive with respect both to Crustacea and Chelicerata. Cisne's (1974) figure (Fig. 11) shows a trichotomous relationship among these groups, forming a major group questionably allied as the sister group of the Uniramia. Cisne (1974, p. 16) cites four "primitive. . . evolutionarily conservative" features which unite trilobites, chelicerates, and crustaceans: "(i) a primitively multiramous limb, (ii) a gnathobasic food ingestion mechanism, (iii) primitively, a posteriorly directed mouth, and (iv) digestive glands in the head region that are often quite extensive". No other known arthropods possess such features, and, remembering that synapomorphic fea-

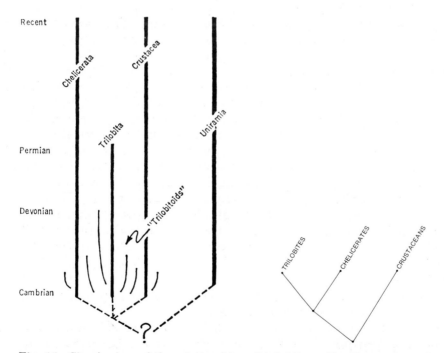

Fig. 11. Cisne's view of the relationships of trilobites with other arthropods. (From Cisne, 1974, p. 16, fig. 3.) See text. Copyright 1974 by the American Association for the Advancement of Science.

Fig. 12. An amalgam of Størmer's (e.g., 1959) and Cisne's (1974) views on the closest relationships of trilobites among Arthropoda. See text.

tures serving to unite a group immediately become plesiomorphic for that group, Cisne's argument seems sound. Moreover, the trichotomy seems ultimately capable of resolution: for the moment I prefer Størmer's view (adopted by Tiegs and Manton, 1958) that chelicerates and trilobites are sister taxa, but accept Cisne's conclusion that they in turn are the sister group of the Crustacea. These relationships are diagrammed in Fig. 12.

Two final points merit discussion. Whittington (e.g., 1971, 1974) and colleagues, in a massive attempt to re-study the fauna of the Middle Cambrian Burgess Shale of British Columbia, while reluctant to speculate freely on the precise phylogenetic affinities of the various different members of the "Trilobitoidea", are providing abundant evidence that "trilobitoids" are not a coherent monophyletic assemblage. Even this wonderful window into early arthropod history merely increases our knowledge of arthropod structural (anatomical) diversity — even when the fossil record is exceptionally good, it is no panacea to problems of phylogenetic reconstruction. Quite the reverse — the additional diversity seems merely to compound the problem! This situation is not unique to the arthropods, for one immediately thinks of the early Palaeozoic echinoderm fossil record as a similar example. It is to be hoped that future studies of the Burgess Shale fauna will more explicitly address themselves to problems of relationship of these remarkable animals.

Finally, largely through the work of Tiegs and Manton (1958; also Manton, 1973) we have the problem of the highest level of relationships among Arthropoda. These workers have meticulously documented the large number of cases of undoubted parallelism in the evolution of the arthropod exoskeleton. For instance, tracheal systems can either be regarded as synapomorphic (in which case terrestrial isopods, spiders, and insects belong to a monophyletic group) or as parallelisms. Clearly they are parallelisms in these groups. Tiegs and Manton (1958) and Manton (1973) therefore opt for a three-fold origin of the arthropod "grade" of organization. Cisne (1974) by linking Crustacea more firmly with Trilobita and Chelicerata, opts for a two-fold origin of arthropods, (as do Hessler and Newman, 1975) while Schram (in press) has concluded that there are no fewer than four arthropod groups of phylum rank that independently reached the "arthropod grade" of organization: Uniramia, Crustacea, Cheliceriformes, and Trilobitomorpha.

A standard pitfall in phylogenetic research is that anatomists engaged in comparing structures they consider homologous among a series of organisms are frequently overly impressed by what they perceive as tremendous morphological gaps between clusters of these organisms. This seems to be the basic reason why some of the early, most outspoken opponents of the very notion of evolution were anatomists: Cuvier and Owen immediately leap to mind. Admittedly, such problems are difficult to surmount (see discussion above concerning problems of relationship among orders of trilobites) but we should never forget that the central problem in a study of phylogenetic reconstruction is the search for sister groups among all known organisms (fossil and Recent) that can be considered as possible relatives at the outset. If the myriapod—hexapod lineage is the sister group of the Onychophora, and *if* the Onychophora are indeed

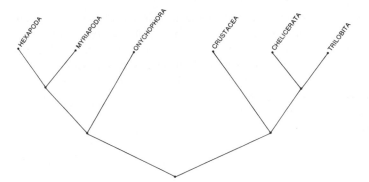

Fig. 13. An hypothesis of relationships among the six major groups conventionally considered to be arthropods. They are alleged to constitute a monophyletic group. See text for discussion.

plesiomorphic with respect to Myriapoda—Hexapoda *and* to Trilobita—Chelicerata—Crustacea, we may still depict their relationships as in Fig. 13. In point of fact, a monophyletic hypothesis of arthropod origins is still the simplest and can only be refuted if one of the components of Fig. 13 is actually shown to be the sister group of some other group either now known, or perhaps yet to be discovered. Only if the Uniramia, for example, can be shown to be the sister group of, e.g., Annelida (or some portion thereof) can we formally reject the hypothesis that Arthropoda are monophyletic.

References

Ashton, J.H. and Rowell, A.J., 1975. Environmental stability and species proliferation in Late Cambrian Trilobite faunas: a test of the niche-variation hypothesis. Paleobiology, 1: 161—174.

Ayala, F.J., Valentine, J.W., DeLaca, T.E. and Zumwalt, G.S., 1975. Genetic variability of the Antarctic brachiopod *Liothyrella notorcadensis* and its bearing on mass extinction hypotheses. J. Paleontol., 49: 1—9.

Beecher, C.E., 1897. Outline of a natural classification of the trilobites. Am. J. Sci., Ser. 4, 3: 89—106; 181—207.

Bergström, J., 1969. Remarks on the appendages of trilobites. Lethaia, 2: 395—414.

Bergström, J., 1973. Organization, life, and systematics of trilobites. Fossils Strata, 2: 1—69.

Bretsky, P.W. and Lorenz, D.M., 1970. Adaptive response to environmental stability: a unifying concept in paleoecology. N. Am. Paleontol. Conv., Chicago, 1969, Proc. E: 522—550.

Campbell, K.S.W., 1977. Trilobites of the Haragan, Bois d'Arc and Frisco Formations. Bull. Okla. Geol. Survey, in press.

Cisne, J., 1974. Trilobites and the origin of arthropods. Science, 186: 13—18.

Cisne, J., 1975. Anatomy of *Triarthrus* and the relationships of the Trilobita. Fossils Strata, 4: 45—63.

Clarkson, E.N.K., 1966a. Schizochroal eyes and vision of some Silurian acastid trilobites. Palaeontology, 9: 1—29.

Clarkson, E.N.K., 1966b. Schizochroal eyes and vision in some phacopid trilobites. Palaeontology, 9: 464—487.

Clarkson, E.N.K., 1966c. The life attitude of the Silurian trilobite *Phacops musheni* Salter 1864. Scott. J. Geol., 2: 76—83.

Clarkson, E.N.K., 1969. A functional study of the Silurian odontopleurid trilobite *Leonaspis deflexa* Lake. Lethaia, 2: 329—344.

Clarkson, E.N.K., 1975. The evolution of the eye in trilobites. Fossils Strata, 4: 7—31.

Clarkson, E.N.K. and Henry, J.-L., 1973. Structures coaptives et enroulement chez quelques Trilobites ordoviciens et siluriens. Lethaia, 6: 105—132.

Cook, H.E. and Taylor, M.E., 1975. Early Paleozoic continental margin sedimentation, trilobite biofacies, and the thermocline, western United States. Geology, 3: 559—562.

Croizat, L., Nelson, G. and Rosen, D.E., 1974. Centers of origin and related concepts. Syst. Zool., 23: 265—287.

Eldredge, N., 1971. The allopatric model and phylogeny in Paleozoic invertebrates. Evolution, 25: 156—165.

Eldredge, N., 1972. Systematics and evolution of *Phacops rana* (Green, 1832) and *Phacops iowensis* Delo, 1935 (Trilobita) from the Middle Devonian of North America. Bull. Am. Mus. Nat. Hist., 146: 45—114.

Eldredge, N., 1973. Systematics of Lower and Lower Middle Devonian species of the trilobite *Phacops* Emmrich in North America. Bull. Am. Mus. Nat. Hist. 151: 285—338.

Eldredge, N., 1974a. Stability, diversity, and speciation in Paleozoic epeiric seas. J. Paleontol., 48: 540—548.

Eldredge, N., 1974b. Revision of the Suborder Synziphosurina (Chelicerata, Merostomata) with remarks on merostome phylogeny. Am. Mus. Novit., 2543: 1—41.

Eldredge, N., 1975. Character displacement in evolutionary time. Am. Zool., 14: 1083—1097.

Eldredge, N., 1977. Alternate approaches to evolutionary theory. Spec. Publ., Carnegie Mus., Pittsburgh, in press.

Eldredge, N. and Gould, S.J., 1972. Punctuated equilibria: an alternative to phyletic gradualism. In: T.J.M. Schopf (Editor), Models in Paleobiology. Freeman, San Francisco, Calif., pp. 82—115.

Eldredge, N. and Tattersall, I., 1975. Evolutionary models, phylogenetic reconstruction, and another look at hominid phylogeny. In: F.S. Szalay (Editor), Contributions to Primate Paleobiology. Karger, Basel, Contrib. Primatol., 5: 218—242.

Fortey, R.A., 1974. The Ordovician trilobites of Spitsbergen. I. Olenidae. Norsk Polarinst., 160: 1—129.

Fortey, R.A. and Owens, R.M., 1975. Proetida — a new order of trilobites. Fossils Strata, 4: 227—239.

Ghiselin, M., 1974. A radical solution to the species problem. Syst. Zool., 23: 536—544.

Gingerich, P.D., 1974. Stratigraphic record of Early Eocene *Hyopsodus* and the geometry of mammalian phylogeny. Nature, 248: 107—109.

Gingerich, P.D., 1976. Paleontology and phylogeny: patterns of evolution at the species level in Early Tertiary mammals. Am. J. Sci., 276: 1—28.

Hahn, G. and Hahn, R., 1975. Forschungsbericht über Trilobitomorpha. Paläontol. Z., 49: 432—460.

Harper, C.W., 1975. Origin of species in geologic time: alternatives to the Eldredge-Gould model. Science, 190: 47—48.

Harrington, H.J., 1959. Classification. In: R.C. Moore (Editor), Treatise on Invertebrate Paleontology. Part O. Arthropoda (1). Geological Society of America and University of Kansas Press, Lawrence, Kansas, pp. O145—O170.

Harrington, H.J. et al., 1959. Part O, Arthropoda 1. In: R.C. Moore (Editor), Treatise on Invertebrate Paleontology. Geological Society of America and University of Kansas Press, Lawrence, Kansas, 560 pp.

Hennig, W., 1950. Grundzüge einer Theorie der phylogenetischen Systematik. Deutscher Zentralverlag, Berlin.

Hennig, W., 1966. Phylogenetic Systematics. University of Illinois, Chicago, Ill., 263 pp.

Henningsmoen, G., 1964. Zig-zag evolution. Norsk Geol. Tidskr., 44: 341—352.

Henry, J.-L. and Clarkson, E.N.K., 1975. Enrollment and coaptations in some species of the Ordovician trilobite genus *Placoparia*. Fossils Strata, 4: 87—95.

Hessler, R.R. and Newman, W.A., 1975. A trilobitomorph origin for the Crustacea. Fossils Strata, 4: 437—459.

Hughes, C.P., Ingham, J.K. and Addison, R., 1975. The morphology, classification and evolution of the Trinucleidae (Trilobita). Philos. Trans. R. Soc. Lond., B, 272: 537—607.

Hutchinson, G.E., 1959. Homage to Santa Rosalia or Why are there so many kinds of animals? Am. Nat., 93: 145—159.

Jaekel, O., 1909. Über die Agnostiden. Z. Dtsch. Geol. Ges., 61: 380—401.

Jell, P.A., 1975. Australian Middle Cambrian eodiscoids with a review of the superfamily. Palaeontographica, A 150: 1—97.

Johnson, J.G., 1974. Extinction of perched faunas. Geology, 2: 479—482.

Kaufmann, R., 1933. Variationsstatistische Untersuchungen über die "Artabwandlung" und "Artumbildung" an der oberkambrischen Trilobitengattung *Olenus* Dalm. Abh. Geol.-Paläontol. Inst. Univ. Greifswald, 10: 1—54.

Lespérance, P.J., 1975. Stratigraphy and paleontology of the Synphoriidae (Lower and Middle Devonian trilobites). J. Paleontol., 49: 91—137.

Lespérance, P.J. and Bertrand, R., 1976. Population systematics of the Middle and Upper Ordovician trilobite *Cryptolithus* from the St. Lawrence lowlands and adjacent areas of Quebec. J. Paleontol., 50: 598—613.

Lespérance, P.J. and Bourque, P.-A., 1971. The Synphoriinae: an evolutionary pattern of Lower and Middle Devonian trilobites. J. Paleontol., 45: 182—208.

Levins, R., 1968. Evolution in changing environments. Monogr. Pop. Biol., 2: 120 pp. (Princeton University).

Lochman-Balk, C., 1970. Upper Cambrian faunal patterns of the craton. Bull. Geol. Soc. Am., 81: 3179—3224.

Manton, S.M., 1973. Arthropod phylogeny — a modern synthesis. J. Zool., Lond., 171: 111—130.

Martinsson, A. (Editor), 1975. Evolution and morphology of the Trilobita, Trilobitoidea and Merosto-
    mata. Proceedings of a NATO Advanced Study Institute held in Oslo 1st—8th July, 1973, organized
    by D. L. Bruton. Fossils Strata, 4: 1—467.
Matthew, W.D., 1915. Climate and Evolution. Ann. N.Y. Acad. Sci., 24: 171—318.
Mayr, E., 1942. Systematics and the Origin of Species. Columbia University, New York, N.Y., 334 pp.
Mayr, E., 1963. Animal Species and Evolution. Harvard University, Cambridge, Mass., 797 pp.
Nelson, G.J., 1970. Outline of a theory of comparative biology. Syst. Zool., 19: 373—384.
Palmer, A.R., 1965a. Biomere — a new kind of biostratigraphic unit. J. Paleontol., 39: 149—153.
Palmer, A.R., 1965b. Trilobites of the Late Cambrian Pterocephaliid Biomere in the Great Basin. Prof.
    Pap. U.S. Geol. Surv., 493: 1—105.
Robison, R.A., 1967. Ontogeny of Bathyuriscus fimbriatus and its bearing on affinities of corynexochid
    trilobites. J. Paleontol., 41: 213—221.
Robison, R.A., 1975. Species diversity among agnostoid trilobites. Fossils Strata, 4: 219—226.
Schaeffer, B., 1952. The Triassic coelacanth fish Diplurus, with observations on the evolution of the Coe-
    lacanthini. Bull. Am. Mus. Nat. Hist., 99: 25—78.
Schopf, T.J.M., 1974. Permo-Triassic extinctions: relation to sea floor spreading. J. Geol., 82: 129—143.
Schopf, T.J.M., Raup, D.M., Gould, S.J. and Simberloff, D.S., 1975. Genomic versus morphologic rates of
    evolution: influences of morphologic complexity. Paleobiology, 1: 63—70.
Schram, F.R., 1977. Arthropods: a convergent phenomenon. Q. Rev. Biol., in press.
Simberloff, D.S., 1974. Permo-Triassic extinctions: effects of area on biotic equilibrium. J. Geol., 82:
    267—274.
Simpson, G.G., 1953. The Major Features of Evolution. Columbia University, New York, N.Y., 434 pp.
Simpson, G.G., 1965. The Geography of Evolution. Chilton, Philadelphia and New York, N.Y., 249 pp.
Stanley, S.M., 1973. Effects of competition on rates of evolution, with special reference to bivalve mol-
    lusks and mammals. Syst. Zool., 22: 486—506.
Stitt, J.H., 1971. Repeating evolutionary patterns in Late Cambrian trilobite biomeres. J. Paleontol., 45:
    178—181.
Stitt, J.H., 1975. Adaptive radiation, trilobite paleoecology, and extinction, Ptychaspidid Biomere, Late
    Cambrian of Oklahoma. Fossils Strata, 4: 381—390.
Størmer, L., 1944. On the relationships and phylogeny of fossil and recent Arachnomorpha. Nor.
    Vidensk.-Akad., 1 (5): 1—158.
Størmer, L., 1959. Arthropoda — general features. In: R.C. Moore (Editor), Treatise on Invertebrate Pale-
    ontology. Part O. Arthropoda 1. Geological Society of America and University of Kansas Press, Law-
    rence, Kansas, pp. O3—O16.
Stubblefield, C.J., 1960. Evolution in trilobites. Q. J. Geol. Soc., Lond., 115: 145—162.
Temple, J.T., 1956. Notes on the Cheiruracea and Phacopacea. Geol. Mag., 93: 418—320.
Tiegs, O.W. and Manton, S.M., 1958. The evolution of the Arthropoda. Biol. Rev., 33: 255—337.
Valentine, J.W., 1973. Evolutionary Paleoecology of the Marine Biosphere. Prentice-Hall, Englewood
    Cliffs, N.J., 511 pp.
Westoll, T.S., 1949. On the evolution of the Dipnoi. In: G.L. Jepsen, G.G. Simpson and E. Mayr (Edi-
    tors), Genetics, Paleontology and Evolution. Princeton University, Princeton, R.I., pp. 121—184.
Whittington, H.B., 1941. Silicified Trenton trilobites. J. Paleontol., 15: 492—522.
Whittington, H.B., 1966. Phylogeny and distribution of Ordovician trilobites. J. Paleontol., 40: 696—737.
Whittington, H.B., 1971. Redescription of Marrella splendens (Trilobitoidea) from the Burgess Shale, Mid-
    dle Cambrian, British Columbia. Bull. Geol. Surv. Can., 209: 1—24.
Whittington, H.B., 1974. Yohoia Walcott and Plenocaris n. gen., arthropods from the Burgess Shale, Mid-
    dle Cambrian, British Columbia. Bull. Geol. Surv. Can., 231: 1—21..
Whittington, H.B., 1975. Trilobites with appendages from the Middle Cambrian, Burgess Shale, British
    Columbia. Fossils Strata, 4: 97—136.

CHAPTER 10

PATTERNS OF EVOLUTION IN THE GRAPTOLITES

R.B. RICKARDS

Introduction

The graptolites are extinct, marine, colonial animals which have most commonly been regarded, in recent works, as comprising a class of the phylum Hemichordata termed the Graptolithina. Kozłowski (1966) persuasively concluded that the rhabdopleurans were the closest relatives of the graptolites, and Rickards (1975) reached the same conclusion notwithstanding the possibility that the latest research may eventually place the graptolites in a new phylum removing them, as in the case of the pogonophorans, from the Hemichordata. Fig. 1 summarizes the main stratigraphical distribution, basic evolutionary framework and suggested mode of life of graptolites.

It can be seen that the pterobranchs occur as far back in time as the Trema-

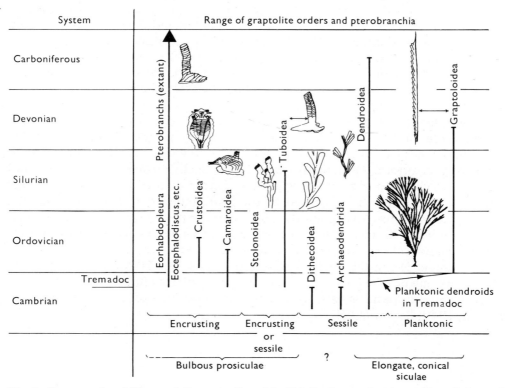

Fig. 1. Gross mode of life, evolution, stratigraphic distribution, and nature of larval stage of graptolites.

333

TABLE I

Comparison between the Dendroidea and Graptoloidea

Order Dendroidea	Order Graptoloidea
(1) roots or basal discs, rarely a nema	almost always a nema
(2) two types of thecae, bithecae and auto-thecae; stolothecae being immature autothecae	one type of theca, the prothecae may be homologous with the stolotheca
(3) black stolons, nodes, and triad divisions of stolons	possibly soft "stolons"
(4) internal tubes present to early parts of thecae	internal tubes absent
(5) thick cortex	relatively thin cortex
(6) many branches and thecae, therefore zooids	relatively few branches and thecae
(7) compound stipes common	compound stipes rarer

doc and a common ancestor, in the low Cambrian or earlier, for these animals and graptolites would not be an unreasonable hypothesis. The encrusting nature and detailed morphology of some tuboids is very close to that of *Rhabdopleura*, whilst further work on the Dithecoidea and Archaeodendrida may reveal yet closer similarities. Within the graptolites themselves the relationships of the orders Crustoidea, Camaroidea, Stolonoidea, Tuboidea, Dithecoidea and Archaeodendrida are obscure: the fossil record is rather poor and neither lineages within these orders, nor suggested links between them, have been worked out.

Consequently the bulk of the following text refers to evolutionary patterns recognizable in the Dendroidea and particularly in the Graptoloidea. The dendroids ranged from the Middle Cambrian to the Upper Carboniferous, were essentially benthonic elements of the inshore, shelly environment, and showed little or no change during this long period of more than 200 m.y. except during the Tremadoc when they gave rise to the planktonic Graptoloidea through the intermediate family Anisograptidae. The anisograptids are arbitrarily retained in the order Dendroidea although the family has many graptoloid morphological characters and exhibits mosaic evolution very well (see Table I).

The orders Dendroidea and Graptoloidea have the following basic skeletal characters, many of which distinguish the one order from the other; soft parts of graptolites have never been found save for possible eggs and embryos (Kozłowski, 1949; Rickards, 1975).

Coloniality

Graptolites were exclusively colonial animals which constructed a scleroproteinaceous skeleton of successive chambers: openings connected each individual zooid of the colony, and each zooid in turn became the "terminal bud" for the growing stipe. In the Dendroidea the zooids were further connected by a sclerotized stolon system. Bulman and Rickards (1966) found partly sclerotized stolons in the Tuboidea, and it is likely that soft, unsclerotized stolons connected all individuals of each graptoloid colony. With such basic coloniality all individuals of the colony would have the same genotype. It is acceptable to talk of the ontogeny of an individual zooid or its thecal sheath; and of the astogeny of the colony as a whole, in which the various thecal shapes are the phenotypic manifestations of the same genotype.

Genetic Polymorphism, Inbreeding and Outbreeding

Kozłowski (1949) concluded that in the dendroids the smaller thecae (bithecae) housed male zooids, and that the larger autothecae housed the females. The planktonic graptoloids, with only one kind of theca, represented a change to hermaphroditism. Within this general framework Skevington (1966, 1967) described what he thought to be an example of genetic polymorphism, and Urbanek (1963, 1970) and Jaanusson (1973) considered that genetic polymorphism may have been more widespread than hitherto supposed. The term genetic polymorphism can only be used or suggested in these contexts if independent evidence indicates that the two or more forms are one and the same palaeospecies: the same subjectivity applies as whenever palaeospecies are defined. Whilst the fact that individuals of the same colony shared the same genotype, and the possibility that graptoloid colonies were hermaphroditic, may suggest inbreeding as the method of reproduction, Urbanek and Jaanusson (1974) concluded that extensive polymorphism indicated outbreeding as the reproductive method because heterosis (superior fitness of the heterozygotes) was the only likely method for maintaining genetic polymorphism in a graptolite population. However, studies on the genetics of wild oats (Allard, 1965) indicate that inbreeding species may in fact maintain high degrees of genetic polymorphism and suggest that this phenotypic variability should not be taken as direct evidence of the reproductive mode of the animals. Jaanusson (1973) further deduced that some of the more significant evolutionary events, such as the origin of uniserial monograptids from the biserial glyptograptids, were initiated by genetic polymorphism. This particular suggestion has been supported by Rickards et al. (1976) who considered that *Atavograptus ceryx*, the earliest known monograptid, originated in a dithyrial population comprising *A. ceryx* and a form which would traditionally be described as a *Glyptograptus* species (Fig. 2). However, the graptoloids may have used a dual "strategy" of breeding (Urbanek, 1973), in times of stress reverting to selfing. Such a flexible approach would have contributed considerably to the evolutionary potential of the graptoloids during the period of the Late Cambrian to Middle Devonian. The dendroids, with their earlier-established and conservative system of having males and females along each stipe of the colony, actually outlasted the planktonic graptoloids but showed distinctly less evolutionary change.

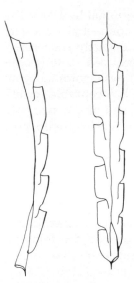

Fig. 2. *Atavograptus ceryx* (Rickards and Hutt) and *Glyptograptus* sp., probable members of dithyrial populations in the *persculptus* Zone, and as such the two morphs should be referred to a single species, in this case *A. ceryx*.

Sequence of Graptolite Faunas

Bulman (1958), in a now classic paper, defined a time sequence of four overlapping faunas based upon the earlier work of Elles (1922). In order of appearance these are: Anisograptid; Dichograptid; Diplograptid and Mono-graptid, the first being a dendroid fauna and the succeeding three graptoloid faunas (see Rickards, 1975, fig. 77). The main features of each of these faunas are described below, and slightly amended by the present author.

Anisograptid fauna

This includes siculate *Dictyonema*, a planktonic dendroid, and pendant and horizontal anisograptids (*Anisograptus, Staurograptus, Kiaerograptus*, etc.) essentially intermediate between dendroids and graptoloids in that they have fewer branches and often incomplete complements of bithecae. The small didymograptids referred to by Bulman (1958, p. 171) are probably referable to *Kiaerograptus*, although possible didymograptids and some tetragraptids have been described more recently (e.g., Jackson, 1974). The fauna is distinguished from the succeeding fauna by its essentially dendroid nature.

Dichograptid fauna

These are mostly dichograptids (*Dichograptus, Loganograptus, Tetragraptus, Didymograptus, Azygograptus*, etc.) with the anisograptids *Clonograptus* and

Bryograptus near the base and some biserial graptoloids towards the top (e.g., *Glyptograptus dentatus*).

Diplograptid fauna

This is divided into four subfaunas by Bulman based upon the association of now abundant biserial graptoloids with some uniserial reclined genera (e.g., *Dicellograptus*), and rare uniserial scandent forms at the very top (e.g. *Atavograptus ceryx*).

Monograptid fauna

This includes elements of the preceding fauna as high as the Upper Llandovery and is divided into several subfaunas by Bulman upon the thecal type of the constituent uniserial scandent monograptids. The fauna is now known to extend into the Devonian (Emsian) and to include many ramose species and genera, though still with the uniserial scandent basic structure.

Elsewhere, Rickards (1975) has related the major structural changes of these four evolving faunas to the broad modes of life adopted as a consequence of their achieving relatively sudden planktonic status: dichograptids had fewer but longer, more slender, and more spread arms, thus tending to counteract sinking, whilst the proximal webs may have actively assisted buoyancy; diplograptids removed the vital siculate region from the more turbulent surface layer of the ocean by becoming scandent (a sharp progression from horizontal and reclined arms), but the resultant heavier biserial rhabdosome needed nemal flotation structures to maintain buoyancy and spines to retard sinking (see Rickards, 1975, fig. 63); monograptids became long, slender and ramose, did not need exaggerated nemal support, nor spinosity to the same extent, whilst the subfaunas based upon thecal types probably reflect the "final" adaptation of the colonies' zooids to the habitat.

Cladogenesis and Anagenesis

George (1962) applied Huxley's (1958) terms to graptolite studies, and these were further elaborated by Bulman (1963). Those evolutionary processes resulting in general biological improvement, or anagenesis, produced what may be termed different grades; the anisograptid, dichograptid, diplograptid, and monograptid faunas just discussed fall naturally into this category. On the other hand cladogenesis, producing genetic divergence, is reflected in thecal elaboration, and the production of divergent lines such as the leptograptids, dicellograptids, lasiograptids, retiolitids and dimorphograptids.

Such a process of classification of evolutionary change hinges on the assumption that rhabdosomal changes are anagenetic or "major" even though, as pointed out by Bulman (1963, p. 406), the biological improvement may be considered obscure. This mode-of-life interpretation I have presented is intended to suggest the actual nature of the improvements, as, indeed, does the rather different interpretation of Kirk (1969), which places the proximal regions of each colony uppermost.

The categorization of thecal changes as cladogenetic and distinct from anagenetic changes may be misleading in the sense that once the planktonic graptoloids had achieved a hydrodynamically stable, uniserially scandent rhabdosome, the remaining rhabdosomal changes possible were few and perhaps limited to cladia formation. Thus, in contrast to the dendroids which remained in one grade through their long history, the planktonic graptoloids underwent a series of grade changes to a more or less entrenched "final" grade in their much shorter and more diverse history.

Origination and Extinction

A great deal of work has been done during the last decade on the stratigraphic range and occurrence of Silurian graptoloids, so that the early work on species abundance in the Silurian (e.g., Bulman, 1933) was necessarily approximate. However, the work by Bulman (1933) on the Ordovician species abundances is as yet unsurpassed: in the British Isles abundance peaks for the total number of species and numbers of new species were remarkably coincident and occurred at the following horizons: *extensus* Zone, *gracilis* Zone and *clingani* Zone. Similar peaks were obtained by analysing North American and Scandinavian data, respectively, taken from Troedsson (1923) and Ruedemann (1904). These peaks, recognizable today on a world-wide scale, might be reasonably termed evolutionary explosions (see later in terms of Silurian occurrences) and they correspond broadly to the sequence of faunas defined by Bulman (1958, p. 170) and to the concept of evolutionary grades also outlined by him (Bulman, 1963, p. 407).

In terms of British Silurian graptoloids I have been able to produce a total range chart of over 300 species (Rickards, 1976). Because the duration of the Silurian is some 40 m.y. and the number of graptoloid zones about the same number, the average duration of each zone is approximately 1 m.y.; in many cases smaller units can be recognized and correlated at least within the confines of the British Isles. Thus it is possible to suggest the mean duration of a Silurian graptoloid species (Fig. 3a): although considerable numbers of species have a very short duration in time (less than 1 m.y.) and a few last as long as 8 m.y. (e.g., *P. dubius* and *D. ramosus*), the mean duration is 1.9 m.y., a figure comparable to other rapidly evolving animal groups.

The number of species at each horizon (Fig. 3b, upper curve) shows very clearly the Llandovery and Ludlow evolutionary explosions which, in my opinion, are in no sense monographic. Immediately prior to the *persculptus* Zone, in the highest Ordovician rocks, graptoloids reach a distinct evolutionary "low" and, as at the level of the *nassa* Zone (or *nassa-dubius* Interregnum of earlier work), came close to becoming extinct with only a small number of species and genera represented on a world-wide scale. The Llandovery saw a spectacular

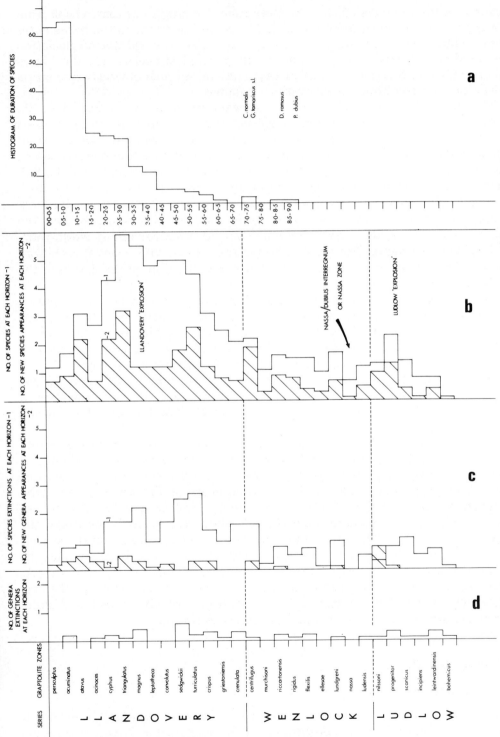

Fig. 3. Histogram of species and generic occurrence in the Silurian of the British Isles and Eire. Full explanation in text.

increase in the numbers of all taxa, the species forming a noticeable peak from the *cyphus* to *turriculatus* Zones, with 59 species occurring in the *triangulatus* Zone. Whilst the Late Ordovician "low" may reflect the widespread glaciation and tendency to lessen the previous extent of tropical oceans, and the Llandovery explosion the amelioration of that climate, no such obvious explanation exists for the *nassa* Zone depauperation of faunas.

The curve for the number of new appearances at each horizon (Fig. 3b, lower curve) is a subdued version of the last curve, but in addition illustrates a distinction between the pre-*acinaces* Zone faunas, essentially composed of relicts from the Ordovician but with the *A. atavus* lineages founded (see later), and the *cyphus* Zone introduction of essentially mid-Llandovery elements. Relatively less innovation takes place following the *triangulatus* Zone peak, until the *sedgwickii* Zone — *turriculatus* Zone fauna appears. Further extensive innovation took place in the *centrifugus* Zone (base of the Wenlock) and the *riccartonensis* Zone, the latter heralding a very distinctive, almost world-wide, Middle Wenlock fauna. The graph showing the number of species at each horizon (Fig. 3b, upper curve) gives only a broad idea therefore of the nature of the evolutionary explosion.

Fig. 3c, the number of species becoming extinct at each horizon, shows a similar succession of peaks to the last two curves but with some of the peaks themselves delayed, and the curve as a whole is flatter. Of interest is the peak of extinctions in the *lundgreni* Zone prior to the *nassa* fauna of world-wide depauperation. What has been said of Fig. 3b (lower curve) and Fig. 3c (upper curve) can also be said of Fig. 3c (lower curve) and Fig. 3d representing, respectively, the introduction and extinction of genera. The mean duration of Silurian graptoloid "genera", based upon 55 such groupings is 4.9 graptolite zones.

Bulman (1933, p. 327) makes the points that these peaks were probably universal in Ordovician faunas (and they certainly are with respect to Silurian faunas) and that they correspond to major structural changes ("sequence of faunas" or anagenetic changes). The Silurian peaks of Fig. 3b similarly correspond to the proliferation of *Atavograptus* (origin of *Monograptus* s.l.); the origin and acme of triangulate monograptids about the *triangulatus* Zone; the appearance of robust *Monograptus* s.s. in the *turriculatus* Zone; the appearance of cladia-bearing forms in the highest Llandovery and low Wenlock, and the proliferation of *Neodiversograptus*, *Lobograptus*, etc. in the low Ludlow. As stated earlier, these cladogenetic changes may well be homologous to the anagenetic changes of earlier faunas but in a situation in which spectacular rhabdosomal changes, other than cladia production, were not possible.

Origin of New Types

Neoteny, paedomorphosis, and recapitulation

Attainment of sexual maturity before the adult stage (neoteny), and retention into the adult stage of youthful or larval biocharacters (paedomorphosis)

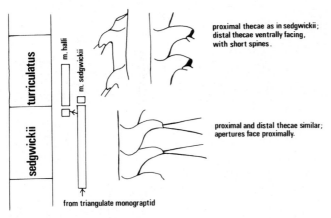

proximal thecae as in sedgwickii;
distal thecae ventrally facing,
with short spines.

proximal and distal thecae similar;
apertures face proximally.

from triangulate monograptid

Fig. 4. Morphology and suggested evolution of *M. halli* from *M. sedgwicki* by incomplete growth of metathecal hook in distal thecae of *M. halli*.

are two concepts not obviously applicable at first sight to the origin of new graptolite types. Neoteny can never, of course, be proposed on direct evidence, but if paedomorphosis is detected in a lineage leading to a new palaeontological species then neoteny as a cause is at least a reasonable hypothesis. Fig. 4 depicts my interpretation of the lineage in which *M. sedgwickii* gives rise to *M. halli*: the latter is characterized by proximal thecae which are closely similar to all the thecae of *M. sedgwickii*, and by distal thecae in which full growth of the hook is retarded so that the resultant tube opens ventrally rather than proximally. In terms of thecal ontogeny *M. halli* may have evolved from *M. sedgwickii* by a form of paedomorphosis perhaps resulting from neoteny. But in terms of rhabdosomal astogeny the argument is more dubious, and it may be that incomplete penetrance of the genotype, reflected in a change of phenotype expression, has resulted in the distal introduction of a "new" character (Urbanek, 1960) to whit the formation of a ventrally facing tube with shorter apertural spines. However, myself and others (Rickards et al., 1976) interpret *M. halli* as the end-member of a long lineage, and in this case the introduction of a new type cannot be claimed as very successful in evolutionary terms: the change from *M. sedgwickii* to *M. halli* would seem to be degenerate in more than morphological expression in which retreat of the dorsal thecal lip takes place after development of retroversion only five zones earlier. Similarly, the end-members of other lineages may exhibit morphological degeneration in the sense that the typical thecal type of the immediate ancestors is only partly completed, and this is particularly true of the distal thecae. Thus the last *Climacograptus*, *C. nebula* from the Upper Llandovery, has a much reduced geniculum and therefore shorter thecae; reduction of overall rhabdosome size may be related in some way.

Other more striking instances of possible paedomorphosis can be cited. Fig. 5 depicts the proposed evolution of *Cephalograptus cometa extrema* from a lineage beginning with the genus *Orthograptus* through species of *Petalo-*

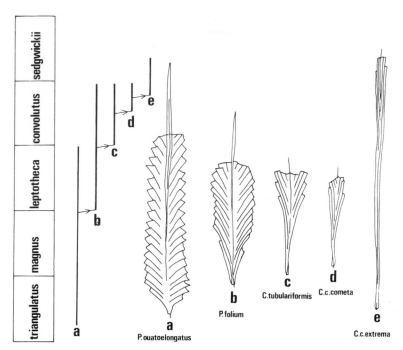

Fig. 5. Range in time and suggested evolution for some *Petalograptus* and *Cephalograptus* species.

graptus. In this lineage the upward growth of elongate early thecae is such that in *C. c. extrema* only the early phases of rhabdosomal astogeny typical of *Petalograptus* are found. Indeed the species may have as few as six thecae and their length (up to 30 mm) may exceed those of any known graptoloids. Again, whether reflecting neotenous development or not, the species *C. cometa* seems not to have given rise to any major new types. The same argument seemingly applies to *Corynoides*, a possible paedomorphic derivative of an isograptid, and to *Nanograptus*, possibly derived in this way from *Glossograptus*. Considering earlier graptolites, such as the multiramous dichograptids, certain types of stipe reduction [one of Elles' (1922) and Bulman's (1933) trends] may reflect paedomorphosis, but not the kind involving an alternation of proximal budding schemes (e.g., *Tetragraptus* to *Didymograptus*). The general conclusion would seem to be that paedomorphic and neotenous processes may well have been widespread in graptolite lineages, but there is as yet no evidence that they gave rise to major grades except possibly in cases of stipe reduction where more investigation is required.

The concept of recapitulation, that ontogeny (or astogeny) reflects phylogeny, has been broadly rejected by most neontologists (Mayr, 1963), and I can find no supporting evidence from the evolution of graptolites. Since the larval stages (pro and metasicular stages) are preserved in most graptolite colonies and since this larval development is remarkably uniform throughout the class and can often be compared with thecal ontogeny and rhabdosomal astogeny, the above remarks can be made with some confidence. The general statement that many unlike adult rhabdosomes, from widely differing ages, go through very similar larval stages of development does not in any way support the concept of recapitulation.

Adaptiveness

It has been pointed out by Mayr (1963) that even quite persistent phenotypes may exist without being strongly selected for, in other words some components, originally selected perhaps, may be permitted to persist during selection pressure provided they do not lower the general fitness of the animal. Thus, the tiny, spike-like virgella of a great many graptolite species seems to confer no great advantages, although the analogous nemal/thecal structures suggest some relationship to the position occupied by the zooid. Such structures are probably better regarded as being on an adaptive peak in the sense that from population to population they must have remained more successful, and therefore continuously selected for, than others produced at the same time.

Preadaption

An animal is said to be preadapted if it can move into a new niche successfully. The most dramatic such change in graptolite evolution is that from the benthonic geographically restricted dendroid to the planktonic cosmopolitan dendroid (*Dictyonema flabelliforme*). The new habitat comprised, of course, the vast, warm surface waters of the tropical "proto-Atlantic" and other oceans. The conical colonies (Fig. 6) may have been apex downwards as suggested by Kirk (1969), or apex upwards (Rickards, 1975); in the latter case secretion of the original holdfast made the slight morphological change to secretion of a nema, or a bundle of fibres, or a nemal vane. In my opinion the benthonic feeding position of a sessile colony, involving more or less passive response to food-carrying currents, would be the same in an essentially planktonic environment and little selection stress would be placed upon pre-existing morphological structures. Eventually the gene pool would change in response to the need for greater buoyancy, retardation of sinking, and necessity to remove the vital proximal region from buffeting in the extreme surface water layers (see earlier). Planktonic conical dendroids were essentially of Tremadoc age, but their immediate descendents, the planktonic graptoloids and more buoyant anisograptids, proliferated rapidly.

Phyletic gradualism and punctuated equilibria

In discussing Palaeozoic invertebrates, which may not, of course, have identical evolutionary patterns to hemichordates, Eldredge (1971) considered that phyletic gradualism, by which a steady transformation of large populations took place, was a less likely process of species formation than was a process of

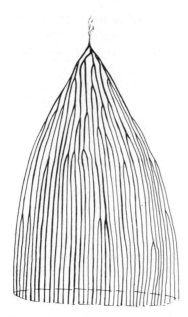

Fig. 6. Schematic, conical, planktonic *Dictyonema* colony, dissepiments omitted.

allopatry. His allopatric model envisaged little or no change in species-specific characters during the range of that species, but a new species arose from it by relatively rapid change in a peripherally isolated part of the whole population. These views were enlarged and generalized by Eldredge and Gould (1972), and the term punctuated equilibria given to the allopatric model. Jaanusson (1973) has also drawn attention to some aspects of discontinuous change in graptolite evolution.

Although in the graptolite lineages that I have examined it is possible to find gradual changes during the time span of a species, such as a gradual shift in the number of thecae in the uniserial portion of *Dimorphograptus erectus* (Fig. 7), or in the thecal spacing of *Monoclimacis flumendosae*, these are the exceptions rather than the rule. More commonly, the new species appear relatively suddenly, and in numbers of cases the reasons are quite clear: thus the change from *Glyptograptus* to *Atavograptus ceryx* is necessarily sudden since the production of a uniserial rhabdosome in this case requires the suppression of dicalycal budding. Jaanusson (1973) has drawn attention to other such striking morphological discontinuities in the evolution of graptolites, and has concluded that at the time of each dramatic change dithyrial populations probably existed. Thus in the above case two polymorphs of the same species existed (Fig. 2), the one biserially scandent (*Glyptograptus* sp.) and the other uniserially scandent (*A. ceryx*). The better adapted of the two would tend to rapidly displace the other.

Other possible examples of such sudden appearances include the following: four-stiped and three-stiped *Tetragraptus fruticosus* (Jaanusson, 1973); four-stiped and three-stiped *Pseudotrigonograptus ensiformis* (Rickards, 1973);

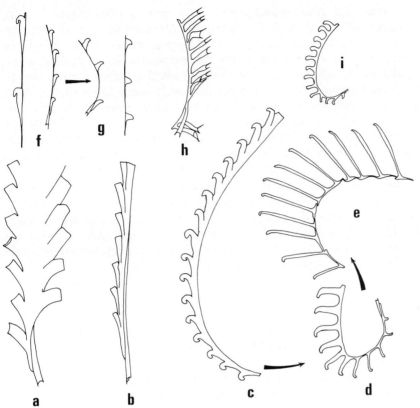

Fig. 7. Thecal isolation in Silurian graptoloids shown diagrammatically: a, *Dimorphograptus decussatus* Elles and Wood; b, *Coronograptus gregarius* (Lapworth); c, *Monograptus difformis* Törnquist, leading to d, *M. triangulatus* (Harkness), leading to e, *Rastrites longispinus* Perner; f, *Monograptus capis* Hutt leading to g, *Rastrites spina* sensu Rickards (1970); h, *Rastrites phleoides* Törnquist; i, *Rastrites richteri*.

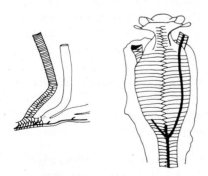

Fig. 8. Comparison of *Rhabdopleura* (left) with a similarly encrusting (camaroid) graptolite: zig-zag fuselli shown; stolons black.

Holmograptus lentus and *Nicholsonograptus fasciculatus* (Skevington, 1966, 1967); *Didymograptus?* sp. A and *D. pakrianus* (Jaanusson, 1960); "*Monograptus*" and *Diversograptus runcinatus*. Indeed such a process of the relatively sudden appearance of a new species seems to be widespread in graptolite evolution, even though, as already stated, some very gradual changes enable one to define arbitrarily successive palaeontological species in some lineages: the *M. t. triangulatus* and *M. t. separatus* lineage would be a good example of this, but the change from *M. difformis* to *M. triangulatus* (Rickards et al., 1976) was a rapid step. In general the process of gradual change results in the palaeontologist being able to define species and subspecies, but the more widespread process of the relatively rapid introduction of a new character after a period of equilibrium enables both the recognition of species and of higher major evolutionary grades in the sense of Huxley (1958) and Bulman (1963). The question of whether plankton clouds of graptolites could develop peripherally isolated populations in the manner envisaged by Eldredge for some invertebrates has not yet been seriously examined, but it is possible that such isolation could take place, for example, in relation to oceanic current gyres.

Mosaic evolution

Mosaic evolution, in which different biocharacters evolve and change at different rates in the same and different lineages, is widely applicable to graptolite evolution of the earliest monograptids (Rickards et al., 1976) (Fig. 9). Here the basic morphological structures of the genus *Atavograptus*, the earliest uniserial scandent graptolite, are as follows: simplified glyptograptid thecae with a very gentle geniculum; a slender, gently dorsally curved rhabdosome; little thecal overlap; gently everted, simpler thecal apertures; a sicula, small (1.3—1.4 mm) in the earliest species *A. ceryx* to moderately long in later species (3 mm in *A. atavus*). *Lagarograptus* originated from *Atavograptus* by the development of apertural processes on the ventral lips of the thecae; the sicular size, shape of rhabdosome, and thecal overlap remained unchanged. In the development of *Coronograptus* from *Atavograptus* the thecal apertures remained more or less unaffected, but the thecal length and overlap (including sicular length and overlap) increased and as a result the rhabdosome became more robust. *Pribylograptus* probably originated from the earliest *Atavograptus* species since the sicula remained small, whilst the other characters also showed little change except for the thecal apertures which became strongly introverted. Strong thecal *extroversion* is the main change accompanying the origin of *Monograptus* s.s. from *Atavograptus* and loss of geniculation in the origin of *Pristiograptus* from *Atavograptus*. Thus the several listed biocharacters change at different rates and in direction in different lineages. This particular plexus of evolving lineages is discussed further in the section on Cope's rule and that dealing with the theory of trends.

Many other evolutionary lineages proposed by Rickards et al. (1976) display mosaic evolutionary patterns to varying degrees. In the *Monograptus communis* lineages, for example, the sicula and fishhook-shaped rhabdosome remained

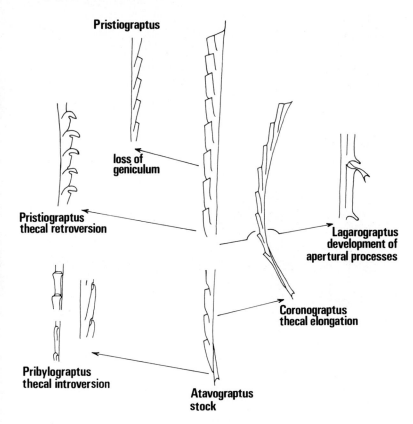

Fig. 9. Outline mosaic evolution involving the genera *Pribylograptus*, *Coronograptus*, *Lagarograptus*, *Monograptus* and *Pristiograptus* from the *Atavograptus* stock: main morphological changes shown diagrammatically. Further explanation in text.

almost the same after the inception of *M. c. communis* from *M. austerus sequens*, but whereas the thecae of *M. millipeda*, *M. clingani* and *M. c. obtusus* became triangulate throughout the colony and heavily hooked, those of *M. c. rostratus* and *M. denticulatus* became higher; and in the latter species rastriform (isolate) at the proximal end, culminating in *Rastrites richteri*.

The lobate *Monograptus* species (Fig. 10) produced excessively retroverted thecae at an early stage (*magnus* Zone) in one lineage, whereas the "main" lineage of this group involving *M. lobiferus* produced robust rhabdosomes with little thecal alterations until the *sedgwickii* Zone upwards when excessively retroverted thecae were again developed. Other groups with enrolled ("streptograptid") thecae either altered the thecae to a dramatic degree (Fig. 11), or else the thecae remained more or less simple hooks whilst the rhabdosome underwent considerable change as in *Diversograptus* or *Sinodiversograptus*.

It would appear that as a rule one or two biocharacters underwent change whilst the remainder were unaffected except by slight dimensional (but not proportional) changes: at a later stage in the same lineage the reverse may ob-

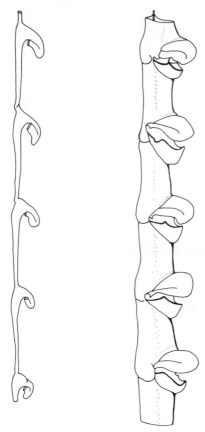

Fig. 10. *Monograptus* sp., showing unusually advanced thecal type as early as the *magnus* Zone.

Fig. 11. *Monograptus nodifer* Törnquist s.l.

tain and those characters which underwent most change earlier on became "stable".

Orthogenesis, programme-evolution, theory of trends

A trend is the occurrence of a character in several independent lineages, in each case undergoing similar development history. Elles (1922) detected several major trends in graptolites: reduction in the number of stipes; change in the direction of growth of stipes relative to the sicula; thecal elaboration; the uniserial scandent rhabdosome coupled with thecal variation. Bulman (1933) amplified these in his concept of a sequence of graptolite faunas, and later (1963) related them to the idea of cladogenesis and anagenesis. Lang (1923) termed the process of trends programme-evolution because each character apparently evolved along a predetermined course. This smacks of vitalistic thought, and can be contrasted with modern ideas on preadaption as defined

by Mayr (1963). Thus in a planktonic environment a new habitat or microhab-
itat may be presented simultaneously to species in quite different lineages but
with not dissimilar basic thecal structure; not surprisingly the gene pools
responded to such pressures in like manner, and the phenotypic reflection
might be expected to be similar in each lineage. For a particular morphological
change in an evolving lineage to be recognized as a trend, that change must take
place at more or less the same time in other definable lineages: orthogenesis
was the term used to describe the apparently "predetermined" course of morphological change with time, and as such is an obsolete term.

Bulman (1933) gave a detailed account of the trend to thecal isolation
affecting lineages of Silurian monograptids, and in particular of Llandovery
genera such as *Rastrites*. Little was known of the morphology of numbers of
Llandovery species and the exercise at that time was difficult: for example
Bulman (1933, p. 332) accepted Elles' (1922) concept of the monograptids
degenerating into Ludlow species with simple thecae. Today we know that
the simple thecae were present through most of the Silurian and at intervals
gave rise to more complex thecal types (Urbanek, 1958; Rickards et al., in
press). The tendency to thecal isolation (Fig. 7) affected far more Silurian
graptoloids than was appreciated by Bulman (1933) or by Sudbury (1958). It
can be summarized as follows:

(1) At the *acinaces* and *cyphus* Zone levels some dimorphograptids and
coronograptids develop more or less isolated thecal apertures, whilst *Monograptus difformis* was the first graptolite with triangular thecae.

(2) In the *triangulatus* Zone, and Middle Llandovery Zones generally, thecal
isolation affected the *M. triangulatus* types, the genus *Rastrites*, some coronograptids, *M. capis*, *M. elongatus* and some undescribed slender monograptids.

(3) Many of the lineages involved in thecal isolation persisted until the higher
Llandovery, such as the triangulate monograptids and *Rastrites*.

Rickards et al. (1976) have identified a considerable number of other trends
affecting Silurian graptoloids of which the following are the more important:

(1) Proximal end protraction affecting the genera *Akidograptus*, *Rhaphidograptus*, *Orthograptus*, *Petalograptus*, and *Cephalograptus*.

(2) Thecal eversion affecting "*Monograptus*", *Monoclimacis*, *Clinoclimaco-
graptus*, *Atavograptus*, *Coronograptus*, *Petalograptus* and others.

(3) Thecal introversion affecting *Metaclimacograptus*, *Pribylograptus*, *M.
sudburiae*, and *Barrandeograptus*.

(4) Thecal hood formation in *P.* (*Pseudoclimacograptus*), *P.* (*Metaclimacograptus*), "*Amplexograptus*", *Pribylograptus*, *Lagarograptus*, *Rhapidograptus*,
and some monograptids of uncertain affinities.

(5) Thecal elongation in *Cephalograptus*, *Akidograptus*, *Orthograptus*,
Petalograptus, *Pribylograptus*, *Rastrites*, *Monograptus*, *Coronograptus*, *Lagarograptus*, *Pristiograptus*, *Pseudoglyptograptus*, and *Cystograptus*.

These and other trends affected different groups at more or less the same time, and affected biserials at the same time as scandent uniserials. Some characters became stabilized, whilst others altered, the whole pattern affording many instances of parallel and mosaic evolution.

Convergence and parallelism

The distinction between parallelism and convergence depends on how close one considers the common heritage needs to be: thus in my mind the tuboid graptolite *Idiotubus*, or a camaroid (Fig. 8a) shows convergence with *Rhabdopleura* (Fig. 8b) in that they had a similar life style as encrusting animals, and the arch of fusellae, the structureless basal layer and position of stolons are closely comparable. On the other hand in the examples of trends taken from my studies of Llandovery graptoloids (see earlier) the lineages can often be traced to a common root a few zones earlier, and the phenotypic responses of the respective genotypes to environmental pressures result in instances of true evolutionary parallelism. There is a difference between the concepts of programme-evolution and parallelism: the former refers to the "predetermined" course or sequence of morphological events; the latter compares or identifies particular similar points of morphology along two or more such sequences. In palaeontology today the term programme-evolution seems happily to have lapsed in usage and most workers discuss such lineages in terms of parallel evolution, using the term descriptively rather than implying "predetermined" sequences. In the lineages *M. t. triangulatus—M. t. extremus—R. longispinus* and *M. convolutus—R. phleoides* (see Rickards et al., 1976) the two rastrites are a result of parallel evolution from monograptids, as indeed are several other *Rastrites* species. The rastritids are clearly polyphyletic in that they originate from several monograptid lineages. The matter is largely one of semantics for the common roots are often but a few graptolite zones earlier; most graptolite genera are polyphyletic in this sense. Thus the examples of trends described above may all be described in terms of parallel evolution, the response of respective gene pools to similar environmental pressures.

Major adaptive trends through time

In his paper on programme evolution Bulman (1933) concluded that there were definite periods of evolutionary activity during which major changes took place (later to be his "sequence of faunas" and "anagenetic grades") followed by periods of quiescence. He made the important observation that all the activity which resulted in what Huxley would term general biological improvement, actually did not seem to result in successful products in that the peaks on his graphs were sharp and many innovations lasted but a short time. This is probably overstating the case for from the Llandovery peak (Fig. 3b) came the longlived major genera *Pristiograptus*, *Monograptus* and *Monoclimacis*, as well as the biserial retiolitids which lasted well into the Ludlow. Bulman was unable at that time to relate major changes of grade to any adaptive strategy and suggested the operation of "some "internal" factor in the evolution of the race". I explained the major changes (Rickards, 1975) as a response to the change from a benthonic to the planktonic mode of life. It is conceivable that dithyrial populations existed for numbers of generations: in some broods the spat settled and in other ones they did not, but in any event the change from benthonic to planktonic was relatively sudden. Whilst the conical dendroid colony must have been pre-adapted to some important degree (see earlier) certain disadvantages ensued, namely that the vital proximal region was in the most buffetted layer of the ocean. Subsequent evolutionary changes, stipe reduction, biserial scandency with floats, uniserial scandency without floats but with cladia in some cases, were the gradual responses of a more or less pre-adapted organism to a hydrodynamically more stable life position. Each of the steps, stipe reduction, etc., affected numerous lineages at more or less the same times and can accurately be termed major adaptive trends.

It is of interest that the pre-adaption of one type of dendroid (*Dictyonema flabelliforme*) should give rise to the whole of the planktonic graptoloid evolutionary plexus, for it is in agreement with Mayr's (1963) view that pre-adapted ecologically specialized forms were more likely to result in successful evolutionary innovations than were unspecialized forms. In this case the bulk of dendroids continued with little change into the Carboniferous and, so far as is known at present, did not again give rise to new graptolite types.

Morphological gradients, penetrance and expressivity

Urbanek (1960) was responsible for the theory of morphophysiological gradients in graptoloids, in which morphogenetic substances generated in the siculo-zooid spread along the rhabdosome and resulted in the regular morphological successions seen on most colonies. The siculo-zooid would be the centre of physiological dominance, and this would not be unreasonable even in the cases of regenerated (bipolar) rhabdosomes lacking a sicula, which are quite common. If this idea is correct it is even possible that the new first zooid was able to pick up where the siculo-zooid left off, and generate and spread the morphogenetic substance. Such a proximally introduced substance could have acted as an inhibitor of distally introduced new biocharacters, which may explain why proximal introduction of new biocharacters is much more common in evolutionary lineages.

Whether or not such a concept would be acceptable to modern geneticists is highly doubtful, and it would probably be more correct to say that differing proximal and distal characters reflect phenotypically the varying expressivity of the genetic factors of the colony. It is at least clear, as has been pointed out by Jaanusson (1973), that information embodied in the genotype was responsible for duplication of the zooids (and thecae) and for the organization of the whole colony. In graptolites, the extent of organization of the colony was probably greater than in any other colonial organisms, for none show the degree of symmetry and balance of the planktonic graptoloids. On the other hand it could be said that the genetic code for a symmetrically organized colony would be simpler and more primitive than that needed to control the astogeny of a "disorganized" colonial association such as that of *Cephalodiscus*.

Cope's Rule

Mayr (1963) considers that "Gigantic forms generally have little leeway left and rapidly become extinct", and Cope took the view that many lineages, after inception, produced increasingly large descendants and that this eventually led to extinction. Graptolite lineages very commonly show a tendency to increased rhabdosomal robustness after a slender beginning, but the last known members of several lineages are diminutive colonies, if not with a small dorso-ventral width then at least having relatively few thecae.

The earliest recognizable *Pristiograptus* species are found in the *magnus* Zone (Rickards, 1974; Hutt, 1975), although a slender species *Pristiograptus fragilis* probably arose from *Atavograptus* as early as the *acinaces* Zone (Rickards et al., 1976). By the *argenteus* Zone the already more robust species *P. jaculum* and *P. concinnus* are found. The Upper Llandovery is then typified by what might loosely be termed the *P. regularis* group of species which are exceedingly long (up to 50 cm) with quite slender proximal regions. Appearing in the latest Llandovery, and then most abundant in the Wenlock and Ludlow, are species of the *P. dubius* group which, whilst shorter in length, have more robust proximal regions and, very commonly, a greater dorso-ventral width distally than representatives of the *P. regularis* group. There was an undoubted overall increase in rhabdosomal robustness (for details of actual lineages illustrated see Rickards et al. 1976). However, in the Late Wenlock and Ludlow a higher proportion of more slender *Pristiograptus* species is found (*P. pseudodubius*, *P. d. ludlowensis* and others now placed in other genera such as *Bohemograptus*) and the rhabdosomes in the Ludlow are commonly shorter than in the Wenlock even for those species with a dorso-ventral width in excess of 2 mm. The last known species are of quite modest proportions. Therefore, the undoubted tendency to increased thecal and rhabdosomal size earlier in the record is not a tendency leading directly to extinction and, indeed, many of the later rather smaller pristiograptids gave rise to numerous Ludlow and Devonian lineages (*Bohemograptus*, *Colonograptus*, *Saetograptus*, etc.).

A very similar situation obtained with that other important Silurian genus *Monograptus* s.s. which includes *M. priodon*. The earliest species is probably *M. sp.* (Rickards, 1974) in the *magnus* Zone, which was rapidly followed in the *argenteus* and *convolutus* Zones by the increasingly robust species *M. undulatus* and *M. lobiferus*. The genus is probably polyphyletic, and later, yet more robust species such as *M. rickardsi*, *M. marri*, *M. priodon*, originated almost certainly from the genus *Monoclimacis*. None the less there is a clear increase in rhabdosomal robustness perhaps reaching an acme in the Wenlock with *M. flemingii* which may be as long as 750 mm and over 4 mm wide. Thereafter the size declined and the Ludlow (and perhaps Devonian) forms are distinctly smaller, in some species very small, and yet they still had a long evolutionary history. The genus *Monoclimacis*, another critical Silurian genus, had exactly the same pattern as *Pristiograptus* and *Monograptus*: slender inception of *Monoclimacis* sp. from *Atavograptus strachani* in the *triangulatus* and *magnus* Zones, increasing robustness through *M. griestoniensis* to *M. vomerina basilica*, then more diminutive Late Wenlock and Early Ludlow species (*M. flumendosae kingi, M. haupti*). This particular, broad pattern seems widely applicable to graptolite evolution, including many Ordovician lines: slender inception, increased robustness and size, decline in robustness and size (possibly giving rise to other groups).

Biogeography

The most recent summary and overall interpretation of graptolite biogeography was that of Skevington (1974) who interpreted the Late Arenig and Llanvirn pronounced graptolite provincialism as a result, primarily, of latitudinal variation in surface water temperatures. Thus the Pacific Province was characterized by stenothermal, tropical forms, and the Atlantic Province by temperate forms. I find the available evidence derived from palaeomagnetic data, tectonic and environmental factors entirely convincing. Earlier, Dewey et al. (1970) noted that the Caledonian oceans had probably been very extensive in the Arenig and Llanvirn, enhancing provincialism, but had decreased in post Llanvirn times as the ocean floor was consumed during the build-up to the Caledonian orogeny. As the available areas decreased so the faunas became uniform, a situation which certainly obtained throughout the Silurian, and possibly the Devonian. However, it is now known that an extensive Late Ordovician glaciation (Beuf et al., 1971; Harland, 1972), with the South Pole positioned in what is now north-west Africa, could have increased the thermal gradient between palaeopole and palaeoequator to such a degree that, not only would provincialism have been lost, but also the essentially tropical zooplankton would have been placed in jeopardy. Very few graptolite species and genera are known from the latest Ordovician strata and it can be stated with some confidence that immediately prior to the Llandovery retreat of the ice sheet, marine transgressions, and explosive evolution, the graptolites very nearly became extinct. Other faunas may have been adversely affected at this time. That the Silurian graptoloid faunas remained uniform suggests that the closing oceans were a factor additional to that of the thermal gradients.

No similar explanation can yet be offered for the near-extinction of the graptoloids in the Late Wenlock *nassa* Zone, nor for the evolutionary explosions about the *gracilis* and *clingani* Zones (Bulman, 1933), although these latter events could, of course, be related to the mixing of Atlantic and Pacific faunas previously so well defined.

Patterns Peculiar to Graptolite Evolution

Bulman (1933) had difficulty in applying neontologists' terminology to the evolutionary patterns he detected, and I have to some extent the same problem

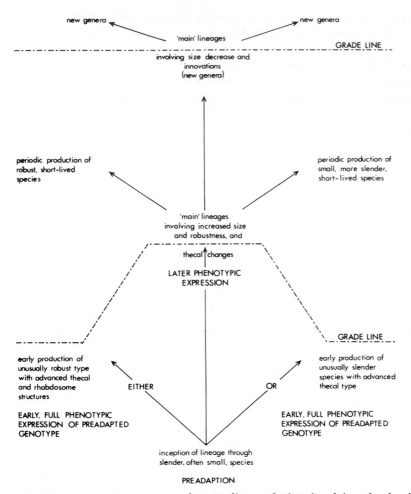

Fig. 12. A common pattern of graptolite evolution involving slender inception, increased robustness at a later stage, and derivation of new genera at a late stage. Full explanation in text.

as is apparent in places earlier in the text. The stratigraphic control available when discussing Silurian graptoloids has enabled some very clear patterns to be deduced. These are described below and, where possible, are defined in terms of modern evolutionary ideas.

When discussing Cope's rule in the preceding section with respect to *Monograptus*, *Pristiograptus* and *Monoclimacis*, a very broad pattern was outlined involving increased rhabdosomal robustness and size. This is a convenient over-simplification, and what seems to be a more real pattern is outlined in Fig. 12. The evolution of *Monograptus* s.s. provides a good example. *Monograptus* sp. (Rickards, 1974; textfig. 1c) probably gave rise to the lobate monograptids like *M. lobiferus*; the lobate thecal structure gradually became accentuated in the "main" lineages (i.e., involving *M. lobiferus* and like forms) to reach extreme forms in *M. knockensis* and *M. singularis* in the Late Llandovery. But a slender *Monograptus* sp. (Rickards et al., 1976; Fig. 10) with a thecal structure as advanced as that of *M. singularis* appears as early as the *magnus* Zone almost at the root of the genus *Monograptus* s.s.

Similarly, in the other main line of *Monograptus* evolution, that in which *Monoclimacis crenularis* probably gave rise to *Monograptus marri* and hence to the *M. priodon* plexus in the *sedgwickii* or *turriculatus* Zones, an unusually robust and advanced species, *M. rickardsi*, was produced in the *turriculatus* Zone, probably from *M. marri*. A similar "grade" of robustness, size, thecal hook, and thecal spinosity is not achieved again until the middle Wenlock and the *M. flemingii* group.

In *Pristiograptus*, *P. fragilis* is possibly a species comparable with *Monograptus* sp. (Fig. 10) in its position in the pattern, in that it is an early extremely slender pristiograptid, probably with relatively advanced thecal structure, whilst *P. concinnus* is an early strikingly robust form, but not perhaps as extreme in its development as *M. rickardsi* in the *Monograptus* lines. The suggested pattern may not hold as well in the genus *Monoclimacis*, itself a more short-lived, less "important" genus than either *Pristiograptus* and *Monograptus*, and the early slender types (*M. griestoniensis*) and robust types (*M. galaensis*) are essentially part of the "main" lineages.

It is tempting to suggest that if preadaption is necessary to specialized successful innovations, and *Pristiograptus* and *Monograptus* certainly fall into this category, then it might reasonably be expected that in a small number of species, perhaps restricted geographically, the gene pool might find full phenotypic expression at a very early stage in direct response to the new environmental pressures. Elsewhere the main bulk of planktonic populations might achieve the same or similar phenotypic expression more gradually.

The above evolutionary pattern may be detectable throughout the known lineages with slight variations probably dependent upon whether the grade introduced is of relatively major or minor nature. In several cases a major innovation, such as the *Glyptograptus* to *Atavograptus* line, may be achieved quite quickly, perhaps through a sequence of dithyrial populations as explained earlier; but following the innovation are more or less abortive attempts to do the same thing. Thus, shortly after the firm establishment of monograptids in the *persculptus* to *atavus* Zones, came the appearance of graptolites with a uniserial proximal end and a biserial distal part (*Dimorphograptus*, "*Bulmanograptus*", *Rhaphidograptus*), as well as genera with a strongly protracted proximal end which may be regarded as simulative (hydrodynamically at least) of the uniserial stipe (*Akidograptus*, etc.). Most of these were short-lived genera.

Numerous similar examples can be found in the evolutionary patterns of graptolites, and it may be that such a process is merely a slight variation on that depicted above for *Pristiograptus* and *Monograptus* in that the pre-adapted genotype found full phenotypic expression at a very early stage, as indicated earlier, but became very successful and itself took up the main lines of evolution.

Summary

(1) The sessile benthonic dendroid graptolites successfully occupied a stable, inshore, shelly environment from the Middle Cambrian to the Carboniferous and showed relatively little change during this span; however, they gave rise to the planktonic graptoloids in the Late Tremadoc through the planktonic dendroid *Dictyonema flabelliforme*.

(2) The change to probable hermaphroditism from an earlier gonochoristic strategy, coupled with outbreeding and, in times of stress, inbreeding, gave the graptoloids a most flexible and competitive mode of reproduction and contributed largely to the evolutionary success of the group, if this factor can be measured in terms of morphological variety.

(3) The genotype of the ancestral dendroid *D. flabelliforme* must have been superbly pre-adapted if we are to explain the Tremadoc/Arenig evolutionary explosion of the Anisograptid and Dichograptid faunas in the new, vast, essentially tropical planktonic habitat.

(4) Subsequent evolutionary explosions could be related to the mixing of Pacific and Atlantic plankton faunas (*gracilis* to *clingani* Zones) during loss of provincialism, whatever its causes, and to the post Ordovician ice-age amelioration and marine transgressions (Llandovery explosion).

(5) I have no explanation to offer at present for the near extinction of graptoloids in the *nassa* Zone, nor for the Ludlow explosion and Devonian extinctions.

(6) The major anagenetic grades probably reflect a change to an increasingly efficient hydrodynamic stability; that is, successively, anisograptid, dichograptid, diplograptid, and monograptid.

(7) The origin of new types may, in some lines, have been by neoteny, but such instances do not seem to have led to very successful lineages. More common was the situation where similarly pre-adapted genotypes, with not too distant roots, responded phenotypically in a like manner to general changes in the environment, so that both parallel evolution and mosaic evolution became established as the typical evolutionary pattern in the graptoloids.

(8) Such preadaption sometimes resulted in full phenotypic expression at an early stage and these were either very successful (*Atavograptus*) followed by later "abortive" changes (*Dimorphograptus*), or were relatively unsuccessful (*Monograptus rickardsi*) followed by more gradual acquisition of the same biocharacters (*M. priodon/M. flemingii* lineages).

(9) Well established major lineages (e.g., *Pristiograptus dubius*) often went through a stage of "giantism", periodically produced robust and slender offshoots, and finally, through quite diminutive end-members, gave rise to strikingly new genera (in this case *Colonograptus* and *Saetograptus*).

(10) Although peripheral isolation of parts of the more or less uniform plankton population is difficult at present to imagine, the concept of punctuated equilibria is certainly applicable at all levels to graptolite evolutionary patterns.

Acknowledgements

I should like to thank Miss Shirley Newman for help with the diagrams; and Professor J.W. Valentine and Miss C. Campbell for their critical reading of the manuscript. Cambridge University Press and the Cambridge Philosophical Society kindly gave permission for the use of Figs. 1, 3 and 4.

References

Allard, R.W., 1965. Genetic systems associated with colonizing ability in predominantly self-pollinating species. In: H.G. Baker and G.L. Stebbins (Editors), The Genetics of Colonizing Species. Academic Press, New York, N.Y., pp. 50—76.

Beuf, S., Biju-Duval, B., Charpal, O. De, Rognon, P., Gabriel, O. and Bennacef, A., 1971. Les Grès du Paléozoique Inférieur au Sahara. Publications de l'Institut Français du Pétrole, Collection "Science et Technique du Pétrole" no. 18. Editions Technip, Paris, 464 pp.

Bulman, O.M.B., 1933. Programme-evolution in the graptolites. Biol. Rev., 8: 311—334.

Bulman, O.M.B., 1958. The sequence of graptolite faunas. Palaeontology, 1: 159—173.

Bulman, O.M.B., 1963. The evolution and classification of the Graptoloidea. Q. J. Geol. Soc. Lond., 119: 401—418.

Bulman, O.M.B. and Rickards, R.B., 1966. A revision of Wiman's dendroid and tuboid graptolites. Bull. Geol. Inst. Ups., 43: 1—72.

Dewey, J.F., Rickards, R.B. and Skevington, D., 1970. New light on the age of Dalradian deformation and metamorphism in western Ireland. Norsk Geol. Tidsskr., 50: 19—44.

Eldredge, N., 1971. The allopatric model and phylogeny in Paleozoic invertebrates. Evolution, 25: 156—165.

Eldredge, N. and Gould, S.J., 1972. Punctuated equilibria: an alternative to phyletic gradualism. In: T.J.M. Schopf (Editor), Models in Paleobiology. Freeman, San Francisco, Calif., pp. 82—115.

Elles, G.L., 1922. The graptolite faunas of the British Isles, Proc. Geol. Assoc. Engl., 33: 168—200.

George, T.N., 1962. The concept of homeomorphy. Proc. Geol. Assoc. Engl., 73: 9—64.

Harland, W.B., 1972. The Ordovician Ice Age. Geol. Mag., 109: 451—456.

Hutt, J.E., 1975. The Llandovery Graptolites of the English Lake District, Part 2. Palaeontogr. Soc. (Monogr.), 129: 57—137.

Huxley, J.S., 1958. Evolutionary processes and taxonomy. Upps. Univ. Årsskr., 6: 21—39.

Jaanusson, V., 1960. Graptoloids from the Ontikan and Viruan (Ordov.) Limestones of Estonia and Sweden. Bull. Geol. Inst. Univ. Upps., 38: 289—366.

Jaanusson, V., 1973. Morphological discontinuities in the evolution of graptolite colonies. In: R.S. Boardman, et al. (Editors), Animal Colonies: Their Development and Function Through Time. Dowden, Hutchinson, and Ross, Stroudsburg, Penn., pp. 515—521.

Jackson, D.E., 1974. Tremadoc Graptolites from Yukon Territory, Canada. In: R.B. Rickards, et al. (Editors), Graptolite Studies in Honour of O.M.B. Bulman. Spec. Pap. Palaeontol., 13: 35—58.

Kirk, N.H., 1969. Some thoughts on the ecology, mode of life and evolution of the Graptolithina. Proc. Geol. Soc. Lond., No. 1659: 273—292.

Kozłowski, R., 1949. Les graptolithes et quelques nouveaux groupes d'animaux du Tremadoc de la Pologne. Palaeontol. Pol., 3: 1—235.

Kozłowski, R., 1966. On the structures and relationships of graptolites. J. Paleontol., 40: 489—501.

Lang, W.D., 1923. Trends in British Carboniferous Corals. Proc. Geol. Assoc. Engl., 34: 120—236.

Mayr, E., 1963. Animal Species and Evolution. The Belknap Press of Harvard University Press, Cambridge, Mass., 797 pp.

Rickards, R.B., 1973. The Arenig graptolite genus *Pseudotrigonograptus* Mu & Lee, 1958. Acta Geol. Pol., 23: 597—604.

Rickards, R.B., 1974. A new monograptid genus and the origins of the main monograptid genera. In: R.B. Rickards et al. (Editors), Graptolite Studies in Honour of O.M.B. Bulman. Spec. Pap. Palaeont., 13: 141—147.

Rickards, R.B., 1975. Palaeoecology of the Graptolithina, an extinct class of the phylum Hemichordata. Biol. Rev., 50: 397—436.

Rickards, R.B., 1976. The sequence of Silurian graptolite zones. Geol. J., 11: 153—188.

Rickards, R.B., Hutt, J.E. and Berry, W.B.N., 1976. The evolution of the Silurian and Devonian graptoloids. Bull. Br. Mus. (Nat. Hist.), 28(1): 1—120.

Ruedemann, R., 1904. Graptolites of New York, I. N.Y. State Mus. Mem., 7.

Skevington, D., 1966. The morphology and systematics of *"Didymograptus" fasciculatus* Nicholson, 1869. Geol. Mag., 103: 487—497.

Skevington, D., 1967. Propable instance of genetic polymorphism in the graptolites. Nature, 213: 810—812.

Skevington, D., 1974. Controls influencing the composition and distribution of Ordovician graptolite faunal provinces. In: R.B. Rickards, et al. (Editors). Graptolite Studies in Honour of O.M.B. Bulman. Spec. Pap. Palaeontol., 13: 59—73.

Sudbury, M., 1958. Triangulate monograptids from the *Monograptus gregarius* Zone (Lower Llandovery) of the Rheidol Gorge (Cardiganshire). Philos. Trans. R. Soc. Lond., (B) 241: 485—555.

Troedsson, G.T., 1923. Førsok till jamforelse mellan Sveriges och Nordamerikas ordoviciska graptolititskiffrar. Geol. Føren. Førh., 45: 227—248.

Urbanek, A., 1958. Monograptidae from erratic boulders of Poland. Palaeontol., Pol., 9: 1—105.

Urbanek, A., 1960. An attempt at biological interpretation of evolutionary changes in graptolite colonies. Acta Palaeontol. Pol., 5: 127—234.

Urbanek, A., 1963. On generation and regeneration of cladia in some Upper Silurian monograptids. Acta Palaeontol. Pol., 8: 135—254.

Urbanek, A., 1970. Neocucullograptinae n. subfam. (Graptolithina); their evolutionary and stratigraphic bearing. Acta Palaeontol. Pol., 15: 163—388.

Urbanek, A., 1973. Organisation and evolution of graptolite colonies. In: R.S. Boardman, et al. (Editors), Animal Colonies: Their Development and Function Through Time. Dowden, Hutchinson and Ross, Stroudsburg, Penn., pp. 441—517.

Urbanek, A. and Jaanusson, V., 1974. Genetic polymorphism as evidence of outbreeding in graptoloids. In: R.B. Rickards, et al. (Editors), Graptolite Studies in Honour of O.M.B. Bulman. Spec. Pap. Palaeontol., 13: 15—17.

CHAPTER 11

Evolution of Trace Fossil Communities

ADOLF SEILACHER

Introduction

Most evolutionary models have their roots in the living world, but it is the privilege of the palaeontologist to test their validity in the dimension of geological time. This applies not only to the phylogeny of individual taxonomic groups, but also to whole communities. These have only recently been viewed as evolving ecosystems whose structure and spatial distribution cannot be explained by the present-day situation alone (Valentine, 1972).

The Present-Day Model

The exploration of modern soft-bottom faunas, which extends with new technologies to the deepest parts of the ocean, has revealed significant trends in radial transects (Fig. 1).

(1) Dredge samples show that *population densities* as well as *body size* decrease with depth (Sanders, 1968; Rex, 1973). Both trends agree with decreasing supply of food away from the zones of primary production, i.e. at increasing distances from the continents and from the photic zone of the sea. The size decrease, however, is not universal, since a number of giant forms have also been reported in deep-sea photographs (Hollister et al., 1975).
(2) *Species diversity* follows a similar trend, but only in the neritic part of the transects. After a subtidal peak, which may be very high in tropical seas (Sanders, 1968), the diversity decreases in a seawards direction (Dörjes and Hertweck, 1975). It came as a surprise that diversity increases again sharply on the continental slope, combined with an increased evenness of species distribution (Sanders, 1968).
The eventual decline in species diversity is equally sharp, but it occurs well out on the abyssal plain. It is particularly noteworthy that this decline is delayed in the Foraminifera, indicating "that productivity affects macrofaunal diversity in some major taxa only when the lower limits of adaptation in size are approached" (Rex, 1973).

This unexpected increase in diversity in the deep-sea environment, which lacks geometric complexity and is adverse not only because of the increasing nutritional depletion but also because of high pressures and low temperatures, has been interpreted as an historical rather than a physiological effect. The *time-stability* hypothesis (Sanders, 1968) accounts for the diversity peaks not

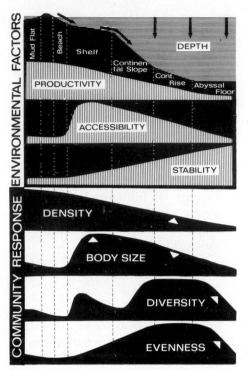

Fig. 1. Environmental and evolutionary trends in space and in time (white arrows). It should be noted that the actual curves could well be step functions.

only in the climatically stable biota of the tropics, but also in deep-sea environments that remain little affected by storms, seasons, sea-level changes, ice ages, plate tectonic events and other perturbances that truncate community evolution in shallow marine habitats, or by lean periods in food supply. In addition, Valentine (1971) has stressed that a decrease in food supply may also be responsible for an increase in diversity.

The time-stability hypothesis has been variously applied to the fossil record (Bretsky and Lorenz, 1970). Thus the diversity of shallow, marine shell assemblages was believed to be a measure of relative water depth (Bretsky, 1969). In contrast, the diversity was found to decrease with depth in modern shelves (Dörjes and Hertweck, 1975); however, the samples were not rarefied. Among fossil bivalves (Levinton, 1974), suspension feeders (which depend on the unstable supply of phytoplankton) and deposit feeders (living on more stable bacteria) show evolutionary differences in agreement with the hypothesis, while the difference in environmental stability between epifaunal and burrowing bivalves seems to be less important (Thayer, 1974).

Translation Into the Palaeontological Record

Palaeontological community studies in general are likely to suffer from the familiar inadequacies of the fossil record: (a) the ratio of fossilizable shell bearers to soft-bodied animals is variable and decreases with depth; (b) fossil assemblages represent time intervals of unknown duration; (c) post-mortem transport of skeletal remains has often admixed elements from alien habitats; (d) diagenetic solution tends to selectively eliminate parts of the original shell assemblages. Therefore, observed diversities may be misleading and original population densities are impossible to reconstruct. These biases are considerably reduced in the *trace fossil* record, which includes the products of soft-bodied as well as shell-bearing organisms, is less affected by diagenesis and not affected at all by sedimentary displacement. On the other hand, trace fossils present problems of their own: (a) they can rarely be identified with animals known from modern dredge samples whose trace production remains unknown, nor with the surface trails that we see in bottom photographs (Hollister et al., 1975), because infaunal burrows have a much higher fossilization potential and therefore dominate in the fossil record; (b) they reflect the behavioural rather than the anatomical adaptation of their unknown producers; (c) they fail to represent organisms too small to have left recognizable traces; (d) they over-emphasize the lower trophic levels, particularly those of deposit feeders; (e) their preservation is facies-dependent, with the optimum in interbedded sands, silts and clays.

In what follows, I have tried to minimize these difficulties by keeping comparisons:

(1) within-fossil-state, i.e. without transgressing the critical "fossilization-barrier";

(2) within-facies, i.e., in soft-bottom assemblages of only two types of deposits that are easily distinguishable by sedimentological as well as palaeontological criteria: (a) the *Cruziana facies*, representing sandy subtidal shelf sediments; (b) *Nereites facies* (Flysch-type), representing distal turbidites of the lower continental slope and the continental rise, brought into the range of our hammer by orogenic uplift.

In the final part of the chapter I shall consider intertidal environments as another frontier for the marine benthos.

Evolution of Flysch Ichnocoenoses Through Phanerozoic Times

Diversity

In Tertiary and Cretaceous ichnofaunas the diversity pattern corresponds to the present-day situation in that they are considerably more diverse in the flysch facies than in contemporary neritic deposits. However, in the perspective of geologic time, this picture changes. While the neritic ichnofaunas, after a steep increase during the Late Precambrian (Webby, 1970), have already reached their level of maximum diversity in the Lower Cambrian, the diversity in the flysch facies is still very low in the Cambrian and rises gradually throughout the Phanerozoic (Fig. 2). Even if we consider the limited number of adequately studied assemblages and our general ignorance about flysch ichnocoenoses of

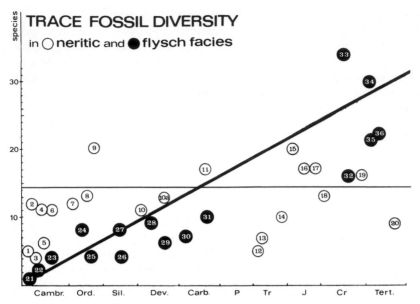

Fig. 2. In comparable trace fossil samples, species diversity shows a strong increase in the deep sea as opposed to level diversity in the shelf during the Phanerozoic (see Seilacher, 1974, fig. 1 for further details).

Permian to Jurassic age, this result suggests that the high diversity of the younger flysches is the result of a gradual build-up in a very stable environment (Crimes, 1974; Seilacher, 1974). Palaeontological evidence thus fully supports the time-stability hypothesis derived from modern dredge samples.

Body size

Deep-sea dredges have shown that photographs of large surface tracks and of bait-attracted amphipods and fish of unusually large sizes are misleading. The majority of the deep-sea benthos is small-sized (Sanders, 1968; Rex, 1973). This fact was interpreted as an adaptation to the food-limited environment.

In trace fossils, body size is not directly recorded; but estimates may be derived from the diameters of trails and burrows, from the size of repetitive patterns and from the areas covered by whole burrow systems. By these measures, some ubiquitous forms like *Gyrochorte* are in fact considerably smaller in the flysch than in the *Cruziana* facies. In the case of the flysch burrow *Granularia* the size difference (0.3 versus 2—6 cm diameter) has so far masked its identity with shallow marine crustacean burrows *(Thalassinoides, Spongeliomorpha, Ophiomorpha)* which is adequately documented by scratch patterns, pellet linings and mode of branching. Others, like *Zoophycos* or *Chondrites*, have smaller, as well as larger, representatives in the flysch facies. As a whole, the size difference is not as great and by no means as consistent as we might expect from the modern dredges.

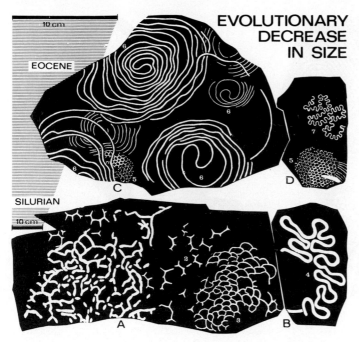

EVOLUTIONARY
DECREASE
IN SIZE

Fig. 3. General size diminution of deep-sea deposit feeders, illustrated by turbidite sole facies from the Silurian of Wales (A, Tübingen catalogue Nr. 1488/1) and New South Wales (from Webby, 1969, pl. 10, fig. 1) and from the Eocene of Pallerstein near Vienna (C, D, both from specimens in the Vienna Collection; D figured by Fuchs, 1895). All traces (*1,2,3,5* = *Paleodictyon*; *4,7* = *Cosmorhaphe*; *6* = *Spirorhaphe*) are graphoglyptid burrows eroded and cast by turbidity currents.

There is, however, a pronounced size gradient in *time* within particular groups. Gradual size reduction can be observed within individual "genera" as well as in larger groups. *Sustergichnus* from the Carboniferous flysch is considerably larger than its Tertiary synonym *Uchirites*, and the same relation exists between *Nereites* and *Helminthoida* (Fig. 4). For graphoglyptid burrows (*Cosmorhaphe, Paleodictyon*, etc.) one needs a tabletop-sized slab in an Ordovician or Silurian flysch to secure a sample of burrow systems equivalent to those contained in a hand-sized specimen from a Tertiary flysch (Fig. 3). Cambrian flysches are characterized by the minute *Oldhamia* burrows (Fig. 4), which are found in the shaly parts of the sequence. Still they do not contradict the general trend: in Argentina, soles of sandy turbidites in the same sequence may be covered with very large burrow casts, which are only too irregular to justify taxonomic determination.

The evolutionary diminution of flysch trace fossils, seemingly contradicting Cope's rule, is by no means universal. *Paleodictyon* nets with a mesh size of 7 cm and other branching graphoglyptids with a 2 cm tunnel diameter ("Pinsdorfer Versteinerung": Abel, 1935, fig. 304) have been found in Tertiary flysches; but such giants are very rare compared to the host of small forms.

In contrast, many cases of a definite size *increase* could be cited from

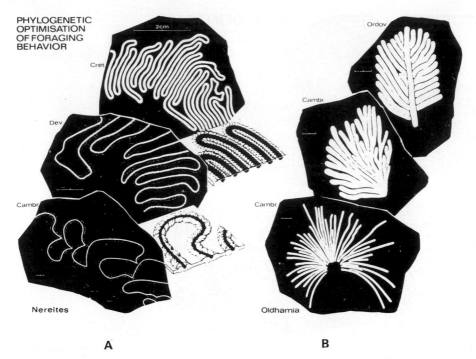

PHYLOGENETIC
OPTIMISATION
OF FORAGING
BEHAVIOR

Fig. 4. Backfilled burrows of sediment-feeding worms from the muddy and silty parts of flysch sections, reflecting the evolutionary increase in fitness by improved foraging programs and sensory monitoring. They also show the size decrease in established lineages. (For details see Seilacher, 1974, fig. 2.)

neritic trace fossils such as *Asteriacites*, *Thalassinoides*, and *Rhizocorallium*. The general trend, however, is much more obscured by variables resulting from changes in climate, salinity and other environmental factors.

Optimization of search patterns

Compared with neritic ichnocoenoses, flysch assemblages show both higher species diversity, and lower ecologic diversity. Feeding burrows of sediment feeders strongly prevail, while resting tracks, trails and habitation burrows of scavengers and suspension feeders are almost absent. Within the burrows and trails of sediment feeders, the occurrence of regular search patterns is also striking. This qualitative difference can in fact be used as a good facies criterion, in addition to the familiar sedimentological features of flysch deposits (Seilacher, 1954).

Distinctive patterns can be explained by highly specialized behavioural programs (Raup and Seilacher, 1969) and include stellate and dendroidal as well as spiral and meandering types. Their "ornamental" character basically reflects selection for uniform and non-duplicating coverage of a food-bearing surface, as has been shown by computer simulation (Papentin and Röder, 1975).

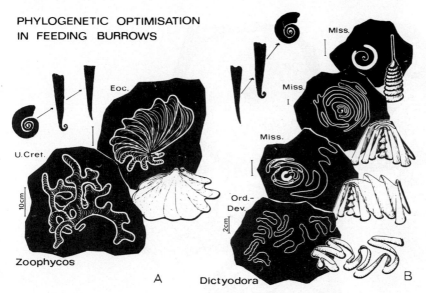

Fig 5. *Zoophycos* (complicated U-shaped tube with backfilled "spreite" structures) and *Dictyodora* (backfilled burrow, connected with the surface by a "spreite"-like structure; see Fig. 6J) reflect sequential behavioural programs that may be related to an ontogenetic sequence. Foraging was phylogenetically improved by "palingenetic" (*Zoophycos*) or "neotenic" shifts (*Dictyodora*) in the behavioural sequence. (From Seilacher, 1974, fig. 3.)

Evolutionary optimization in this direction can be observed at various levels.

(a) Circular *scribbles* provide better coverage than irregular trails; but because of many overcrossings they are less efficient than spirals and meanders based on more complex programs. Accordingly, scribble patterns occur in Early Palaeozoic, but disappear completely in later flysches (Fig. 6).

(b) Within particular lineages, the evolution from open to tight meanders (Fig. 4A), from radial to pinnate branching (Fig. 4B) or from tongue-like to strip mining types of spreite burrows (Fig. 5A) expresses improved coverage and resource utilization of foragers.

Introduction of new feeding strategies

Not all observed trends in pattern evolution conform to the paradigm for simple foraging. This paradigm is most perfectly approximated by the trace fossil *Helminthoida labyrinthica* (Fig. 4, top left), whose meanders are in close contact with each other and free enough to snugly fit into corners left between previous patterns. The imperfections of its Palaeozoic ancestors (Fig. 4) are clearly a primitive feature resulting from a poorer sensory control of the search behaviour.

In other, more regular and more complex patterns, the deviation from the foraging paradigm cannot be explained as an imperfection, but requires a modification of the paradigm itself.

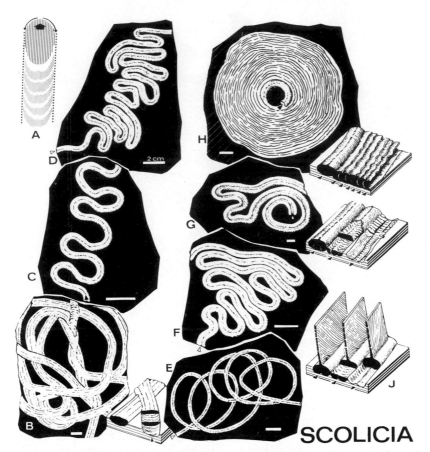

SCOLICIA

Fig. 6. *Scolicia* burrows result from a conveyor belt motion of infaunal sediment feeders
(A). Their search patterns have evolved from ineffective "scribbles" to increasingly tight
meanders in neritic (B—D) as well as in flysch habitats (E—H). However, the tight spiral
of H systematically violates the paradigm for simple foraging, indicating the introduction
of a new feeding strategy within the group. B. Lower Cambrian; Simrishamn, Sweden (=
Psammichnites gigas; Geology Department, Lund University). C. Lowermost Cambrian
(Kessjusinkaja Fm.), Siberia (Tubingen catalogue No. 1488/2; courtesy of Dr. Rozanov,
Moscow). D. Upper Carboniferous (Auernigg Beds); Hermayor, Austria (Naturhist. Museum,
Vienna). E. Ordovician; Barrancos, Portugal (Delgado, 1910, pl. 34, fig. 2). F. Devonian;
Aljustrel, Portugal (Delgado, 1910, pl. 14, fig. 2). G. Eocene; Zarauz, Spain (= *Taphrhel-
minthopsis;* from field photograph). H. Upper Cretaceous; Unterpurkersdorf near Vienna
("*Helminthoida*" *zumayensis* Llarena; Tübingen catalogue No. 1488/3). J. The backfill struc-
tures of *Dictyodora* (see Fig. 5) suggest a similar origin, but with a long "schnorchel"
connecting the burrowing animal constantly with the sediment surface.

(1) "*Scolicia*" is a comprehensive name for trails of a conveyor-belt type of
motion through the sediment. As in *Nereites* and *Helminthoida* the track
became backfilled with concave lamellae behind the animal, but without clear
separation of rejected sand and digested clay fraction. Whether *Scolicia* burrows are the products of burrowing gastropods or irregular echinoids, or both,

remains an open question in view of their long time span (Precambrian to Tertiary). *Scolicia* was probably the first group to develop spiral and meandering
search patterns, and in this particular case meandering is not restricted to the
very deep environments (Fig. 6).

In contrast to the *Nereites* line, not all *Scolicia* forms approach the tight foraging paradigm. One Late Cretaceous species [*S. zumayensis* (Llarena); Fig. 6H]
strongly violates this paradigm in a highly systematic way: its spiral trail consistently cuts into previous whorls, thus producing a spreite-like structure. This
means that the animal did predominantly rework sediment that it had already
ploughed through in earlier turns. While being extremely inefficient for a sediment feeder, such behaviour could be advantageous if new nutrients became
available in the ploughed sediment between successive passages.

(2) *Graphoglyptid* burrow systems (Fuchs, 1895), in contrast to *Nereites* and
Scolicia, did not become backfilled but remained open, probably supported
by a mucus lining. This explains why they became preferentially eroded and
cast by turbidity currents, so that now almost complete replicas of the burrow
systems can be observed on turbidite soles. Their meandering, spiral, net- and
antler-like patterns are not only too complicated, but also too widely spaced,
to conform to the simple foraging paradigm. As far as the spaces between the
tunnels are concerned, the spiral tunnels of the worm *Paraonis* in modern beach
sands (Fig. 7) provide a functional model. The *Paraonis* spirals seem to trap
mobile micro-organisms which are then harvested during later passages. The
complex meanders of *Cosmorhaphe* (Fig. 3), the spirals of *Spirorhaphe* (Figs.
3 and 7) and the majority of *Paleodictyon* (Fig. 3) would then represent alternative pathways towards a net-like paradigm with constant mesh size (Röder,
1971).

The *Paraonis* model may apply to all Palaeozoic and some later graphoglyptids; however, it does not explain why the majority of graphoglyptid burrow systems in the Upper Cretaceous and Tertiary flysches had an excessive
number of outlets (Fig. 8). Multiple openings hamper active ventilation, which
is best achieved in a U-shaped burrow with only two openings, but they could
have trapped fine detritus from the surface, and have facilitated diffusive aera-
tion of the tunnel system. This would be essential if the tunnels were used as
"mushroom gardens" in which bacteria or other micro-organisms grew on
detrital cellulose and other refractory materials (Seilacher, in press).

Such strategy would seem very appropriate in this environment, since food
not only diminishes, but also deteriorates on the way to the deep-sea floor.
Eventually the particulate food resources consist mainly of refractory materials
that animals can digest only with the help of micro-organisms. Such symbiosis,
which also provides fixed nitrogen of which this food is deficient, usually operates in the animals' guts and intestines (Carpenter and Culliney, 1975). But
there is no reason why it should not also work outside the body if the animals
provide the space and mucus secretions that favour the metabolism of micro-
organisms which is otherwise extremely slow in the deep sea (Jannasch and
Wirsen, 1973). This assumption also supports the fact that the radiation of
"farming" burrow types in the Cretaceous coincides with the conquest of ter-

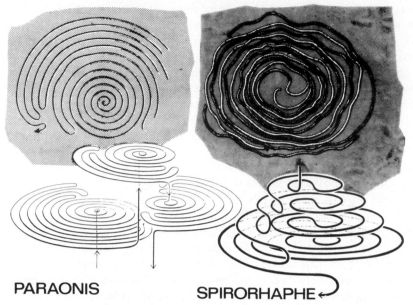

PARAONIS   **SPIRORHAPHE**

Fig. 7. The foraging paradigm is also violated by open graphoglyptid burrows like *Spiro-rhaphe* (others see Figs. 8 and 9) that leave large areas of sediment between the loops untouched in a systematic way (not like in the "poor" meanders of early *Nereites* and *Scolicia*, Figs. 4 and 6). Analogy with the spiral tunnels of the living worm *Paraonis fulgens* suggests trapping of mobile micro-organisms as an alternative feeding strategy. The multi-level reconstruction of the *Spirorhaphe* tunnel (see Fig. 3C), derived from surplus loops in the center of erosional casts (top right), also facilitates the behavioural interpretation of the structure (From Seilacher, 1967.)

restrial and subtidal environments by angiosperms, which must have greatly increased the influx of detrital cellulose.

Dissolved organic matter should also be considered as a possible food resource. It could become preferentially adsorbed from the open sea water by mucus secretions, as well as from the pore solutions ascending in suddenly deposited turbidite sediments. The spirals of *Scolicia zumayensis* (Fig. 6H) would be particularly useful for this feeding method if we assume a very slow rate of burrowing.

Whether or not these speculations are sound, the cited examples do indicate the evolution of new trophic niches, perhaps in response to evolutionary changes in very distant parts of the biosphere.

Graphoglyptid speciation

Trace fossil taxonomy (or parataxonomy) in general suffers from the fact that the taxa tend to be monospecific. Even very distinctive forms, such as *Gyrochorte*, do not justify subdivision into subgroups, despite world-wide distribution and very long time span (Mississippian to Tertiary). This must not

FOOD EXTRACTION FROM
DEEP SEA MUD

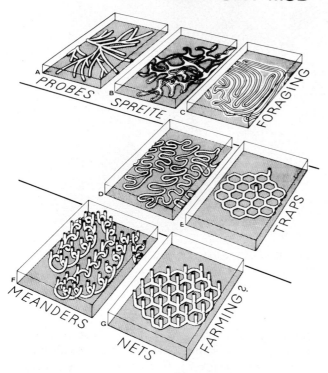

Fig. 8. In contrast to the burrows of foraging sediment feeders, which are typically back-filled and tend to cover a given surface as completely as possible (A—C), graphoglyptid burrow systems (D—G) are designed to leave gaps and remain unfilled, allowing repeated passages. Passive trapping of mobile and dissolved food within the sediment would explain the simpler forms (D, E), active farming the forms with multiple openings (F,G). A. *Chondrites* with terminal backfilling. B. *Phycosiphon* with "spreite" structures (white) inside loops. C. *Helminthoida labyrinthica* (see Fig. 4) with separate backfilling of rim zone (white) and central tunnel. D. *Cosmorhaphe* (see Fig. 3). E. Simple *Paleodictyon* (see Fig. 3, 1—3). F. *Helicolithus* with multiple openings (see Fig. 9, g). G. *Paleodictyon* with multiple openings (see Fig. 3, 5).

imply that the producers remained unaltered, but that trace morphology tends to be a conservative feature shared by many related genera and species and often also by unrelated, but ecologically and ethologically similar organisms. As a modern example we may again cite the paraonid spirals (Fig. 7) which are indistinguishable in German and Pacific beaches, although the producers belong to different genera and species.

In flysch ichnocoenoses the situation is less extreme. In many ichnogenera, such as *Chondrites* and *Zoophycos*, several well-defined ichnospecies can be distinguished. This may in part be due to the fact that the underlying behavioural patterns are less generalized than in shallower environments, thus allow-

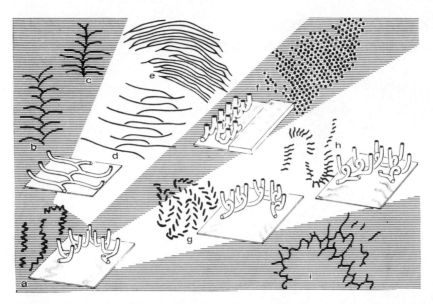

Fig. 9. Graphoglyptid diversification. The modification of one (out of five) basic behavioural program into contemporaneous (Upper Cretaceous to Eocene) and largely sympatric species is explained by co-evolution rather than by trophic niche partitioning. (From Seilacher, 1974, fig. 4.)

ing a finer grade of ichnotaxonomic distinction. But since most observable differences reflect different solutions of the same biological task (exploitation of a limited, but evenly distributed food resource), they represent a relatively clear case of within-habitat, and perhaps sympatric, speciation.

Graphoglyptid burrows are the most remarkable example of this. Not only can we distinguish several basic types of behavioural programs, but each of them can be subdivided into a large number of well-defined ichnospecies (Fig. 9). Most of them range through Upper Cretaceous and Tertiary flysches and have world-wide distribution. They may also be found associated on the same bedding planes, which suggests potential co-existence in the same habitat. This is all the more surprising since related species present only minor modifications of the same program. In terms of palaeontological discrimination, they are ecologically equivalent, without visibly improving the approximation to the supposed functional model. Their diversification may be understood, however, if we assume a shift in the evolutionary bonus from the efficiency in food utilization to the distinctiveness of the pattern itself. In open traps or "mushroom gardens" complex codification of the burrow pattern could have acted as a safety key against usurpers, while the tunnels were left undisturbed between subsequent visits by the owner. The variations in the pattern could also reflect different symbionts. In this sense, graphoglyptid speciation could be viewed as a final level of community evolution in which faunal diversity increases less by trophic niche partitioning than by co-evolutionary diversification.

Intertidal Communities

As shown in the previous section, the increase in diversity of the deep-sea benthos resulted in large part from within-habitat (though not necessarily sympatric) speciation. This process was probably very slow owing to low geometric complexity, adverse conditions and slow metabolic rates compared with shelf environments (Jannasch and Wirsen, 1973). The fact that the immigration of new forms from the shelf did still not prevail, emphasizes the low accessibility which the deep sea shares with intertidal environments (Fig. 1).

Modern beach communities

Accessibility decreases not only on the oceanward side of the shelf but also on the landward side. In intertidal environments obvious barriers for immigration are created by drastic changes in turbulence, temperature and salinity, as well as by intermittent air exposure. The sudden drop in diversity (Dörjes and Hertweck, 1975) reflects these immigration barriers of the beach environment, in which food resources may support extreme population densities.

In contrast to the opportunistic hordes of land-based beach combers searching the zone during low tide, and of nektonic high-tide visitors, true beach animals are provided with very highly specialized feeding mechanisms. Swash feeders in particular have evolved most complicated filtering devices and behavioural responses to their ever-changing environment (Seilacher, 1953, 1959, 1961). The same is true for beach animals feeding within the sediment. The spiral tunnel systems of *Paraonis* (Fig. 7), used and constructed only during high tide, compete in regularity with the "ornamental" flysch forms. In a way, beach animals combine extreme physiological tolerance with very high trophic specialization.

Taken as a measure of evolutionary maturity, the high specialisation of the beach community seems to contradict the extreme physical instability of the environment. This dilemma disappears if we replace the term "stability" by "predictability" (Slobodkin and Sanders, 1969). For an organism that has to face waves, storms and emersion at short intervals, long-term transgressions, regressions and climatic changes are no longer "catastrophic" events. In this situation the physical instability rather favours further evolutionary adaptation, as it fences off biological perturbance by new invaders. But since at the same time food resources are too unpredictable to become a limiting factor, co-evolutionary speciation did not become effective. In this sense, the highly adapted, but low diversity communities of modern beaches do not contradict the time-stability hypothesis.

Modern mud-flat communities

In this context it is particularly tempting to compare sandy beaches with intertidal sand and mud flats, which combine similar physical hazards with a more predictable food resource. In mud flats, particularly those of the tropical

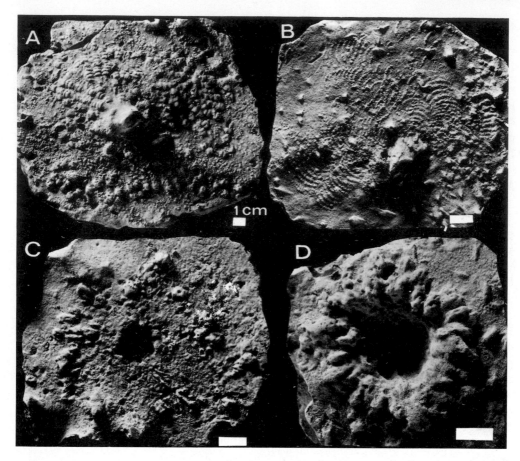

1 cm

zone, the activity of stationary sediment feeders is concentrated on the low-tide phase of the tidal cycle. During this period a rich algal film develops on the exposed sediment surface, which is exploited by a highly diverse fauna of crustaceans, bivalves, gastropods, echinoderms, polychaetes and other worm-like creatures.

In some way the even food distribution in intertidal mud flats resembles the situation on the deep-sea floor. Still we do not find the same meandering or spiral feeding patterns. This is because of the constant threat from birds and other terrestrial predators, which does not allow leisure for systematic foraging. Instead, the grazing animals remain within easy reach of the protective burrow opening or — if not provided with an effective warning system — emerge only partly and only for quick radial excursions from their central burrow. Like meander trails in the deep sea, stellate patterns are therefore produced in this environment by representatives of many different phyla (Fig. 10).

As in the deep sea, however, we find an extreme case of speciation in a group that has the sensory prerequisite for effective co-evolutionary niche partition-ing: the different species of fiddler crabs (*Uca*) feed on the algal film around

Fig. 10. Star-shaped tracks exposed on a tropical tidal flat (El Salvador, Central America) reflect foraging on an algal film around a central sheltering burrow. (Photographs by W. Wetzel from plaster casts made in the field, except for field photograph-fig. H.) A. Bite marks of a small gobiid fish repeatedly jumping out of a water-filled hole in the center. B. Delicate claw marks of a small *Uca* species. C-D. Radial tracks of other brachyuran crabs scraping the surface mud into their burrows. E. Branching probes of large annelid worm. F. Probes of a sipunculid worm. G. Markings left by annelid that intermittently catapults the annulated end of its body out of its tube. H. Siphuncular probes of a tellinid bivalve.

their burrows in very much the same way. But they differ greatly in their fiddling behaviour, which may be very distinctive in species that are hardly distinguishable by morphological features. Most of these forms are also smaller than other brachyuran crabs, and many are found co-existing in the same habitat. Very likely, the speciation of the fiddler crabs was predominantly a co-evolutionary effect in a very mature community, comparable in a broad sense to the speciation of the graphoglyptid burrows in the deep sea.

Ancient mud-flat communities

Owing to the low fossilisation potential of intertidal sediments, beach and mud-flat ichnocoenoses are very rarely preserved in the geologic record. One fossil example, however, is particularly interesting in the context of community evolution.

During a 1972 helicopter excursion in the Hedjaz, made possible by the Saudi-Arabian government and Dr. Glen Brown (U.S. Geological Survey), I was

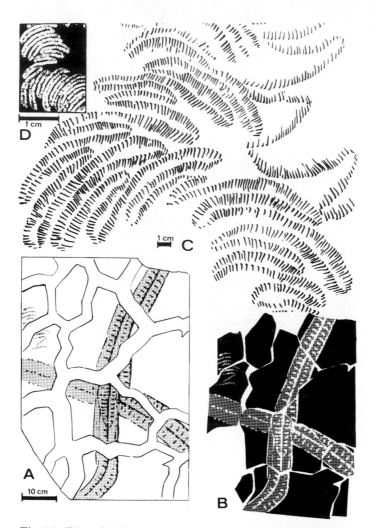

Fig. 11. Trace fossils from an Early Palaeozoic (probably Late Cambrian) tidal flat in Saudi Arabia (U.S. Geol. Survey locality No. 1803). A—B. Desiccation cracks disrupt a trilobite burrow (*Cruziana*) produced while the underlying mud was still soft and covered by sea water. A gastropod grazing track (upper left) is also disrupted (Tübingen catalogue No. 1488/4). C. Another specimen (Tübingen catalogue No. 1488/5) showing all the characteristics of a gastropod grazing track with meandrically arranged radula bites, but giant in comparison to Recent counterparts (D). By this time the mud surface was enough consolidated by drying and by the algal mat not to yield under the weight of the giant gastropod. D. Radular markings of a Recent limpet (*Helcion pellucidum*) for comparison. The width of the track roughly corresponds to the diameter of the animal.

able to study Early Palaeozoic sandstones onlapping over the basement of the Arabian Shield. At one place the sandstones contained trace fossils of presumably intertidal origin (Fig. 11): trilobite burrows (*Cruziana*), indicating a submerged marine stage, which later became disrupted by large desiccation cracks. On the *Cruziana*-bearing surface, as well as in a higher bed, I found meandering radular bite marks of what seems to have been a truly gigantic gastropod. The heavy animal had left no other trail, which indicates that the mud had already become rather firm by the time the gastropod grazed on the exposed and presumably algal covered surface. Possibly this was one of the earliest, large-sized pioneers that left the water to graze in an environment that was otherwise still uninhabited by higher organisms and not yet endangered by terrestrial predators. It is probably no coincidence that the only other trails referable to gastropod-like creatures of similar size (*Climactichnites*) come from the Potsdam Sandstone of New York, which is comparable not only in age (Late Cambrian), but also in its assumed littoral origin, with this Saudi-Arabian example.

General Conclusions

The structure and diversity of soft-bottom communities, as expressed in the trace-fossil record, is suggested to have evolved in the following successive phases.

(1) *Pioneer phase* with emphasis on trophic flexibility and large body size.

(2) *Saturation phase* with emphasis on trophic diversification, behavioural optimization and body size diminution, in order to accommodate more species and increase their evenness within the limits of a given and predictable food resource.

(3) *Co-evolution phase* with emphasis on the behavioural diversification of trophically equivalent species.

This succession, however, can become effective only in environments that provide enough time and evolutionary stability by high physical and food predictability and effective immigration barriers, as, for example, the deep-sea bottom and tropical mud flats. In shelf biotopes, unpredictable perturbations by climatic or tectonic events, and by sudden immigration of alien species, tend to inhibit the maturation of the communities and make them oscillate at lower levels of diversity, evenness and adaptation.

In a stable biotope, the picture may be further obscured by the fact that maturation may be retarded in particular niches, or at higher trophic levels with less predictable food supply (deep-sea scavengers). On the other hand, the beach-sand example shows that high immigration barriers may allow a limited number of pioneer species to gradually optimize without permanent saturation of the biotope and without co-evolutionary diversification, and in spite of seemingly low stability of the environment.

Acknowledgements

This article ("Fossil-Vergesellschaftungen" Nr. 54; Nr. 53 *Neues Jahrb. Geol. Paläontol., Abh.*, in press) grew out of the work of an integrated research group (Sonderforschungsbereich 53, "Palökologie") in Tübingen, supported by the Deutsche Forschungsgemeinschaft. At earlier stages the theme was fruitfully discussed with Tübingen colleagues and with Dr. G. von Wahlert and his group (Stuttgart—Ludwigsburg).

References

Abel, O., 1935. Vorzeitliche Lebensspuren. Fischer, Jena, 644 pp.

Bretsky, P.W., 1969. Evolution of Paleozoic benthic marine invertebrate communities. Palaeogeogr., Palaeoclimatol., Palaeoecol., 6: 45—59.

Bretsky, P.W. and Lorenz, D.M. 1970. Adaptive response to environmental stability: a unifying concept in paleoecology. In: E.L. Yochelson (Editor), Proceedings of the North American Paleontological Convention, I. Allen Press, Lawrence, Kansas, pp. 522—550.

Carpenter, E.J. and Culliney, J.L., 1975. Nitrogen fixation in marine ship-worms. Science, 187: 551—552.

Crimes, T.P., 1974. Colonisation of the early ocean floor. Nature, 248: 328—330.

Delgado, J.F.N., 1910. Etude sur les fossiles des schistes à néréites de San Domingos et des schistes à néréites et à graptolites de Barrancos. Commission Serv. Geol. Portugal, Lisbonne.

Dörjes, J. and Hertweck, G., 1975. Recent biocoenoses and ichnocoenoses in shallow-water marine environments. In: R.W. Frey (Editor), The Study of Trace Fossils, pp. 459—491.

Fuchs, T., 1895. Studien über Fukoiden und Hieroglyphen. Denkschr. Math. Naturwiss. Kl. Akad. Wiss., Vienna.

Hollister, C.D., Hollister, C.D. and Nafe, K.E. 1975. Animal traces on the deep-sea floor. In: R.W. Frey (Editor), The Study of Fossils. Springer Verlag, Heidelberg, pp. 493—510.

Jannasch, H.W. and Wirsen, C.O. 1973. Deep-sea microorganisms: In situ response to nutrient enrichment. Science, 180: 641—643.

Levinton, J.S., 1974. Trophic group and evolution in bivalve molluscs. Palaeontology, 17: 579—585.

Papentin, F. and Röder, H., 1975. Feeding patterns: The evolution of a problem and a problem of evolution. Neues Jahrb. Geol. Paläontol., Monatsh., 1975: 184—191.

Raup, D.M. and Seilacher, A., 1969. Fossil foraging behavior: Computer simulation. Science, 166: 994—995.

Rex, M.A., 1973. Deep-sea species diversity: Decreased gastropod diversity at abyssal depths. Science, 181: 1051—1053.

Röder, H., 1971. Gangsysteme von *Paraonis fulgens* Levinsen 1883 (Polychaeta) in ökologischer, ethologischer und aktuopaläontologischer Sicht. Senckenberg. Marit., 3: 3—51.

Sanders, H.L., 1968. Marine benthic diversity: A comparative study. Am. Nat., 102: 243—281.

Seilacher, A., 1953. Der Brandungssand als Lebensraum in Gegenwart und Vorzeit. Nat. Volk, 83: 263—272.

Seilacher, A., 1954. Die geologische Bedeutung fossiler Lebensspuren. Z. Dtsch. Geol. Ges., 105: 214—227.

Seilacher, A., 1959. Schnecken im Brandungssand. Nat.Volk, 89: 359—366.

Seilacher, A., 1961. Krebse im Brandungssand. Nat. Volk, 91: 257—264.

Seilacher, A., 1967. Vorzeitliche Mäanderspuren. In: W. Hediger (Editor), Die Strassen der Tiere. Viehweg, Braunschweig, pp. 294—306.

Seilacher, A., in press. Pattern analysis of *Paleodictyon* and related trace fossils. In: T.P. Crimes (Editor), Trace Fossils II. Geol. J., Spec. Iss.

Seilacher, A., 1974. Flysch trace fossils: Evolution of behavioural diversity in the deep-sea. Neues Jahrb. Geol. Paläontol., Monatsh., 1974: 233—245.

Slobodkin, L.B. and Sanders, H.L., 1969. On the contribution of environmental predictability to species diversity. In: G.M. Woodwell and H.H. Smith (Editors), Diversity and Stability in Biological Systems. Brookhaven, New York, N.Y., pp. 82—95.

Thayer, C.W., 1974. Environmental and evolutionary stability in bivalve mollusks. Science, 186: 828—830.

Valentine, J.W., 1971. Resource supply and species diversity patterns. *Lethaia*, 4: 51—61.

Valentine, J.W., 1972. Conceptual models of ecosystem evolution. In: T.J.M. Schopf (Editor), Models in Paleobiology. Freeman, San Francisco, Calif., pp. 192—215.

Webby, B.D., 1969. Trace fossils (Pascichnia) from the Silurian of New South Wales, Australia. Paläontol. Z., 43: 81—94.

Webby, B.D., 1970. Late Precambrian trace fossils from New South Wales. Lethaia, 3: 79—109.

CHAPTER 12

THE PATTERN OF DIVERSIFICATION AMONG FISHES

KEITH STEWART THOMSON

Introduction

Any attempt to review and discuss the overall patterns of evolution of so diverse a group as "the fishes" is bound to be incomplete and unsatisfactory. "The fishes" comprises too complex an assemblage of organisms with too imperfect a fossil record to allow us to examine many of the interesting questions current in evolutionary theory. However, there is some usefulness in bringing together an overview of the group if only because the fishes, taken as a unit, give us a very broad sample of phenomena in aquatic systems as a whole.

In this paper I have used the word diversity simply as a term denoting numbers of taxa present. I have drawn data from two main sources — Romer's *Vertebrate Paleontology* (1966) and Harland et al.'s *The Fossil Record* (1967: chapter by Andrews, Gardiner, Miles and Patterson). In addition I have drawn useful information from Obruchev's volume from the Russian treatise *Fundamentals of Paleontology* (1967). The data are, however, not really very satisfactory. Romer gives a very complete listing of genera within each higher taxon but generic counts and some for families are unavailable from Harland's compilations and different higher groupings are used. Romer uses only a 27-fold division of the Phanerozoic, while the less comprehensive data in Harland et al. are given with a 73-fold division. In certain cases the results of using the different systems vary widely; in others (e.g. Fig. 11) the two are remarkably consistent. I have preferred to work with genera throughout because of the larger number of taxa thus available for quantitative analysis.

I have largely followed the classification of higher categories used by Romer (1966) because, being highly conservative, it suffers from a minimum of the problems to be encountered in trying to collate the competing classifications. At the same time, the scheme presented by Moy-Thomas and Miles (1971, fig. 11.1) is the best current summary of the phylogenetic relationships of the various higher groups of fishes. There are no major problems in reconciling the subclasses of fishes recognized by Moy-Thomas and Miles and the subclasses used by Romer.

There are many problems in the taxonomy and classification of fishes. For example, there appears to be a marked turnover in genera at the end of each period because most palaeoichthyologists have been unwilling to admit that any genus could cross a period boundary. Work at the species level with fossil fishes is virtually impossible. In the early part of the record, the characters used to distinguish "species" would probably be sufficient to separate genera of living fishes. In very few cases have large enough samples of taxa within single

faunas been available for quantitative analysis of inter- and intra-specific variation. Where fossil materials are referred to extant genera, caution often deters
an author from making specific diagnoses. Among the works of different
authors there are many inconsistencies in the use of higher taxonomic designations and in the groupings to which the terms are applied. Cladistic analysis
(e.g., Greenwood et al., 1974) has not yet clarified the situation.

I have not followed the practice of treating the various subdivisions of the
geological column as being equal in length throughout the Phanerozoic but
have made greater allowance for the absolute measure of time by treating
only the subdivisions of each period as equal. In most cases I have not included
data from the Pleistocene and Holocene because they are available on a quantitatively different scale from those for the rest of the record. In most cases
throughout this paper, results that should strictly be plotted as histograms have
been given a series of points (middle of the relevant time intervals) joined by
lines to form curves. These are much easier to follow and compare.

It would be highly desirable to make separate analyses for the records of
freshwater and marine fishes. However, although Pitrat (1973) has made a
limited study along these lines, for the vast majority of fossil occurrences
detailed environmental information is not readily available. For very many
families it is virtually impossible to assign an exclusively fresh-water or marine
status.

Perhaps to make any further listing of the problems of the fossil record of
fishes would deter the reader entirely. Therefore, apart from noting that none
of the familiar problems encountered in treating the fossil record has failed to
turn up in this study, I will add no further *caveat*.

There have, of course, been many treatments in the literature of changing
diversity, differential rates of origination and extinction, and questions of
periodicity in the fossil record. There is no need to review them here, although
we must note the special contributions of Newell (e.g. 1952, 1967) and Simpson (particularly *Tempo and Mode in Evolution*, 1944 and *The Major Features
of Evolution*, 1953). There have been few studies that have dealt in detail with
the diversifications of fishes as a whole. Gregory (1955) included some data in
a general review of the vertebrates. Flessa and Imbrie (1973) made a major set
of analyses in which fossil fishes contributed in a small way to the identification of "diversity associations". That paper also gives a useful review of the
ideas and literature relevant to the study as does Cisne's (1974) study of arthropod evolution.

Here I will treat three subjects separately: a presentation of the raw data on
overall diversity of fishes through time; rates of origination and extinction and
the course of adaptive radiation among higher groups of fishes. The results are
then discussed in a final section.

Overall Diversity

We will describe first the simple numerical data on taxonomic diversity
through the Phanerozoic. The basic data are summarized in Figs. 1 and 4. Here

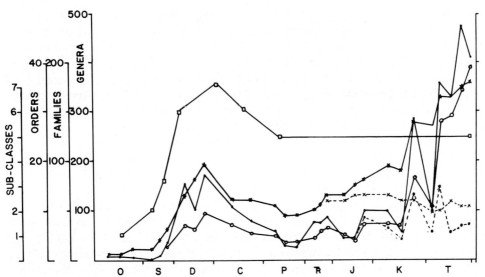

Fig. 1. Numbers of genera (solid circles), families (open circles), orders (crosses) and sub-classes (squares) of fishes from Ordovician to Pliocene (data from Romer, 1966). Dotted lines show counts exclusive of teleosts.

are shown the numbers of genera, families, orders and subclasses of "fishes" at intervals through the Phanerozoic; the numbers of first appearances and last appearances of families and orders; and rates of origination and extinction of families.

The fishes first appear in the Ordovician, in sediments now universally con-

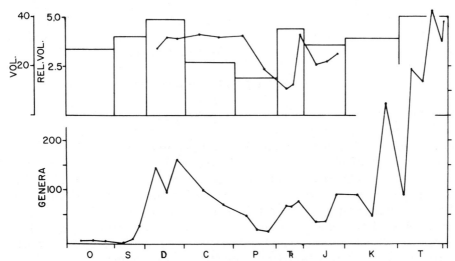

Fig. 2. Comparison of the generic diversity of fishes through time with two estimates of the volume of the geological record. The histograms show "relative sedimentary volume" from the data of Garrells and Mackenzie (1971), the superimposed curve is from the data of Ronov (1959), showing volume in millions of cubic kilometers.

sidered to be marine in origin (Denison, 1956; Spjeldnaes, 1967; and Thomson, 1971 for review). During the Ordovician only Agnatha — the jawless fishes — have been found, and only at low diversity. This low diversity continues until the end of the Silurian when, rather abruptly, larger numbers of taxa, including the gnathostome subclasses Acanthodii, Osteichthyes, and Placodermi all appear, a large proportion being fishes of freshwater origin. This corresponds to the first extensive occurrences of freshwater sediments. There continues a general rise in numbers of all taxa through to the end of the Devonian, but with a relatively low count of both genera and families for the Middle Devonian. In the remainder of the Palaeozoic there is a general decline in the numbers of genera, families and orders present. Counts of genera show a small peak in the Triassic and a general increase in the number of all categories begins at the end of the Triassic, continuing (accelerating) to the present. The number of "fish" subclasses reaches its peak in the Early Mississipian and, with the extinction of the Placodermi and Acanthodii, reaches its present level in the Late Palaeozoic. As Fig. 1 shows, the great increase in numbers of taxa from the Middle Cretaceous onwards arises almost wholly from the adaptive radiation of the Teleostei. If the teleost counts are subtracted, the resulting subtotals (particularly at the ordinal level) show a relatively constant diversity from the Early Carboniferous to the Recent.

Depending on the subdivision of the geological time frame used the history of first and last appearances of families differs somewhat. The general impression, however, is that first appearances of families are distributed rather regularly throughout the history of the fishes and there are no major peaks and valleys (see Cutbill and Funnell, 1967). Last appearances rise at the beginning and end of the Devonian but otherwise also show an even amplitude. Last appearances and thus the turnover of families are high at the end of each period but that is almost certainly an artifact of palaeontologists' usage. The data on first and last appearances of orders (not shown) is equally random after the peak in the Devonian; the only major peak comes in the Cretaceous.

Major factors affecting the nature of the fossil record of fishes

Newell (1967), among others, has documented the following pattern of overall taxonomic diversity of organisms in the Phanerozoic: an early rise in diversity in the Palaeozoic, followed by a decline to the end of the Permian; there is a "crash" in numbers of most taxa in the Permo-Triassic, followed by a rise in diversity to the Late Palaeozoic levels and then a pronounced climb from the end of the Mesozoic to the Recent, when diversity levels two or three times greater than the Palaeozoic peak are reached. Valentine (1969, 1973; Valentine and Moores, 1970, 1972) has proposed a model to explain this changing pattern in terms of global tectonics, changing sea-levels and changing environmental conditions. He shows that since the first major diversifications of the Palaeozoic, there has been a gradual process of "replacement" of high taxonomic categories by lower categories. Raup (1972) has presented an alternative model according to which diversity levels have remained constant since the

Late Palaeozoic and he explains the fluctuating numerical counts in terms of artifacts caused by the nature of the geological record and sampling problems.

Given the extremely small samples we have, our raw numerical data seem to fit the pattern suggested by Newell and Valentine, but it is clear that there is no Permo-Triassic crash in numbers and the higher taxa do not show a general decline after the Carboniferous.

It is necessary, therefore, to interpret the raw data in terms of the nature of the geological record and sampling problems in order to derive a more accurate view of the course of diversity of fishes through the Phanerozoic. The problems of the nature of the geological record itself, particularly the different rates of accumulation of sediment at different time periods, have been discussed elsewhere (see review in Raup, 1972, who shows that the effect of variations in the record should be most apparent at the lower taxonomic levels), but estimates of the volumes of deposits included in the geological periods are surprisingly hard to come by. Ronov (1959) has provided some very useful data for the interval from the Devonian to the Jurassic and other authors have given less detailed estimates for the whole Phanerozoic. Garrells and Mackenzie (1971) have combined the various data and their estimates are given in Fig. 2 with Ronov's data and a summary of generic diversity of fishes through time. It will immediately be obvious that there is a general coincidence of the data on diversity with those on "quantity" in the geological record. The decline in taxonomic diversity (families and genera) from the Carboniferous to the Triassic seems to coincide with smaller volumes of deposit and also lower sea levels (see Hallam, 1971). The Triassic peak in generic diversity coincides with a small peak in volume in the geological column.

We can take the analysis one step further by calculating the relationship between the number of taxa and the relative volume of deposit at each period (Fig. 3). This procedure has the effect, first of all, of smoothing out the curves because of the smaller numbers of data points available. The curve of genera per unit volume still shows a Palaeozoic peak but it is shifted from the Devonian to the Carboniferous. Similarly, the peak of familial diversity is moved from the Devonian to the Carboniferous. The curve of orders is little modified except that its peak is moved from the Carboniferous to the Permian. Allowance for the volumes of geological deposits removes both the Triassic sub-peak in the pooled data and also (except in the generic counts) largely smooths out the Permo-Triassic decline in diversity. The shift to the right of the various peaks may itself be a new artifact of the method, but it is significant that the procedure does not eliminate the Palaeozoic peaks in diversity at all ranks. Similarly, it is not possible thereby to eliminate the increase in the Tertiary shown by all ranks except the subclasses that is in large part attributable to the adaptive radiation of one group of fishes — the Teleostei. The adaptive radiation of the Teleostei occurs within what are at present considered to be the lower taxonomic ranks. In the modified data, the diversity of non-teleost fishes, instead of remaining steady during the Tertiary, would show a marked decline.

These general observations on the correlations between apparent numerical

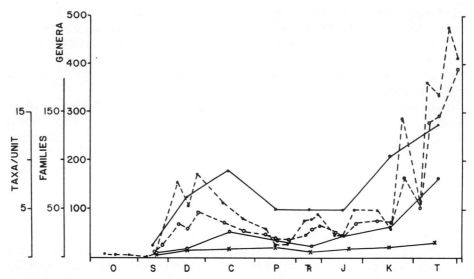

Fig. 3. Plots comparing the raw numerical diversity of genera (solid circles) and families (open circles) of fishes with the same respective values divided by the factors of relative sedimentary mass (Garrells and Mackenzie, 1971) shown in Fig. 2: dashed lines, genera (solid circles), families (open circles) and orders (crosses).

diversity and the nature of the geological record are amplified by the separate consideration of certain groups of fishes. The record of generic and familial diversity for the actinopterygian grades Chondrostei and Holostei, when compared with Ronov's data (1959) on the accumulation of continental deposits, shows a good correlation between a Triassic peak in diversity with a peak in Triassic continental deposits. The data on the Chondrichthyes (shark-like fishes, rays and holocephalans), when compared with Ronov's data on marine clastics and marine carbonate deposits again show a good general correlation especially between a Late Carboniferous decline, a Permo-Triassic low and a small peak in the later Triassic.

Rates of Origination and Extinction

In Fig. 4, I present the two rates of taxonomic evolution: new families appearing per million years and taxa becoming extinct per million years. The data are subject to the usual vagaries of such estimates (see, for example, Simpson, 1953). It will be seen that origination rates are rather constant with the exception of a major peak in the Triassic and large surges at the end of the Cretaceous and in the Early Tertiary. Extinctions are concentrated in Devonian, Permian, Early Jurassic and Late Cretaceous peaks, with some smaller peaks and one exceptionally low point in the mid-Permian. While the data show a Late Devonian mass extinction, this extinction is not greater than other major extinctions. The Permo-Triassic phase of high extinction rates that has been

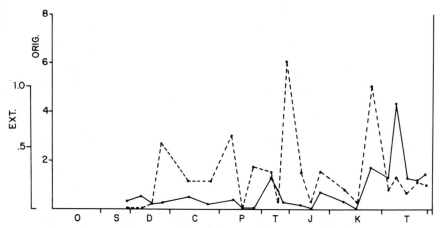

Fig. 4. Rates of origination (new families of fishes per million years, solid line) and rate of extinction (last appearance of families per million years, dashed line).

seen in many other groups (e.g. Newell, 1967) and has even been claimed for fishes is not apparent, but the third major occurrence of high extinctions in most organisms, the Late Cretaceous extinction, is easily recognized.

In Figs. 5 and 6, the raw data on extinction and origination are modified somewhat, being plotted as the numbers of taxa originating as a proportion of the number of taxa present in the immediately preceding time interval and the taxa becoming extinct as a proportion of those present. This has the effect of partially screening out the errors due to differences in the "quantity" of the fossil record for different time intervals and gives us a measure of rates that is more meaningful biologically than a record of taxa appearing or disappearing as a proportion of the total numbers ever existing or of maximum numbers known for any given time interval. In these modified plots, the rates appear more uniform through time, with the peaks of activity more even in height. The rate of origination, plotted this way, is highest in the Devonian and then is high again in the Late Cretaceous; in between, peaks of origination occur rather regularly (see later). The changed measure of extinction rates eliminates the very pronounced Late Triassic peak and, indeed, no single time interval stands out as a major time of extinction.

A plot of relative turnover of taxa (not shown; see the compilations of Cutbill and Funnell, 1967, fig. 13) shows that origination rates and extinction rates combine to give periods of maximum taxonomic turnover at the Devonian—Carboniferous and Permian—Triassic boundaries, while compilations of net profit and loss (from the same source) show even fluctuations throughout the Palaeozoic and major peaks of profit in Late Jurassic, mid-Cretaceous, and Early Tertiary.

The data on rates of origination and extinction of fishes, taken overall, do not therefore fit the notion of three major mass extinctions, in the Late Devonian, Permian—Triassic and Late Cretaceous, that seems to apply to most inver-

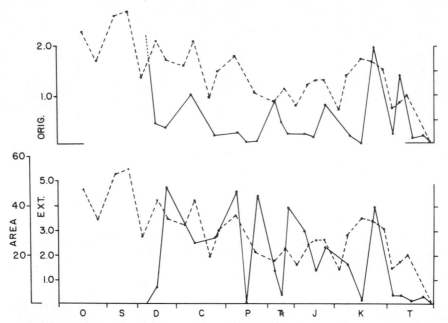

Fig. 5. Comparisons of the rates of origination of fishes (new families at each interval divided by number of families present in preceding interval), rate of extinction (families missing divided by families present in preceding interval) and Hallam's (1971) data on the area of marine transgressions (dotted lines). Note that the same dotted line is superimposed on each plot but the scale is given only once.

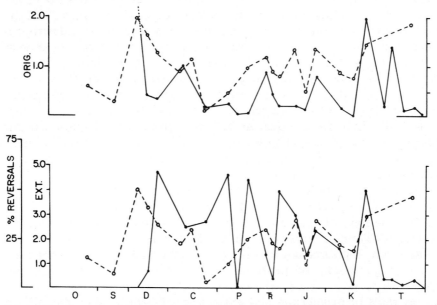

Fig. 6. Similar comparison as in Fig. 5 but including McElhinny's (1971) data on the rate of geomagnetic reversal — dashed line (scale, given only once, is percentage of mixed polarity measurements at each time interval).

tebrates. Rather, they suggest a more complicated and more evenly distributed pattern of changing rates of origination and extinction through time.

The data may tend to show an even pattern in part because both marine and fresh-water fishes are included. Pitrat (1973) concluded, for example, that the Permo-Triassic major extinction is shown only in the record of marine fishes, and not in fresh-water fishes. Schaeffer, however, stated that the Permo-Triassic boundary "has little significance for the fishes in relation to major extinctions or origination bursts" (1973, p. 493).

While the nature of the raw numerical data on taxonomic diversity may be shown to be affected by the nature of the geological column, there is obviously a possibility of a correlation between evolutionary rates and physical factors in the history of the earth. In Fig. 5, I compare Hallam's data on the area of marine transgressions and regressions in the Phanerozoic with the data on rates of origination and extinction of families of fishes in the Phanerozoic. Another factor that has intrigued many students of diversity is the rate of change in polarity of the earth's magnetic field, possibly in part because it is a more readily quantifiable subject than "climatic changes" or "provinciality". The best compilation of data seems to be that of McElhinny (1971). Crain (1971) used McElhinny's data to show a good positive correlation between the rate of reversal and general biological extinction (see also J.F. Simpson, 1966, and Hays, 1971). In Fig. 6, the rate of geomagnetic reversal is compared with the rates of origination and extinction for fishes.

A strict comparison using correlation coefficients shows that none of the comparisons in Figs. 5 and 6 has statistical validity, although, interestingly, there is a weak negative ($P = 0.1$) correlation between the area of transgression and the rate of geomagnetic reversal. The lack of correlation is interesting because there is the superficial *appearance* of a positive correlation between originations and reversals and perhaps also originations and area. The correlation between extinction and area *appears* negative. It could be argued that the precision of the data is not sufficient to allow strict comparisons, but that is a weak position. The best case that can be made for the data seems to be as follows: if we assume a normally low origination rate, then all significant *increases* in the rate seem to fall near increases in area of marine transgression and high rates of magnetic reversals (but this does not apply completely vice versa). Similarly, low points in the extinction rate seem to come *near* high points in the area of transgression. No amount of special pleading will produce a consistent relationship between extinctions of fishes and magnetic reversals.

The Course of Adaptive Radiation

From a treatment of the pooled data of all fishes, we may continue to an examination of the contributions of the various major groups to overall diversity. In Fig. 7, the data are presented, broken down in terms of the numbers of genera of Agnatha, Acanthodii, Placodermi, Sarcopterygii, Chondrostei, Holostei, Teleostei and Chondrichthyes. From this we can see that the Devo-

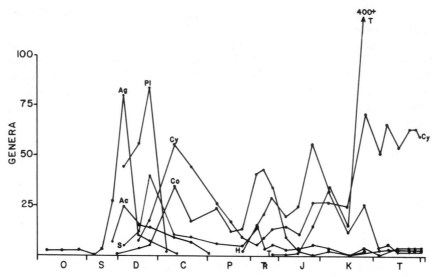

Fig. 7. Counts of generic diversity for separate groups of fishes through time. *Ac* = Acanthodii; *Ag* = Agnatha; *Cy* = Chondrichthyes; *Co* = Chondrostei; *H* = Holostei; *Pl* = Placodermi; *S* = Sarcopterygii and *T* = Teleostei.

nian peaks in diversity are due largely to the agnathan and placoderm fishes, that the Carboniferous to mid-Cretaceous diversity is varyingly made up of the chondrichthyan, chondrostean and holostean fishes, and that at the end of the Cretaceous only the teleosts and chondrichthyans are at all important, the former outnumbering the latter four- or five-fold. Fig. 7 also shows that the rise and fall in numbers of genera per higher group is rather precipitous, whether the group has a single peak of diversity, as in the Agnatha, Placodermi and Acanthodii (and Dipnoi and Rhipidistia among the lobe-finned fishes), or whether, as in the case of Chondrostei, Holostei and Chondrichthyes, there are several clear separate peaks in diversity at widely separated points in time. No group maintains a high numerical diversity for a long period of time. Also, the major and subsidiary peaks, while of different magnitude, seem to appear in a rather regular order through the time scale and peaks may be coincident in time.

Newell (1952) looked at possible periodicities in invertebrate evolution and found peaks in generic diversity in the Devonian, Early Carboniferous, Permian, Triassic, Jurassic, Cretaceous and Tertiary that seem to correspond (given the lack of fine detail in the time scale he used) to the major peaks of generic diversity among fishes. Simpson (1952) also noted that "there is a well-marked periodicity in vertebrate evolution. . . (the) successive peaks. . . of high. . . proliferation of new groups, in one class after another or sometimes within one class at different times. . . are rather evenly distributed in geological time" (1952, p. 370). However, he also observed (1952, p. 363) that "for the Vertebrata as a whole, however, there is certainly no tendency for their times of most rapid diversification to coincide". This last conclusion seems not to be

supported by the data presented in here. These results are interesting enough
to be worth a fuller investigation.

The most difficult problem is to decide whether or not the observed fossil
record is an accurate or even useful reflection of events in the adaptive radiations of the separate groups of fishes that actually occurred through time. It is
probably significant that for each of the major points in the geological record
where groups show a sharp decline in diversity, there are marked changes of
facies that could at least contribute to that decline.

Similarly, the coincident peaks of abundance of fish groups seem to reflect
the peak development of particular environmental conditions. The Late Devonian record of fishes is largely the record of a world-wide "Old Red Sandstone"
that can be seen from Scotland to Antarctica. The assemblages associated with
Carboniferous coals, the Triassic fresh-water faunas and Jurassic and Cretaceous
(especially chalk) marine faunas all obviously reflect times of optimum conditions for the diversification of particular groups. The Placodermi are an example to note here. When they first appear in our fossil record, there are (for the
Lower Devonian) 43 genera arranged in 9 families; 3 orders are present in the
Gedinnian, 4 more appear in the Siegenian. Even given taxonomic artifacts it
seems obvious that the Placodermi had been radiating first in the Silurian under
conditions of which we know nothing. However, it would take a complete revolution of the fossil record, and enormous discoveries of new taxa, to change the
general *shape* of the diversification of placoderms. It is unlikely that new discoveries will, in fact, do more than modify the leading and trailing "tails" on
the curves. The Holostei present a different picture. The first order to appear is
coincident with the last radiation of Chondrostei (Triassic). Next, the Jurassic
peak is actually made up of the radiation of three groups — Amiiformes, Pholidophoriformes and Pycnodontiformes. Did these Jurassic radiations stem from
known fish groups, or did they arise from Triassic radiations of which we have
no record at all? Here we must turn to the evidence of phylogenetic study
(which is, of course, based on available forms). According to Patterson (1974)
the best evidence allows the proposition that the holostean lines arose "para-
phyletically" from known chondrostean groups. The postulation of missing
radiations is therefore unnecessary, although this is not the same as saying that
they did not exist!

We can never be sure about radiations of groups for which an extensive record
is absent, for it is easy to put any "aberrant" forms into "wastebasket" groups
without assessing their significance. Nonetheless, I am inclined to the view that
the fossil record is a reasonably representative, if very small, sample of total
fish diversity. I suggest that the major bursts of radiations, assigned to different
taxonomic groups in Fig. 7, are more the *products* than the artifacts of world-
wide environmental conditions through time.

Accepting the above premise, we may take an analysis of the course of diversification among the fishes one step further by means of a simple modification
of the data. In order to compare the separate groups independently of the absolute numbers of taxa preserved, and in order to minimize some of the other imperfections of the fossil record, we may plot the raw data as follows. For each

time interval the number of taxa present is plotted as a percentage of the maximum ever recorded for a similar time interval (not, it must be emphasized, of the total number ever recorded). Such plots, expressed as curves, show us the course of diversification of each group in a common frame. Whether or not some included taxa are missing from the record should not affect the overall shape of the curve, particularly the position in time and shape of the peak in diversification, because it comes when the fossils are most common. The tails of the curve are, of course, more suspect, but probably less subject to error than plots of absolute numbers of taxa.

Comparison of Figs. 7 and 8 will show that this modification of the data does not alter the timing and sequence of diversification in the eight major groups of fishes. The Agnatha appear first in the fossil record, but their maximum diversification occurs in the Lower Devonian, at the same time as the maximum diversification of the Acanthodii whose first appearance in the record is only Late Silurian. The next peaks of diversification occur in the Upper Devonian and belong to the Placodermi and Sarcopterygii, both of which appear first in the Lower Devonian. Following this there is a gap during which the chondrostean and chondrichthyan fishes show important diversifications. Then in the Early Triassic we find the peak appearance of the Chondrostei, a major peak of the Holostei and of lobe-finned fishes (the coelacanths). The Holostei reach their maximum diversification in the Late Jurassic. The Chondrichthyes reach their maximum in the Late Cretaceous. Finally the teleosts peak in the Neogene.

Such an analysis draws attention to a very interesting fact. In the adaptive radiations of actinopterygians, the teleosts occupy a position rather like the mammals among tetrapods. That is, they appear in the Triassic but the major Mesozoic radiations of fishes are chondrostean and holostean (as the major tetrapod radiations are reptilian). Further, rather than teleosts and holosteans completely excluding each other, the first radiations of teleosts are coincident

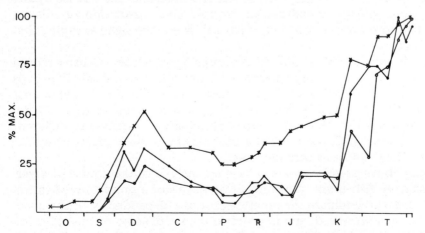

Fig. 8. Plots of percentage maximum diversity of genera for groups shown in Fig. 7.

with those of the holosteans. Similarly, the first radiation of holosteans, rather than excluding the chondrosteans, is coincident with the last chondrostean radiation.

From the figures we see that the overall longevity of groups (in the fossil record at least) may be widely different. The Agnatha and Teleostei survived at low generic diversity in the fossil record for a long time before their major bursts of diversification. The Chondrichthyes and Holostei diversified soon after their appearances in the record.

We can carry the analysis a step further by attempting to break down the curves of those groups showing multiple peaks, such as the Chondrostei, Holostei, and Teleostei, into separate group diversity curves. For the Holostei this is easy and, as shown in Fig. 9, a simple curve of diversification with a single peak is given by the data on each of the following recognized holostean constituents: Semionotiformes, Amiiformes and Pholidophoriformes. The Pycnodontiformes retain a curve with a double peak. Although the data are very poor, similar constituent curves can be drawn for the Teleostei (Fig. 10). There is, however, no current taxonomic division of the Chondrostei or Chondrichthyes that will give us such a separation.

Several interesting points emerge from an examination of Figs. 8—10: (1) We can see clearly the succession of different higher groups of fishes through time. (2) The breakdown of each higher group into subgroups (roughly at the ordinal level), each of which shows only a single peak of diversification, may reflect a situation of major biological importance. (3) The peaks of the separate curves tend to be remarkably similar to each other. Finally, (4) there is some indication that peaks of diversity occur in a regular manner through the Phanerozoic, rather than randomly. These points will be discussed in some detail later.

First we need to test the accuracy of the "curves" and "peaks" produced

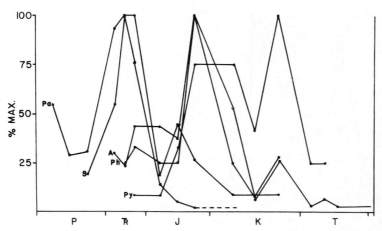

Fig. 9. Detailed breakdown of part of the data for Chondrostei and the four major groups of the Holostei (cf. Fig. 8). A = Amiiformes; Pa = Palaeonisciformes; Ph = Pholidophoriformes; Py = Pycnodontiformes; S = Semionotiformes.

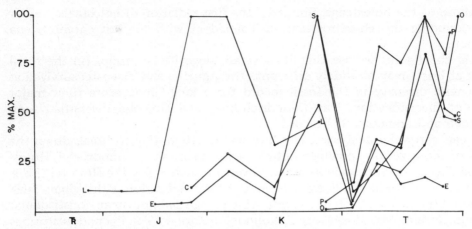

Fig. 10. Detailed breakdown of the generic data for six subdivisions of the Teleostei: orders C = Clupeiformes; E = Elopiformes; L = Leptolepiformes; O = Osteoglossiformes; P = Perciformes; S = Salmoniformes.

by this simple method. It could be argued that the generic counts used here, based on a three-fold division of each time period, are insufficient for an accurate analysis. For most groups, the alternative data in Harland et al. (1967), although broken down by stages, are insufficient for analysis (see Introduction). However, for one group, the Palaeonisciformes, there are sufficient genera *and* families for us to compare two sets of data. This is done in Fig. 11. The two curves are essentially identical. Although there is a lot more information in the data from Harland, there is also a lot more noise. For example, in the first peak, the plot of Romer's generic data smooths out an hiatus in the

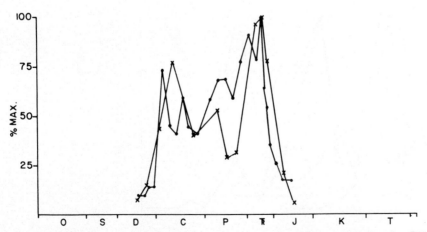

Fig. 11. Comparison of percentage maximum diversity curves for the Palaeonisciformes. Crosses mark the plot of generic data from Romer (1966) and solid circles the familial data from Harland et al. (1967). Note the different subdivisions of the time axis.

Harland data which is due to a poor record in the mid-Mississippian. The positions of the peaks of diversity are essentially the same in both plots and the slopes of the ascending and descending curves are the same. This shows that the data and the method yield a satisfactory level of information. The example of the Palaeonisciformes is also useful for another reason. While it is easy to divide the multiple-peak curve for the Holostei into a series of single curves, there is no taxonomic division of the Palaeonisciformes that will accomplish this, although it is quite clear (Fig. 12) that two, and possibly three, separate stages of diversification (Carboniferous, Permian and Triassic) occur in the history of the group. This shows that the diversifications are not always due to adaptive radiations that are exactly mirrored in the taxonomy and the diversity curves produced are therefore not mere artifacts of taxonomy. Further, that the shape of the group diversity curves is not seriously affected by the "sampling" problem of the fossil record is also shown by the fact that curves for genera and families are essentially identical for groups that give a simple group diversity curve. If, for example, low generic counts were simply due to patchiness in the fossil record we would expect family and generic curves to be different.

A more important possibility (discussed in part before) is that the pattern of diversity for separate groups seen in the fossil record, whether in the raw data or the percentage maximum diversity curves, might be artifacts of the volume of deposits in the geological record. In Fig. 13 I compare two sets of percentage maximum diversity curves for Chondrichthyes, Chondrostei and Holostei. One curve is plotted from the raw data, the other from the generic counts corrected for the volume of geological deposits (using Ronov's detailed breakdown for the period Devonian—Jurassic, 1959). For the Chondrostei, adjusting the data diminishes the Carboniferous peak in diversity and almost eliminates the Permian sub-peak. The record for the Holostei is completely unchanged. For

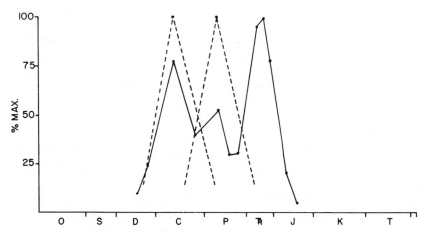

Fig. 12. Percentage maximum diversity curve for genera of Palaeonisciformes (solid line) and a superimposed hypothetical set of curves indicating three separate adaptive radiations within the group (dashed line).

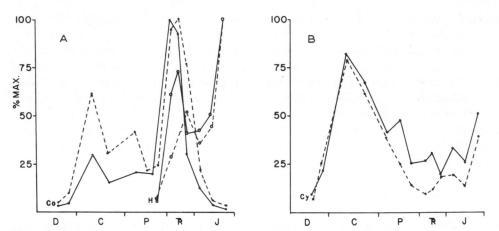

Fig. 13. Comparison of percentage maximum diversity curves plotted for genera of fishes (dashed lines) and for the same data modified by dividing the number of genera present at each time interval by the total volume of deposits (km³) estimated by Ronov (1959) for that interval (solid lines). *Co* = Chondrostei; *Cy* = Chondrichthyes; *H* = Holostei. Note that such comparisons are only possible from Devonian to Jurassic.

the Chondrichthyes a new minor peak appears in the Permian. In every case, the position of the major peak and the slopes of the ascending and descending curves around the main peak are unaltered. One might add, further, that there are no variations in the volume of sedimentary deposits that can account for the sudden disappearance of the Agnatha, Placodermi, and Holostei, or the equally sudden rises of the Agnatha, Placodermi, Sarcopterygii, and Teleostei. Thus it seems likely that although variations in the volume of sedimentary deposits through time may cause minor artifacts in the diversity of higher groups of fishes, the method of plotting percentage maximum diversity to chart the course of diversification is basically a reliable technique with which to compare groups.

For each separate group of fishes we can plot a simple curve of diversification through time, with a single peak and roughly symmetrically steep ascending and descending slopes. Each seems to represent a single adaptive radiation, but the possibility must also be borne in mind that (because we start with data sorted into taxonomic categories) any curve might reflect more than one exactly synchronous radation within a single group. For example, the diversity curves for all agnathan sub-groups are the same; therefore there is a simple curve for agnathans as a whole. At approximately the ordinal level of traditional classifications, however, the conclusion is probably safe that one curve equals one adaptive radiation.

The properties of the group diversity curves are remarkable. Not only do they tend to be synchronous, but they also tend to have a very similar shape. In Fig. 14A, curves for various groups of fishes have been superimposed. It will be seen that the similarity between the curves extends between approximately the 25% points on the ascending and descending arms. Between these points

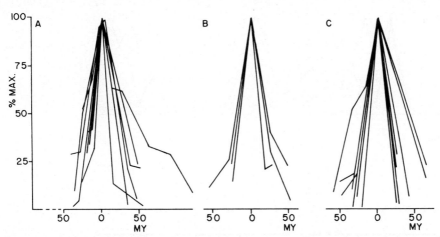

Fig. 14. Group diversity curves of (A) fishes, (B) amphibians and (C) reptiles (cf. Fig. 15) superimposed to show common features.

the two slopes are essentially linear, suggesting that growth in numbers of taxa is more exponential than logistic. Of course, in certain cases there are only few data points available for each curve, but a check with the data in Fig. 11 will show that where there are more data points, the picture is not greatly changed. Given the imperfections of the fossil record, the major point to be made concerning these curves is that their span is extremely short (roughly 50 million years from 50% increasing to 50% decreasing diversity) and that they are all similar. (In two cases, a square-topped curve can be plotted: Pycnodontiformes (Holostei), Fig. 9; and Leptolepiformes (Teleostei), Fig. 10. In both cases the sample size is very small, a *total* of 18 and 15 genera, respectively. They are also contemporaneous. I suggest that these are artifacts.)

The common shape of these curves is extremely interesting. It shows that one can identify, in retrospect, a point in the history of each group at which diversity rises suddenly. As this point represents a very different number of taxa for different groups the coincidence of the curves is striking. In addition, while it is intuitively obvious that the increase in diversification of a group might be very rapid once it has acquired a "toe-hold" through some particular set of adaptations and therefore reached the "point of critical mass" allowing such diversification, it is less obvious why the group should so quickly enter an equally rapid (symmetrical) decline; see discussion below.

The similarity between group diversity curves in fishes is sufficiently interesting that two other major groups of lower vertebrates — Amphibia and Reptilia — were tested to see if a similar situation prevailed. As shown in Fig. 15, a completely comparable set of group diversity curves may be plotted at the ordinal level. The mean span of the amphibian curves is possibly less than that for fishes, while those for the Reptilia are essentially the same as for fishes (Figs. 14B,C). The plots of percentage maximum diversity thus indicate that the time course and pattern of diversification is essentially constant within, and pos-

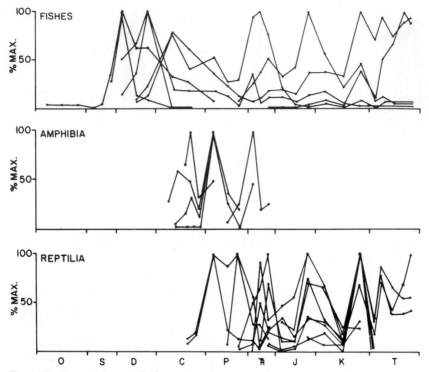

Fig. 15. Comparison of the sets of percentage maximum diversity curves for fishes and orders of amphibians and reptiles. For simplicity the individual curves are not labelled.

sibly also characteristic of, major groups of vertebrates. If this is proved correct by further analyses, it must demonstrate a fundamental property of the adaptations of such groups.

The group diversity curves for the Amphibia and Reptilia also show a similar property to those of fishes in the coincidence of peaks at a small number of apparently regularly spaced points in time. There are only eight such points for the combined lower tetrapods; all but two of these are points of peak diversification for fishes. For the whole of the lower vertebrates there are only 12 points in time at which peak diversities occur (more than 50 peaks).

We have already noted that the distribution of points of peak diversity for fishes in the time scale appears to be regular. The data shown in Fig. 15 indicate that this regularity applies over the whole of the lower vertebrates. After the initial bursts of diversification, the peaks occur regularly at an average spacing of 62 m.y. The record of the Amphibia is rather short and both the Amphibia and Reptilia show a similar pattern of increase in spacing between peaks in diversity. In all three cases, the early spacing has an interval of roughly 30 m.y., half that of the final spacing.

In Fig. 16 a set of group diversity curves based on generic counts is given for ten major groups of marine invertebrates whose occurrence overlaps with that

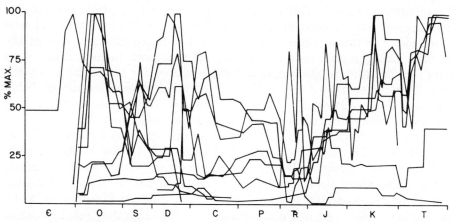

Fig. 16. Percentage maximum diversity curves plotted for generic counts within the following marine invertebrate groups: ammonites, bivalves, brachiopods, corals, cystoids, echinoids, graptolites, nautiloids, ostracods and trilobites. Data generously supplied by Stephen Jay Gould.

of fishes and which were major components of shallow water ecosystems. Similar results to those for the fishes obtained. The curves for each group show very sharply rising and falling slopes and the peaks tend to be coincident around a relatively small number of points in time. These points, particularly in the Devonian, Early Carboniferous, Triassic, Jurassic and Cretaceous, are broadly but not precisely coincident with the similar positions of peaks for fishes.

Discussion

During the Ordovician and the first half of the Silurian, fishes existed at a relatively low diversity, measured at the generic and family level, but the major groups of gnathostome fishes, Acanthodii, Actinopterygii, Sarcopterygii, and Placodermi, were probably differentiated before or at the end of the Silurian. This low diversity at the family level probably reflects the intense competition that the fishes were encountering from invertebrate groups within relatively confined adaptive zones. At a certain critical point in the (mid?) Silurian, differentiation of the major adaptive morphological complexes that we recognize as the bases of major taxonomic groupings produced an effective increase in the available adaptive zones. Then, increasing competitive superiority over certain benthic invertebrates resulted in the sudden rise in diversity at all levels that began in the Late Silurian and continued to a peak in the Carboniferous.

That the initial bursts of diversification involved very different sorts of fishes — both agnathan and gnathostome — in parallel suggests that the availability of the particle/detritus feeding niches, leading to the later evolution of scavenging feeding and active predation, was the product not merely of factors intrinsic to the separate groups of fishes (their morphological adaptations) but also of extrinsic factors experienced in common. Thereafter, the emergence of new sorts of organisms in the aquatic ecosystem caused major changes, one of the principal ones being a complete re-ordering of carnivore niches: these same fishes being, by the Middle Devonian, a food resource for other fishes. By the end of the Upper Devonian there had occurred a full diversification of fishes living both as active predators in open water (fast and slow swimming) and bottom-living forms eating a broad range of hard-shelled prey or scavenging.

I suggest that the history of the fishes can best be explained by the following model. At the end of the Devonian or the beginning of the Carboniferous, fishes had come to occupy all available carnivorous niches in the aquatic environment. Thereafter, fishes remained at (declined to) an *apparent* equilibrium in diversity demonstrated as an effective upper limit on diversity (at whatever rank) which they could not/did not exceed. Although they can be seen to have undergone significant morphological "advancement" (for example from the chondrostean to holostean grade among Actinopterygii, or the "cladodont" to "hybodont" grade in Chondrichthyes), these improvements turned out only to be sufficient to allow the fishes to keep their heads above water (to use a metaphor perhaps more appropriate to the Sarcopterygii) in competition in evolving aquatic ecosystems. Throughout the turnover of fish groups from the Carboniferous to the Early Jurassic, there was no change in adaptation that allowed the exploitation of significantly greater dimensions of the aquatic environment until the diversification first of the holosteans and then of the teleosts, the adaptations of the latter permitting entry into a whole new range of adaptive zones, including increased predation on small food items through suction feeding and improved locomotion.

This history of changing diversity among fishes occurred in the context of changing world conditions, as outlined in the model of Phanerozoic diversity of Valentine (see earlier). The development of a first great peak of diversity among fishes occurs near the end of the general organismic diversification of the Early Palaeozoic. The more stable diversity of the Late Palaeozoic is maintained through the assembly of Pangea. With the break-up of Gondwana and the opening of the Atlantic came new diversifications of fishes and the beginning of a steady increase in diversity, the major contributors to which are the teleosts. It is this last group that is able to exploit the possibilities of a major increase in the number of different faunal provinces available.

Schopf et al. (1975) have suggested that the high levels of diversification within such groups as mammals, ammonites and teleosts are artifacts of the apparent greater morphological complexity of these groups. In addition, Raup (1972) contends that the Mesozoic—Tertiary increase in diversity of organisms is due to biases in sampling and that diversity levels have remained in equilibrium since the Late Palaeozoic. The data presented here suggest that the Mesozoic—Tertiary increase is real for fishes (*pace* Schopf et al., 1975). It is interesting that no measure of the diversity of fishes indicates a major "crash" at the Permo-Triassic boundary.

I suggest that the results presented here concerning the common course of diversification within separate groups of fishes may be explained by analogy with the evolution of *species* within a single habitat. Even though we can only deal with the generic level, in each diversification the individual taxa become more and more specialized, giving the effect of species packing (see, for example, MacArthur, 1972). Then any shift in the conditions defining niches would cause rapid extinctions of taxa. If this is correct, the coincident shapes of the group diversity curves reflect the common nature of the process of adaptation of any group in an unstable environment. Once the set of adaptations that defines the higher group is entrained and broad diversification at lower taxonomic levels is possible, the course of the diversity curve is essentially inevitable. Only the absolute numbers of taxa produced may be different. The group will rapidly reach whatever peak in numbers of specialized lower ranks is possible and then decay at a similar standard rate (some 4—8% of the maximum diversity per million years) regardless of the *immediate* cause of the decay — see later. Extinction is also independent in respect of taxonomic rank, families and genera decaying at the same rate. This invites comparison with the ideas of Van Valen (1973), and tends to support the validity of higher taxa.

The question of the cause or causes of the temporal sequence of origination and diversification among fishes is most difficult to discuss. The fact that the peaks in diversity come in waves rather than being scattered more randomly through the record and that the timing of these waves seems to be common to more than one group of organisms (fishes, amphibians, reptiles, many invertebrates) suggests that some extrinsic factor is important in determining their occurrence. Sea-floor spreading, plate motions, changes in sea level, marine transgressions and regressions, and particularly the resulting changes in area of environments and changes in provinciality have been prime candidates for explaining evolutionary patterns (see, *inter alia*, Newell, 1952; Moore, 1954, 1955; Rutten, 1955; House, 1967; Kauffman, 1970; Jardine and McKenzie, 1972; Hays and Pitman, 1973; Schopf, 1974, and Simberloff, 1974, for varying views on the subject). Another possibility is that some physical factor directly affects evolutionary (mutation) rates; geomagnetic reversals are an obvious candidate here (e.g., J.F. Simpson, 1966; Crain, 1971; Reid et al., 1976). The comparisons in Figs. 5 and 6 do not wholly support such conclusions. Particularly, there is an inverse correlation in the Palaeozoic between diversification and marine transgressions but a more or less positive correlation in the Mesozoic.

A diametrically opposite explanation of the pattern of diversification arises from the particular shape of the group diversity curves and their juxtaposition in time. It is possible that the rise to peak diversity of new groups is not possible until the decline of preceding groups has been entrained. This would fit with the notion of an upper limit to overall diversity. One given group could be replaced in the record by a numerically large group of specialists or a small group of generalists, but in the sequence of group diversity curves the rise of any group is prevented until the preceding group makes way for it. In addition, it is also possible, especially during the very early history of fishes when the full range of adaptive zones is being realized for the first time, that the adaptive radiation of a succeeding group actually contributes to the quality and quantity of instability in the environment that drives an existing specializing group to extinction. Such interactions would provide a good explanation of an "intrinsic" mechanism producing regular spacing of diversity peaks in time. In Fig. 17 a set of curves each with the mean value for the group diversity curves of fishes is superimposed on the combined data. It is seen that with one exception (the mid-Devonian peak) such a set of curves will give a regular progression of peaks in diversity. However, the curves interact between the 30–42% points which does not prove a simple replacement of one group by another. The difference is such that it is unlikely to be explained simply by errors in estimating the shapes of the curves. Only the interactions of the earliest curves (Devonian and Carboniferous) are close enough to suggest such an interaction between the histories of successive adaptations of fishes. The data for amphibians and reptiles show the same situation.

The following explanation seeks to combine the possible causal factors already suggested. The Lower Devonian, Upper Devonian, and Lower Carboniferous diversifications of fishes (and possibly the Lower Carboniferous and Permian actinopterygian diversifications) follow each other sufficiently closely that one may postulate that each successive wave is a function of decay of its predecessor (see also the cephalopods below). Only in these cases could a major component of the *instability* of niches be the rapid origin of superior competitors in the filling of these adaptive zones for the first time (*cf.* Hutchinson, 1957). The fact that the Middle Devonian low point corresponds to a marine transgression and that there is a general correspondence between Palaeozoic peaks of diversity and the regressions may be connected with the fact that many of these early fishes were fresh-water or estuarine in habit. The small Permian peak is the first to be directly contemporaneous with a marine transgression.

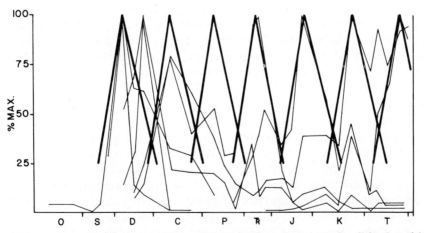

Fig. 17. The data shown in Figure 8 (group diversity curves for fishes) — thin lines — with a set of identical curves drawn from the mean value of these curves (see Fig. 14A) — heavy lines superimposed to show the apparent regularity in the peaks of diversification.

After the Lower Carboniferous (except for the apparent waves of radiations within the chondrostean order Palaeonisciformes) the overall diversity of fishes remained rather stable, with the emergence of no *major* adaptive complexes until the Triassic. It is difficult to sort out a possible circularity in the explanation of this stability. On the one hand, production of new adaptations may have been reduced under stable environmental conditions. On the other hand, new adaptations may have constantly been arising but did not lead to new diversifications because of the absence of significant changes in the environments that would allow them the foothold necessary to diversify (specialize) at low taxonomic ranks. Tappan and Loeblich (1973) point out that a general decrease in the productivity of marine systems through the Late Palaeozoic would cause extinction of "most suspension feeders and higher trophic levels; only the most cosmopolitan and efficient grazers, carnivores and detritus feeders persisted" (1973, p. 465). Such conditions would favour generalists among fish carnivores; generalists would be more immune to environmental fluctuations.

Origination rates for fishes rose sharply in the Early Triassic with the coincident radiations of the old Chondrostei and the first group of the new holostean grade (Semionotiformes). Possibly this event began among estuarine and freshwater fishes (the ancestral group for Semionotiformes is the Lower Permian Aduellidae; Patterson, 1974), but the full course of the diversifications was greatly facilitated by the *subsequent* mid-Triassic marine transgression which hypothetically produced conditions favouring the diversification of both marine taxa (through the creation of new zones for marine colonization) and estuarine taxa (by fragmentation and multiplication of marginal environments). This hypothesis needs to be tested by a rigorous investigation of the environments of deposition for Permian and Triassic fishes. The next diversifications of fishes show their peaks, and rates of origination are highest, in the Upper Jurassic, and the Jurassic transgression was at its maximum in the Oxfordian and Kimmeridgian (Hallam, 1975). This may suggest that the next phase in the history of predatory fishes was not possible without the full development of new trophic patterns at lower levels in aquatic ecosystems. The same situation would apply for the Cretaceous diversifications.

This pattern of diversification is exemplified by a comparison of fishes and cephalopods, which are in many ways analogous predators (Packard, 1972). The Lower Devonian peak in fishes follows immediately after (effectively intersects) the decline of the penultimate peak in the generic diversity of nautiloids (Fig. 18). It is immediately followed by (and again intersects) the last diversification of nautiloids and the coincident first radiation of ammonoids. Through the Late Palaeozoic the three groups appear to decline (remain stable, allowing for preservational bias) together. The Triassic diversification of fishes is roughly coincident with that of ammonoids, but the Jurassic and Cretaceous peaks in ammonoid diversity clearly precede those of fishes. There is no good evidence that the fishes and ammonoids tended to exclude each other in the Mesozoic, although they probably did in the Palaeozoic. This seems to confirm a theory of increasingly available quantitative trophic resources after the Permian. The

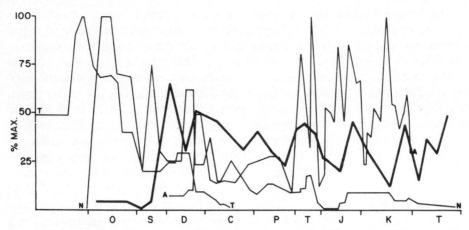

Fig. 18. Percentage of maximum diversity curves for (*A*) ammonoids, (*N*) nautiloids and (*T*) trilobites compared with the average diversity for all fishes shown in Fig. 8. Superimposed (heavy line).

decays of the fish and ammonoid radiations seem to reflect jointly the retreat of seas after each transgression. Overall, there is a strong suggestion that the diversifications of cephalopods, and possibly other invertebrate competitors, may have played a role in the maintenance of equilibrium in fish diversity. Marine reptiles were probably also important (Packard, 1972).

There is a great deal that remains unexplained in this account. For example, the regular succession of peaks in the diversity of fishes largely shows correlation with *regressions* of the seas in the Palaeozoic and *transgressions* in the Mesozoic. It may simply be that in the former cases we are dealing with radiations based upon fresh-water and estuarine fishes and in the latter with largely marine groups. If this were true it would be demonstrated in the relationship between evolutionary rates and rates of transgression and regression. High origination rates for fresh-water fishes in the Palaeozoic would be expected to follow regressions, and high extinction rates would follow transgressions. Our present levels of data are not sufficient to test this. But also how would we account for the shift from one mode to another? Perhaps it is connected with the emergence of the freshwater Amphibia and Reptilia as direct competitors of fresh-water fishes. Although precise identification of cause and effect is not possible from the data on fishes, all the evidence suggests that the controlling feature underlying the history of diversification is a combination of results of global tectonic changes. Probably geomagnetic reversals, rather than being proximal causes of evolutionary changes, merely reflect the same tectonic processes.

While the data for fishes fit current equilibrium models of Phanerozoic diversity — with the exception that a new level of equilibrium seems to be reached in the Tertiary — the data do not constitute *prima facie* evidence for such equilibria. If the apparent regularity in the course of diversification of fish groups were real, then it would indicate an underlying oscillatory mechanism that, whether it were intrinsic to any group of organisms or extrinsic, might produce the same effect as an equilibrium condition but by different means.

Finally, a somewhat surprising conclusion must be mentioned. If, as we have seen, there is a common pattern of diversification among organisms — a rapid rise in diversity at low taxonomic levels and an equally sudden decline — then we have an explanation for the famous apparent mass extinctions of the Devonian, Permo-Triassic and Cretaceous that have been the subject of so much discussion by palaeobiologists. If an identically sudden fall to extinction of all groups is an inevitable result of diversification and specialization, then mass extinctions are explained. *Mass* extinctions occurred when the immediately preceding time spans were particularly favourable to the essentially contemporaneous diversifications of large numbers of groups. We should concentrate our attentions therefore on finding the reasons for these mass *increases* in diversity, not for their declines, in order to explain the phenomena (Thomson, 1976).

Summary of Conclusions

(1) Numbers of fossil taxa of fishes reach a major peak in the Devonian. Thereafter, numbers of all taxa decline through the beginning of the Triassic, the decay of higher taxa necessarily being slower than that of lower taxonomic ranks. In the Triassic a new increase in number occurs, but only of the lower taxonomic ranks (order and below), leading to very high levels in the Late Tertiary.

(2) The fossil record of fishes and the taxonomy of many groups is inadequate. The numbers known from the record represent a very small fraction of those presumed to have lived at any time.

(3) The numbers of fossil occurrences is strongly modified by the volume of deposition in the geological record. After allowing for this bias it seems that, after the Devonian to Lower Carboniferous peak in diversity, the numbers of all taxa remained at a constant level until the mid-Mesozoic.

(4) Rates of taxonomic evolution and records of first and last occurrences of genera and families show rather even fluctuations in evolutionary activity except for the Devonian and possibly the Tertiary. There is little evidence among fishes for a Permo-Triassic extinction. There is no simple and direct statistical correlation between rates of origination or extinction and rates of marine transgressions and geomagnetic reversals. However, a more complicated relationship does exist.

(5) A modification of the data is used to compare the pattern of diversification among higher groups of fishes. It is discovered (a) that each group (at approximately the ordinal level) gives a simple curve of change in diversity through time, corresponding to a single adaptive radiation, and (b) that these curves are quite uniform among all groups measured between the 25% points of relative diversity on the ascending and descending slopes of the curve. This suggests that the diversifications of higher groups of fishes have major points in common, beyond the factors intrinsic to each separate radiation. (c) The

diversifications of the individual groups of fishes tend to be synchronous, several curves showing their peaks at the same time interval. This coincidence applies also to the Amphibia and Reptilia, so that there are only some 12 points in time upon which the diversifications of more than 50 orders converge. (d) These points of peak diversification are regularly distributed through the time scale, with an average interval of some 60 m.y. between peaks.

(6) The growth in diversification of a single group is compared with species packing in a niche, and the subsequent rapid decay of the diversification is compared with the resultant extinction of taxa due to progressive specialization and an unstable environment. In addition to the physical factors in such an environment, other organisms, including other groups of fishes, may be important in causing the decay of diversifications.

(7) The development of peak diversity among fishes in the Devonian and Early Carboniferous is compared with crash filling of a series of niches. Thereafter, it is hypothesized, the development of an equilibrium in numbers of taxa reflects a ceiling on diversity imposed largely by trophic resources. In the Late Palaeozoic fishes tended to become generalized carnivores and new specializations (new diversification of higher groups) did not emerge until the Triassic when apparently diversifications of estuarine and fresh-water fishes began, later extending into the environments produced by the Triassic marine transgression. In the Jurassic and Cretaceous major radiations follow the marine transgressions. In general, Palaeozoic diversifications are contemporaneous with regressions and Mesozoic diversifications with transgressions.

(8) The apparent regularity in diversifications of fishes cannot be accounted for by any single factor but a combination of biological and physical factors, all of which in the final analysis are probably to be traced to global tectonic events affecting the quality and quantity of fossil environments.

(9) While the data for fishes fit the equilibrium hypothesis it is possible that other mechanisms, perhaps involving the factors producing the apparent regularity in diversification, may be involved in the control of diversity of fishes.

(10) No special cause need be sought for periods of multiple extinctions in the fossil record. Rather, explanations must be found for the periods of multiple occurrences of high diversity.

Acknowledgements

I am grateful to Professor G.E. Hutchinson, Drs. J.P. Finnerty and D.S. Webb, and S.P. Rachootin, D.M. Schankler and D.E. Simanek for valuable advice and assistence. Dr. S.J. Gould very generously made available his compilations of data on invertebrate diversity. The illustrations were prepared by Linda Price Thomson. Studies supported by NSF Grant GB 28823X.

References

Cisne, J.L., 1974. Evolution of the world fauna of aquatic free-living arthropods. Evolution, 28: 337—366.

Crain, I.K., 1971. Possible direct causal relation between geomagnetic reversals and biological extinctions. Geol. Soc. Am. Bull., 82: 2608—2606.

Crain, I.K. and Crain, P.L., 1970. New stochastic models for geomagnetic reversals. Nature, 228: 39—41.

Cutbill, J.L. and Funnell, B.M., 1967. Numerical analysis of the fossil record. In: W.B. Harland et al., The Fossil Record. Geological Society of London, pp. 791—820.

Denison, R.H., 1956. A review of the habitat of the earliest vertebrates. Fieldiana Geol., 11: 361—357.

Flessa, K.W. and Imbrie, J., 1973. Evolutionary pulsations: evidence from Phanerozoic diversity patterns. In: D.H. Tarling and S.K. Runcorn, Implications of Continental Drift to the Earth Sciences, 1. Academic Press, London, pp. 247—285.

Garrells, R.M. and Mackenzie, F.T., 1971. Evolution of Sedimentary Rocks. Norton, N.Y., 397 pp.

Greenwood, P.H., Miles, R.H. and Patterson, C., 1974. The Interrelationships of Fishes. Academic Press, London, and Linnean Society of London, 536 pp.

Gregory, J.T., 1955. Vertebrates in the ecologic time scale. In: A. Poldervaart (Editor), Crust of the Earth. Geol. Soc. Am. Spec. Pap., 62: 593—608.

Hallam, A., 1971. Re-evaluation of the palaeogeographic argument for an expanding earth. Nature, 232: 180—182.

Hallam, A., 1975. Jurassic Environments. Cambridge University Press, Cambridge.

Harland, W.B. et al., 1967. The Fossil Record. Geological Society of London, 828 pp.

Hays, J.D., 1971. Faunal extinctions and reversals of the earth's magnetic field. Geol. Soc. Am. Bull., 82: 2433—2447.

Hays, J.D. and Pitman III, W.C., 1973. Lithospheric motion, sea level changes and climatic and ecological consequences. Nature, 246: 18—22.

House, M.R., 1967. Fluctuations in the evolution of Palaeozoic invertebrates. In: W.B. Harland (Editor), The Fossil Record. Geological Society of London, pp. 41—54.

Hutchinson, G.E., 1957. Concluding remarks. Cold Spring Harbor Symp. Quant. Biol., 22: 415—427.

Jardine, N. and McKenzie, D., 1972. Continental drift and the dispersal and evolution of organisms. Nature, 235: 20—24.

Kauffman, E.G., 1970. Population systematics, radiometrics and zonation — a new biostratigraphy. North Am. Paleontol. Conv. Proc., F: 612—666.

MacArthur, R.H., 1972. Geographical Ecology: Patterns in the Distribution of Species. Harper and Row, New York, N.Y., 269 pp.

McElhinny, M.W., 1971. Geomagnetic reversals during the Phanerozoic. Science, 172: 157—159.

Moore, R.C., 1954. Status of Invertebrate Palaeontology, 1953, X. Evolution of Late Palaeozoic invertebrates in response to major oscillations of shallow seas. Bull. Mus. Comp. Zool. Harvard Coll., 112: 259—286.

Moore, R.C., 1955. Expansion and contractions of shallow seas as a causal factor in evolution. Evolution, 9: 482—483.

Moy-Thomas, J.A. and Miles, R.S., 1971. Palaeozoic Fishes. Saunders, Philadelphia, Del., 259 pp.

Newell, N.D., 1952. Periodicity in invertebrate evolution. J. Paleontol., 26: 371—385.

Newell, N.D., 1967. Revolutions in the history of life. Geol. Soc. Am., Spec. Pap., 89: 63—91.

Obruchev, D.V., 1967. Fundamentals of Paleontology, Vol. XI, Agnatha, Pisces. Israel Program for Scientific Translations. Jerusalem, 825 pp. (Translation from the Russian Edition of 1964.)

Packard, A., 1972. Cephalopods and fish: the limits of convergence. Biol. Rev., 47: 241—307.

Patterson, C., 1974. Interrelationships of holosteans. In: P.H. Greenwood, R.S. Miles and C. Patterson, Interrelationships of Fishes. Academic Press, London, Linnean Society of London, pp. 233—305.

Pitrat, C.W., 1973. Vertebrates and the Permo-Triassic extinction. Palaeogeogr., Palaeoclimatol., Palaeoecol., 14: 249—264.

Raup, D.M., 1972. Taxonomic diversity during the Phanerozoic. Science, 177: 1065—1071.

Raup, D.M., Gould, S.J., Schopf, T.J.M. and Simberloff, D.S., 1973. Stochastic models of phylogeny and the evolution of diversity. J. Geol., 81: 525—542.

Reid, G.C., Isaksen, I.S.A., Holzer, T.E., and Critzen, P.J., 1976. Influence of ancient solar-proton events on the evolution of life. Nature, 259: 177—179.

Romer, A.S., 1966. Vertebrate Paleontology. University of Chicago Press, Chicago, Ill., 468 pp.

Ronov, A.B., 1959. On the post-Cambrian geochemical history of the atmosphere and hydrosphere. Geochem. U.S.S.R., 1959: 493—506.

Rutten, M.G., 1955. Evolution and oscillations of shallow shelf seas. Evolution, 9: 481—482.

Schaeffer, B., 1973. Fishes and the Permian—Triassic Boundary. In: A. Logan and L.V. Hills, The Permian and Triassic Systems and Their Mutual Boundary. Can. Soc. Pet. Geol., Calgary, Alta., pp. 493—497.

Schopf, T.J.M., 1974. Permo-Triassic extinctions: relation to sea-floor spreading. J. Geol., 82: 129—143.

Schopf, T.J.M., Raup, D.M., Gould, S.J. and Simberloff, D.S., 1975. Genomic versus morphologic rates of evolution: influence of morphologic complexity. Paleobiology, 1: 63—70.

Simberloff, D.S., 1974. Permo-Triassic extinctions: effects of area on biotic equilibrium. J. Geol., 82: 267—274.

Simpson, G.G., 1944. Tempo and Mode in Evolution. Columbia University Press, New York, N.Y., 237 pp.

Simpson, G.G., 1952. Periodicity in vertebrate evolution. J. Paleontol., 26: 359—370.

Simpson, G.G., 1953. The Major Features of Evolution. Columbia University Press, New York, N.Y., 434 pp.

Simpson, G.G., 1969. The first three million years of community evolution. In: G.M. Woodwell and H.H.

Smith, Diversity and Stability in Ecological Systems. Brookhaven Symp. Biol., 22: 162—177.
Simpson, J.F., 1966. Evolutionary pulsations and geomagnetic polarity. Geol. Soc. Am. Bull., 77: 197—204.
Spjeldnaes, N., 1967. The paleoecology of the Ordovician vertebrates of the Harding Formation (Colorado, USA). Coll. Int. Cent. Rech. Sci., 168: 11—20.
Tappan, H. and Loeblich, A.R., 1973. Smaller protistan evidence and explanation of the Permian—Triassic crisis. In: A. Logan and L.V. Hills, The Permian and Triassic Systems and Their Mutual Boundary. Can. Soc. Pet. Geol. Calgary, Alta., pp. 465—480.
Thomson, K.S., 1971. The adaptation and evolution of early fishes. Q. Rev. Biol., 46: 139—166.
Thomson, K.S., 1976. Explanation of large scale extinctions of lower vertebrates. Nature, 261 (5561): 578—580.
Valentine, J.W., 1969. Patterns of taxonomic and ecological structure of the shelf benthos during Phanerozoic time. Palaeontology, 12: 684—700.
Valentine, J.W., 1973. Phanerozoic taxonomic diversity: a test of alternate models. Science, 180: 1078—1079.
Valentine, J.W. and Moores, E.M., 1970. Plate tectonic regulation of biotic diversity and sea level: a model. Nature, 228: 657—659.
Valentine, J.W. and Moores, E.M., 1972. Global tectonics and the fossil record. J. Geol., 180: 167—184.
Van Valen, L., 1973. A new evolutionary law. Evol. Theory, 1: 1—30.

Addendum

Since this chapter was written, A.G. Fischer and M.A. Arthur have shown me a preprint of their paper "Secular variations in the pelagic realm", in which they document a ca. 60-m.y. periodicity in evolution and give a full review of literature not cited here. Another paper mentioning such periodicity is D.V. Ager (Proc. Geol. Assoc. Lond., 87: 131—159 (1976)).

CHAPTER 13

PATTERNS OF AMPHIBIAN EVOLUTION: AN EXTENDED EXAMPLE OF
THE INCOMPLETENESS OF THE FOSSIL RECORD

ROBERT L. CARROLL

Introduction

Amphibians present a curiously bimodal pattern of diversity (Fig. 1). The
group emerged in the Late Devonian and radiated extensively in the Carboniferous, Permian and Triassic. By the end of the Triassic all of the ancient
groups had become extinct. The second radiation is first evident in the Jurassic and has continued to the present. Not only are the two radiations distinct in
time, but the basic structure and biology of the ancient and modern amphibian
groups are very different from one another.

The fossil record of the initial terrestrial adaptation and radiation is unfortunately very incomplete. No forms are known that can be considered as intermediate between rhipidistian fish and amphibians. Although primitive in some
respects, the ichthyostegids (the earliest known amphibians) are unquestionably terrestrial in the nature of the girdles and limbs. The skull is clearly
distinct from the trunk; the great reduction of the opercular series suggests that
gill respiration was no longer possible. It may be assumed that the elaboration
of the girdles, limbs, vertebrae and ribs for support of the body on land must
have occurred very rapidly, with corresponding modification of the respiratory
and circulatory systems, etc.

There is no direct evidence that more than a single lineage of rhipidistian fish
gave rise to Palaeozoic amphibians. There is, however, an extensive radiation of
amphibians within the Lower Carboniferous which might be interpreted as evidence of the differentiation of several distinct lineages at the level of the rhipidistians.

As the only tetrapods present in the Devonian and Lower Carboniferous, amphibians had the entire spectrum of terrestrial environments in which to diversify. Differentiation of the various taxonomic groups during the Palaeozoic
constitutes a major part of this chapter. Each of the major groups diversified
into a spectrum of habitats. All include some families that were presumably
highly adapted to a terrestrial way of life, some that were semi-terrestrial or
semi-aquatic, and others that were obligatorily aquatic or even neotenic. Close
convergence is not a common feature, however, for both aquatic and terrestrial
adaptations have typically followed different patterns in the various groups.

During the Carboniferous several groups approached the reptilian grade of
development in the elaboration of strong vertebral support and proportionately
longer limbs. One of these groups also evolved the definitive reptilian reproductive pattern and subsequently gave rise to the great diversity of amniotes.

Although the evolution of reptiles will not be considered in detail in this

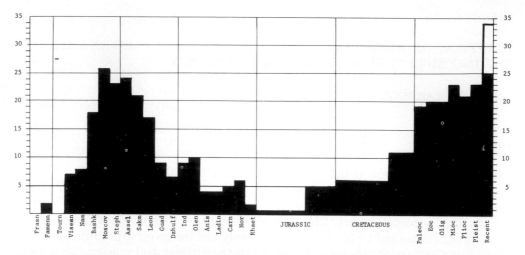

Fig. 1. Total number of amphibian families from successive geological horizons. Families are counted as occurring in a particular horizon, even if their remains have not been reported, if the family is known to occur in both lower and higher horizons. The records are not otherwise extrapolated. Names of geological horizons follow *The Fossil Record* except for the substitution of Steph(anian) for Upper Carboniferous, a term which is usually applied to a much longer time interval. The small number of occurrences in the Jurassic and Cretaceous makes it impractical to employ the numerous stages. Disagreement as to the division between the Asselian and Sakmarian, and the difficulty of assigning species to one horizon or the other, give an erroneous impression of longevity of many taxa.

chapter, it is necessary to comment on the effect of reptilian diversification on the major patterns of amphibian radiation. Reptiles remain rare and relatively small until near the end of the Pennsylvanian. Up to that time, their presence probably affected only a few groups of small amphibians. By the end of the Carboniferous, however, with the emergence of the pelycosaurs as dominant carnivores, a large spectrum of amphibians were subject to both predation and competition from reptiles. Because of the initial adaptation of reptiles to a primarily terrestrial way of life, the more terrestrial amphibians were the first groups to be strongly affected. Few survived the Lower Permian. Aquatic and semi-aquatic amphibian groups continued to diversify during the Permian, and more strictly aquatic groups underwent a major radiation in the Late Permian that continued to the end of the Triassic. With the expansion of reptiles into fresh-water and marine environments in the Late Permian and Triassic, the last labyrinthodonts were finally eliminated from these environments as well.

In the Late Carboniferous or Early Permian, one or more groups of small amphibians became adapted to a quite different way of life than that of other Palaeozoic forms. Taking advantage of a large surface-to-volume ratio, they could use the general body surface for gas exchange. Although making them more subject to desiccation, their small size made it easier to adapt to burrowing, aquatic or otherwise cryptic habits that made them relatively safe from predation by the more terrestrial vertebrates, and enabled them to take advantage of food sources and environmental extremes unavailable to reptiles. From this group or groups arose the ancestors of the modern amphibian orders. Their small body size and cryptic habits led to a remodelling of the skeletal system, producing patterns that were very different from those of their Palaeozoic antecedents. These changes formed the basis for a renewed radiation of amphibians which has continued to the present.

It was originally expected that some measure of the rates of various evolutionary processes might be determined for the Class Amphibia: origination, extinction and longevity of groups at various taxonomic levels. To that end, data on all described fossil amphibians were collected. It soon became apparent that a strictly quantitative treatment of this information would be very misleading in view of the incompleteness of the fossil record. Nevertheless, it was felt that availability of the raw data was necessary in order to evaluate the general patterns discussed. A list of all amphibian species known from fossils and the time of their occurrence has been prepared for this purpose *.

Labyrinthodontia

Three subclasses of amphibians are currently recognized: Labyrinthodontia, Lepospondyli and Lissamphibia (Romer, 1966). The labyrinthodonts are known from the Upper Devonian to the end of the Triassic. They are generally large forms (up to 3 m in length), heavily scaled, and characterized by having the vertebral centra ossified in more than a single unit. The lepospondyls are known from the Lower Carboniferous to the end of the Lower Permian. Like the labyrinthodonts, they are heavily scaled, but generally of smaller size and typically have a single central ossification. The term Lissamphibia is applied to the three living orders — Urodela, Anura and Apoda (Gymnophiona, Caecilia) — whose fossil record begins, respectively, in the Upper Jurassic, Lower Jurassic and Paleocene. These forms are generally small, and always have only a single central ossification. All three groups exhibit pedicellate teeth in which the base and crown of the teeth are separated by a band of fibrous tissue. All three practise cutaneous respiration. Anurans and salamanders lack body scales, but they are present in some apodans. The interrelationships among the three subclasses have not been established. There is no convincing evidence as to whether or not either lepospondyls or the Lissamphibia are natural groups.

Labyrinthodonts are clearly derived from rhipidistian fish, with which they share a similar pattern of the bones of the skull, labyrinthine infolding of the dentine of the teeth, and presence and pattern of large palatal tusks. The pattern of the girdles and limbs of amphibians can be readily derived from that of the rhipidistians. The closest affinities of amphibians appear to be with the rhizodontid rhipidistians on the basis of skull proportions and details of limb and girdle structure. No rhipidistians are known that can be specifically cited as labyrinthodont ancestors. On the basis of the degree of differentiation of

---

* This list has been printed separately, and will be sent by the author on request.

Upper Devonian and Lower Mississippian amphibians, it has been assumed that divergence must have occurred by the Middle Devonian.

Three major groups of labyrinthodonts are recognizable in the Early Carboniferous, but they are sufficiently similar to support the assumption that all could have evolved from a single group of rhipidistians.

Ichthyostegids

Romer (1966) recognized three orders of labyrinthodonts: Ichthyostegalia, Temnospondyli and Anthracosauria. The Ichthyostegalia are only definitely known from the Upper Devonian of East Greenland. Two monotypic families are recognized: Ichthyostegidae and Acanthostegidae, although it has been suggested that other taxa are present (Säve-Söderbergh, 1932; A.L. Panchen, personal communication, 1976). Taxa from later horizons have been assigned to the Ichthyostegalia (*Colosteus, Erpetosaurus* and *Otocratia*) (Romer, 1947), but all appear to be related to the temnospondyl stock. Panchen (1973a) has assigned the Lower Carboniferous genus *Crassigyrinus* to a monotypic order of primitive labyrinthodonts, but the remains are too incomplete to preclude its relationship to either the ichthyostegids or the Rhipidistia. The ichthyostegids are clearly the most primitive of labyrinthodonts, retaining the break between the two parts of the braincase characteristic of rhipidistians, as well as two elements of the opercular series between the cheek and the pectoral girdle, and a vertebral construction of essentially rhipidistian grade; but they are clearly amphibian in the structure of the limbs and girdles. The pattern of the bones of the skull roof makes them unlikely ancestors of either temnospondyls or anthracosaurs, and they are considered an early specialization among labyrinthodonts that left no descendants.

Since ichthyostegids are unquestionably labyrinthodonts and yet not ancestral to either of the more advanced orders, one must assume that at least one other group of labyrinthodonts was present in the Upper Devonian, but left no fossil record. The presence of Devonian footprints in Australia (Warren and Wakefield, 1972) indicates a wide geographic distribution as well. With the rich and widespread distribution of rhipidistians in the Middle and Upper Devonian in North America, Europe, Asia, Australia and Antarctica, it is surprising to find so few of their descendants in the Upper Devonian and Mississippian.

Temnospondyls (Fig. 2)

Temnospondyls are the dominant amphibians from the Lower Carboniferous to the end of the Triassic. They are clearly differentiated from anthracosaurs on the basis of the pattern of the dermal bones of the skull roof, as emphasized by Romer (1947). In the temnospondyls, the tabular is small and separated from the parietal by the supratemporal. In the anthracosaurs, the tabular is large and in contact with the parietal. Primitive members of both groups retain the intertemporal, a bone missing in ichthyostegids. According to Panchen (1970), the anthracosaur pattern probably evolved from a condition resembling

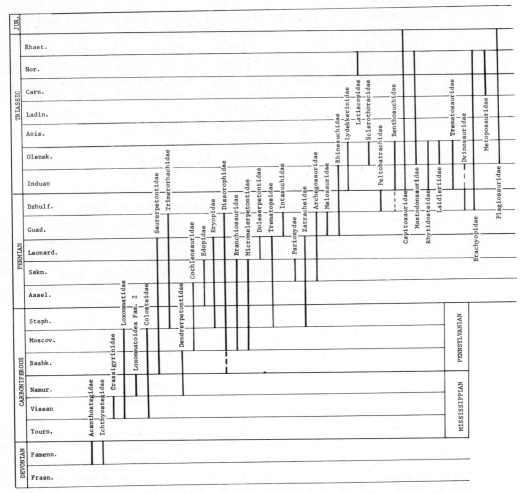

Fig. 2. Fossil record of Ichthyostegalia and Temnospondyli. Dashed lines indicate range extension based on doubtful records or doubtful generic assignments. Records are indicated as if they extended the entire length of the stage in which they are found. Most taxa are known from only a single very restricted horizon. The Edopidae should be limited to the Asselian.

that seen in temnospondyls, but no intermediates have been described. Anthracosaurs appear more primitive in retaining mobility between the cheek region and the skull roof, resembling that seen in rhipidistians, but temnospondyls have the cheek and skull roof strongly attached. The pattern of sculpturing of the dermal bones of the skull roof serves to separate most members of the two orders, with nearly all temnospondyls exhibiting a pattern of distinct pits and grooves, while these are generally (but not always) more diffuse among anthracosaurs.

Romer (1947) emphasized the pattern of vertebral centra as serving to differentiate anthracosaurs and temnospondyls, with the pleurocentra (the true centra of more advanced tetrapods) dominant in the former group and the intercentra (an anterior, usually crescentic element) in the latter. Romer went further (1964a) and elaborated a scheme for the derivation of the anthracosaur pattern from that of temnospondyls, arguing that the latter was basic to all labyrinthodonts. Subsequent information (Panchen, 1975; Holmes and Carroll, 1977) indicates that the pleurocentra were fully developed in even the most primitive anthracosaurs. There is currently no evidence to determine the origin of the vertebral pattern in any of the three labyrinthodont orders, nor is there any evidence of the specific relationship or relative time of differentiation of temnospondyls and anthracosaurs. An anthracosaur has been identified from the lowermost Mississippian on the basis of femora (Carroll et al., 1972, p. 21) and temnospondyls from the Visean.

The fossil record for temnospondyls in the Mississippian or Lower Carboniferous is clearly very incomplete and certainly not indicative of their actual diversity or distribution. Most genera can be grouped into two categories: trimerorhachoids and loxommatoids. The loxommatoids (Tilley, 1976), known in both North America and Europe and extending into the Late Pennsylvanian, are characterized by the presence of a peculiar anterior extension of the orbit, giving a keyhole shape to the combined orifice. Postcranial material is very poorly known, but they were apparently aquatic, with poorly developed limbs. The trimerorhachoids are known from complete skeletons of two genera from the Upper Mississippian: *Pholidogaster* (Panchen, 1975) from Europe and *Greererpeton* (Romer, 1969) from North America. These forms are approximately a metre in length, with 40—42 presacral vertebrae. The limbs are very small, presumably precluding terrestrial activities. The orbits are small and circular in shape. The family to which these forms are allocated — the Colosteidae — continues into the Pennsylvanian. The Upper Pennsylvanian and Permian families Saurerpetontidae and Trimerorhachidae are presumably descendants of the same stock, and retain a primarily aquatic way of life. It is possible that trimerorhachoids and loxommatoids evolved from a common ancestor in the Early Mississippian. Neither are appropriate ancestors to the more common terrestrial temnospondyls of the Late Carboniferous and Permian.

Judging from the remains of ichthyostegids and the earliest anthracosaurs, the early labyrinthodonts probably had well-developed limbs, a relatively small number of presacral vertebrae (25—32) and were adapted to spend a considerable portion of their adult life on land. The majority of Lower Carboniferous temnospondyls are not of this pattern, however. The earliest temnospondyl that appears to exhibit terrestrial habits is *Caerorhachis* (Holmes and Carrol, 1977) from the Namurian. One must assume that such forms were present in the earlier Carboniferous, but that the nature of the deposits from which most amphibians have been discovered was biased in favour of more aquatic forms. *Caerorhachis* may be placed among the edopoid amphibians that became dominant in the Upper Carboniferous and Lower Permian. These forms do not have lateral line canal grooves, and have relatively well-developed limbs; none are known to have a large number of presacral vertebrae. From among the edopoids has evolved the major assemblage of Upper Carboniferous and Lower Permian amphibians, the eryopoids. These forms are distinguished by the loss of the intertemporal bone (also missing in some trimerorhachoids and loxommatoids) and by the sutural attachment of the palate and base of the braincase (a movable articulation is retained in some forms). The earliest known eryopoids are from the Middle Pennsylvanian, but the groups may be considerably older. Eleven families are here classified within the Eryopoidea, including highly terrestrial, semi-terrestrial, semi-aquatic and secondarily completely aquatic lineages. Bolt (1969) recognized two superfamilies at this level of evolution: a closely knit Dissorophoidea, including the dissorophids, trematopsids and doleserpetontids; with the remaining retained in the Eryopoidea s.s. [Boy (1972) subsequently included the neotenic "branchiosaur" families Branchiosauridae and Micromelerpetontidae in the Dissorophoidea]. This latter assemblage has not been redefined, and contains a very heterogeneous assemblage of families. The aquatic zatrachiids seem as close to the dissorophids as do the trematopsids; the highly terrestrial parioxyids seem only slightly more distant. The archegosaurids, on the other hand, are quite unlike either dissorophids or eryopids. The family Intasuchidae is suggested by Olson (1962) as being closely related to trematopsids. Melosaurids alone seem very closely related to eryopids. Until this entire assemblage can be considered as a unit, it seems premature to subdivide the Eryopoidea as it was conceived by Romer. It does make a large, diverse assemblage, difficult to define rigorously in terms of either anatomy or habits, but there is no convincing evidence that these families did not evolve from a single assemblage, above the level of the known edopoids.

The Lower Permian sees the climax of the eryopoids in Europe and North America, as well as the last of the edopoids and trimerorhachoids. The Middle Permian marks a hiatus in the record of most tetrapod groups. In the later Middle and Upper Permian of Russia and South Africa there is a lingering record of eryopoids, but primarily the initiation of derived lineages.

Above the eryopoid level, the Temnospondyli underwent a major radiation in the Upper Permian and Triassic. Unfortunately, the fossil record is such that the interrelationships of the seven or more superfamilies have not been established, nor can their derivation from among the Carboniferous and Lower Permian forms be specified. Watson used the term stereospondyl to apply to the Triassic labyrinthodonts and Romer formalized the term in a taxonomic sense to apply to genera which developed the intercentra to more or less complete cylinders, with the corresponding reduction or elimination of the pleurocentra. As Säve-Söderbergh indicated in 1934, and Welles and Cosgriff (1965) more recently emphasized, the "stereospondyls" are an unnatural assemblage, rather like the holosteans among the osteichthyes, of diverse and poorly established ancestry. They are here considered as temnospondyls, with no attempt being made to assemble suborders within that group.

An assemblage of medium- and large-sized, semi-aquatic forms, readily derivable from the Lower and Middle Permian eryopids are grouped in the Rhinosuchoidea. *Peltobatrachus* is a highly terrestrial form that probably evolved from this basic stock and represents a terminal point for such adaptive types among labyrinthodonts. There is little evidence of its being antecedent to later, specialized aquatic forms, as suggested by Panchen (1959).

The rhinesuchids, known from the Middle Permian to Lower Triassic, primarily from South Africa, represent a continuation of the pattern seen among the eryopids and melosaurids; most genera being medium- to large-sized, presumably semi-terrestrial forms. There is a general tendency for the skull profile to be reduced within this assemblage, a factor which is accentuated in some specialized lineages.

The rhinesuchids are succeeded in time by the capitosauroids, with which they form an evolutionary continuum. The capitosauroids are limited to the Triassic, but span the length of that period and have a cosmopolitan distribution. In this group the skulls are very much flattened, the limbs small and ossification reduced. Most if not all were obligatorily aquatic. These forms reached the greatest dimensions of any labyrinthodonts, with skulls approaching a metre in length. Remains are common in many parts of the world, giving rise to a plethora of generic and specific names, most of which have been reduced to synonymy in the latest review by Welles and Cosgriff (1965).

The metoposaurs are of similar size and body proportions to the capitosaurs, but are differentiated by the anterior position of the orbits. This group is common and widespread in the Upper Triassic, but their antecedents have not been described. The group is known from both eastern and western North America, Europe and India, but is unrepresented in South America, Africa and Australia. The remains of metoposaurs are sometimes found in large numbers in what are presumably the last remnants of shallow lakes.

Another isolated group is the Trematosauroidea, restricted to the Lower Triassic, but common in such distant regions as Greenland, Spitsbergen and Madagascar. Remains have also been reported from western North America, South Africa, Australia and Russia. Trematosaurs are typically long-snouted forms, with well-developed lateral line canal grooves and rather high narrow skulls. They retain the vertebral pattern of most primitive temnospondyls, with the persistence of paired pleurocentra. The nature of the deposits in which they are found indicates that some were marine in habitus — the only group of amphibians to invade that environment. It is conceivable that they evolved from such long-snouted aquatic forms as *Archegosaurus*, from the Lower Permian, but there is no convincing evidence and no known intermediates. Long-faced rhinosuchoids, benthosuchids and primitive capitosaurs have also been suggested as possible ancestors.

The rhytidosteids were included among the trematosauroids by Romer (1966), but Cosgriff (1965) had elevated them to equal rank, as an assemblage distinguished by the distinctly triangular shape of the skull, the small orbits and the pustular ornamentation seen in some genera. Their rare remains are reported from the Lower Triassic of South Africa, Australia and Spitsbergen.

Once suggested as having evolved from the trimerorhachoids, the brachiopoids are a rare group, extending throughout the Triassic, and with plausible antecedents in the Upper Permian. The skull is short and flat, but not bizarrely so. Remains have been described from North and South America, Australia, South Africa, Europe and Antarctica. The large neotenic form from the Upper

Permian of Russia, *Dvinosaurus*, is suggested as a collateral of this group.

The plagiosaurs are the most peculiar of the temnospondyls, with extremely short, wide skulls and pustular ornamentation. Some, at least, were perennibranchiate, and reach a skull width of 0.5 m. They are represented by rare fossils throughout the Triassic of Europe. Although they resemble the brachyopids somewhat in skull proportions, they are totally different in cranial details, and the vertebrae are complete cylinders that have been homologized with the pleurocentra, in contrast with the condition in all other temnospondyls. Their ancestry is totally unknown. The isolation of the few species of this group in an order or suborder of their own does not seem justified despite the distinct nature of their vertebrae. Superfamily rank appears more appropriate.

One species that deserves special mention is *Latiscopus disjunctus* (Wilson, 1948) described from the Upper Triassic of Texas. It is the only amphibian of its age that is not a metoposaur, a plagiosaur, a capitosauroid or a brachyopoid. It is small, yet not obviously immature, with the orbits near the middle of the skull length. It is presumably a relict of one of the Late Permian or Early Triassic families, such as the lydekkerinids or rhytidosteids. Its discovery emphasizes the absence of other small amphibians, of any sort, from the Late Triassic.

Although most of the major groups of Triassic temnospondyls are without known phyletic connection with one another, Howie (1972a), has suggested that skull proportions indicate two major groups. On one hand, the trematosauroids, brachyopoids and metoposauroids have the orbits relatively anterior in position. The rhinesuchoids and capitosauroids, in contrast, have the orbits more posteriorly located. A new genus, *Rewena*, she suggests is an isolated offshoot of the short-faced line.

No labyrinthodonts are known from the Jurassic. A mandible reported from that period in Australia is apparently reworked from the Upper Triassic (Colbert, 1967). The rarity of continental beds in the Lower Jurassic may conceal the last phases in the evolution of the group, but it is unlikely that lineages survived long into the later Mesozoic without leaving some fossil record. The large size and distinctive sculpturing of the dermal bone would render identification of even small fragments possible. The late surviving labyrinthodonts certainly left no descendants.

The Triassic temnospondyls illustrate several of the problems inherent in making any quantitative measures regarding rates of origination, extinction and longevity. Neither the time of origin or the particular ancestral group is known for the majority of the superfamilies. At the level of the genus, the problem is compounded by the use of the vertebrate fossils to establish the age of the beds in which they are found. Without an independent means of dating the beds, it is meaningless to discuss the possible time span of the fossils. When the age of the beds is known with more assurance, this factor is used to establish the identity of fossils in which the anatomy is poorly known. The reliance on vertebrate fossils for dating of beds tends to ignore the possibility of slow migration, and obscures its detection. This practice assumes rapid and more or less similar rates of evolution throughout the world. If these assumptions were correct, it would suggest that Triassic amphibians exhibit a much more consistent pattern than that seen in Tertiary amphibians, or Permian and Triassic reptiles.

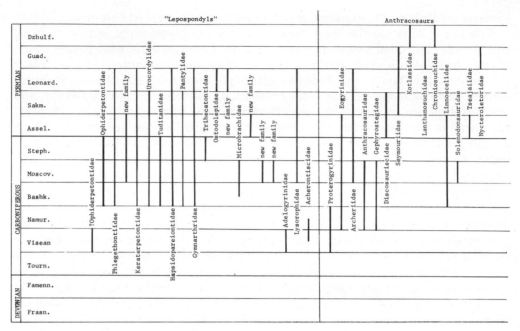

Fig. 3. Fossil record of "Lepospondyls" and Anthracosaurs. The "new family" between Phlegethontiidae and Keraterpetontidae is the Scincosauridae.

Anthracosaurs (Fig. 3)

The remaining labyrinthodonts are here grouped as the Order Anthracosauria. Various names have been used to refer to the higher taxa of non-temnospondyl labyrinthodonts (Panchen, 1970, 1975; Kuhn, 1972; Tatarinov, 1972). The usage here follows that of Romer (1947, 1966) except for the accommodation of newly recognized groups.

The anthracosaurs were never as numerous or diverse as the temnospondyls. They presumably originated from a common ancestor in the Upper Devonian or Lower Mississippian. Panchen (1975) has recently coined the term Herpetonspondyli for the primitive anthracosaurian stock. This group is represented in the Late Mississippian by *Eoherpeton* in Scotland and *Proterogyrinus* in North America. These forms have well-developed limbs and girdles, approximately 32 presacral vertebrae, and must have been capable of active life on land, although traces of lateral line canals remain in the adult. The vertebrae have large, nearly cylindrical pleurocentra and small crescentic intercentra. This is essentially the pattern of amniotes, and it is probable that the line leading to reptiles diverged at only a slightly earlier stage in anthracosaur evolution. Panchen (1972a) suggests, primarily on the basis of the structure of the ear, that reptiles did not evolve from a labyrinthodont ancestry. No non-labyrinthodont amphibians are known that even vaguely approach the anatomy probable for reptilian ancestors, however, and among the labyrinthodonts, anthracosaurs are certainly much closer to the reptilian pattern than are temnospondyls.

Forms similar to the known herpetospondyls could have given rise to all the subsequent anthracosaur lines, although no intermediate forms are known. In the Carboniferous, several divergent groups can be recognized. The embolomeres represent a major aquatic adaptation based on elongation of the trunk region, with approximately 40 presacral vertebrae. The individual vertebrae have both the intercentra and pleurocentra elaborated as subequal cylinders, presumably allowing a great deal of flexibility in the column associated with an undulatory swimming mode. The limbs, however, retain a pattern similar to that of their more terrestrial antecedents. Some genera have reevolved a dorsal caudal fin. Lateral line canal grooves are well developed. Panchen (1970) suggests that the embolomeres, which are best represented in the Carboniferous of Great Britain, inhabited fairly deep bodies of water. In the Permian, however, their remains are restricted to deltaic deposits in southwestern North America. The group did not survive the Lower Permian.

The remaining Carboniferous anthracosaurs were apparently much more terrestrial in their habits. The gephyrostegids, suggested as relicts of the group which gave rise to reptiles (Carroll, 1969a), are relatively small forms, with only 24 trunk vertebrae and relatively great limb-to-trunk length ratio. The earliest known genus, *Bruktererpeton* (Boy and Bandel, 1973) has body proportions approaching those of primitive captorhinomorph reptiles.

Forms allied to the known gephyrostegids may have given rise to the seymouriamorphs of the Permian. This group is not known in the Carboniferous, but the retention of some primitive features suggests that it may have diverged from the primitive anthracosaur stock prior to the emergence of the known gephyrostegids. Unlike other anthracosaurs, but in common with temnospondyls, the skull and cheek region are solidly attached and the dermal bone is marked by distinct pits and grooves.

The Seymouriidae, Discosauriscidae and Kotlassiidae seem to belong to a single phylogenetic assemblage with relatively conservative skeletal patterns, but the differences in habits or known life stages make detailed comparison difficult. Discosauriscids, all from central and eastern Europe, are known only from larval or neotenic forms. The Seymouriidae are primarily North American, with only questionably assigned material from Russia. All are fully terrestrial as adults with stout limbs and massively swollen neural arches. The kotlassiids, known only from the Upper Permian of Russia, like the terminal temnospondyls, have reverted to an aquatic way of life. Like some dissorophids and plagiosaurs, they have developed dermal armour, in the form of close-set sculptured plates, covering the trunk region.

Other anthracosaurs, from the Upper Permian of Russia, are difficult to associate with any of the previously mentioned groups. They have been classified among the seymouriamorphs because they may have diverged from the same general stock, and certainly show few similarities with more primitive groups. Members of the Chroniosuchidae have very long narrow skulls (rather similar to those of some embolomeres in their general proportions) and large plates of dermal armour. Members of the Lanthanosuchidae have very wide, low skulls, fenestrate in the temporal region.

None of these Upper Permian groups shows any evidence of continuing into the Triassic, although beds of this age in Russia have abundant remains of temnospondyls.

A number of other families may be affiliated with the Anthracosauria, but their specific position is now difficult to assess. In the Upper Carboniferous and Lower Permian are several groups that approach the reptilian grade of development in some skeletal features, but retain others that are typically amphibian. Members of the Limnoscelidae, Solenodonsauridae and Tseajaiidae may not be particularly closely related to one another, but all retain an otic notch and have a primitive configuration of the occiput and temporal region that bar them from close relationship to unquestioned reptiles — the captorhinomorphs and pelycosaurs. Whether one accepts the arguments concerning the configuration of the occiput and otic region in regard to reproductive habits (Carroll, 1970) or simply considers their general cranial anatomy, these forms are anthracosaur derivatives and not close allies to true reptiles. All are relatively large, and their skeleton suggests that they were primarily terrestrial in habits, at least as adults.

The diadectids are another group of Permo-Carboniferous forms that have been classified as both reptiles (Romer, 1956) and amphibians (Romer, 1964b; Olson, 1966), but are here considered to be phylogenetically related to the reptilian stock.

At the other end of anthracosaur phylogeny is a further group of questionable affinities, the Nycteroleteridae. The anatomy of the occiput again suggests anthracosaurian affinities, but not close relationship to either seymouriamorphs or more primitive suborders. The pattern of the skull roof has suggested affinities with the Procolophonoidea (Efremov, 1940; Colbert, 1946). Since procolophonoids otherwise are without obvious antecedents, this is a tempting suggestion, but it is difficult to defend rigorously. The position of these forms remains undetermined.

A glance at the known ranges of the various families of anthracosaurs (Fig. 3) emphasizes the absence of intermediate forms, and the tendency for a given morphological pattern to be long perpetuated without significant change. In the absence of any well-established ancestor-descendant relationships between any of the major groups, it is impossible to establish probable times of origination. Since most genera and species are confined to a single horizon, no estimation of average longevity has much biological significance.

"Lepospondyls"

The term lepospondyl is used in reference to a variety of small Palaeozoic amphibians which differ from labyrinthodonts in lacking the distinctive enfolding of the dentine, and in the absence of an otic notch and palatine fang pairs. The vertebral centra are typically ossified as a single cylindrical element, presumably homologous with the pleurocentra of labyrinthodonts and amniotes. Three orders are typically recognized: Microsauria, Nectridea and Aïstopoda.

All are currently being reviewed. It is becoming increasingly clear that these orders are not closely related to one another, and that relationships within the Microsauria and Aïstopoda are also uncertain. The groups are known only from North America and Europe.

Aïstopods (Baird, 1964; McGinnis, 1967) are snake-like in having a very long body with up to 230 vertebrae, and in the complete absence of limbs and girdles. A degree of differentiation into cervical, dorsal and caudal series implies the presence of limbs in an ancestral form, but this cannot be substantiated. Most of our information on the group comes from the Upper Carboniferous families Ophiderpetontidae and Phlegethontiidae. The skulls of these forms are highly specialized, implying a long separate history for each family, as well as from any basic amphibian stock. The presence of relatively numerous specimens in a variety of Upper Pennsylvanian coal swamp deposits suggests that the fossils may provide a fairly accurate sample of the group as a whole, at least during this time span, in this particular environment, in Europe and eastern North America. A single specimen from the early Lower Carboniferous has been referred to this order, but not adequately described. Its affinities with the later forms are not known. The latest aïstopods are known in the Lower Permian.

Work by Beerbower (1963), Bossy (1976) and Milner (1977) provides a good basis for understanding the nectrideans. Three families can be recognized, all of which are aquatic forms, with the caudal vertebrae specialized to form a swimming organ. The group appears in the Lower Pennsylvanian, is known from diverse forms in the later Pennsylvanian and Lower Permian, and is absent from the fossil record thereafter. It has been suggested that nectrideans are related to anthracosaurs (Thomson and Bossy, 1970), but there is no convincing evidence. K. Bossy (personal communication, 1975) suggests that the ancestral forms were probably terrestrial in habit, with large and well ossified limbs.

The terminal members of the group — *Diploceraspis* and especially *Diplocaulus* — are large forms, with the skulls greatly flattened and the tabular region expanded posteriorly in the form of "horns". *Diplocaulus* is common in the later Lower Permian redbeds of Texas. It may be said to parallel the general habitus of the aquatic temnospondyl amphibians of the Triassic.

Microsaurs have recently been reviewed by Carroll and Gaskill (1977). Eleven families are recognized, several of which are monotypic. Many are already clearly distinguished in the Lower Pennsylvanian and the major diversification may have been completed by that time, although other families are known only from the later Pennsylvanian or Lower Permian. Two basically distinct patterns of the skull roof support separation of the order into suborders, but both may have had a common ancestry in the Lower Carboniferous. Each suborder exhibits much variability in the body proportions, with vertebral counts ranging from 19 to 44. The limbs may be small, but they are always present, and the tail is never specialized in the manner of nectrideans. Several microsaurs have intercentra in the trunk region in addition to the normal spool-shaped centra. Such elements have never been reported in nectrideans and aïstopods. Habits range from perennibranchiate forms to lizard-like, with a variety of large and small burrowing genera. Despite their diversity, none of the microsaur lines is known to survive the Lower Permian.

Usually allied with microsaurs are two additional families: the lysorophids and the adelogyrinids. In both, the pattern of the skull appears to preclude close affinities with typical microsaurs. The lysorophids have a very open skull, with large orbits, and the maxilla and premaxilla are quite freely movable (Bolt and Wassersug, 1975). There are up to 72 presacral vertebrae and very small limbs. The short tail exhibits haemal arches, but there are never trunk intercentra. The family is most common in the Upper Pennsylvanian and Lower Permian of North America, and then vanishes from the fossil record. Possibly ancestral forms are known from the Lower Pennsylvanian of Ireland.

Much more restricted in forms and geography are the adelogyrinids, with four genera described from the Upper Mississippian of Scotland (Watson, 1929; Carroll, 1967b; Brough and Brough, 1969). Since only six specimens are known, this degree of taxonomic diversity is subject to question. The skull is solidly roofed, but lacks one of the elements present in typical microsaurs. The eyes are very far forward. The trunk is apparently quite long, but the limbs are well developed. Neither trunk intercentra nor haemal arches have been identified.

A final lepospondyl family, Acherontiscidae (Carroll, 1969b), is represented by a single specimen from the Mississippian of Scotland. The centra are formed from two subequal cylinders, grossly resembling the pleurocentra and intercentra of embolomeres. The skull is poorly known but is solidly roofed and has lateral line canal grooves. The trunk region is long and the limbs little developed.

Lysorophids, adelogyrinids and acherontiscids appear like the isolated orders of Mesozoic mammals. They have no obvious time or phylogenetic position of origin. They hint at an even greater diversity of small amphibians in the Palaeozoic.

Unlike the labyrinthodonts, the lepospondyls show no obvious affinities with the rhipidistians. The characteristics by which they differ from both these groups — the lack of labyrinthodont enfolding of the dentine, an otic notch and palatine fang pairs — may all be related to the small size of these forms and do not necessitate close relationship among them. The factor of size may be related to the common appearance of cylindrical vertebral centra in the group as well.

It may be suggested that the various lepospondyl groups arose during the Upper Devonian and Lower Carboniferous from various stocks of labyrinthodonts. It is conceivable that one or more might have evolved directly from rhipidistians, making the times of their origins even more uncertain.

In addition to the uncertainty of the origin of various lepospondyl groups, there is also a question as to the time of their extinction. Their complete absence in the Middle and Upper Permian is difficult to explain. There are, in fact, almost no small amphibians of any sort between the Lower Permian and the Upper Jurassic. It would be rather surprising to have what would appear to be a fairly wide adaptive zone completely empty until the differentiation of Lissamphibia in the Late Mesozoic. It seems more probable that preservational or collecting bias has left these elements of the amphibian fauna unsampled. A conceivable exception to the absence of small amphibians in the Lower Mesozoic is provided by a skull described by Von Huene and Bock (1954), from the Upper Triassic of Pennsylvania, as a late lysorophid with urodele connections.

"Lissamphibia"

Structure and biology of the living orders

One of the most profound gaps in all of vertebrate phylogeny separates the Palaeozoic and Triassic labyrinthodonts and "lepospondyls" from the modern amphibian orders. Not only are the structure and way of life of the modern orders totally distinct from those of the earlier forms, but no intermediates are known, and no specific interrelationships have been established.

The generally amphibian nature of the Palaeozoic orders was recognized soon after the first adequate descriptions of their skeletons were published. By the time that biologists had understood the basic distinction between amphibians and reptiles in the modern fauna, it became obvious that labyrinthodonts and, presumably, lepospondyls, belonged to the former category. The recognition of the phyllospondyls or branchiosaurs as larval labyrinthodonts confirmed the common presence of terrestrial and aquatic life stages in both groups. It is almost certain that most, if not all, Palaeozoic amphibians practiced external fertilization, and reproduced via anamniotic eggs. These similarities, definitely of a primitive nature, should not, however, obscure the basic biological and structural differences between archaic and modern amphibians.

Modern amphibians depend on cutaneous respiration for gas exchange with the environment. Judging from the generally greater body size and the common presence of a continuous covering of heavy overlapping scales, this capacity was not a significant factor in the biology of Palaeozoic forms. The lack of scales and the necessity of keeping the body surface moist and readily permeable restricts the modern forms to damp environments and relatively small size. Body length approaching that of moderate-sized labyrinthodonts (up to 1800 mm) is achieved by the cryptobranchid *Andrias* (Meszoely, 1966) but this genus is permanently aquatic, and very exceptional in its size.

Despite the tendency of herpetologists to accept the inclusion of the three living amphibian orders within a single group, the "Lissamphibia" (Parsons and Williams, 1963), it is necessary to discuss the structure and general biology of frogs, salamanders and apodans separately, for each group differs markedly from the others throughout their known history.

Frogs are the most readily characterized, for all adhere to a single skeletal pattern that is among the most specialized of any vertebrate order. The vertebral column is greatly reduced, with only 5—9 trunk vertebrae and no tail in the adult stage. The immediately postsacral region is modified to form a longitudinal rod, the urostyle. The rear limbs are emphasized, with both terrestrial

and aquatic locomotion typically being produced by their symmetrical movement. The skull is typically, and presumably primitively, specialized in being
very open, in contrast with the labyrinthodont condition, with the main support being provided by the braincase which forms a stout longitudinal bar.
The margin of the skull is typically in the form of an anteriorly pointing arch.
Posteriorly the otic region of the braincase extends laterally to reach the dorsal
portion of the cheek. The palate exhibits very large interpterygoid vacuities. In
most genera, there is a large tympanum supported by a tympanic annulus set
into a notch in the squamosal. Vibration of the tympanum is transmitted to the
inner ear via a thin stapes, or columella. This is almost certainly a primitive condition for frogs and is almost universally accepted as being directly derived
from the pattern ascribed as typical for labyrinthodonts. Frogs also possess the
operculum-opercularis system, common to salamanders. This will be considered
with the latter group.

Among the most striking specializations of frogs is the evolution of a larval
stage that is radically different in structure and biology from the adult (Wassersug, 1973). Except in the few genera where direct development occurs, the tadpole characterizes all frog groups. Their way of life and relationship to the
environment is so different from those of the adult that they may be treated as
if they were distinct organisms, with an adaptive history quite separate from
that of the adults. Tadpoles are well known in the Miocene deposits in Czechoslovakia (Špinar, 1972) and pipid larvae from the Lower Cretaceous similar to
those of living species are being described by Estes et al. (1977). It seems
probable that the presence of such a specialized larval stage is as old a feature as
the specialization of the axial skeleton, and the two characteristics may in some
way be related. Among the specializations of the tadpole is that of diet, for
most are herbivorous suspension feeders and algae grazers, in contrast with all
adult amphibians.

Despite the high degree of specialization of the larval stage, anurans remain
primitive in other aspects of reproduction, for nearly all species practice external fertilization.

Salamanders may be considered the least specialized of living amphibians in
retaining conservative body proportions, with limbs present in most forms and
with most of the power for locomotion resulting from waves of contractions of
the axial musculature. Locomotor specialization within the group includes limb
reduction and loss of the pelvic limbs in one superfamily, and elongation of
the trunk.

The skull of salamanders is modified in a manner analogous and possibly
homologous with that of frogs, with the main support provided by the braincase, the otic capsules of which attach posterolaterally to the dorsal portion of
the cheek. In contrast with most frogs, the parabolic outline of the skull is not
complete posteriorly in the area of the temporal musculature.

Salamanders lack a tympanum. A short stapes is frequently present and
sound is said to be transmitted from the substrate through the lower jaw to the
inner ear by bone conduction. The physiology of this system has never been
confirmed, however. Many salamanders also have a second system for trans-

mission of external vibrations to the inner ear, including a second ear ossicle, the operculum. This is an element forming part of the wall of the otic capsule that is attached via a muscle to the shoulder girdle (Monath, 1965). It is assumed, but again without experimental confirmation, that the system transmits vibrations from the substrate through the forelimbs. This system seems to be of use primarily on land, for in permanently aquatic genera and the larval stages of terrestrial forms it is not evident. Frogs have a similar arrangement to that of adult terrestrial salamanders. The complexity of this peculiar hearing system is a strong argument for the close relationship or common ancestry of frogs and salamanders.

The reproductive system of primitive salamanders presumably resembles that of Palaeozoic amphibians, for external fertilization is the rule and the larvae are similar to the adult in general anatomy, except for the presence of external gills. In advanced forms, both aquatic and terrestrial, the male deposits a spermatophore that is picked up by the cloacal lips of the female so that fertilization is internal. Complicated courtship coordinates the sexual behaviour of the males and females.

Within an advanced salamander family, the Plethodontidae, many genera lay their eggs on land and development is direct, with the young miniature replicas of the adults, with no specifically larval features. For their body size, the plethodontids have the largest eggs of all salamanders, and the young are one fourth to one half the length of the adults. It has been suggested that the reproductive pattern seen in plethodontids is analogous with that which led to the development of the amniotic egg of ancestral reptiles. The European genus *Salamandra* gives birth to live young.

Apodans are the least well known of the living amphibians, being restricted to the damp tropics, and are aquatic or burrowing in habit. They are without trace of girdless or limbs, and although the trunk region may have more than 200 vertebrae, the tail is very short or absent. In contrast with frogs and salamanders, there are well-developed ribs throughout the column. Unlike the condition in frogs and salamanders, the skull roof is solidly ossified as in Palaeozoic forms. The orbital opening is small, and may be completely covered with bone. A specialized tactile or chemosensory organ, the tentacle, extends out from the skull anterior to the area of the orbit. Apodans all practice internal fertilization; in contrast with frogs and salamanders, the males possess a copulatory organ.

Fossils of frogs essentially like modern genera are known as early as the Lower Jurassic (Estes and Reig, 1973) and remains of fairly specialized salamanders from the Upper Jurassic (Hecht and Estes, 1960). The only fossil apodan is a single vertebra from the Paleocene (Estes and Wake, 1973) indicating an essentially modern pattern. The only fossil to suggest an earlier appearance of any of the modern orders is a single skeleton of *Triadobatrachus* from the Lower Triassic of Madagascar (Estes and Reig, 1973). The skull is frog-like, with a squamosal suggesting support of a tympanum. There are 14 trunk vertebrae and 6 caudals. The ilium is specialized in a manner like that of frogs, but to a lesser degree. This fossil is a plausible ancestor to frogs, but shows no specific affinities with any of the advanced families. Unfortunately, *Triadoba-trachus* provides no clues of affinities with any of the Palaeozoic amphibian groups, or with apodans or salamanders.

It is very surprising for there to be such a profound gap between the ancient and modern groups of amphibians. The apodan and salamander habitus seems to be matched by various lepospondyl groups, which suggests the presence of a similar adaptive zone in the Late Palaeozoic and Early Mesozoic. The high degree of specialization of the oldest known frogs and salamanders suggests a considerable period of prior evolution, but it would be very surprising if these forms had been at all common and yet left absolutely no fossil record. One might suppose a long "dark age" in amphibian evolution, such as that which occurred between the Upper Triassic and the Late Cretaceous for mammals, during which advanced features were perfected, but when neither extensive radiation occurred nor large populations developed. Tihen (1968) suggested that salamanders may not have diversified extensively until late in the Cretaceous, perhaps in association with the diversification of angiosperms and holometabolous insects.

The living amphibian orders must have been distinct from one another since at least the Late Palaeozoic. In the absence of any known common ancestor, it is at least possible that the ancestors of the various orders are to be found among different Palaeozoic groups, which opens up the possibility that the lineages separated as early as the Late Devonian or Early Mississippian. The specialized nature of the skeleton of salamanders and frogs separates them markedly from all Palaeozoic groups, and it is reasonable to think that the definitive pattern was not established until at least the end of the Palaeozoic. Hence the age of all these orders may be taken as Late Palaeozoic or earliest Mesozoic, whatever the specific ancestry.

Fossil Record

Apoda

There are approximately 34 genera and 160 species of living Apoda. None has a fossil record. A single vertebra from the Upper Paleocene of Brazil is the only known fossil. The strictly tropical distribution of the group is probably a major factor in reducing both the likelihood of fossilization and the subsequent recovery of their remains.

There is no basis for considering the age of any other constituent genera or species. Most authors have placed all the genera in a single family, but Taylor (1969) recognizes four families. There is no basis for estimating the time of differentiation of these groups.

The geographical distribution of the order does not appear to conform with the modern arrangement of continents. This may be interpreted as reflecting an original distribution prior to the establishment of the modern continental pattern, thus suggesting a Palaeozoic origin for the group. If apodans share a common ancestry with frogs and urodeles, the divergence of each is presumably late in the Palaeozoic. If the apodans (as suggested most recently by Carroll and Currie, 1975) have evolved separately from these groups, the Late Devonian or Early Carboniferous may mark the time of phylogenetic divergence, although not the establishment of the definitive morphological pattern.

Anurans

The classification of frogs, both the family designation of modern genera and the question of interrelationships of major lineages, is subject to continuing dispute. Various proposals are presented in the recent symposium volume, *Evolutionary Biology of the Anurans* (Vial, 1973). The number of families accepted by recent authors varies from 17 (Porter, 1972) to 23 (Savage, 1973). The number of genera affected by the differences in family assignment is, however, relatively small. The rare genera or groups of genera whose family assignment are most contentious have essentially no fossil record. In Porter's classification, only 33 genera (of a total of 218) belong to families with no pre-Pleistocene fossil record. In the classification used by Lynch (1973), only 11 living genera would be included in families without a fossil record. In terms of numbers, a major difference depends on whether the Rhacophoridae and Hyperoliidae are accepted as distinct families (Liem, 1970) or are included in the Ranidae. Whatever system of classification is accepted, most major groups of frogs do have a long, if spotty, fossil record (Fig. 4). Families that have been considered primitive on the basis of their skeletal morphology — As caphidae, Discoglossidae, Pipidae and Palaeobatrachidae — have an ancient record, going back to the Jurassic or the base of the Cretaceous. Rhinophrynids, also considered primitive on the basis of the living forms, are known from the Paleocene. Pelobatids (possibly including the pelodytids) are generally accepted as intermediate in terms of the morphology of the living genera. The record of this group extends to the top of the Cretaceous. Microhylids, although specialized in some features, are considered to have diverged from the other frog families at a comparatively early stage. Orton (1957) and Starrett (1973) suggest that the morphology of the tadpole shows the most primitive pattern known among living frogs. In contrast with other primitive frog groups, the fossil record for microhylids goes back no further than the Lower Miocene.

Leptodactylids, bufonids and hylids, all structurally more modern frogs, are known from the Paleocene, where they are represented by members of modern species groups (Estes and Reig, 1973). None of these groups has been reported from the Mesozoic. It should be emphasized that the earliest records of all these families (as well as that of a living pipid genus) come from a single locality, a fissure filling, in Brazil (Estes, 1975b; Estes and Reig, 1973). Certainly these families were present in the Cretaceous, but there is no way of estimating how early in the Cretaceous, or whether they might have been distinguishable even in the Jurassic.

Ranids have not been reported prior to the Eocene. Almost all known fossils can be placed in the genus *Rana*. Forms difficult to distinguish from living

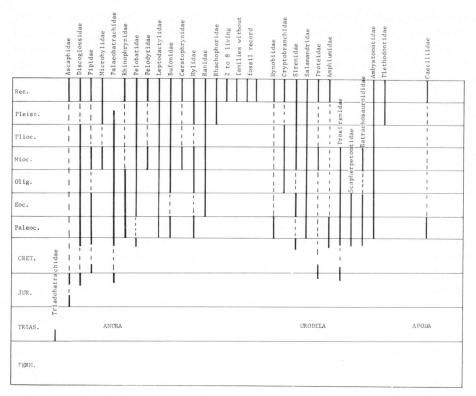

Fig. 4. Fossil record of the "Lissamphibia". Dashed lines within range indicate the absence of a fossil record.

members of this genus are known in Europe from the Oligocene (R. Estes, personal communication, 1976). In general, there is little if anything to distinguish fossil ranids from living species. Both *esculenta* and *temporaria* species groups can be recognized in the Miocene of Europe (Sanchiz, personal communication, 1976). Modern species of *Rana* appear in North America at the base of the Miocene (Estes, 1970b). Despite the relatively late appearance of ranids in the fossil record, Griffiths (1963) argues that the group had probably diverged from the more archaic frogs as early as the Jurassic. Such an early divergence is suggested by the low probability of any of the advanced frogs having evolved from the groups currently known from the Jurassic or Cretaceous. On the basis of their geographical distribution, Savage (1973) suggested that both ranids and microhylids appeared in the southern continents by the end of the Cretaceous.

Except for the Microhylidae, the fossil record generally confirms the systems of classification based on living genera, to the extent that the groups considered most archaic appear earliest in the record. Although there is general acceptance for the ranking of anuran families according to their relative degree of modernity, the possible interrelationships of the various archaic and advanced groups are subject to a bewildering number of possible arrangements (listed or cited in Vial, 1973). Some schemes accord fairly well with the fossil record, in terms of the time of appearance of most families. Others (notably Griffiths, 1963), who suggest a Jurassic origin for advanced frogs, would suppose a large gap in the fossil record for several of the families.

The evidence from both fossil and living forms suggests at least three episodes of diversification at the family level among anurans: ascaphids, discoglossids, pipids and palaeobatrachids in the Late Triassic through the Early Jurassic; pelobatids, leptodactylids, bufonids and hylids in the Cretaceous, subsequent to the divergence of the more archaic forms; ranids and microhylids apparently did not differentiate in the northern continents until the Early Tertiary.

Once a particular pattern was achieved, the basic anatomy of each of the major frog families has remained essentially constant. In as far as can be determined from the skeleton, modern genera and even species (or species groups) are in appearance from the Middle or even Early Tertiary. From the skeletal anatomy, one has the impression of at least three "quantum" jumps of general modernity. Unfortunately, there is no obvious association between the observed modernization of the skeleton in successive levels of anuran radiation and the evolutionary success of the various families. It is quite possible that these features of anuran adaptation are not clearly reflected in the skeleton.

Nevo (1968) notes that all the Jurassic and Early Cretaceous frogs are primarily aquatic forms, to judge both by the habits of their living descendants and the nature of the deposits in which they have been found. More terrestrial forms only appear later in the fossil record. The terrestrial habit among frogs might have been established at an early date, but the natural bias of the fossil record would not favour the preservation of such forms.

In order to appraise the pattern of anuran evolution, one should also consider the relative number of individuals and species at successive intervals in time. It does not seem possible yet to differentiate the obvious bias in the fossil record which progressively decreases the probability of the discovery of fossils from successively older horizons, from a real decrease or increase in population size or species numbers in successive epochs. Raup (1975) has applied a mathematical model to the data of fossil echinoderms in an effort to determine how well the fossil record may be expected to reflect actual diversity. The information for that group, however, is far richer than is that for anurans.

Wassersug (1975) has suggested that anurans (particularly the larval stage) may be adapted specifically to the seasonal fluctuations in water level and aquatic food resources which have been intensified since the breakup of the unified Palaeozoic and early Mesozoic world land mass. This would accord with a gradual increase in numbers and species diversity throughout the Late Mesozoic and Tertiary.

A striking feature of anuran evolution is that only a single family, the Palaeobatrachidae (Špinar, 1972) has become extinct. The fossil record at the level of the genus and species, compared with the number of living forms, is clearly so inadequate as to make any estimation of total number, time of origination and extinction or longevity meaningless.

Geographical distribution

The history of the geographical distribution of the anurans is a subject of continuing dispute. Quite different views have been expressed by Darlington (1957) and Savage (1973). Among the more advanced frog groups, the question of distribution is so complicated by taxonomic disputes that it is impossible to give a short summary that accurately reflects current opinion. The primitive groups certainly originated and diversified prior to the formation of the Atlantic Ocean as a major barrier. Savage views major events in subsequent anuran evolution as being associated with the separation of the northern Laurasian land mass from the southern Gondwanaland supercontinent, and subsequent separation of the southern continents.

Urodeles

The fossil record of salamanders has been discussed recently by Estes (1965, 1970b) and Tihen (1968) and a list of species given by Brame (1967). The fossil record (Fig. 4) is extremely incomplete and casts very little light on the origin of the group as a whole, the interrelationships of the various major groups, or the general evolutionary trends.

Unlike the situation for frogs, the earliest records of the various salamander families do not correlate well with the relative antiquity suggested by the degree of modernity of the living forms. Most of the forms known from the Cretaceous are paedomorphic or neotenic. Although the tendency toward neotenic habits may be one of the basic adaptations of urodeles as a group, it is felt by Estes (1969a) that the predominance of neotenic forms in the early fossil record of salamanders is a result of geological conditions, rather than indicating the dominance of this way of life for urodeles in general.

If, as is frequently suggested, a mountain brook habitat is primitive for salamanders, it would explain the rarity of their fossil remains, for such an environment is rarely preserved in the fossil record.

Two other features differentiate the fossil record of salamanders from that of frogs: the larger ratio of fossil to extant families; and the possible relationship of a fossil family (the Prosirenidae) to living groups (Plethodontidae and Ambystomatidae) (Estes, 1969b).

One aspect which shows similarity to the pattern in frogs is the relatively early appearance of most living groups. The basic pattern of sirenids and amphiumids was certainly established in the Upper Cretaceous, although these forms are placed in extinct genera. Cryptobranchids are essentially "modern" in the Oligocene. Salamandrids have a fairly extensive record in the Tertiary of Europe. A long list of extinct genera and species is almost certainly misleading. This group is currently being reviewed by Estes, who finds that most of the fossils can be assigned to modern genera.

In an appraisal of the entire order, Tihen (1968, p. 144) has stated, "Although some Miocene specimens do undoubtedly represent extinct genera (e.g. *Batrachosauroides*) it is likely that the great majority of Miocene forms represent extant genera. Although this does not of itself necessarily imply the converse, i.e., that most modern genera were already in existence during the Miocene, I think that this also is likely to prove true. In other words, most evolutionary differentiation at the generic level was probably completed before or during the Miocene."

Among the urodeles, there is little question as to the number and content of any of the families, but some variety of opinions as to their interrelationships. Being limited almost entirely to the northern temperate zone, and being much less numerous than frogs, the basic taxonomy is much better known. The fossil record, however, provides much less evidence as to the general pattern of evolution of the group.

Discussion

Patterns of radiation

The fossil record of frogs and salamanders in the Late Mesozoic and Cenozoic suggests the early establishment of many distinct lineages which have then continued for 50 to 100 m.y. with little morphological modification.

Among the "lepospondyls" of the Palaeozoic, there is at least the impression of a similar pattern. At least there is little evidence of gradual but significant transformation in a single lineage, or a series of successive radiations. This may, however, be only an artifact of the inadequacy of the fossil record.

Among the temnospondyls, clearly the most diverse of all amphibian groups, the fossil record is too incomplete to determine any particular pattern. There has been a tendency to assume a "tree-like" phylogeny, with a sequence of successive radiations. The fossil record does not support a pattern of extensive early radiation followed by long-term stability, but it is probably not sufficiently complete to preclude such a pattern. There are few families whose specific relationships to others can be documented, or whose time of origin can be established with any reasonable degree of accuracy.

The nature of the interrelationships among the various groups of anthracosaurs is too poorly established to discern any pattern.

Incompleteness of the fossil record

Eldredge and Gould (1972) have recently emphasized how one's phylogenetic conceptions are influenced by prior assumptions regarding the probable mode or pattern of evolution. The concepts of phylogenetic gradualism, in particular, have tended to exaggerate the incompleteness of the fossil record. Eldredge and Gould suggest that gradual phyletic changes in species or genera are very unlikely to be documented in the fossil record (particularly within one small geographic area). At higher taxonomic levels (previously discussed at length by Simpson, 1953) there is unlikely to be a good fossil record linking any of the major levels of organization. In addition to these expected gaps (which are observed in presumably well-represented assemblages such as the Upper Pennsylvanian, Permian and Triassic temnospondyls), there are intervals, including the whole of the Mississippian and much of the Upper Mesozoic, for which there is almost no fossil record of amphibians, despite the occurrence of numerous lineages in preceding or succeeding periods.

On the basis of known fossils, obvious gaps are evident in the record of most major amphibian groups. There is little basis for estimating the amount of radiation in the missing lineages and their longevity cannot be accurately estimated; nevertheless, they must represent a significant proportion of the total amphibian diversity.

Devonian: The ancestors of anthracosaurs and temnospondyls are unrecorded, as are those of the known ichthyostegids. Other groups, related to the ancestry of reptiles and the various "lepospondyl" groups, may have been distinct in the Upper Devonian, but have not been discovered.

Mississippian: The record of fossil amphibians is distressingly poor during this period, lacking representatives of all the following groups: the ancestors of reptiles and reptilian-like anthracosaurs; loxommatoids; trimerorhachoids; terrestrial edopoids; and an unknown number of lineages ancestral to the various "lepospondyl" groups.

Pennsylvanian: The immediate ancestors of the various "lepospondyl" families and the ancestors of seymouriamorphs are either unknown or unrecognized.

Permian: The Early Permian record gives the impression that no major groups are unrepresented, although some Pennsylvanian families that appear to become extinct may have remained in areas or environments that have not been adequately sampled. It may be that the ancestors of frogs and salamanders were distinct by this time, and yet no unequivocal evidence has been found. *Doleserpeton* (Bolt, 1969) is a possible exception. In the Middle and Upper Permian, the immediate ancestors of most of the Triassic temnospondyl superfamilies are either unknown or unrecognized. The immediate ancestors of anurans and urodeles must have been present by this time. Apodans may have been a recognizable distinct group as well, but no "lissamphibian" remains have been found.

Triassic: The majority of the temnospondyl lineages may be represented in the fossil record, but only a single fossil is known of the major anuran radiation that must have been occurring at this time. Of urodeles and apodans, the subsequent record is too incomplete to know whether these groups had begun to radiate in the Triassic, or only later. Their ancestors must have been recognizable by this time, however.

Jurassic: Several primitive frog groups are represented by the end of the period, by which time many of the surviving families must have evolved. The ancestors of even the advanced families may have been distinguishable by the end of the Jurassic, but are not represented in the fossil record until the Tertiary. Presumably, most of the salamander families had emerged by the end of the Jurassic, but the fossil record consists of a single femur and two atlas vertebrae.

Cretaceous: The record of frogs is very incomplete, lacking the ancestors of

TABLE I

Fossil record of living amphibian groups

	Apoda	Urodela	Anura	Total
Living families	1—4	8	16—22	25—34
Living families known as fossils	1	8	14	23
Percentage of living families known as fossils	25—100%	100%	64—87%	68—92%
Families known only as fossils	0	3	1	4
Living genera	34	54	218	306
Living genera known as fossils	0	16	18	34
Percentage of living genera known as fossils	0%	34%	8.3%	11%
Genera known only as fossils	1	28	26	55
Living species	158	316	2600	2974
Living species known as fossils	0	9	57	66
Percentage of living species known as fossils	0	2.8%	2.2%	2.2%
Species known only as fossils	1	63	86	150

all the intermediate and advanced families known in the Cenozoic. There is almost no record of the more terrestrial salamander groups and no trace of apodans, which must have been diversifying by the end of the period.

Comparison of the fossil record of the living amphibian groups with the number of living taxa provides some basis for estimating the probability of representation at particular taxonomic levels. Records for families, genera and species are indicated in Table I. Of the living species, slightly more than 2% are known as fossils. Eleven percent of the living genera are represented as fossils. In the few cases where modern species are known as fossils, their range may extend nearly the entire length of the Cenozoic. Such great longevity is even more evident at the level of the genus. Although extrapolation is extremely hazardous, one may assume that many living amphibian genera and species have similar ranges. Whatever their longevity, it is obvious that the vast majority of genera and species living in the Tertiary has left no fossil record.

Most genera known only as fossils are reported from only one or two localities, and appear to have a very short range. Other genera, such as *Eopelobates* (Cret.—Mioc.) and *Palaeobatrachus* (Eoc.—Mioc.) are probably more representative of the actual time range as this taxonomic level.

At the level of the family, approximately 80% of the living groups are represented in the fossil record. We may expect that during the Tertiary there is a fairly high degree of probability that most common or widespread families (aside from apodans) did leave a fossil record. This is particularly true of those living in North America and Europe.

Can we extrapolate these observations of the Tertiary groups to their Palaeozoic predecessors? It has been suggested that salamanders and apodans may have occupied the same general environmental zones as some of the lepospondyls. For these forms, comparison seems justified. It is probable that the num-

ber and ranges of genera and species of most "lepospondyl" groups indicated by
the fossil record very much underestimate the actual diversity and longevity at
these taxonomic levels. Judging from Tertiary forms, the actual diversity may
be 20 to 100 times that indicated by the fossil record. The longest well-established ranges for genera and species may be suggested as typical. This, of
course, does not provide a valid basis for comparison with the longevity of
other groups.

Among lepospondyls the large number of monotypic families suggests that
even at this level there may have been significantly more taxa than have been
reported.

The large size of most labyrinthodonts suggests that individual specimens
had a much greater probability of being preserved and subsequently identified.
Although the known ranges of genera and species are almost certainly shorter
than the actual time span of these groups, the number of recorded families may
be close to that which really existed. On the basis of phylogenetic considerations, it is probable that the actual range of most families was significantly
greater than is recorded by the known fossils.

To judge from the modern orders, it is clear that the family provides a much
better basis for comparing the diversity of fossil and living amphibians than
does the genus or species. For both fossil and living orders, the family appears
to be a much more stable taxonomic unit. Above the level of the family, various problems are encountered that make taxonomic comparison difficult. The
content and nature of superfamilies and suborders is variable from group to
group and from author to author. It is hence awkward to estimate either their
number or their longevity. At the level of the order, the numbers are so small
as to make comparisons of limited value. Furthermore, the longevity of the
amphibian orders is extremely difficult to establish, as has been previously
indicated.

Although one could tabulate the time of initial occurrence of groups at various taxonomic levels, this would clearly have little biological significance.

Summary

Amphibians have undergone two separate episodes of evolution. Palaeozoic
amphibians, divisible into labyrinthodonts and lepospondyls, radiated into a
wide spectrum of terrestrial, semi-terrestrial and aquatic habitats from the
Upper Devonian into the Triassic. With the emergence of reptiles, these groups
were successively eliminated. None survived the Triassic. Modern amphibians
— frogs, salamanders and apodans — emerged from an unknown ancestry at the
end of the Palaeozoic, remained extremely rare until the end of the Mesozoic,
but were varied and widespread elements in the Cenozoic fauna.

On the basis of the number of living forms, the fossil record, even in the Late
Cenozoic, is very incomplete at the level of the genus and species. Although
most living genera and species are without a fossil record, the fossils that are
found indicate that many genera have persisted throughout the entire Ceno-

zoic. Several modern species of frogs and salamanders are known as early as the Miocene. The great longevity of the modern taxa suggests that the very brief records of Palaeozoic genera and species are a result of geological or collecting biases rather than a true indication of more rapid origin, evolution and extinction. The majority of living families have a record extending at least to the Early Tertiary. It is probable that most Palaeozoic labyrinthodont families are represented in the fossil record. The large number of monotypic families among the Palaeozoic lepospondyl groups suggests that many other families might have existed without any record yet being found. In very few cases do the known records of family longevity appear to reflect accurately the total span of the groups. The most important gaps in the fossil record of amphibians occur in the Mississippian (Lower Carboniferous), during which ancestors of the major Palaeozoic groups were evolving, and during the Triassic and Jurassic, when the ancestors of the modern amphibians were undergoing their initial radiation. As a result of these gaps, the nature of the interrelationships of all the orders remains unknown. It is likewise impossible to establish the longevity of any of the major groups.

Acknowledgements

Ms. Lise Winer was responsible for the accumulation and tabulation of much of the taxonomic data, as well as typing the manuscript and drafting the charts. Her extensive contribution to this paper is gratefully acknowledged.

Drs. Eileen Beaumont, Oxford, Katherine Bossy, Brown University, and Angela Milner, the British Museum, were most generous in contributing information from unpublished works bearing on the diversity of Palaeozoic amphibian groups. Drs. Richard Estes and Borja Sanchiz assisted in clarifying the taxonomic position of many genera of Cenozoic amphibians. Dr. Angela Milner provided a long list of labyrinthodont species missed in the initial search of the literature.

This work was supported by grants from the Faculty of Graduate Studies and Research, McGill University, the Merrill Trust Fund and the National Research Council of Canada.

References

Baird, D., 1964. The aïstopod amphibians surveyed. Breviora, (206): 1—17.
Baird, D., 1970. Type specimen of the Oligocene frog *Zaphrissa eurypelis* Cope 1866. Copeia 1970(2): 384—385.
Baird, D. and Carroll, R.L., 1967. *Romeriscus*, the oldest known reptile. Science, 157(3784): 56—59.
Beerbower, J.R., 1963. Morphology, paleoecology, and phylogeny of the Permo-Pennsylvanian amphibian *Diploceraspis*. Bull. Mus. Comp. Zool., 130(2): 34—108.
Berman, D.S., 1973. A trimerorhachid amphibian from the Upper Pennsylvanian of New Mexico. J. Paleontol., 47(5): 932—945.
Bolt, J.R., 1969. Lissamphibian origins: possible protolissamphibian from the Lower Permian of Oklahoma. Science, 166: 888—891.
Bolt, J.R. and Wassersug, R.J., 1975. Functional morphology of the skull in *Lysorophus*: a snake-like Paleozoic amphibian (Lepospondyli). Paleobiology, 1(3): 320—332.

Bonaparte. J.F., 1963. *Promastodonsaurus bellmanni,* n.g. et n.sp., capitosáurido del Triásico Medio de Argentina (Stereospondyli—Capitosauroidea). Ameghiniana, 3(3): 67—78 (English abstract).

Bossy, K., 1976. Morphology, Paleoecology and Evolutionary Relationships of the Pennsylvanian Urocordylid Nectrideans (Subclass Lepospondyli, Class Amphibia). Thesis, Yale University, New Haven, Conn.

Boy, J., 1971. Zur Problematik der Branchiosaurier (Amphibia, Karbon-Perm). Paläontol. Z., 45(3/5): 107—119.

Boy, J., 1972. Die Branchiosaurier (Amphibia) des saarpfälzischen Rotliegenden (Perm, SW-Deutschland). Abh. Hess. Landesamt. Bodenforsch., 65: 1—137.

Boy, J. and Bandel, K., 1973. *Bruktererpeton fiebigi,* n. gen., n.sp. (Amphibia: Gephyrostegida) der erste Tetrapode aus dem Rheinischwestfälischen Karbon (Namur B; W.-Deutschland). Palaeontographica, 145: 39—77.

Brame, A.H., 1967. A list of the world's recent and fossil salamanders. Herpeton, 2: 1—26.

Brough, M.C. and Brough, J., 1967. Studies on early tetrapods. Philos. Trans. R. Soc. Lond., 252: 107—165.

Carroll, R.L., 1964a. Early evolution of the dissorophid amphibians. Bull. Mus. Comp. Zool., 131(7): 1—250.

Carroll, R.L., 1964b. The relationships of the rhachitomous amphibian *Parioxys.* Am. Mus. Novit., (2167): 1—11.

Carroll, R.L., 1967a. Labyrinthodonts from the Joggins Formation. J. Paleontol., 41: 111—142.

Carroll, R.L., 1967b. An adelogyrinid lepospondyl amphibian from the Upper Carboniferous. Can. J. Zool., 45(1): 1—16.

Carroll, R.L., 1967c. A limnoscelid reptile from the Middle Pennsylvanian. J. Paleontol., 41(5): 1256—1261.

Carroll, R.L., 1969a. Problems of the origin of reptiles. Biol. Rev., 44: 393—432.

Carroll, R.L., 1969b. A new family of Carboniferous amphibians. Palaeontology, 12(4): 537—548.

Carroll, R.L., 1970. Quantitative aspects of the amphibian—reptilian transition. Forma Functio, 3: 165—178.

Carroll, R.L., 1972. Gephyrostegida, Solenodonsauridae. Handbuch der Paläoherpetologie, teil 5B: Batrachosauria (Anthracosauria), Gephyrostegida—Chroniosuchida, pp. 1—19 (Fischer, Stuttgart).

Carroll, R.L. and Currie, P.J., 1975. Microsaurs as possible apodan ancestors. Zool. J. Linn. Soc., 57(3): 229—247.

Carroll, R.L. and Gaskill, P., 1977. The Order Microsauria. Mem. Ser. Am. Philos. Soc., in press.

Carroll, R.L., Belt, E.S., Dineley, D.L., Baird, D. and McGregor, D.C., 1972. Excursion A59: Vertebrate Paleontology of Eastern Canada. Int. Geol. Congr., 24th, Ottawa, 113 pp.

Casamiquela, R.M., 1961. Un pipoideo fósil de Patagonia. Rev. Mus. La Plata, Sec. Paleontol., N.S., 4: 35—69 (English abstract).

Chakravarti. D.K., 1968. Fauna and stratigraphy of the *Gangamopteris* beds. J. Paleontol. Soc. India, (5—9): 9—15.

Chantell, C.J., 1964. Some Mio-Pliocene hylids from the Valentine Formation of Nebraska. Am. Midi. Nat., 72: 211—225.

Chantell, C.J., 1970. Upper Pliocene frogs from Idaho. Copeia 1970 (4): 654—664.

Chantell, C.J., 1971. Fossil amphibians from the Egelhoft local fauna in north-central Nebraska. Mich. Univ. Mus. Paleontol. Contrib. 23(15): 239—246.

Chase, J.N., 1965. *Neldasaurus wrightae,* a new rhachitomous labyrinthodont from the Texas Lower Permian. Bull. Mus. Comp. Zool., 133(3): 156—225.

Chatterjee, S. and Chowdhury, T. Roy, 1974. Triassic Gondwana vertebrates from India. Indian J. Earth Sci., 1(1): 96—112.

Chowdhury, T. Roy, 1965. A new metoposaurid amphibian from the Upper Triassic Maleri Formation of central India. Philos. Trans. R. Soc. Lond., B, 250(761): 1—52.

Colbert, E.H., 1946. *Hypsognathus,* a Triassic reptile from New Jersey. Bull. Am. Mus. Nat. Hist., 86: 225—274.

Colbert, E.H., 1967. A new interpretation of *Austropelor,* a supposed Jurassic labyrinthodont amphibian from Queensland. Mem. Qld. Mus., 15(1): 35—41.

Colbert, E.H. and Cosgriff, J.W., 1974. Labyrinthodont amphibians from Antarctica. Am. Mus. Nov., (2552): 1—30.

Cosgriff, J.W., 1965. A new genus of Temnospondyli from the Triassic of western Australia. J. R. Soc. W. Austr., 48(3): 65—90.

Cosgriff, J.W., 1972. *Parotosaurus wadei,* a new capitosaurid from New South Wales. J. Paleontol., 46(4): 545—555.

Cosgriff, J.W., 1974. Lower Triassic Temnospondyli of Tasmania. Geol. Soc. Am., Spec. Pap., 149: 134 pp.

Cosgriff, J.W. and Garbutt, N.K., 1972. *Erythorbatrachus noonkanbahensis,* a trematosaurid species from the Blina Shale. J. R. Soc. W. Austr., 55(1): 5—18.

Daly, E., 1973. A Lower Permian vertebrate fauna from southern Oklahoma. J. Paleontol., 47: 562—589.

Darlington, P.J., 1957. Zoogeography: the Geographical Distribution of Animals. Wiley, New York, N.Y., 675 pp.

DeMar, R., 1968. The Permian labyrinthodont amphibian *Dissorophus multicinctus* and adaptations and phylogeny of the family Dissorophidae. J. Paleontol., 42(5), part 1: 1210—1242.

Dutuit, J.M., 1975. Notions de spécialisation et de stade évolutif chez les Almasauridés et les metoposauri-
   dés. Coll. Int. C.N.R.S., No. 218, Prob. Act. Paléontol. (Evol. Vertébr.), pp. 325—330.
Eaton, T.H., 1973. A Pennsylvanian dissorophid amphibian from Kansas. Occ. Pap. Mus. Nat. Hist. Univ.
   Kansas, (14): 1—8.
Efremov, I.A., 1940. Die Mesen-Fauna der permischen Reptilien. Neues Jahrb. Min. Geol. Paläontol., B,
   84: 379—466.
Eldredge, N. and Gould, S.J., 1972. Punctuated equilibria: an alternative to phyletic gradualism. In:
   T.J.M. Schopf (Editor), Models in Paleobiology. Freeman, San Francisco, Calif., pp. 82—115.
Estes, R., 1963. Early Miocene salamanders and lizards from Florida. Q. J. Fla. Acad. Sci., 26: 234—256.
Estes, R., 1964. Fossil vertebrates from the late Cretaceous Lance Formation, eastern Wyoming. Univ.
   Calif. Publ. Geol. Sci., 49: 1—180.
Estes, R., 1965. Fossil salamanders and salamander origins. Am. Zool., 5: 319—334.
Estes, R., 1969a. The Batrachosauridae and Scapherpetontidae, late Cretaceous and early Cenozoic sala-
   manders. Copeia, 1969(2): 225—234.
Estes, R., 1969b. Prosirenidae, a new family of fossil salamanders. Nature, 224(5214): 87—88.
Estes, R., 1969c. The fossil record of amphiumid salamanders. Breviora, 332: 1—11.
Estes, R., 1969d. A new fossil discoglossid frog from Montana and Wyoming. Breviora, 328: 1—7.
Estes, R., 1970a. New fossil pelobatid frogs and a review of the genus Eopelobates. Bull. Mus. Comp.
   Zool., 139(6): 293—339.
Estes, R., 1970b. Origin of the recent North American lower vertebrate fauna: an inquiry into the fossil
   record: Forma Functio, 3: 139—163.
Estes, R., 1975a. Xenopus from the Palaeocene of Brazil and its zoogeographic importance. Nature, 254:
   48—50.
Estes, R., 1975b. Fossil Xenopus from the Paleocene of South America and the zoogeography of pipid
   frogs. Herpetologica, 31: 263—278.
Estes, R., 1977. Relationships of the South African fossil frog Eoxenopoides reuningi (Anura, Pipinae).
   Ann. S. Afr. Mus., in press.
Estes, R. and Hoffstetter, R., 1976. Les Urodeles de Miocène de la Grive-St. Alban (Isère, France). Bull.
   Mus. Nat. Hist. Nat., Paris, in press.
Estes, R. and Reig, O., 1973. The early fossil record of frogs: a review of the evidence. In: J. Vial (Edi-
   tor), Evolutionary Biology of the Anurans. University of Missouri Press, Columbia, Mo., pp. 11—64.
Estes, R. and Tihen, J., 1964. Lower vertebrates from the Valentine Formation of Nebraska. Am. Midl.
   Nat., 72: 453—472.
Estes, R. and Wake, M., 1973. The first fossil record of caecilian amphibians. Nature, 239: 228—231.
Estes, R. and Wassersug, R., 1963. A Miocene toad from Colombia, South America. Breviora, 193: 1—13.
Estes, R., Hecht, M. and Hoffstetter, R., 1967. Paleocene amphibians from Cernay, France. Am. Mus.
   Novit., (2295): 1—25.
Estes, R., Nevo, E. and Špinar, Z., 1977 Early Cretaceous pipid tadpoles from Israel (Amphibia, Anura).
   Herpetologica, im press.
Goin, C.J. and Auffenberg, W., 1955. The fossil salamanders of the family Sirenidae. Bull. Mus. Comp.
   Zool., 113: 498—514.
Goin, C.J. and Auffenberg, W., 1957. A new fossil salamander of the genus Siren from the Eocene of
   Wyoming. Copeia, 1957(2): 83—85.
Griffiths, I., 1963. The phylogeny of the Salientia. Biol. Rev., 38(2): 241—292.
Guilday, J.E., Martin, P.S. and McCrady, A.D., 1964. New Paris No. 4: a late Pleistocene cave deposit in
   Bedford County, Pennsylvania. Bull. Natl. Speleol. Soc., 26: 121—194.
Harland, W.B., Holland, C.H., House, M.R., Hughes, N.F., Reynolds, A.B., Rudwick, M.J.S., Satter-
   thwaite, G.E., Tarlo, L.B.H. and Willey, E.C. (Editors), 1967. The Fossil Record. Geological Society,
   London, 828 pp.
Haughton, S.H., 1931. On a collection of fossil frogs from the clays at Banke. Trans. R. Soc. S. Afr., 19:
   321—346.
Hecht, M.K., 1959. Amphibians and reptiles. In: P. McGrew et al. (Editors), The Geology and Paleonto-
   logy of the Elk Mountain and Tabernacle Butte Area, Wyoming. Bull. Am. Mus. Nat. Hist., 117: 117—
   176.
Hecht, M.K. and Estes, R., 1960. Fossil amphibians from Quarry Nine. Postilla, 46: 1—19.
Herre, W., 1935. Die Schwanzlurche der mitteleocänen (Oberlutetischen) Braunkohle des Geiseltales und
   die Phylogenis der Urodelen unter Einschluss der fossilen Formen. Zoologica, 33: 1—85.
Heyler, D., 1969a. Vertébrés de l'Autunien de France. Editions de Centre National de la Recherche
   Scientifique, Paris, 255 pp.
Heyler, D., 1969b. Un nouveau stégocephale du Trias inférieur des Vosges, Stenotosaurus lehmani. Ann.
   Paléontol., Vertébr., 55: 73—80.
Holman, J.A., 1963. Anuran sacral fusions and the status of the Pliocene genus Anchylorana Taylor. Her-
   petologica, 19(3): 160—166.
Holman, J.A., 1965a. Early Miocene anurans from Florida. Q. J. Fla. Acad. Sci., 28(1): 68—82.
Holman, J.A., 1965b. A small Pleistocene herpetofauna from Houston, Texas. Texas J. Sci., 17(4): 418—
   423.
Holman, J.A., 1965c. A late Pleistocene herpetofauna from Missouri. Trans Ill. Acad. Sci., 58(3): 190—
   194.

Holman, J.A., 1966. A small Miocene herpetofauna from Texas. Q. J. Fla. Acad. Sci., 29(4): 267—275.
Holman, J.A., 1967a. Additional Miocene anurans from Florida. Q. J. Fla. Acad. Sci., 30: 121—140.
Holman, J.A., 1967b. A Pleistocene herpetofauna from Ladds, Georgia. Bull. Ga. Acad. Sci., 25(3): 154—
    166.
Holman, J.A., 1968. Lower Oligocene amphibians from Saskatchewan. Q. J. Fla. Acad. Sci., 31(4): 273—
    289.
Holman, J.A., 1969a. The Pleistocene amphibians and reptiles of Texas. Publ. Mus. Mich. State Univ.,
    Biol. Ser., 4(5): 163—193.
Holman, J.A., 1969b. A small Pleistocene herpetofauna from Tamaulipas. Q. J. Fla. Acad. Sci., 32(2):
    153—158.
Holman, J.A., 1969c. Pleistocene amphibians from a cave in Edwards County, Texas. Texas. J. Sci.,
    21(1): 63—67.
Holman, J.A., 1970. A Pleistocene herpetofauna from Eddy County, New Mexico. Texas J. Sci., 22(1):
    29—39.
Holman, J.A., 1972a. Herpetofauna of the Kanopolis local fauna (Pleistocene: Yarmouth) of Kansas.
    Mich. Acad., 5(1): 87—98.
Holman, J.A., 1972b. Herpetofauna of the Calf Creek local fauna (Lower Oligocene: Cypress Hill Forma-
    tion) of Saskatchewan. Can. J. Earth Sci., 9(12): 1612—1631.
Holman, J.A., 1973a. New amphibians and reptiles from the Norden Bridge fauna (Upper Miocene) of
    Nebraska. Mich. Acad., 6(2): 149—163.
Holman, J.A., 1973b. Herpetofauna of the Mission local fauna (Lower Pliocene) of South Dakota. J.
    Paleontol., 47(3): 462—464.
Holmes, R. and Carroll, R.L., 1977. A temnospondyl amphibian from the Mississippian of Scotland.
    Bull. Mus. Comp. Zool., 147(12): 489—511.
Hotton, N., 1959. Acroplous vorax, a new and unusual labyrinthodont amphibian from the Kansas Per-
    mian. J. Paleontol., 33: 161—178.
Howie, A.A., 1970. A new capitosaurid labyrinthodont from East Africa. Palaeontology, 13(2): 210—
    253.
Howie, A.A., 1972a. On a Queensland labyrinthodont. In: K.A. Joysey and T.S. Kemp (Editors), Studies
    in Vertebrate Evolution. Oliver and Boyd, Edinburgh, pp. 51—64.
Howie, A.A., 1972b. A brachyopid labyrinthodont from the Lower Triassic of Queensland. Proc. Linn.
    Soc. N.S. W., 96: 268—277.
Iskokova, K.I., 1969. Fossil amphibians of Irtysh region. Akad. Nauk Kazak. S.S.R., Izv., Ser. Biol., 1:
    48—52 (in Russian).
Ivakhnenko, M.F., 1971. New data on early Triassic labyrinthodonts of the Russian platform (on the spe-
    cific composition of the genus Thoosuchus). Mosk. O.-Va. Ispyt. Prir. Biul. Btd. Geol., 46(6): 145—
    146 (in Russian).
Ivakhnenko, M.F., 1972. A new benthosuchid from the Lower Triassic of the Upper Volga region. Paleon-
    tol. Zh., 4: 93—99 (in Russian).
Jux, V., 1966. Entwicklungshöhe und stratigraphisches Lager des Mechernicher Stegocephalen. Neues
    Jahrb. Geol. Paläontol., Monatsh., 6: 321—325.
Konjukova, E.D., 1955. Platyops stukenbergi Trautsch. — An archegosaurid of the lower zone of the
    Upper Permian of the Ural region. In: Materials on the Permian and Triassic Terrestrial Vertebrates of
    the U.S.S.R. Tr. Inst. Paleontol., Acad. Sci. U.S.S.R., 49: 89—117 (in Russian).
Kuhn, O., 1941. Die eozänen Anura aus dem Geiseltale nebst einer Ubersicht über die fossilen Gattungen.
    Nova Acta Leopoldina, N.F., 10: 345—376.
Kuhn, O., 1972. Seymouria. Handbuch der Paläoherpetologie, teil 5B: Batrachosauria (Anthracosauria),
    Gephyrostegida—Chroniosuchida, pp. 20—69 (Fischer, Stuttgart).
Langston, W., 1953. Permian amphibians from New Mexico. Univ. Calif. Publ. Geol. Sci., 29(7): 349—416.
LaRivers, I., 1966. A new frog from the Nevada Pliocene. Occ. Pap. Biol. Soc. Nev., (11): 1—7.
Lehman, J.P., 1967. Remarques concernant l'évolution des labyrinthodontes. Colloq. Int. C.N.R.S., Prob.
    Act. Paléontol., (163): 215—279.
Lehman, J.P., 1971. Nouveaux vertébrés fossiles du Trias de la série de Zarzaïtine. Ann. Paleontol.,
    Vertébr., 57(1): 71—113.
Lewis, G.E. and Vaughn, P.P., 1965. Early Permian vertebrates from the Cutler Formation of the Placer-
    ville Area, Colorado. Contrib. Paleontol., Geol. Surv. Prof. Pap. 503-C, U.S. Government Printing
    Office, Washington, D.C., 50 pp.
Liem, S.S., 1970. The morphology, systematics and evolution of the old world treefrogs (Rhacophoridae
    and Hyperoliidae). Fieldiana, Zool., 57: 1—145.
Lozovskiy, V.R. and Shishkin, M.A., 1974. First labyrinthodont find in the Lower Triassic on Mangy-
    shlak. Dokl. Acad. Sci., U.S.S.R., Earth Sci. Sect., 214(1—6): 42—44.
Lynch, J.D., 1965. The Pleistocene amphibians of Pit II, Arredondo, Florida. Copeia, 1965: 72—77.
Lynch, J.D., 1971. Evolutionary relationships, osteology, and zoogeography of laptodactyloid frogs. Univ.
    Kansas Publ. Mus. Nat. Hist., 53: 1—238.
Lynch, J.D., 1973. The transition from archaic to advanced frogs. In: J. Vial (Editor), Evolutionary
    Biology of the Anurans. University of Missouri Press, Columbia, Mo., pp. 133—182.
McGinnis, H.J., 1967. The osteology of Phlegethontia, a Carboniferous and Permian aïstopod amphibian.
    Univ. Calif. Publ. Geol. Sci., 71: 1—46.
Meszoely, C., 1966. North American fossil cryptobranchid salamanders. Am. Midl. Nat., 75(2): 495—515.

Milner, A.C., 1977. Morphology and Taxonomy of Carboniferous Nectridea (Amphibia). Thesis, University of Newcastle upon Tyne.

Młynarski, M., 1961. Amphibians from the Pliocene of Poland. Acta Paleontol. Polon., 6: 261—282 (in Polish).

Monath, T., 1965. The opercular apparatus of salamanders. J. Morphol., 116: 149—170.

Moss, J.L., 1972. The morphology and phylogenetic relationships of the Lower Permian tetrapod *Tseajaia campi* Vaughn (Amphibia: Seymouriamorpha). Univ. Calif. Publ. Geol. Sci., 98: 1—63.

Nevo, E., 1968. Pipid frogs from the early Cretaceous of Israel and pipid evolution. Bull. Mus. Comp. Zool., 136: 255—318.

Nevo, E. and Estes, R., 1969. *Ramonellus longispinus*, an early Cretaceous salamander from Israel. Copeia, 1969 (3): 540—547.

Noble, G.K., 1928. Two new fossil Amphibia of zoogeographic importance from the Miocene of Europe. Am. Mus. Nov., 1928 (303): 1—13.

Ochev, V.G., 1972. Capitosauroidea labyrintodonty yugo-vostoka yevropeiskoi Chasti S.S.S.R. Saratov University Press, Saratov, 209 pp.

Olson, E.C., 1952. Fauna of the Upper Vale and Choza: 6. *Diplocaulus*. Fieldiana, Geol., 10: 147—166.

Olson, E.C., 1955. Fauna of the Vale and Choza: 10. *Trimerorhachis*: including a revision of pre-Vale species. Fieldiana, Geol. 10(21): 225—274.

Olson, E.C., 1956a. Fauna of the Vale and Choza: 11. *Lysorophus*: Vale and Choza. *Diplocaulus, Cacops* and Eryopidae: Choza. Fieldiana, Geol., 10(25): 313—322.

Olson, E.C., 1956b. Fauna of the Vale and Choza: 12. A new trematopsid amphibian from the Vale Formation. Fieldiana, Geol., 10(26): 323—328.

Olson, E.C., 1957. Catalogue of localities of Permian and Triassic terrestrial vertebrates of the territories of the U.S.S.R. J. Geol., 65(2): 196—226.

Olson, E.C., 1962. Late Permian terrestrial vertebrates, U.S.A. and U.S.S.R. Trans. Am. Philos. Soc., N.S., 52(2): 1—224.

Olson, E.C., 1965. New Permian vertebrates from the Chickasha Formation in Oklahoma. Circ. Okla. Geol. Surv., 70: 1—70.

Olson, E.C., 1966. Relationships of *Diadectes*. Fieldiana, Geol., 14(10): 199—227.

Olson, E.C., 1970a. New and little known genera and species of vertebrates from the Lower Permian of Oklahoma. Fieldiana, Geol., 18(3): 359—434.

Olson, E.C., 1970b. *Trematops stonei* sp. nov. (Temnospondyli: Amphibia) from the Washington Formation, Dunkard Group, Ohio. Kirtlandia, (8): 1—12.

Olson, E.C., 1972a. *Fayella chickashaensis*, the Dissorophoidea and the Permian terrestrial radiations. J. Paleontol., 46(1): 104—114.

Olson, E.C., 1972b. *Diplocaulus parvus* n. sp. (Amphibia: Nectridea) from the Chickasha Formation (Permian: Guadalupian) of Oklahoma. J. Paleontol., 46(5): 656—659.

Ortlam, D., 1970. *Eocyclotosaurus woschmidti* n.g., n.sp., ein neuer Capitosauridae aus dem Oberen Buntsandstein des nördlichen Schwarzwaldes. Neues Jahrb. Geol. Paläontol., Monatsh. 9: 568—580.

Orton, G.L., 1957. The bearing of larval evolution on some problems in frog classification. Syst. Zool., 6(2): 79—86.

Panchen, A.L., 1959. A new armoured amphibian from the Upper Permian of East Africa. Philos. Trans. R. Soc. Lond., B, 242(691): 207—281.

Panchen, A.L., 1967. Amphibia. In: W.B. Harland et al. (Editors), The Fossil Record. Geological Society of London, London, pp. 685—694.

Panchen, A.L., 1970. Batrachosauria. Anthracosauria. Handbuch der Paläoherpetologie, teil 5A, pp. 1—84.

Panchen, A.L., 1972a. The interrelationships of the earliest tetrapods. In: K.A. Joysey and T.S. Kemp (Editors), Studies in Vertebrate Evolution. Oliver and Boyd, Edinburgh, pp. 65—87.

Panchen, A.L., 1972b. The skull and skeleton of *Eogyrinus attheyi* Watson: Labyrinthodontia). Philos. Trans. R. Soc. Lond., B, 263(851): 279—326.

Panchen, A.L., 1973a. On *Crassingyrinus scoticus* Watson, a primitive amphibian from the Lower Carboniferous of Scotland. Palaeontology, 16(1): 179—193.

Panchen, A.L., 1973b. Carboniferous tetrapods. In: A. Hallam (Editor), Atlas of Palaeobiogeography. Elsevier, Amsterdam, pp. 117—125.

Panchen, A.L., 1975. A new genus and species of anthracosaur amphibian from the Lower Carboniferous of Scotland and the status of *Pholidogaster pisciformis* Huxley. Philos. Trans. R. Soc. Lond., B, 269(900): 581—640.

Parsons, T. and Williams, E., 1963. The relationship of modern Amphibia: a re-examination. Q. Rev. Biol., 38: 26—53.

Paton, R., 1974. Capitosaurid labyrinthodonts from the Trias of England. Palaeontology, 17(2): 253—289.

Paton, R., 1975. A Lower Permian temnospondylous amphibian from the English midlands. Palaeontology, 18(4): 831—845.

Peabody, F.E., 1954. Trackways of an ambystomatid salamander from the Paleocene of Montana. J. Paleontol., 28(1): 79—83.

Porter, K.R., 1972. Herpetology. Saunders, Philadelphia, Pa., 524 pp.

Price, L.I., 1948. Um anfibio labirintodonte da formacao pedra de fogo, estado do Maranhao. Bol. Det. Nac. Prod. Min., Div. Geol. Min., (24): 1—32.

Raup, D.M., 1975. Taxonomic diversity estimation using rarefaction. Paleobiology, 1(4): 333—342.

Riabinin, A.N. and Shishkin, M.A., 1962. On the Upper Permian labyrinthodont *Jugosuchus*. Paleontol. Zh., 1: 140—145 (in Russian).

Richmond, N.D., 1964. Fossil amphibians and reptiles of Frankstown Cave, Pennsylvania. Ann. Carnegie Mus., 36: 225—228.

Romer, A.S., 1930. The Pennsylvanian tetrapods of Linton, Ohio. Bull. Am. Mus. Nat. Hist., 59: 77—147.

Romer, A.S., 1939. Notes on branchiosaurs. Am. J. Sci., 237: 748—761.

Romer, A.S., 1947. Review of the Labyrinthodontia. Bull. Mus. Comp. Zool., 99(1): 1—366.

Romer, A.S., 1952. Late Pennsylvanian and early Permian vertebrates of the Pittsburgh—West Virginia region. Ann. Carnegie Mus., 33: 1—112.

Romer, A.S., 1956. Osteology of the Reptiles. University of Chicago Press, Chicago, Ill., 772 pp.

Romer, A.S., 1964a. The skeleton of the Lower Carboniferous labyrinthodont *Pholidogaster pisciformis*. Bull. Mus. Comp. Zool., 131(6): 129—159.

Romer, A.S., 1964b. *Diadectes* an amphibian? Copeia, 1964 (4): 718—719.

Romer, A.S., 1966. Vertebrate Paleontology. University of Chicago Press, Chicago, Ill., 3rd ed., 468 pp.

Romer, A.S., 1969. A temnospondylous labyrinthodont from the Lower Carboniferous. Kirtlandia, (6): 1—20.

Savage, J.M., 1973. The geographic distribution of frogs: patterns and predictions. In: J. Vial (Editor), Evolutionary Biology of the Anurans. University of Missouri Press, Columbia, Mo., pp. 351—445.

Säve-Soderbergh, G., 1932. Preliminary note on Devonian stegocephalians from East Greenland. Medd. Grønland, 94(7): 1—107.

Säve-Soderbergh, G., 1934. Some points of view concerning the evolution of the vertebrates and the classification of this group. Ark. Zool., 26A(17): 1—20.

Seiffert, J., 1969. Urodelen-Atlas aus dem obersten Bajocien von SE-Aveyron (Südfrankreich). Palaeontol. Z., 43(1/2): 32—36.

Seiffert, J., 1972. Ein Vorläufer der Froschfamilien Palaeobatrachidae und Ranidae im Granzbereich Jura-Kreide. Neues Jahrb. Min. Geol. Paläontol., 2: 120—131.

Shishkin, M.A., 1960. *Inflectosaurus amplus*, a new Triassic trematosaurid. Paleontol. Zh., 1960 (2): 130—148 (in Russian).

Shishkin, M.A., 1967. Plagiosaurs from the Triassic of the U.S.S.R. Paleontol. J., Am. Geol. Inst., (1): 86—92.

Shishkin, M.A., 1968. On the cranial arterial system of the labyrinthodonts. Acta. Zool. (Stockh.), 49(1—2): 1—22.

Shishkin, M.A., 1973. The morphology of the early Amphibia and some problems of the lower tetrapod evolution. Tr. Paleontol. Inst., Akad. Nauk. S.S.R., 137: 1—256 (in Russian).

Simpson, G.G., 1953. The Major Features of Evolution. Columbia University Press, New York, N.Y., 434 pp.

Špinar, Z.V., 1952. Revision of some Moravian Discosauriscidae. Rov. Ustav. Geol., 15: 1—159 (in Czechoslovakian).

Špinar, Z.V., 1972. Tertiary Frogs from Central Europe. Junk, The Haque, 346 pp.

Špinar, Z.V., 1975a. A new representative of the genus *Neusibatrachus* Seiffert, 1972 (Anura) from the Miocene at Devinska Nova Ves and some considerations on its phylogeny. Casopis Min. Geol. 20(1): 59—68.

Špinar, Z.V., 1975b. *Miopelobates fejfari* n.sp., a new representative of the family Pelobatidae (Anura) from the Miocene of Czechoslovakia. Vestn. Ustred. Ustav. Geol., 50: 41—45.

Špinar, Z.V., 1976a. Hyoid skeleton and vomeri of paleobatrachids (Anura). Vestn. Ustred. Ustav. Geol., 51: 179—183.

Špinar, Z.V., 1976b. *Opisthocoelellus hessi*, a new representative of the family Bombinidae Fitzinger, 1826 — from Oligocene of Č.S.R. Sb. Ustred. Ustav. Geol., 51: 285—290.

Starrett, P.H., 1973. Evolutionary patterns in larval morphology. In: J. Vial (Editor), Evolutionary Biology of the Anurans. University of Missouri Press, Columbia, Mo., pp. 251—271.

Swinton, W.E., 1956. A neorhachitome amphibian from Madagascar. Ann. Mag. Nat. Hist., 12th Ser., 9: 60—64.

Tatarinov, L.P., 1972. Seymouriamorphen aus der Fauna der U.S.S.R. Handbuch der Paläoherpetologie, 5B: 70—80 (Fischer Verlag, Stuttgart).

Taylor, E.H., 1941. A new anuran from the Middle Miocene of Nevada. Univ. Kansas Sci. Bull., 27: 61—69.

Taylor, E.H., 1968. The Caecilians of the World: A Taxonomic Review. University of Kansas Press, Lawrence, Kansas, 848 pp.

Taylor, E.H., 1969. Skulls of Gymnophiona and their significance in the taxonomy of the group. Univ. Kansas Sci. Bull., 48: 585—687.

Tewari, A.P., 1962. A new species of *Archegosaurus* from the Lower Gondwana of Kashmir. Rec. Geol. Surv. India, 89(2): 427—434.

Thomson, K.S. and Bossy, K.H., 1970. Adaptive trends and relationships in early Amphibia. Forma Functio, 3: 7—31.

Tihen, J.A., 1942. A colony of fossil neotenic *Ambystoma tigrinum*. Univ. Kansas Sci. Bull., 28(9): 189—198.

Tihen, J.A., 1952. *Rana grylio* from the Pleistocene of Florida. Herpetologica, 8: 107.

Tihen, J.A., 1960. Notes of late Cenozoic hylid and leptodactylid frogs from Kansas, Oklahoma and Texas. Southwest Nat., 5(2): 66—70.

Tihen, J.A., 1962. A review of new world fossil bufonids. Am. Midl. Nat., 68: 1—50.

Tihen, J.A., 1968. The fossil record of salamanders. J. Herpetol., 1(1—4): 113—116.

Tihen, J.A., 1974. Two new North American Miocene salamandrids. J. Herpetol., 8(3): 211—218.

Tilley, E.H., 1971. Morphology and Taxonomy of the Loxommatoidae (Amphibia). Thesis, University of Newcastle upon Tyne.

Vaughn, P.P., 1958. On the geological range of the labyrinthodont amphibian Eryops. J. Paleontol., 32(5): 918—922.

Vaughn, P.P., 1969a. Further evidence of close relationship of the trematopsid and dissorophid labyrinthodont amphibians with a description of a new genus and new species. Bull. S. Calif. Acad. Sci., 68(3): 121—130.

Vaughn, P.P., 1969b. Upper Pennsylvanian vertebrates from the Sangre de Cristo Formation of central Colorado. Los Angeles County Mus. Contrib. Sci., (164): 1—28.

Vaughn, P.P., 1971. A Platyhystrix-like amphibian with fused vertebrae, from the Upper Pennsylvanian of Ohio. J. Paleontol., 45(3): 464—469.

Vergnaud-Grazzini, C., 1966. Les amphibiens du Miocène de Beni-Mellal. Notes Mém. Serv. Géol. Maroc., 194: 43—74.

Vergnaud-Grazzini, C., 1968. Amphibiens pléistocènes de Bolivie. Soc. Géol. Fr. Bull., 10(6): 688—695.

Vergnaud-Grazzini, C. and Hoffstetter, R., 1972. Présence de Palaeobatrachidae (Anura) dans des gisements tertiaires français: caractérisation, distribution et affinités de la famille. Palaeovertebrata, 5(4): 157—177.

Vergnaud-Grazzini, C. and Wenz, S., 1975. Les discoglossidés du Jurassique supérieur du Montsech (Province de Lérida, Espagne). Ann. Paléontol. (Vertébr.), 61(1): 1—20.

Vial, J.L. (Editor), 1973. Evolutionary Biology of the Anurans. University of Missouri Press, Columbia, Mo., 470 pp.

Von Huene, F. and Bock, W., 1954. A small amphibian skull from the Upper Triassic of Pennsylvania. Bull. Wagner Free Inst. 29: 27—33.

Warren, J.W. and Wakefield, N.A., 1972. Trackways of tetrapod vertebrates from the Upper Devonian of Victoria, Australia. Nature, 238: 469—470.

Wassersug, R.J., 1973. Aspects of social behavior in anuran larvae. In: J.R. Vial (Editor), Evolutionary Biology of the Anurans. University of Missouri Press, Columbia, Mo., pp. 273—297.

Wassersug, R.J., 1975. The adaptive significance of the tadpole stage with comments on the maintenance of complex life cycles in anurans. Am. Zool., 15: 405—417.

Watson, D.M.S., 1929. The Carboniferous Amphibia of Scotland. Paleontol. Hung., 1: 219—252.

Watson, D.M.S., 1956. The brachyopid labyrinthodonts. Bull. Br. Mus. (Nat. Hist.), Geol., 2(8): 317—391.

Welles, S.P. and Cosgriff, J., 1965. A revision of the labyrinthodont family Capitosauridae: and a description of Parotosaurus peabodyi, n.sp. from the Wupatki member of the Moenkopi Formation of northern Arizona. Univ. Calif. Publ. Geol., 54: 1—148.

Welles, S.P. and Estes, R., 1969. Hadrokkosaurus bradyi from the Upper Moenkopi Formation of Arizona: with a review of the brachyopid labyrinthodonts. Univ. Calif. Publ. Geol. Sci., 84: 1—56.

Wilson, J.A., 1948. A small amphibian from the Triassic of Howard County, Texas. J. Paleontol., 22(3): 359—361.

Wilson, R.L., 1968. Systematics and faunal analysis of a Lower Pliocene vertebrate assemblage from Trego County, Kansas. Mich. Univ. Paleontol. Contrib., 22: 75—126.

Young, C.C., 1965. On the first occurrence of fossil salamanders from the Upper Miocene of Shantung, China. Acta Paleontol. Sin., 13: 457—459.

Young, C.C., 1966. On the first discovery of capitosaurids from Sinkiang. Vertebr. Palasiat., 10: 58—63.

CHAPTER 14

TETRAPOD MASS EXTINCTIONS — A MODEL OF THE REGULATION OF
SPECIATION RATES AND IMMIGRATION BY CYCLES OF TOPOGRAPHIC
DIVERSITY

R.T. BAKKER

Introduction

The description of mass extinctions, one of the unique gifts of palaeontology
to evolutionary and ecological theory, is a rich source of opportunities for
generating and testing hypotheses about the regulation of speciation, faunal
diversification, and the fate of clades. The two best known mass extinctions are
those at the Permian—Triassic boundary and at the Cretaceous—Tertiary
boundary. These two extinction events have two outstanding characteristics:
(1) in a relatively short interval of stratigraphic time, a few tens of millions of
years or less, many groups of high taxonomic rank, which had been dominant
and diverse in various trophic-habitat guilds for a very long time, disappear
from the fossil record and are never seen again; after the mass extinction other
groups tend to evolve to fill the vacated trophic-habitat roles; thus mass extinctions result in times of accelerated turnover in community composition; (2)
during mass extinctions, the standing diversity in many habitats decreases; the
number of species, genera, and higher taxa present worldwide and in each local
fauna falls.

For terrestrial vertebrates the most discussed extinctions are those at the end
of the Pleistocene and at the Cretaceous—Tertiary boundary. The Pleistocene
extinctions were concentrated in island faunas and among very large mammals,
and the agency of man is suspected; the Cretaceous extinctions were far more
comprehensive, wiping out all terrestrial tetrapods with a body weight of over
10 kg. The terminal Cretaceous tetrapod extinctions have attracted much speculation. Most proposed explanations are *ad hoc* — search is made for a climatic
or other environmental change during the extinction event, and when evidence
for such change is found, the change itself is indicted as the agent which
directly or indirectly caused the extinctions. A more satisfying approach, I
believe, is to make a quantitative survey of tetrapod diversity through several
periods, plot standing diversity and extinction rates, and see if extinction
events have a common pattern, a *modus operandi*, which might suggest that
one process or cycle of processes controls mass extinctions all through tetrapod
history. This present study began when I was collecting data for calculating
predator—prey ratios. I sought out large collections from single formations,
made with little conscious field bias by collectors who tended to bring back
nearly all identifiable specimens, and counted and measured every individual.
Calculations of live weight and biomass ratios followed. Community composition patterns were thus drawn for the Early Permian of North America, the

Late Permian and Triassic of South Africa, and the Early Cretaceous and Late
Jurassic of North America. The literature supplied data for other times and
places. When this survey of community composition through time was completed, it seemed to suggest that diversity of large, fully terrestrial tetrapods did
not fluctuate randomly, but rather followed a repeated pattern of sudden, mass
extinctions followed by gradual rediversification. Five or six extinction events
seem to have occurred in the Permian—Cretaceous interval. This paper presents
a summary description of these diversity—extinction cycles and offers a single
explanation for all of them, based on cycles of topographic diversity, driven by
plate tectonics, regulating speciation and immigration rates.

Measuring Diversity in Fossil Tetrapods

Advantages of the Permian—Cretaceous interval

The interval from the Early Permian to the Cretaceous/Tertiary transition is
ideal for studying mass extinctions because of the lack of major zoogeogra-
phical barriers among tetrapods. Interpretation of competition and extinction
among Cenozoic mammals is complicated by delayed dispersal of isolated
faunas. For the first half of the Tertiary, South America, Africa and Australia
were separated by major oceanic barriers from each other and from Laurasia;
during the later Tertiary these barriers broke down, first for Africa, then for
South America, lastly for Australia. Europe and North America had intermittent contact — sometimes their faunas were nearly identical, sometimes each
went its separate way. Sudden changes in faunal diversity and composition can
be the result of breakdown of barriers and intermingling of previously isolated
faunas. The Permian—Cretaceous interval is simpler — the distribution of most
tetrapod groups was limited only by local climate and habitat. The Pangaea
supercontinent was welded together during the Permo-Carboniferous and
remained more or less intact through the first half of the Jurassic. Among Permian, Triassic and Jurassic tetrapods little or no provinciality can be detected
(Cox, 1974; Kalandadze, 1974). A few groups of large ectothermic tetrapods
were restricted to the tropics and subtropics and hence are absent from high
latitudes in the Permian and Triassic (Bakker, 1975b). Other groups favoured
either wet or dry conditions (edaphosaurids and caseids, respectively, in the
Permian; Olson and Vaughn, 1970). But for any given climate and habitat, the
faunal composition in the Permian to Jurassic is quite astoundingly homogeneous from one modern continent to another. The vertical sequence of faunas
in individual depositional basins also varies little from continent to continent,
so that we can be assured that most new species, genera, and families dispersed
throughout the world, wherever suitable local habitats existed, at rates which
are practically instantaneous by the standards of stratigraphic resolution.
    Sea-floor spreading and continental dispersal opened some major oceanic
barriers to tetrapod dispersal in the Cretaceous (Cox, 1974), but the differences
among even Late Cretaceous faunas can be explained mostly by local habitat

differences. For example, the dominant dinosaurs of the northern Western Interior of North America are ceratopsians (horned dinosaurs) and hadrosaurs (duckbills); sauropods are very rare or absent. Contemporary faunas from Mongolia have no large ceratopsids and a hadrosaur fauna of limited diversity, but sauropods are present. In South American faunas of the same age sauropods are dominant, hadrosaurs rare, ceratopsians rare or absent. Mammal faunas also show contrasts: marsupials are common in North America; absent from Mongolia. The climate of the northern Western Interior was wet — lignites and coals are common, and floodplain deposits are poorly oxidized. Mongolia was seasonally very arid; aeolian and reworked aeolian sediment is common and the oxidation state of most facies is high (Gradzinski, 1969; Lefeld, 1971). The South American sauropod beds are also oxidized (Price, 1947; Feruglio, 1949—50). The distribution of Cretaceous sauropods thus can be explained merely by their preference for seasonally dry climates. The existence of major oceanic barriers in the Cretaceous can be tested by tetrapod distribution only when we obtain both wet and dry facies faunas from all the modern continents.

Sampling problems — basin quality versus species flow

The terrestrial fossil record is far spottier in time and space than that of shallow-water marine invertebrates. Sedimentation in an epeiric sea can sample a fair fraction of the ecological diversity and total habitat area. But terrestrial sedimentation is severely limited to lowland basins adjacent to continental shelves, or along the shorelines of epeiric seas, or in rift valleys and basins deep within the continental interior. These lowland and valley locales represent the parts of the continents which are least diverse topographically and often least diverse in flora and fauna as well. The species diversity of Recent mammals, birds, angiosperms, and other groups has a positive correlation with topographic diversity (Simpson, 1964; Ricklefs, 1973; Nel, 1975): the most diverse vertebrate faunas occur in the mountains and adjacent high plains, not in the peripheral lowland basins. The terrestrial fossil record cannot be construed as a *direct* sampling of the entire range of the terrestrial tetrapod diversity.

If speciation and adaptive radiation are so rapid that most habitats are kept full to capacity with species and higher taxa, then the diversity of fossil tetrapod faunas would reflect merely the quality of the basin at the time of deposition — the habitat area, the rainfall and productivity, floral diversity, seasonality, etc. Diversity of large, fully terrestrial tetrapods correlates positively with rainfall in arid and semi-arid habitats (Nel, 1975); in the forest and dense forest habitats the correlation is reversed, since the understory becomes too dense to permit easy movement by large-bodied terrestrial species. Diversity of small tetrapods generally increases in forests with increasing rainfall, since the floral-structural diversity increases (Ricklefs, 1973), adding room for progressively more arboreal species. Positive correlation of megafaunal diversity with rainfall can be found in fossil communities. The highest diversity among Late Cretaceous dinosaur communities is from the Old Man Formation of the northern Western Interior; the habitat was well-watered floodplain, dotted with gallery forests. Contemporary faunas from Mongolia have much less variety; the fully oxidized clastic sediments show that rainfall was lower and more seasonal. The late Late Permian faunas of southern Africa (Lower Beaufort) are diverse — five big herbivores are about equally common; the habitat was a temperate, well-watered floodplain. The contemporary faunas of the Pri-Uralian Russian Zone IV are very poor by comparison — only one big herbivore is common; the Pri-Uralian region was intensely arid throughout the last half of the Permian; thick evaporites are present and the floodplain deposits are often fully oxidized (Efremov, 1950).

The species diversity in local basins may reflect not only the quality of the basin, its carrying capacity for species, but also the diffusion rate of new species into the basin. Since topographic and biotic diversity in lowland basins are low, speciation rates should be low; most basin species probably are not the result of speciation *in situ*, but rather of speciation and migration from the more diverse environments of the continental interior. A fundamental observation of stratigraphy is that most species appear suddenly, already possessing their diagnostic characters, and the amount of morphological change within most species during their observed life span is quite modest compared to the morphologic change separating successive species in lineages. Eldredge and Gould (1972) and Stanley (1975) argue forcefully that this pattern is not simply the product of gaps in sedimentation but rather reflects the concentration of morphological change during relatively short-lived speciation events occurring in small, peripherally isolated populations. Since the speciation event is brief and occurs in a limited geographic area, the probability of recording the quick burst of morphological change in the fossil record is small. For terrestrial basins, one could argue that species tend to appear suddenly in the sedimentary record not only because the speciation event was short-lived and occurred in small isolates, but also because most speciation does not occur in the basin at all, but in more diverse continental habitats.

The usual fate of a species in a basin is extinction within a few million years or less. If the diffusion rate of new species equals the extinction rate, a steady-state diversity is maintained within the basin. If the diffusion rate exceeds the extinction rate, diversity rises until the "carrying capacity" of the basin for species is reached. Whether such a saturation number of species of large tetrapods has ever been reached in large continental areas is a moot point. As we shall see below, the fossil record suggests that evolution can continue to increase the number of species and families of large tetrapods until a mass extinction empties the habitat.

If the diversity in local basins is determined by the diffusion rate of new species into the basin, then for any given category of basin quality, the changes in diversity observed in the Permian–Cretaceous fossil record reflect the speciation and migration rates over the entire continental surface which supplied the potential migrants into the basin of deposition. Since major oceanic barriers were few or absent, and diffusion of new species, or at least genera, was stratigraphically instantaneous, each basin during the Permian–Cretaceous interval received the products of speciation and migration from all of Pangaea.

Sampling problems — family statistics

One difficulty in the study of terrestrial tetrapod diversity is fluctuating and often small sample sizes; many important faunal stages are represented by only a few dozen large specimens; others produce tens of thousands. Another problem is that facies vary from one faunal stage to another. Simple species lists, such as are commonly published, are of little value if sample size and relative frequency are not given. One stratagem to avoid the sample-size problem is to compile statistics on the number of families for various size and habitat categories. Fluctuating sample size should have less effect on the number of families recorded than on the number of species. Taphonomic processes vary from one depositional environment to another; we may get a more complete sampling of swamp vertebrates than of those inhabiting seasonally well-drained floodplains. And taphonomy generally discriminates according to the absolute size of the bones; small bones may be winnowed out of a channel bar, but may collect in local depressions lacking megafauna (Behrensmeyer, 1975). Comparison of species lists without data on body size and environment can be misleading. Tracking the number of families of large-bodied (>10 kg adult weight) and small-bodied (<10 kg) families separately for fully terrestrial, fresh-water aquatic, and marine aquatic tetrapods should yield a fair representation of the changes in family diversity, even though the family list will most probably underestimate the absolute number present for any one time. I have plotted (Figs. 2, 3) the families present for each faunal stage and for several size/habitat categories. The terrestrial and fresh-water record for the first 4/5ths of the Jurassic is virtually non-existent; the few records are not worth plotting. I have omitted those families known from only one formation or local rock unit equivalent to a formation. Many of these single-formation records come from highly unusual local facies, which are not consistently represented in most faunal stages. These single formation records may indicate short-lived families but more probably represent sporadic sampling of habitats which usually are not trapped by fossilization. Families which are known from only one faunal stage but are present in several formations in different areas are included in the figures. This compilation will omit families which were in fact short-lived and restricted to one basin, but the low provinciality of the Permian—Cretaceous interval ensures that such short-lived endemics should be few.

Sampling problems — Simpson's D for single formations

Using family statistics should indicate changes in world-wide family diversity of tetrapods. A measure of species diversity in each local basin would also be useful. In the Permian—Cretaceous interval most genera of large terrestrial tetrapods have only one common species in one basin at one time; the number of large-bodied genera is quite close to the number of large-bodied species for any local fauna, with few exceptions. Diversity measures are available which suppress the effect of sample size. Simpson's D is such a measure, easy to compute and widely used (MacArthur, 1972, p. 197). It asks this useful question: if the individual specimens are randomly distributed through the sample area, or sample rock volume, what is the probability of drawing out two individuals of the same species consecutively? This probability is given by the expression:

$$D = \sum_i 1/(P_i)^2$$

where P_i is the relative frequency of the ith species. In a fauna with two species, each making up 50% of the sample, $D = 2$; D_{max} is equal to the total number of species. If the two species were distributed 90% and 10%, D would be about 1.2. Use of D minimizes the importance of rare species which are picked up in increasingly large samples. In practice, in tetrapod faunas, species with less than a 5% relative frequency contribute very little to the D value.

Simpson's D can be calculated for biomass; P_i becomes the biomass of species i divided by the total biomass in the sample. Since the rate at which a species contributes carcasses to the fossil record should be controlled by the productivity rate (in a steady-state population, yearly productivity is equal to yearly mortality) the calculated biomass contributed by a fossil species should indicate productivity, energy flow, and resource consumption. I believe that biomass D is a better measure of tetrapod diversity than a simple species-frequency D. And the biomass D suppresses the effect of small specimens (under 10 kg), since in most local faunas nearly all the biomass represented comes from large species (10 kg and over). This is a valuable quality, since the presence or absence of abundant microfauna depends more on taphonomy than on life frequency. In some faunas all the small specimens were removed by scavengers or current winnowing. Use of a simple species frequency D would make diversity in faunas lacking microvertebrates appear artificially low. In practice, biomass D for Permian—Cretaceous terrestrial faunas is usually close to species frequency D for specimens representing animals of 10 kg and larger. The fluctuating taphonomic bias against small specimens is a difficult problem; the best measure of small tetrapod diversity through time probably would be long-term trends in family-level statistics. An impoverished microfauna at any one formation or faunal stage may have no biologic meaning.

Taphonomic prudence must be used in censusing tetrapod collections. The census must be based on skeletal elements which vary least in robustness and resistance to destruction and current transport for any given live weight class.

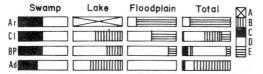

Fig. 1. Diversity of potential prey for large *Dimetrodon* in four successive formations from the Early Permian of Texas. From bottom to top: *Ad* = Admiral Fm.; *BP* = Belle Plains Fm.; *Cl* = Clyde Fm.; *Ar* = Arroyo Fm. Three biofacies are shown for each (swamp, lake, drained floodplain). The rectangle for each biofacies represents 100% of the preserved biomass. Key to genera: *A* = *Secondontosaurus*; *B* = *Ophiacodon*; *C* = *Edaphosaurus*; *D* = *Eryops*; *E* = *Diadectes*. Each genus has only one common species in any one biofacies.

In Early Permian tetrapods, the head size varies enormously for a given body weight — edaphosaurs are microcephalic; *Eryops* is hypermacrocephalic. A skull count would overestimate the *Eryops*/edaphosaur ratio. Femora and humeri vary much less and make excellent census indices. For thecodonts and dinosaurs limb bones are also the most reliable index. In therapsids (mammal-like reptiles) skull size varies little relative to body size, and the heavy-snouted, akinetic skulls are often the most abundant skeletal elements preserved.

The Patterns of Diversity

Dynasty I. Carboniferous — early Late Permian (Fig. 1)

As interpreted in this section, and in Figs. 2, 3 and 8, the Permian—Cretaceous record of land tetrapods is a succession of seven or eight complexes: each complex lasted from 20 to 60 m.y., was composed, at its peak, of about a half dozen to a dozen families of large herbivores, and was truncated by a mass extinction of most of the characteristic lineages. I use the term "dynasty" for these complexes.

Dynasty I has roots in the latest Carboniferous (Olson, 1962; Olson and Vaughn, 1970), and was composed of a cluster of primitive reptiles, near-reptiles, and both terrestrial and aquatic amphibians. The pole-ward temperature gradient was steep; continental glaciation occurred in south Gondwanaland and probably in northern Siberia (Frakes et al., 1975). All the tetrapods were probably ectothermic; nearly all are restricted to the warm tropics and subtropics of Laurasia and northern Gondwanaland (Bakker, 1975a). Throughout the Permian, the overall climatic trend in the tropics was towards increasingly severe aridity (Olson and Vaughn, 1970). This trend controlled changes in the relative frequencies of the tetrapod faunas — genera linked to swamps and large lakes became increasingly rare; genera characteristic of oxidized floodplains became more common. Relatively little morphological evolution occurred within the families, except for phyletic size increase.

Dynasty II. The first therapsids (Fig. 2, faunal stage *1*)

The important dry-habitat families of the Early Permian survive into the early Late Permian (*Dimetrodon*, caseids, captorhinids, dissorophids) and overlap in time with the first therapsids (mammal-like reptiles: Olson, 1962). The first therapsids were probably already fully endothermic (Bakker, 1975b) and quickly radiated into an extraordinary variety of large and small terrestrial herbivores and carnivores. The ectothermic families died out early during this initial therapsid radiation. Here is excellent evidence for progressive evolution producing a fundamental improvement (endothermy), and the improved clade radiating rapidly, causing the extinction of clades lacking the improvement through competition and predation.

Diversity within any local fauna of the all-ectothermic Dynasty I was low;

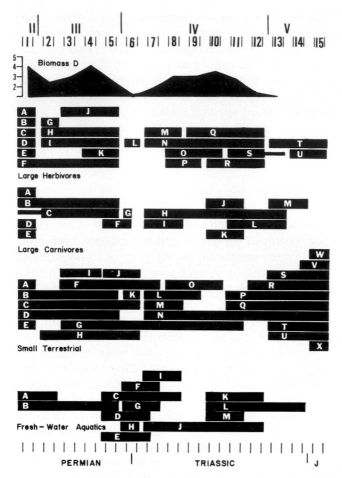

Fig. 2. Diversity of non-marine tetrapods, early Late Permian to Early Jurassic. Each bar represents one family. Narrow extensions of bars indicate that the family is present but very rare. Families known from only one formation are omitted. Roman numerals at top show the successive "dynasties" as described in the text. Biomass D has been calculated for each faunal level from the formation with the largest number of identifiable specimens. The faunal levels are represented by the following formations (the first named formation supplies the data for D, except for no. 8, where two have been combined because of small sample size): 1, Tap Zone, Russian Zone II; 2, Ruhuhu, low *Kistecephalus* Zone; 3, upper *Kistecephalus* Zone, Madumabisa Mudstone; 4, low and middle *Daptocephalus* Zone, Kawinga; 5, upper *Daptocephalus* Zone; 6, *Lystrosaurus* Zone (South Africa); 7, *Cynognathus* Zone, ?Ur-Ma-Ying; 8, Omingonde Mudstone, Russian Zone VI; 9, Manda; 10, Santa Maria; 11, Ischigualasto, lower Red Beds (South Africa); 12, Lower Chinle, lower Los Colorados, Wolfville; 13, upper Los Colorados, Stuben Sandstone; 14, Lufeng Beds, Knollenmergl, upper Red Beds (South Africa); 15, Forest Sandstone, Cave Sandstone, Portland Arkose. The South African and Russian "zones" correspond in rank to formations or series.

Family abbreviations: *Large Herbivores*: A = titanosuchids; B = struthiocephalids; C = tapinocephalids; D = moschopids; E = styracocephalids; F = pareiasaurids; G = endothiodontids; H = oudenodontids; I = aulacephalodontids; J = whaitsiids; K = daptocephalids; L = lystrosaurids; M = diademodontids; N = kannemeyeriids; O = stahleckeriids; P = shansiodontids; Q = rhynchosaurids; R = traversodontids; S = aetosaurids; T = melanorosaurids; U = plateo-

usually one or two large species make up most of the specimens (Fig. 1). The low diversity is especially marked among big herbivores: in early faunas, there is only one common species in the swamp facies (*Lupeosaurus* or *Edaphosaurus*) and one in the drained floodplain facies (*Diadectes*): in the later, more arid faunas two common herbivores can be found together (*Angelosaurus* and *Cotylorhynchus*: Olson, 1962). Large herbivore diversity is much higher in the first therapsid radiation, Dynasty II (Fig. 2, faunal stage *1*, families *A—F*): in the floodplain facies of the Tap Zone of the South African Karroo, five families and six genera are rather widespread and common; biomass *D* is about four. High large herbivore diversity was achieved by the radiations of the succeeding therapsid, therapsid—archosaur, dinosaur, and mammal dynasties (Figs. 2, 3). I believe that the low herbivore diversity of Dynasty I was not controlled by local environmental quality, since diversity was low in all types of habitats, but rather by a low rate of evolution of trophic divergence. Dynasty I was probably all-ectothermic; low-energy budgets and food consumption kept trophic competition to a minimum. The evolutionary conversion rate of carnivores and omnivores into herbivores was low, and the herbivore clades themselves showed only feeble rates of diversification. The therapsids, dinosaurs, and mammals were probably endothermic, with about twenty-fold increase in the energy consumption for a given body weight; trophic competition and diversification was much more vigorous than in Dynasty I.

Dynasty III. Therapsids (Fig. 2, faunal stages *2—5*)

All five fully terrestrial large herbivore families of Dynasty II (*A—E* in Fig. 2), the only family of very large predators (anteosaurids, *A* in Fig. 2) and the two most common and diverse medium-to-large carnivore families (*D* and *E* in Fig. 2) disappear abruptly together in the fossil record of the Karroo Basin in South Africa. This apparently massive extinction marks the division between the Tap Zone and the *Kistecephalus* Zone in the Karroo, and between Dynasty II and III of this chapter. In the earliest known post-Tap Zone fauna of southern Africa, new groups of big herbivores (dicynodonts) begin to appear, descended from families which were common but restricted to small body sizes during the Tap Zone. Initial diversity of big dicynodonts may have been low, with one genus, *Endothiodon*, dominating some early local faunas (*G* in Fig. 2; Cox,

---

saurids. *Large Carnivores*: *A* = anteosaurids; *B* = hipposaurids; *C* = gorgonopsids; *D* = lycosuchids; *E* = pristerognathids; *F* = moschorhinids; *G* = proterosuchids; *H* = erythrosuchids—rauisuchids; *I* = cynognathids; *J* = herrerasaurids; *K* = chiniquodontids; *L* = ornithosuchids; *M* = procompsognathids (including halticosaurs and dilophosaurs). *Small Terrestrial*: *A* = dissorophids; *B* = dikopsids; *C* = scaloposaurids; *D* = emydopsids; *E* = nycteroleterids; *F* = kingoriids; *G* = procolophonids; *H* = kistecephalids; *I* = procynosuchids; *J* = galesaurids; *K* = prolacertids; *L* = trirachodontids; *M* = bauriids; *N* = sphenodontids; *O* = gracilosuchids; *P* = pedeticosaurids; *Q* = heterodontosaurids; *R* = anchisaurids; *S* = tritylodontids; *T* = fabrosaurids; *U* = ictidosaurids; *V* = icarosaurids; *W* = khuneotheriids; *X* = morganucondontids. *Fresh-Water Aquatics*: *A* = archegosaurids; *B* = rhinesuchids; *C* = brachyopoids; *D* = benthosuchids; *E* = uranocentrodontids; *F* = rhytideosteids; *G* = sclerothoracids; *H* = lydekkerinids; *I* = trematosaurids; *J* = capitosaurids; *K* = metoposaurids; *L* = phtyosaurs; *M* = cerritosaurids.

1964). But more big dicynodont families were added, and at the end of Dynasty III (faunal stages *4* and *5* in Fig. 2) four fully terrestrial big families were common, plus pareiasaurs, big aquatic herbivores. Biomass *D* rises during the first part of Dynasty III, and appears to reach maximum some time before the end of the dynasty. In the latest Dynasty III fauna (late *Daptocephalus* Zone, stage *5* in Fig. 2) one genus, *Daptocephalus*, increases in relative frequency at the expense of the other big herbivores; thus local diversity, measured by *D*, decreases although all of the genera and families seem to be present right through to the end of the zone.

One family dominates the top predator role in Dynasty III, the gorgonopsians, making up nearly all the specimens known (*C* in Fig. 2). Large gorgons were present but very rare in the preceding Tap Zone. The medium-size hipposaurids, also present in the Tap Zone, were wide-spread but uncommon all through Dynasty III (Kitching, 1972). Late in the dynasty two more large predator families appear — the proterosuchian thecodonts and the therapsid moschorhinids. The latter make up about half the predator specimens in the latest *Daptocephalus* Zone faunas at the end to Dynasty III (Kitching, 1972).

Dynasty IV. Therapsids, thecodonts, early dinosaurs (Fig. 2, stages *6—12*)

At the end of the *Daptocephalus* Zone, about at the Permian—Triassic boundary, all of the big herbivore families of Dynasty III, including the pareiasaurs, disappear (faunal stage *5*, fig. 2). The gorgons and hipposaurids also disappear. The next faunal zone, considered here to be the beginning of Dynasty IV, is the *Lystrosaurus* Zone; only one genus of big herbivore is present, the dicynodont *Lystrosaurus*, with only a very few large species present in any one locale (Cluver, 1971; *L* in Fig. 2). *Lystrosaurus* occurs in astounding numbers and in a wide range of facies and geographic locations. In the early *Lystrosaurus* Zone in South Africa, top predators are holdovers from the late Dynasty III — moschorhinids and proterosuchians. In the mid and late *Lystrosaurus* Zone, top predators are absent entirely; although big lystrosaurs are common, no carnivore larger than 1 kg adult weight has been found (Kitching, 1972). The empty top predator role here is surprising, especially so because lystrosaurs seem to occur in a variety of facies including overbank, floodplain sediments. Evidently the early *Lystrosaurus* Zone big carnivores had become extinct and were not immediately replaced by the evolution and migration into the basin of new big carnivores.

As recognized here, Dynasty IV begins at about the Permian—Triassic boundary and lasts until early Late Triassic; it is the best sampled of all the dynasty units recognized in this chapter, with many local faunas scattered through all the continents. Increase in herbivore diversity occurred in a series of steps: (1) in the *Lystrosaurus* Zone, one big herbivore family was present, the lystrosaurids, probably descended from daptocephalids (Cluver, 1972); (2) in the next Karroo Zone, the *Cynognathus* Zone (stage 7 in Fig. 2), lystrosaurids become extinct and are replaced by a big dicynodont group, the kannemeyerids, also probably daptocephalid descendents, and by diademodontid cynodonts, descendents of small omnivorous cynodonts; (3) in the next stage, (8 in Fig. 2), two more big dicynodont families, again daptocephalid descendents, are added — shansiodonts and stahleckeriids; (4) in the following stage (9), diademodontids disappear but are replaced by big rhynchosaurs, descendents of small rhynochosaurs of stage 7. By faunal stage 8 of Fig. 2 a standing diversity of four big herbivore families was reached, with biomass D between 2 and 4 in local basins; turnover continued in big herbivores in stages 9 through 12, but the number of big herbivore families remained at about 4. In one local fauna at the very end of Dynasty IV (Chinle, stage 12) only one big herbivore genus is common, the kannemeyerid *Placerias*.

Top carnivore turnover is brisk in Dynasty IV. Two groups appear in stage 7, cynognathid cynodonts and erythrosuchid thecodonts, both descendents of groups which were probably of small size in preceding stages. Erythrosuchids die out shortly but give rise to the very long lived rausiuchid family. Cynognathids are replaced by chiniquodont cynodonts (K in Fig. 2), big predators also descended from some small cynodont family of the Early Triassic. Other thecodonts and, finally, early dinosaurs enter the top predator role in the Middle and Late Triassic.

Dynasty V. Prosauropod dinosaurs (stages 13 and 14 in Fig. 2)

All or nearly all of the late Dynasty IV big herbivores become extinct within a narrow stratigraphic range in the Late Triassic — the rhynchosaurs, dicynodonts, and big, armoured aetosaur thecodonts disappear. In the faunal zone following these extinctions (13 in Fig. 2) prosauropod dinosaurs of two closely related families (T and U in Fig. 2) make up nearly all big herbivore specimens. In any one local basin, the diversity appears to be very low, reminiscent of that of the *Lystrosaurus* Zone; usually one genus of big prosauropod dominates the collections, although various growth stages sometimes have been recognized as distinct genera (Rozdestvenski, 1965). Surprisingly, the top predators of these early prosauropod zones are holdovers from the mid-Triassic — ornithosuchid and rausiuchid thecodonts (H and L in Fig. 2). Advanced theropod dinosaurs take over this trophic role at the Triassic—Jurassic boundary (M in Fig. 2).

Dynasty VI. Sauropod and stegosaur dinosaurs (Jurassic in Fig. 3)

The first three quarters of the Jurassic is extremely poorly sampled for terrestrial faunas. Latest Jurassic show a splendid diversity of giant sauropod dinosaur herbivores and the big, armoured stegosaur herbivores ($A-E$, Fig. 3), with several other, less common families. The transition from the prosauropod faunas of Dynasty V to the sauropod/stegosaur faunas of Dynasty VI is not at all clear; there are not enough data to decide whether a sudden extinction terminated the prosauropod radiation, or whether this radiation gradually produced the sauropod radiation, without the intervention of an extinction event. I suspect the latter alternative is correct, which would make the distinction between Dynasties V and VI artificial.

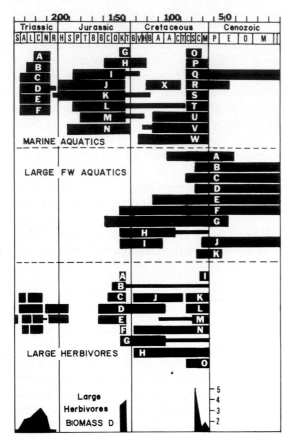

Fig. 3. Diversity of tetrapods, Early Triassic to Cretaceous. Standard marine stages are indicated by intitials in boxes at top. Families indicated as in Fig. 2. Biomass D for large herbivores for Triassic taken from Fig. 2; D calculated for a few large Jurassic and Cretaceous samples from the following formations: *1*, Tendaguru (Kimmeridgian); *2*, Morrison (Tithonian); *3*, Old Man (Campanian); *4*, Lower Edmonton A (Campanian/Maastrichtian); *5*, Lower Edmonton B (Maastrichtian); *6*, Lance—Hell Creek—Frenchman (latest Maastrichtian).

Family abbreviations: *Marine Aquatics*: A = henodontids; B = pachypleurosaurids; C = mixosaurids; D = placocheliids; E = nothosaurids; F = shastasaurids; G = pliosaurids; H = metriorhynchids; I = rhomaliosaurids; J = ichthyosaurids; K = plesiosaurids; L = stenopterygiids; M = teleosaurids; N = rhamphorhynchids; O = protostegids; P = mosasaurids; Q = cheloniids; R = toxocheliids; S = ichthyornids; T = hesperornids; U = elasmosaurids; V = polycotylids; W = ornithocheirids (pteranodontids); X = ornithodesmids. *Large Fresh-Water Aquatics*: A = baenids; B = varanids; C = pelomedusids; D = dermatemyids; E = crocodylids; F = trionychids; G = pholidosaurids; H = goniopholids; I = glyptopids; J = emyids; K = champsosaurids. Triassic fresh-water aquatics shown in Fig. 2. The sampling of non-marine tetrapods during the first eight stages of the Jurassic is so poor that the records are not worth plotting on this compilation. *Large Terrestrial Herbivores*: A = camarasaurids; B = diplodocids; C = stegosaurids; D = brachiosaurids; E = cetiosaurids; F = camptosaurids; G = hypsilophodontids; H = panoplosaurids; I = ceratopsids; J = iguanodontids; K = hadrosaurids; L = protoceratopsids; M = titanosaurids; N = pachycephalosaurids; O = euoplocephalids. Explanation of Trias-

Dynasty VII. Duck bill, ankylosaur, ceratopsid dinosaurs (Cretaceous in Fig. 3)

The sauropod/stegosaur faunas continue to be diverse well into the Upper Morrison Formation of the Western Interior of the U.S., recently dated by pollen as middle or late Tithonian (latest Jurassic; Bakker et al., 1976). But all the stegosaurs and most of the sauropod lineages become extinct or greatly reduced in number before the earliest Cretaceous faunas from Germany and England (Wealden Facies; Koken, 1887). In these earliest Cretaceous faunas a cluster of new big herbivore families appear (*J, N, H* in Fig. 3), all probably descended from ornithischian dinosaurs of small or medium size during the Jurassic. One of these Early Cretaceous ornithischian families, the iguanodonts, becomes extinct in mid-Cretaceous, but gave rise to the hadrosaurs, a dominant Late Cretaceous group (*K* in Fig. 3). The Cretaceous terrestrial record is spotty, without large sample sizes for much of the period. Extinction and addition of big herbivores seems to have continued during most of the period, with more families present in the latest Cretaceous than at any time during the Early Cretaceous (Fig. 3).

The largest carnivores of the Early Cretaceous seem to be derived from big Late Jurassic types; new top predators, the tyrannosaurs, take over in the Late Cretaceous. A new group of medium-size but very heavily armed carnivores (deinonychids) appear in the Early Cretaceous and are the most common predators in the Cedar Mountain and Cloverly Formation (Bakker et al., 1976). Tyrannosaurs and deinonychids may well have a common ancestry in some small-bodied Jurassic theropod.

In the Late Cretaceous of the Western Interior, a sequence of three faunal stages with good sample sizes is known from sediments deposited on a well-watered deltaic plain (Russell, 1967b; Bakker, 1972). The earliest fauna, that of the Old Man Formation, is very rich, with five or six common big herbivore genera (Fig. 3; biomass *D*): the next, from the Lower Edmonton, is noticeably impoverished, with only two or three common herbivores; the last, from the latest Cretaceous Hell Creek—Lance—Upper Edmonton Formations, has a very low diversity; in most of Montana, South Dakota and Wyoming one genus, the great horned dinosaur *Triceratops*, makes up between 60 and 90% of the specimens. Thus local diversity, measured by *D*, in the northern Western Interior seems to have decreased in a series of steps during the Late Cretaceous, although none of the families of big herbivores seem to have disappeared until the very end of the period.

---

sic terrestrial families as in Fig. 2. Figs. 2, 3 and 4 include data from the following sources: Anderson and Anderson (1970); Appleby et al. (1967); Bonaparte (1971). Boonstra (1969); Camp and Welles (1956); Carroll et al. (1972); Chatterjee and Roy-Chowdhury (1974); Cosgriff (1974); Cox (1964, 1965, 1974); Cruickshank (1972); Estes (1970); Estes and Reig (1973); Gaffney (1975); Galton (1971); Hopson and Kitching (1972); Kalandadze (1970, 1974); Kalin (1955); Keyser (1973); McGowan (1970, 1972a, b); McKenna (1969); Olson (1962); Panchen (1967); Persson (1963); Russell (1967a, b); Steel (1967, 1970, 1973); Von Huene (1932, 1942, 1935—42); Young (1951 with corrections from Rozdestvenski, 1965); Zangerl (1953, 1960, 1971).

End of Dynasty VIII — beginning of the age of mammals

Although local diversity may have decreased towards the end of the period, all of the Late Cretaceous big herbivore families seem to have lasted, in one basin or another, until the latest Cretaceous. But then, apparently within a very short stratigraphic interval (Van Valen and Sloan, 1976), all dinosaurs became extinct. The only surviving terrestrial tetrapods were mammals, lizards, and other groups with a body size of under 10 kg. This great mass extinction has attracted much attention (Russell, 1971; Van Valen and Sloan, 1976), but it should be pointed out that extinction of large terrestrial tetrapods was nearly as complete at the end of Dynasty II and Dynasty III.

Therian mammals had begun an adaptive radiation within the Late Cretaceous (Crompton and Kielan-Jaworowska, 1976), but were restricted to small body sizes until after the dinosaur extinction had occurred. Therian mammal diversification was explosive in the Early Cenozoic, but most of the variety produced was in insectivorous and omnivorous forms. In most Paleocene and Eocene faunas only one or two big herbivore families are common; standing diversity of big mammalian herbivores did not reach the level of the Late Cretaceous dinosaur faunas until the Oligocene. The Late Cretaceous faunas had included two groups of big dinosaur herbivores equipped with powerful shearing batteries of continuously replaced cheek teeth (ceratopsids and hadrosaurs); these dental batteries were capable of dealing with tough fibrous plant tissues. Mammals with analogous batteries, made up of rootless, continuously growing cheek teeth, did not become common until the Oligocene or Miocene.

World-wide synchroneity of diversity cycles

The fluctuations in large terrestrial herbivore diversity, measured by the world-wide number of families and by D for individual formations, and as plotted in Figs. 2 and 3, faithfully reflect the diversity preserved in museum collections made with little conscious field bias. The best sampled dynasties, III, VII, and especially IV, seem to be asymmetric through time: they begin with a total or near-total extinction of all big herbivores; diversity of big herbivores is then increased gradually, in a series of steps, as introduction rate of new families exceeds the extinction rate (see stages *6—9* in Fig. 2); end of the dynasty comes with extinction of most or all of the big herbivores in what appears to be a narrow stratigraphic interval.

The single most important methodological question in this study is this: are the diversity fluctuations in the observed Permian—Cretaceous record merely the result of a combination of taphonomic bias, variable sample size, and differences in the local environmental quality of the depositional basins? Or do the diversity patterns actually reflect changes in the world-wide pool of potential immigrants which supplies most new species into the basins? I believe that the latter interpretation is correct, for the following reasons.

None of the faunas used to calculate D in Figs. 2 and 3 come from facies with evidence of severe aridity (aeolian clastics; abundant evaporites), so that low diversity at the beginning of Dynasties III, IV and V is probably not the result of low rainfall and primary productivity. The *Lystrosaurus* faunal zone is very widespread — South Africa, India, China, and Antarctica — and occurs in red/green mottled floodplain sediments (Karroo) and in clastic units associated with coal beds (India, Antarctica: Colbert, 1974; Chatterjee and Roy-Chowdhury, 1974). Everywhere the large herbivore diversity is very low. Mid-Triassic faunas are also widespread — Brazil, Argentina, Southwest Africa, Tanzania, Zambia, India, and Pri-Uralian Russia — and everywhere even small samples yield three or four families of big herbivores (Kalandadze, 1970; Keyser, 1973). Late Triassic faunas from Argentina, South Africa, China, and Germany all seem to be dominated by a very few species of big prosauropod dinosaurs (Bonaparte, 1971; Rozdestvenski, 1965). The two large Late Jurassic samples come from quite different facies. The Morrison was seasonally arid floodplain deep in the continental interior; the Tendaguru Beds were a delta building out into the proto-Indian Ocean. But diversity is high in both and the family composition is rather similar.

The overall patterns of family diversity and biomass *D* do not seem to be controlled by local facies or faunal provinces.

Does a saturation number exist?

Five fully terrestrial families of big herbivores are recognized for Dynasty II, but structural differences among some of these are certainly a grade less than that commonly considered necessary for family distinction among later therapsids and dinosaurs. The moschopids and tapinocephalids could be merged into one family. Dynasty II, and III, and IV then have the same maximum number of fully terrestrial big herbivore families — four. These dynasties were of short duration — 15 to 30 m.y. each. The number of big families is much higher at the end of the Cretaceous and Jurassic — seven to nine (Fig. 3). The Cretaceous was a long dynasty, 70 m.y. of radiation without a major extinction, and the Jurassic dynasty may have been as long. The pattern suggests that diversity would have increased in Dynasties II, III and IV if the dynasty-terminating extinctions had been postponed, and that the maximum diversity observed in these short dynasties does not reflect a saturation of the environment.

Diversity patterns — small terrestrial tetrapods

Small (adult weight <10 kg) tetrapods do not show patterns of mass extinction. Microvertebrate sampling is spotty, but marked decreases in diversity do not occur at any of the dynasty terminations (Figs. 2, 3, 4). Mid-Triassic microvertebrates are very poorly known, and the apparent decrease in number of families here probably is the result of poor sampling. At the end of the Triassic the apparent number of families increases. Webb (1969) suggests that when megafauna diversity declines, more resources become available to small vertebrates, and the microfaunal diversity may increase; perhaps the Late Triassic bloom of small families is such a case. Jurassic, Early and early Late Cretaceous

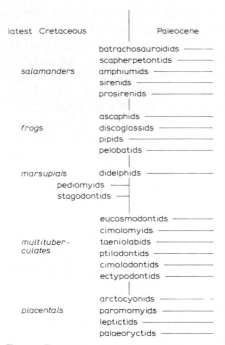

Fig. 4. Families of small tetrapods at the Cretaceous—Tertiary boundary.

microfaunas are poorly sampled. In the northern Western Interior, microverte-brate faunas have been studied intensely across the Cretaceous—Tertiary boundary (Fig. 4). Two families of marsupials become extinct at the systemic boundary, possibly from competition with radiating placental mammals. But all or nearly all of the other mammals and small lizards survived across the boundary in undiminished numbers and diversity. The size-specificity of this extinction is remarkable: all terrestrial families with an adult weight of 10 kg or more became extinct; nearly all below 10 kg survived.

Diversity patterns — fresh-water aquatics

Fresh-water aquatic tetrapods do not show mass extinction. The number of families appears to fall at the end of the Triassic (Fig. 2), but this decline is probably caused by facies peculiarities of the known tetrapod sites at this time — fully oxidized clastics and aeolianites are very common, and most sites seem to represent quite dry habitats. The most diverse fresh-water aquatic family of Dynasty IV, the phytosaurs (L in Fig. 2), survives into the latest Triassic (early Dynasty V). Fresh-water crocodiles and turtles probably existed in the Late Triassic. The number of families increases at the Permo-Triassic boundary and at the Cretaceous—Tertiary boundary (Figs. 2, 3). Wet facies are abundant and widespread across these transitions, and the data probably do reflect real diversity increments. Sampling is especially good across the Cretaceous—Tertiary boundary, and virtually all the large and small fresh-water families survive; most are still with us in the Recent (Figs. 3, 4). More families are added in the Paleocene (chelydrids; Estes, 1970). The most important fresh-water families survive across the Jurassic—Cretaceous boundary (goniopholid crocodiles; soft shell and glyptopid turtles).

Pareiasaurs are listed as big herbivores (*F* in Fig. 2) but are segregated in swampy facies and are rarely found with the dinocephalians in the Karroo; in Russia pareiasaurs are most common in channel deposits (Efremov, 1950). These animals most probably were strongly aquatic. And they are the only big herbivores to survive from Dynasty II through Dynasty III.

Diversity patterns — marine aquatics

Marine aquatic tetrapods suffer mass extinctions in phase with the big terrestrial herbivores (Fig. 3). Permian marine aquatics include only little *Mesosaurus* of periglacial Gondwanaland. Marine extinctions are most nearly complete in the Late Triassic and at the Cretaceous—Tertiary boundary; at the Jurassic—Cretaceous boundary a number of families survive but only with greatly reduced numbers (Fig. 3). The only two marine families crossing the Cretaceous—Tertiary boundary are the toxochelyid and cheloniid sea turtles. These families seem to be the least fully pelagic of all the big Cretaceous marine families; both retained relatively unmodified hindlimbs and probably could move over marshlands. Protostegid sea turtles had fore and hind flippers far more specialized for an open-sea existence (Wieland, 1900; Zangerl, 1953, 1960), and protostegids did not survive the systemic boundary. Facies distributions also indicate that toxochelyids and cheloniids were less strongly linked to open-sea conditions than were protostegids: the latter are common in the off-shore facies of the broad epeiric sea sediments in Kansas and Mississippi (Zangerl, 1953, 1960) but are absent from the narrow shelf sediments of the Cretaceous Atlantic coast, deposits which yield a mixed fauna of terrestrial, estuarine, and near-shore marine vertebrates (Russell, 1967a); toxochelyids and cheloniids are common in the Atlantic facies. One of the few Jurassic marine families to remain common in the early Cretaceous were the metriorhynchid sea crocodiles; these animals retained a sacral rib/ilium contact, a long femur, and a hindlimb generally less modified for swimming and more capable of terrestrial locomotion than that of the various ichthyosaurs and plesiosaurs which were greatly depleted by the terminal Jurassic extinctions. It appears that marine aquatics had a probability of surviving mass extinction events inversely proportional to the degree of specialization for fully pelagic habits.

Suggested Causes of Mass Extinctions

Climatic change

A climatic change towards cooler, wetter conditions has been suggested as the cause of the terminal Cretaceous terrestrial extinctions (Russell, 1971;

Worsley, 1971). The climate-change paradigm for explaining evolutionary events has peculiar properties: when the stratigraphic record at a mass extinction is examined intensively by palynologists, sedimentologists, and micropalaeontologists it is almost certain that some evidence for a climatic trend will be found, simply because climate is never really stable. If a climatic trend is to be accepted as a cause of an extinction of many long-lived clades, then at the very least the climatic change must be demonstrated to be of greater amplitude than the climatic fluctuations which occurred previous to the mass extinctions and which had no visible effect on diversity or turnover rates of taxa. Some evidence does exist for a slightly steeper poleward temperature gradient at the Cretaceous—Tertiary boundary, but winters were still mild enough for crocodylines to survive in Saskatchewan in the Early Paleocene (Sternberg, 1932). Vast continental areas must have remained tropical and warm year round. And the terrestrial extinctions occurred in hot, dry continental interior basins (Mongolia) as well as in the wet Western Interior.

Climatic fluctuations had occurred throughout the Cretaceous (Krassilov, 1973; Vachrameev, 1975), and it is not obvious that the terminal Cretaceous changes were unusually severe or sudden. In the Western Interior the most dramatic climatic-sedimentological change was in the mid-Cretaceous. During the Jurassic marine evaporites and non-marine evaporites with red beds and aeolianites are common; the Morrison contains thick caliches and non-marine evaporites; the overlying Clovery (Aptian—Albian) also has saline-lake sediment and oxidized floodplain facies (Bakker et al., 1976). In the Late Cretaceous the climate shifted towards wetter conditions — floodplain sediments become drab and coal and lignite become prominent. This sedimentological shift is more striking in the field than that of the Cretaceous—Tertiary boundary, where facies change little. Most of the Cretaceous dinosaur lineages continued to diversify throughout the period; the mid-Cretaceous climate change seems to have had little effect on extinction rates. Dinosaurs were the dominant megafauna in all known habitats — arid and wet basins, deltas near epicontinental seas; floodplains deep within continents and far from any seaway, and in both high and low palaeolatitudes (Russell, 1973). Clearly the Late Cretaceous dinosaur radiation encompassed the full range of terrestrial habitats; it is difficult if not impossible to invoke a climatic change which would exterminate every species in every locale.

Similar objections can be raised to climatic hypotheses for the other terrestrial extinctions; the dominant large herbivores and carnivores of each dynasty were distributed through the full range of contemporary palaeogeography and local habitats; climatic change might favour one family over another locally but it is unlikely that all members of all families would suffer. The probable presence of endothermy in therapsids, thecodonts, and dinosaurs makes climate-induced mega-extinction very improbable; minor evolutionary change in thermo-conductance can adapt a tropical endotherm to arctic habitats and *vice versa* (Scholander et al., 1950). Among living large herbivores the most diverse group is the ruminant artiodactyls; a world-wide climatic shift towards wetter conditions would favour browsing ruminants (cervids and primitive antelopes) over obligate grazing ruminants (large, advanced antelopes) but could not adversely effect all ruminants simultaneously.

Regression and decreased equability

A striking correlation exists between major regressions of epeiric seas and tetrapod mass extinctions (see Fig. 8). Extinctions at the end of the Cretaceous, Jurassic, Triassic and Permian coincide with the most extensive regressions in the Permian—Cretaceous interval. The Permian regression was long (beginning in the Late Carboniferous) and pulsatory; extinction of Dynasty II (the Tap Zone fauna) may coincide with a major regressive pulse at the mid Late Permian, but stratigraphic correlation between terrestrial Beaufort Series and the marine stages is too imprecise to test this possibility.

Shallow sheets of water buffer fluctuations in temperature and humidity on the adjacent terrestrial habitat (the "Great Lakes" or "Mediterranean Effect"). Regression should decrease climate equability in the cratonic areas previously covered by transgression. It has been argued that a decrease in equability was the direct cause of the Cretaceous dinosaur extinctions (Worsley, 1971). Objections to this mechanism are similar to those for the simple climate change paradigm. Dinosaurs were dominant in continental climates (Mongolia) far from any buffering seaway. And large ectothermic reptiles, big crocodylines, champsosaurs, big aquatic turtles, groups which would require mild, frost-free winters, remained common and diverse across the Cretaceous—Tertiary boundary (Fig. 4). If decrease in equability did not affect the distribution of these big ectotherms in the Western Interior, it is difficult to argue that dinosaurs were victims.

Floral change

Major changes in terrestrial flora do not coincide with the tetrapod mass extinctions (see Fig. 8). Floral change is best known for a dynasty-ending extinction at the Cretaceous—Tertiary boundary; change was very modest (Brown, 1962; Leffingwell, 1971). The systemic boundary records merely the continuation of the gradual and progressive angiosperm diversification. Moreover, the increase in angiosperm diversity and dominance during the Cretaceous is in phase with the increase in diversity of families of big ornithischian dinosaur herbivores. The Cretaceous radiation of big herbivores was successfully co-evolving with the angiosperms, and no good evidence can be found for floral change at the end of the period causing the mass extinction. Floral sequences in the Permian—Jurassic are best documented for Gondwanaland (Anderson and Anderson, 1970). The shift from a *Glossopteris* flora to a *Dicroidium* flora began in the *Lystrosaurus* Zone and was completed in the mid-Triassic. The shift from *Dicroidium* to a conifer-dominated flora occurred in the Late Triassic and Early Jurassic. The last great floral shift was to an angiosperm flora; the first flowering plants appear in the mid Early Cretaceous and gradually increase in diversity and relative frequency throughout the rest of the period and the Early Tertiary (Wolfe et al., 1975; and Doyle, Ch. 16). All these floral shifts began *after* mass extinctions and replacements of big herbivores, suggesting that changes in the selective pressure exerted by browsers on the flora induces evolutionary response from the plants.

Regression-habitat area/habitat diversity

Fresh-water aquatics. Regression of shallow seas leaves behind fresh-water wetlands. In the Cretaceous Western Interior, the terminal regression increased the area of fresh-water swamps as deltas prograded into the regressing shoreline (Asquith, 1974). Major regressions are marked in the pollen record by an increase in the relative frequency of triletes, the cryptogams, especially ferns, lycopods, and horsetails (Bharadwaj, 1971); these plants are characteristic of wet, fresh-water habitats. The increase in area and diversity of fresh-water wetlands probably causes increase in diversity and numbers of fresh-water aquatic tetrapods. The greatest Recent diversity of fresh-water aquatic tetrapods occurs in the extensive areas of wetlands (Kiester, 1971), where resources are most abundant and the constantly shifting patterns of rivers and lakes provides a dynamic mosaic of local geography which should maximize the speciation rate of aquatics. The lack of extinctions and the apparent increase in diversity and abundance of fresh-water aquatics during terrestrial extinctions is almost certainly linked to increasing wetland area.

Marine aquatics. Regression reduces the habitat area and diversity of shallow marine environments (Schopf, 1974); extinction of marine tetrapods during regression is not surprising. The survival of toxocheliids, cheloniids, and the sea crocodilians is also not surprising — these groups were less dependent upon broad, open shelf seas and more capable of exploiting near-shore and estuarine resources than were most of the other marine aquatics.

Terminal Cretaceous extinctions included micro-invertebrates which lived in open water over deep ocean basins as well as those in epeiric seas (Bramlette and Martini, 1964). Some marine tetrapods, mosasaurs and plesiosaurs especially, are widely distributed throughout Laurasia down to New Zealand and may have been capable of travelling across the high seas. The extinction of high-seas aquatics indicates that cratonic regressions were accompanied by changes in the qualities of the open sea (perhaps changes in circulation), or that the high-seas groups were linked by part of their life cycle to the cratonic seas, perhaps through breeding sites.

Cratonic regression increases the area of terrestrial habitat, though the increment is modest (10—20% at most). The area of lowland basins increases several fold during regression. Hence the extinction of big terrestrial families cannot be explained by a simple decrease in total habitat area.

Haug Effect Control of Diversity

Are most extinctions clumped?

When the sequence of faunas in one basin is examined in detail, much of the turnover in species and genera between mass extinctions seems to be clumped. For example, after the Permo-Triassic boundary extinctions, standing diversity in the Karroo System increased in a series of steps (Fig. 2): *Lystrosaurus* disappears before the first *Kannemeyeria* and *Diademodon* appear; shansiodonts and stahlecheriids appear abruptly together. In the Cretaceous Western Interior, most of the Old Man genera disappear without descendents (*Centrosaurus, Monoclonius, Kritosaurus, Parasaurolophus, Brachylophosaurus*). In the next faunal level (Lower Edmonton) several genera appear abruptly, without recognizable ancestors among the Old Man fauna (*Edmontosaurus, Arrhinoceratops*). The Tertiary record is full of cases where clumps of species and genera appear together or disappear together. One could invoke the incompleteness of the fossil record — the clumped appearances and extinctions are produced by discontinuous deposition and fossilization. But, I believe, the clumped pattern also occurs in sections where major gaps cannot be recognized (the Paleocene—Eocene boundary in North America, for example).

The intensive analysis of Pleistocene climate shows that the short-term trends in temperature and humidity often are not smooth but abrupt, with sudden reversals of direction and sharp changes in rate (Emiliani, 1964), a pattern which may hold for the Phanerozoic as a whole. Clumped appearances and extinctions could well reflect sudden minor climatic changes, inducing speciation, permitting migration of new taxa and changing the rules of competition and survival in the local basin.

At the Cretaceous—Tertiary boundary a half dozen dinosaur genera seem to disappear together (*Triceratops, Ankylosaurus, Edmontosaurus, Thescelosaurus, Leptoceratops, Tyrannosaurus*). These genera seem to have been present through much if not all the Lance—Hell Creek faunal zone. This final dinosaur fauna of the Western Interior was a complex of species which coexisted for a substantial stratigraphic interval and then became extinct together in a very short time, but this pattern is probably not unique to the last dinosaur fauna; the Old Man, Lower Edmonton, and many other faunal zones had clumped appearances and disappearances. The mean stratigraphic longevity of dinosaur genera is about the same as that of Tertiary mammals — a few million years. The unique aspect of the last dinosaur fauna was not that all its members became extinct, but rather that new species, produced in the continental interior, did not migrate into the basin to replace those going extinct.

The terminal Cretaceous extinctions, and those terminating the other dynasties, do not necessarily indicate that the extinction rate in each basin increased (see Fig. 7). Decrease in diversity and complete extinction of the megafauna is possible if the immigration rate of new species falls below the extinction rate, even if the extinction rate remains constant or decreases. Species diversity in the Western Interior, as measured by biomass D, decreased during the Old

Man—Lance interval; if the extinction rate in the basin was controlled by, among other things, the intensity of competition, the local extinction rate would decrease with decreasing diversity. But the immigration rate of new genera remained below the extinction rate in the Old Man/Lower Edmonton and the Lower Edmonton/Upper Edmonton—Lance transitions (Fig. 6).

Control of speciation by transgression/orogeny cycles

Most of the new species introduced into any given basin were not derived from speciation *in situ*, but rather immigrated. What controls the immigration rate? Species diversity in Recent mammals and birds is highest in regions of vigorous topographic relief (Ricklefs, 1973; Simpson, 1974; Nel, 1975); Regions of vigorous topographic relief have high habitat diversity and can accommodate high species density; high topographic relief also increases the probability of speciation, since there is an abundance of geographic barriers which might isolate a small population during a speciation event.

Are there cycles of geographic—topographic diversity which could control speciation rate and immigration rates? Transgression itself increases geographic diversity by cutting up cratons into partially isolated fragments. An additional and very powerful geographic cycle is associated with transgressions — the Haug Effect. Opinion about the periodicity of orogeny has fluctuated between belief in widely spaced, short, intense orogenies synchronized world-wide, and belief in more or less continuous orogenic activity throughout the stratigraphic column. Johnson (1971) has argued that although the notion of a world-wide cycle of short, intense episodes of mountain building is obsolete, it is equally incorrect and misleading to suppose that orogeny has been spread evenly through time. Some evidence for earth movements probably can be found somewhere for every stratigraphic stage, but Johnson presents formidable documentation that orogenic intensity has peaked during maximum transgression. During regression, orogeny may continue but at a reduced rate. Evaluating the magnitude of orogeny is difficult; the size of the clastic wedge shed onto the craton is one indicator of the rate of uplift. Johnson shows that the building of clastic wedges on the North American craton are maximized during maximum orogeny. For the Permian—Cretaceous interval this correlation seems to hold: clastic wedges and transgression were at a maximum during the transition from Carboniferous to Permian (the Hercynian and Appalachian events); earth movements were few and mild at the Permo-Triassic boundary; mid-Triassic wedges are present in the Karroo System (the Molteno clastics). In North America a thick clastic wedge was built out eastward into the Western Interior in Late Jurassic time, culminating in the Morrison Formation. After Morrison deposition ceased in latest Jurassic, a phase of non-deposition and erosion followed; clastic wedges again began to be built out in the Aptian and Albian (Bakker et al., 1976). Cretaceous earth movements were wide-spread and intense, but the Cretaceous—Tertiary boundary was relatively quiet, except in the Western Interior (Worsley, 1971). Orogeny did continue across the systemic boundary here but not necessarily with undiminished intensity — in Saskatche-

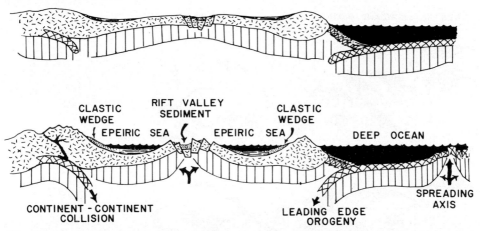

Fig. 5. One possible plate-tectonic mechanism for the synchroneity of transgression and oro-geny. Schema modified from Dewey and Bird (1970), Veevers and Evans (1973) and Hays and Pitman (1973). Below: Active sea-floor spreading axes expand in volume from increased heat flow; water is displaced from the oceanic basins onto the cratonic interior; activity of many spreading axes increases the number and intensity of orogeny and rifting. Above: Rela-tive decrease in the number and activity of spreading axes reduces the volume of the axes, draining water off the cratons and reducing the number and intensity of orogeny. Erosion becomes faster than uplift in continental mountain belts.

wan and Montana the boundary locally is marked by a decrease in palaeoslope and extensive coal deposition (Sternberg, 1932; Russell, 1971).

The synchroneity of maximum orogeny with maximum transgression is known as the Haug Effect (Johnson, 1971). A plate-tectonic mechanism can be proposed (Fig. 5). Other mechanisms are possible. The Haug Effect provides a high, steady-state supply of geographic barriers during transgression; the pulsa-tory pattern of epeiric seas constantly changes shorelines; uplift is as fast or faster than erosion, and new highlands are produced as old ones are worn down. The high topographic—geographic diversity should give a high speciation rate and a high standing diversity within the continental highlands and adjacent plains. The pool of potential immigrants to the lowland basins should be large. Immigration rates will be high. Basin diversity will grow until saturation is reached, or until a major regression, coupled to a diminution of orogenic uplift, reduces continental habitat diversity, reduces the speciation rate and the spe-cies diversity in the continental pool, and decreases the immigration rate into the basin below the extinction rate (Figs. 6, 7, 8).

Size specificity of the Haug Effect

The magnitude of the geographical barrier necessary for speciation varies with the size and mobility of the animal; for dinosaurs and elephants barriers must be of considerable size and duration; for mice a much smaller feature is sufficient. During regression the supply rate of major topographic features is

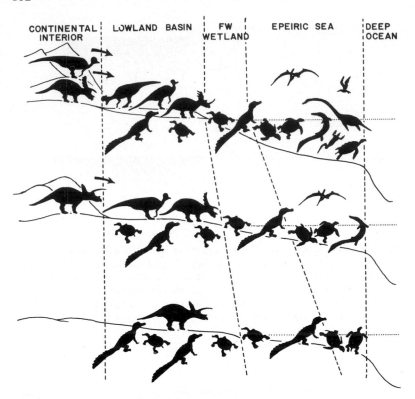

Fig. 6. A model for mass extinction, using the Late Cretaceous of the American northern Western Interior as an example. The three profiles represent, from top to bottom the changing topography and diversity in the Campanian (Old Man), Early Maastrichtian (Lower Edmonton), and latest Maastrichtian (Lance—Hell Creek). Above: High world-wide rates of orogeny and transgression cause high rates of speciation among large terrestrial tetrapods; new species diffuse into the lowland basin at rates as high or higher than the extinction rates in the basin; a large, steady-state diversity is maintained. Middle: World-wide decrease in orogeny and transgression reduces the speciation and diffusion rates to a level below the extinction rate of large terrestrial tetrapods in the lowland basin; regression causes extinction among marine families, but increases the habitat diversity and area for fresh-water aquatics. Bottom: Regression at maximum, orogeny at minimum. The last species in the basin becomes extinct and is not replaced. Fresh-water families increase in number and diversity. Of the marine families, only two types of sea turtle, with limbs capable of progression on land, survive.

cut severely, but since even subdued highlands have a surface texture of minor topographic features, the supply of barriers for small tetrapod speciation should be reduced far less or not at all. The speciation rate and the size of the pool of potential immigrants should be reduced far less for small tetrapods than for large species. The basin megafauna is more vulnerable to extinction than the microfauna; the total number of large species is usually small, and if the immigration rate falls below the extinction rate for only a short interval, the local megafauna can be reduced to one or two species.

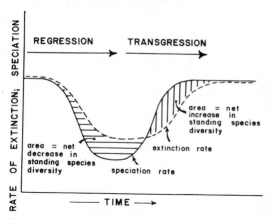

Fig. 7. Speciation and extinction during a major regression. The decreasing standing diversity may decrease the extinction rate during regression, but the speciation rate falls below the extinction rate.

Some problems of complete megafauna extinction

It is not difficult to imagine how decreased topographic—geographic diversity during a regression will decrease the diversity on the continents at large and in the lowland basins dependent upon the continental habitats for immigrants. But the terminal Cretaceous extinctions eliminated all of the megafauna. It might be supposed that as the standing diversity decreases, the population sizes of some of the surviving species will increase. How are these last survivors eliminated? I believe that no special mechanism is required. The extinction of the last dinosaurs probably came from one of the agents which cause extinction all through tetrapod history — abrupt but minor climate change, floral change, introduction of parasites, predator/prey cycles, or a combination of these. Dominant, common tetrapod species can suffer sudden and complete extinction. There are two levels of causation for mass extinction: (1) an immediate cause which kills off the last species cluster in each local basin and habitat — this agent may be different from basin to basin; (2) the cause of mass termination of families of big tetrapods is the decrease in speciation rate and in the size of the pool of potential immigrants on the continents.

Was the final extinction of the last dinosaur species cluster in each habitat simultaneous? Stratigraphic resolution indicates that the final extinctions were concentrated into a few million years or less (Van Valen and Sloan, 1976), but a more precise statement cannot be made. For the other mass extinctions, the time—space geometry of the final extinctions is not well known.

Concluding remarks

The model for mass extinction presented here is a working hypothesis. I believe that the most useful hypotheses for major extinction events must include the following considerations: (1) species and higher taxa must be segregated into body-size, trophic and habitat guilds; (2) measures of diversity and extinction must suppress the effect of varying sample size in the fossil record;

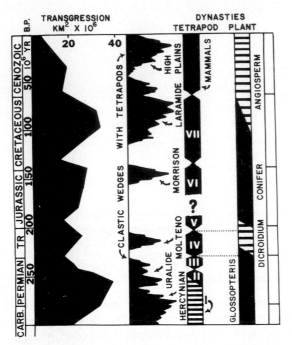

Fig. 8. The Haug Effect and terrestrial dynasties. Left: Black pattern shows world-wide transgression across continents; data from Hallam (1963, 1971), Hays and Pitman (1973), Schopf (1974). Middle: Black pattern shows a very schematic record of the major tetrapod-bearing clastic wedges. Right: Successive dynasties of tetrapods and land plants. In the tetrapods, the transitions between dynasties, except between the first and second, are marked by mass extinctions of families and a zone with low diversity. The floral transitions appear to lack mass extinctions of higher taxa and are more gradual, with new clades progressively expanding in numbers and diversity as the older clades decrease.

(3) since the diversity in any local basin is controlled by a combination of the basin quality and the size of the pool of potential immigrant species, local diversity events must be examined to separate the effects of change in local environment from change in the larger regions which support the immigrant pool; (4) all extinction events should be compared to search for iterative patterns. Many *ad hoc* explanations for particular extinction events, especially some of those proposed for the Cretaceous—Tertiary boundary, do not include these considerations. The correlation of epeiric-sea regression with terrestrial megafaunal extinction and marine extinction is extremely good, and it seems likely that the overall control of mass extinction lies in some eustatic—orogenic cycle which changes the geometry of biological geography, rather than in a simple, world-wide climatic shift. Schopf's (1974) species-area argument is one successful application of ecological theory to regression. Since terrestrial habitat area increases during regression, the mechanism for land megafaunal extinction must be a little different; the Haug Effect, as suggested here, may well be the geographic mechanism which controlled land diversity cycles.

Summary

(1) In the Permian—Cretaceous interval few oceanic barriers existed; newly evolved tetrapod species spread quickly all over Pangaea, wherever suitable local habitats existed.

(2) Most morphological change is concentrated in short-lived speciation events in small, isolated populations; species appear in depositional basins abruptly and show only modest change during their stratigraphic range ("punctuated equilibria").

(3) Throughout the faunal stages of any one basin species appear and disappear often in clumps, perhaps because of sudden, minor climatic change and migrations between previously separate habitats.

(4) Topographic—geographic diversity is low in basins. Species diversity and speciation rates are highest in the continental regions of high relative relief. Most basin species are products of speciation and migration from these regions.

(5) Diversity in the basin is controlled by the quality of the basin (its size, rainfall, productivity and habitat diversity) and by the combination of the immigration rate of new species from the continental areas of high speciation and the local extinction rate.

(6) The Permian—Cretaceous has six or seven cycles of diversification and extinctions ("dynasties"); each begins with a sudden, mass extinction of most of the families of big terrestrial and marine aquatic tetrapods; families of small terrestrial and fresh-water aquatics show no clear pattern of mass extinction and may increase in number during the big terrestrial—marine extinctions. After the extinction, big terrestrial diversity is gradually built up by radiation from the surviving large tetrapods and from the small-bodied groups.

(7) Large herbivore diversity was greatest at the end of the longest diversification cycles (60 m.y. between mass extinctions), suggesting that the shorter cycles had not reached an ultimate saturation of diversity.

(8) Climatic change during mass extinctions is not of greater magnitude or suddenness than the fluctuations occurring all through the diversification cycles.

(9) Terrestrial plants do not show clear patterns of mass extinction. Major new plant groups increase in diversity and dominance gradually, generally beginning an adaptive radiation shortly after a mass extinction and replacement of big herbivores, suggesting that change in browsing pressure induces an evolutionary response from the browsed plants.

(10) Mass tetrapod extinctions coincide with major regressions of shallow seas and probably with a decrease in the world-wide level of orogeny (the Haug Effect).

(11) Regression decreases the habitat area and diversity of marine aquatics, causing extinctions; the surviving groups often are the least specialized for pelagic life and retain the capacity to move over wetlands and exploit nearshore habitats.

(12) Regression increases the habitat area and diversity of fresh-water aquatics.

(13) Diversity of big terrestrial species falls in several steps in any one basin during regression; since most families survive until the end of the cycle in one basin or another, the final extinctions of families world-wide tend to be concentrated into a short interval when the last cluster of species in each basin becomes extinct.

(14) The immediate cause of the extinction of the last cluster of species in each basin does not have to be the same nor does it have to differ from the agents which caused extinction of clumps of species all through the diversification cycle.

(15) The overall control of the mass extinctions is the decreased geographic—topographic diversity in the continents. The combination of regression of shallow seas and reduction of relief on land reduces the supply of geographic barriers for speciation and reduces the diversity of habitats for newly evolved species. Fewer new species are formed, and fewer of these survive long enough to disperse widely. As the pools of potential immigrants for the basins or any other local habitat shrink, the immigration rate falls below the extinction rate for each habitat.

(16) Since small tetrapods can be isolated by small topographic—geographic features, the speciation rate of small species decreases much less than that of large species. Diversity of small tetrapods may increase if they begin to exploit resources made available by the decreasing diversity of megafauna.

References

Amalitsky, V., 1922. Diagnoses of the new forms of vertebrates and plants from the Upper Permian on the North Dvina. Izv. S.S.R. Nauk., 1922: 329—340 (in Russian).

Anderson, H.M. and Anderson, J.M., 1970. A preliminary review of the biostratigraphy of the Uppermost Permian, Triassic, and Lowermost Jurassic of Gondwanaland. Palaeontol. Afr., 13: 1—22.

Appleby, R.M., Charig, A.J., Cox, C.B., Kermack, K.A. and Tarlo, L.B.H., 1967. The Reptilia. In: W.B. Harland (Editor), The Fossil Record. Geol. Soc. Lond., Spec. Publ., 2: 695—732.

Asquith, D.O., 1974. Sedimentary models, cycles, and deltas, Upper Cretaceous, Wyoming. Bull. Am. Assoc. Pet. Geol., 58: 2274—2283.

Bakker, R.T., 1972. Anatomical and ecological evidence of endothermy in dinosaurs. Nature, 238: 81—85.

Bakker, R.T., 1975a. Experimental and fossil evidence for the evolution of tetrapod bioenergetics. In: D. Gates and R. Schmerl (Editors), Perspectives of Biophysical Ecology. Springer Verlag, New York, N.Y., pp. 365—395.

Bakker, R.T., 1975b. Dinosaur renaissance. Sci. Am., April: 57—78.

Bakker, R.T., Behrensmeyer, A.K., Dodson, P. and McIntosh, J., 1976. Second Report of the Morrison Dinosaur Ecology Group. Nat. Geogr. Res. Grant Rep., 1976, pp. 1—47.

Behrensmeyer, A.K., 1975. The taphonomy and paleoecology of Plio-Pleistocene vertebrate assemblage East of Lake Rudolf, Kenya. Bull. Mus. Comp. Zool., 146: 473—578.

Bharadwaj, D.C., 1971. Palynological subdivisions of the Gondwana Sequence in India. Proc. Second Gondwana Symposium, Pretoria, Scientia, pp. 531—536.

Bonaparte, J., 1971. Los Tetrapodos del Sector Superior de la Formacion Los Colorados, La Rioja, Argentina (Triasico Superior). Opera Lilloana, 22: 1—183.

Boonstra, L.D., 1969. The Fauna of the *Tapinocephalus* Zone (Beaufort Beds of the Karroo). Ann. S. Afr. Mus., 56: 1—73.

Bramlette, M.N. and Martini, E., 1964. The great change in calcareous nannoplankton fossils between the Maestrichtian and Danian. Micropaleontology, 10: 291—328.

Brown, R.W., 1962. Paleocene flora of the Rocky Mountains and Great Plains. U.S. Geol. Surv. Prof. Pap., 375: 1—55.

Camp, C.L. and Welles, S.P., 1956. Triassic dicynodont reptiles, Part I: The North American genus *Placerias*. Mem. Univ. Calif., 13: 255—304.

Carroll, R.L., Belt, E.S., Dineley, D.L., Baird, D. and McGregor, D.C., 1972. Vertebrate Paleontology of Eastern Canada. 24th Int. Geol. Congr., J.D. McAra Ltd., Calgary, Alta., pp. 1—113.

Chatterjee, S. and Roy-Chowdhury, T., 1974. Triassic Gondwana vertebrates from India. Indian J. Earth Sci., 1: 96—112.

Cluver, M.A., 1971. The cranial morphology of the dicynodont genus *Lystrosaurus*. Ann. S. Afr. Mus., 56: 155—274.

Colbert, E.H., 1974. *Lystrosaurus* from Antarctica. Am. Mus. Novit., 2535: 1—44.

Cosgriff, J.W., 1974. Lower Triassic Temnospondyli of Tasmania. Geol. Soc. Am. Spec. Pap., 149: 1—134.

Cox, C.B., 1964. On the palate, dentition, and classification of the fossil reptile *Endothiodon* and related genera. Am. Mus. Novit., 2171: 1—25.

Cox, C.B., 1965. New Triassic dicynodonts from South America, their origins and relationships. Philos. Trans. R. Soc. Lond., Ser. B., 248: 457—517.

Cox, C.B., 1974. Vertebrate palaeodistribution patterns and continental drift. J. Biogeogr., 1: 75—94.

Crompton, A.W. and Kielan-Jaworowska, Z., 1976. Molar structure and occlusion in Cretaceous therian mammals. Int. Congr. Dental. Morphol., in press.

Cruickshank, A.R.I., 1972. The proterosuchian thecodonts. In: K.A. Joysey and T.S. Kemp (Editors), Studies in Vertebrate Evolution. Oliver and Boyd, London, pp. 89—119.

Dewey, J.F. and Bird, J.M., 1970. Mountain belts and the new global tectonics. J. Geophys. Res., 75: 2625—2647.

Efremov, I.A., 1950. Taphonomy and the geological record. Acad. Sci. U.S.S.R., Publ. Paleontol. Inst., 24: 3—176 (in Russian).

Efremov, I.A. and Vjushkov, B.P., 1955. Catalogue of Permian and Triassic terrestrial vertebrate localities in the territory of the U.S.S.R. Tr. Paleontol. Inst., 46: 1—185 (in Russian).

Eldredge, N. and Gould, S.J., 1972. Punctuated equilibria: an alternative to phyletic gradualism. In: T.M. Schopf (Editor), Models in Paleobiology. Freeman, San Francisco, Calif., pp. 82—115.

Emiliani, C., 1964. Paleotemperature analysis of the Caribbean Cores A254-BR-C and CP-28. Bull. Geol. Soc. Am., 75: 129—144.

Estes, R., 1970. Origin of the Recent North American lower vertebrate fauna: an inquiry into the fossil record. Forma Functio, 3: 139—163.

Estes, R. and Reig, O.A., 1973. The early fossil record of frogs: a review of the evidence. In: J.L. Vial (Editor), Evolutionary Biology of the Anurans. University of Missouri Press, St. Louis, Mo., pp. 10—63.

Feruglio, E., 1949—50. Descripcion Geologica de la Patagonia, I. Coni, Buenos Aires, 334 pp.

Frakes, L., Kemp, E.M. and Crowell, J.C., 1975. Late Paleozoic Glaciation: Part VI, Asia. Bull. Geol. Soc. Am., 86: 454—464.

Gaffney, E.S., 1975. A phylogeny and classification of the higher categories of turtles. Bull. Am. Mus. Nat. Hist., 155: 389—436.

Galton, P.M., 1971. A primitive dome-headed dinosaur (Ornithischia: Pachycephalosauridae) from the Lower Cretaceous of England and the function of the dome of pachycephalosaurids. J. Paleontol., 45: 40—47.

Gradzinski, R., 1969. Sedimentation of dinosaur-bearing Upper Cretaceous deposits of the Nemegt Basin, Gobi Desert. Palaeontol. Pol., 21: 147—229.

Hallam, A., 1963. Major epeirogenic and eustatic changes, since the Cretaceous, and their possible relationship to crustal structure. Am. J. Sci., 261: 397—423.

Hallam, A., 1971. Reevaluation of the palaeogeographic argument for an expanding earth. Nature (Lond.), 232: 180—182.

Hays, J.D. and Pitman, W.C., 1973. Lithospheric motion, sea level changes and climatic and ecological consequences. Nature, 246: 18—22.

Hopson, J.A. and Kitching, J.W., 1972. A revised classification of cynodonts (Reptilia; Therapsida). Palaeontol. Afr., 14: 71—85.

Janensch, W., 1927. Material und Formangehalt der Sauropoden in der Ausbeute der Tendaguruexpedition. Palaeontographica, Suppl., 7: 1—34.

Johnson, J.G., 1971. Timing and coordination of orogenic, epeirogenic and eustatic events. Geol. Soc. Am. Bull., 82: 3263—3298.

Kalandadze, N.N., 1970. New Triassic kannemeyerids from the Southern Pri-Urals. Material for the Evolution of Land Vertebrates, Akad. Nauk, Moscow, pp. 51—57.

Kalandadze, N.N., 1974. Intercontinental connections of tetrapod faunas in the Triassic Period. Palaeontol. Zh., 3: 75—86.

Kalin, J., 1955. Crocodilia. In: J. Piveteau (Editor), Traité de Paléontologie, Masson, Paris, pp. 696—784.

Keyser, A.W., 1973. A New Triassic vertebrate fauna from South West Africa. Palaeontol. Afr., 16: 1—15.

Kiester, A.R., 1971. Species density of North American amphibians and reptiles. Syst. Zool., 20: 127—137.

Kitching, J.W., 1967. On the examination of the Beaufort Beds within the flood area of the Hendrik Verwoerd Dam, Orange River Project. S.-Afr. Tydskr. Wet., 10: 386—388.

Kitching, J.W., 1972. Fossil Localities of the Beaufort Succession. Thesis, University of the Witwatersrand, Johannesburg.

Koken, E., 1887. Die Dinosaurier, Crocodiliden und Sauropterygiens des Norddeutschen Wealden. Paläontol. Abh., 5: 311—419.

Krassilov, V.A., 1973. Climatic changes in Eastern Asia as indicated by fossil floras, I. Early Cretaceous. Palaeogeogr., Palaeoclimatol., Palaeoecol., 13: 261—274.

Kutty, T.S., 1971. Two faunal associations from the Maleri Formation of the Pranhita—Godavari Valley. J. Geol. Soc. India, 12: 63—67.

Lefeld, J., 1971. Geology of the Djadokhta Formation of Bayn Dzak (Mongolia). Palaeontol. Pol., 25: 101—127.

Leffingwell, H.A., 1971. Palynology of the Lance and Fort Union Formations of the Type Lance Area, Wyoming. Geol. Soc. Am. Spec. Pap., 127: 1—64.

Lillegraven, J.A., 1972. Ordinal and familial diversity of Cenozoic mammals. Taxon, 21: 261—274.

MacArthur, R.H., 1972. Geographical Ecology. Harper and Row, New York, N.Y., 269 pp.

McGowan, C., 1970. The distinction between latipinnate and longipinnate ichthyosaurs. Life Sci. Contrib. R. Ont. Mus., 20: 1—8.

McGowan, C., 1972a. Evolutionary trends in longipinnate ichthyosaurs with particular reference to the skull and forefin. Life Sci. Contrib. R. Ont. Mus., 83: 1—38.

McGowan, C., 1972b. The systematics of Cretaceous ichthyosaurs with particular references to the material from North America. Contrib. Geol. Univ. Wyo., 11: 1—29.

McKenna, M.C., 1969. The origin and early differentiation of therian mammals. Ann. N.Y. Acad. Sci., 167: 317—340.

Nel, J.A.J., 1975. Species density and ecological diversity of South African mammal communities. S. Afr. J. Sci., 71: 168—170.

Olson, E.C., 1962. Late Permian terrestrial vertebrates; U.S.A. and U.S.S.R. Trans Am. Philos. Soc., 52: 3—224.

Olson, E.C. and Vaughn, P.P., 1970. The changes of terrestrial vertebrates and climates during the Permian of North America. Forma Functio, 3: 113—138.

Panchen, A., 1967. Amphibia. In: W.B. Harland (Editor), The Fossil Record. Geol. Soc. Lond. Spec. Publ., 2: 685—694.

Persson, P.O., 1963. A revision of the classification of the Plesiosauria with a synopsis of the stratigraphical and geographical distribution of the group. Acta. Univ. Lund., 59: 1—60.

Price, L.I., 1947. Sedimentos mesozoicos na Baia De Sao Marcos Estado do Maranhao. Not. Prelim. Estud., Min. Agric. Brazil, 40: 1—7.

Ricklefs, R.E., 1973. Ecology. Chiron Press, Newton, Mass., 861 pp.

Rozdestvenski, A.K., 1965. Growth changes in Asian dinosaurs and some problems of their taxonomy. Paleontol. Zh., 1965: 95—109 (in Russian).

Russell, D.A., 1967a. Systematics and morphology of American mosasaurs. Bull. Peabody Mus. Nat. Hist. Yale Univ., 23: 1—240.

Russell, D.A., 1967b. A census of dinosaur specimens collected in Western Canada. Nat. Hist. Pap. Natl. Mus. Can., 36: 1—13.

Russell, D.A., 1971. The disappearance of the dinosaurs. Can. Geogr. J., 83: 204—215.

Russell, D.A., 1973. The environments of Canadian dinosaurs. Can. Geogr. J., 87: 4—11.

Scholander, R.F., Hock, R., Walters, V., Johnson, F. and Irving, L., 1950. Heat regulation of some Arctic and Tropical mammals and birds. Biol. Bull., 99: 225—236.

Schopf, T.J.M., 1974. Permo-Triassic extinctions: relation to sea floor spreading. J. Geol., 82: 129—143.

Simpson, G.G., 1964. Species density of North American Recent mammals. Syst. Zool., 13: 57—73.

Stanley, S.M., 1975. A theory of evolution above the species level. Proc. Natl. Acad. Sci., 72: 646—650.

Steel, R., 1967. Ornithischia. Handbuch der Palaeoherp., 15: 1—84.

Steel, R., 1970. Saurischia. Handbuch der Palaeoherp., 14: 1—87.

Steel, R., 1973. Crocodilia. Handbuch der Palaeoherp., 16: 1—116.

Sternberg, C.M., 1932. A new fossil crocodile from Saskatchewan. Can. Field Nat., 46: 128—133.

Vachrameev, V.A., 1975. The basic boundaries of phytogeographic provinces of the terrestrial world in the Jurassic and Early Cretaceous time. Paleontol. Zh., 1975: 123—132 (in Russian).

Van Valen, L. and Sloan, R., 1976. Ecology and the extinction of the dinosaurs. Evol. Theory, in press.

Veevers, J.J. and Evans, R.R., 1973. Sedimentary and magnetic events in Australia and the mechanism of world-wide Cretaceous transgressions. Nature, 245: 33—36.

Von Huene, F., 1932. Die Fossile Reptil-Ordnung Saurischia, ihre Entwicklung und Geschichte. Monogr. Geol. Paläontol., Ser. 1., 4: 1—368.

Von Huene, F., 1935—42. Die Fossilen Reptilien des Südamerikanischen Gondwanalandes an der Zeitenwende. Heine, Tübingen, 332 pp.

Von Huene, F., 1942. Die Anomodontier des Ruhuhu-Gebietes in der Tübinger Sammlung. Palaeontographica, 94: 154—184.

Webb, S.D., 1969. Extinction—origination equilibria in Late Cenozoic land mammals of North America. Evolution, 23: 688—702.

Wieland, G.R., 1900. Some observations on certain well marked stages in the evolution of the testudinate humerus. Am. J. Sci., 9: 413—424.

Wolfe, J.A., Doyle, J.A. and Page, V.M., 1975. The bases of angiosperm phylogeny: paleobotany. Ann. Mo. Bot. Garden, 62: 801—824.

Worsley, T.R., 1971. Terminal Cretaceous events. Nature, 230: 318—320.

Young, C.C., 1951. The Lufeng saurischian fauna in China. Palaeontol. Sin., N.S.C., 134: 19—96.

Young, C.C., 1964. The Pseudosuchians in China. Palaeont. Sin., N. S. C, 19: 1—134.

Zangerl, R., 1953. The vertebrate fauna of the Selma Formation of Alabama. Part III: The turtles of the Family Protostegidae. Fieldiana Geol. Mem., 3: 59—133.

Zangerl, R., 1960. Vertebrates of the Selma Formation. Part V: An advanced cheloniid sea turtle. Fieldiana Geol. Mem., 3: 281—312.

Zangerl, R., 1971. Two toxochelyid sea turtles from the Landenian Sands of Erquelinnes (Hainaut) of Belgium. Inst. R. Sci. Nat. Belg. Mem., 169: 1—32.

CHAPTER 15

PATTERNS OF EVOLUTION IN THE MAMMALIAN FOSSIL RECORD

PHILIP D. GINGERICH

Introduction

Mammals are the most successful and most intelligent of land vertebrates. While most mammals today remain terrestrial like the ancestral mammalian stock, one progressive group has invaded the air and others have invaded the sea. Living mammals differ rather sharply from other living vertebrates in bearing their young alive, in suckling their young (hence the class name Mammalia), and in sometimes "educating" their young. Mammals are warm-blooded, or endothermic, and their high metabolic rates make possible a sustained high level of continuous activity. Osteologically, living mammals also differ sharply from most other vertebrates in possessing a double occipital condyle on the skull, a single mandibular bone (the dentary) which articulates directly with the squamosal bone of the cranium, an eardrum supported by an ossified ectotympanic bone, three auditory ossicles, only two tooth generations — deciduous and permanent, cheek teeth of complicated morphology, a single bony nasal opening in the skull, and epiphyses on the long bones. While living mammals are well separated from other groups of vertebrates today, the fossil record shows clearly their origin from a reptilian stock and permits one to trace the origin and radiation of mammals in considerable detail.

If one could choose any one anatomical system of mammals for preservation in the fossil record, the system yielding the most information about the animals would undoubtedly be the dentition, and it is fortunate that this *is* the most commonly preserved element of fossil mammals. Dental enamel is the hardest mammalian tissue, and it thus has the best chance of being preserved in the fossil record. The teeth of different families and genera of mammals have a characteristic, genetically determined pattern which makes them ideal for systematic identifications. Teeth are involved in the mastication of food, and the pattern of cusps and crests characteristic of the teeth of different mammalian groups reflects the dietary preference of the group as well as its heritage, offering insight into the ecological adaptations of each group. Finally, the fact that there is a single definitive set of permanent teeth which form within the jaw before they erupt is of great importance for detailed evolutionary studies. Within related groups of mammals, body size is highly correlated with tooth size, which varies within recognized limits. Furthermore, this variance in tooth size has a demonstrated high additive genetic component (i.e., high heritability; see Bader, 1965, Alvesalo and Tigerstedt, 1974), meaning that tooth size does respond to natural selection. The fact that teeth do not continue to grow after they erupt greatly simplifies estimation and comparison of the definitive body size of individuals in different samples because it is not necessary to correct for ontogenetic size increase in dental dimensions. This makes mammalian teeth ideal for microevolutionary studies, offering insight into the relative body size of related species, body size being one of the most important components of an animal's adaptation.

In the following pages I have outlined some of the major features of the radiation of mammals, including examples of major adaptive trends, rates of origination and extinction, and taxonomic longevity. Specific mammalian adaptations are also discussed, including some remarkable examples of convergence, mosaic evolution, and small-scale evolutionary reversals. This is followed by a consideration of speciation in mammals, including discussion of the origin of higher taxonomic groups of mammals.

Mammalian Radiations

Mammals are first known from Upper Triassic (Rhaetic) strata in England, Wales, Switzerland, southern China, and South Africa. Most of these earliest mammals, including some known from skulls and skeletons, are prototherians of the family Morganucodontidae (e.g., *Eozostrodon*), with relatively simple triconodont molars (see below). Others, placed in the prototherian family Haramiyidae, have multicusped teeth and are possibly closely related to the origin of the Multituberculata, an important group of extinct rodent-like Mesozoic and Early Tertiary mammals. However, one Late Triassic mammal is also known, *Kuehneotherium*, which has the three major cusps on the upper and lower molars rotated to form interlocking triangles as in the more advanced "therian" mammals (see Fig. 1).

Morganucodontids, with their simple triconodont molars, are thought to represent an early prototypical stage in the evolution of the mammalian molar. *Kuehneotherium* from the Rhaetic represents an advance over the triconodont pattern in having the cusps rotated so that upper and lower cheek teeth form a row of interlocking triangles. The next stage, addition of a shearing heel onto the back of the lower molar triangles, is represented by the Early Cretaceous genus *Aegialodon*. By the mid-Cretaceous (Albian) two forms with fully developed tribosphenic dentitions characteristic of modern mammals are known, *Holoclemensia* and *Pappotherium*, which represent respectively the earliest marsupial and placental mammals. The mid-Cretaceous was the time of the initial major radiation of angiosperm plants, with a correlated radiation of insects and other terrestrial invertebrates, and it is not surprising that these changes in plant and insect communities were accompanied by a modest radiation of insectivorous mammals.

Near the end of the Cretaceous, placental mammals began the first of their major radiations, leading to a characteristic fauna in the Paleocene that was dominated by archaic primates, proteutherian insectivores, and a diverse assemblage of archaic ungulates, the Condylarthra. Multituberculates were also important elements of virtually all known Paleocene faunas. Other groups of placental mammals making their first appearance in the Paleocene were the

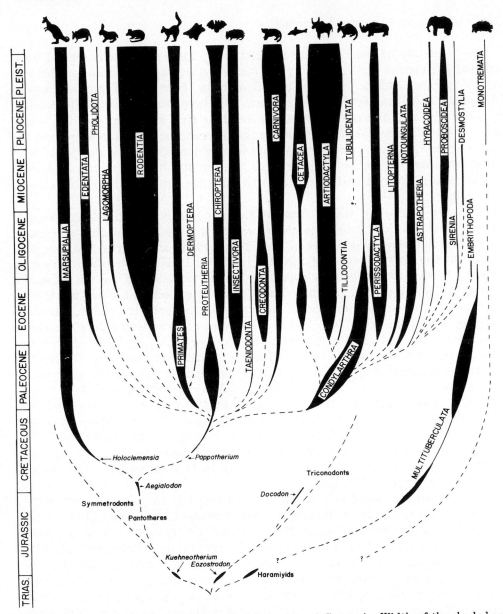

Fig. 1. Radiation of mammals during the Mesozoic and Cenozoic. Width of the shaded area gives a very general estimate of the relative number of genera in the fossil record for each order. See Figs. 2—5 for quantitative data on diversity.

Dermoptera, the Insectivora (sensu Lipotyphla), the Taeniodonta, archaic true Carnivora (Viverravinae), the Pantodonta, and several predominantly South American orders, the Litopterna, Notoungulata, and Astrapotheria. Most groups of archaic mammals typical of the Paleocene survived into the Eocene, but became extinct during or shortly after that epoch.

Origin of modern orders of mammals

The beginning of the Eocene is marked in the fossil mammal faunas of western North America and Europe by the sudden appearance of mammals belonging to modern orders. For example, Rodentia, Primates of modern aspect (Adapidae and Omomyidae), Chiroptera, primitive true Carnivora (Miacinae), Artiodactyla, and Perissodactyla all make their first appearance at the beginning of the Eocene. We do not have a fossil record actually documenting the origin of any of these major groups, but a consideration of the morphology of the most primitive forms together with the climatic history of the Paleocene and Eocene in Europe and western North America provides a possible clue to their origin.

The Late Paleocene was a time of climatic cooling, with the subtropical climate of the Middle Paleocene giving way to a warm temperature climate in the Late Paleocene, which was in turn followed by a return to a subtropical climate in the Early Eocene (Wolfe and Hopkins, 1967). Probable ancestors of several of the modern groups appearing in the Eocene are known from the Middle Paleocene of North America and possible ancestors are known for the others, but no connecting forms are yet known from the Late Paleocene (Gingerich, 1976b, table 13). This hiatus in an otherwise rich fossil record is correlated with Late Paleocene climatic deterioration, and was almost certainly a result of the temporary decline in average temperatures. During this climatic deterioration the geographic ranges of many mammals formerly inhabiting western North America (or Europe, or Asia) probably contracted, following the northern border of the subtropical climatic zone as it retreated southward. When the subtropical zone expanded again at the beginning of the Eocene, highly evolved descendants of the former North American Middle Palaeocene fauna (which had remained in Central American refuges) reinvaded North America (Sloan, 1969). It is probable that a similar phenomenon occurred in Europe and in Asia.

The climatic warming in the Early Eocene not only brought new, highly evolved mammalian forms northward, but it made high-latitude land connections between the Holarctic continents accessible to many mammalian groups. The result was a high level of faunal interchange and rapid dispersal of modern mammals between North America, Europe, and Asia (McKenna, 1975). Thus the Early Eocene dispersal was perhaps as much a result of climatic change as it was of continental positions, although breaking up the land connection between Europe and North America and the final opening of the North Atlantic ocean created a permanent barrier to further mammalian migration between Europe and North America early in the Eocene.

The appearance of modern orders was sudden in the fossil record, but it is probable that their evolutionary origin was gradual and continuous in areas (such as Central America) where we do not yet have an adequate Early Tertiary fossil record. It is also probable, in view of the structural changes involved in the origin of the characteristic ever-growing incisors present in the earliest known rodents, or the double pulley astragalus characteristic of artiodactyls, that the evolution leading to differentiation of the modern orders was relatively rapid in the phyletic lineages involved. Rapid but continuous evolution of this sort can be traced in Early Tertiary primates in the transition from *Plesiadapis* to *Platychoerops*, where the incisors were considerably reorganized morphologically and functionally in the space of only 2—3 m.y. (Gingerich, 1976b).

These relatively rapid rates of change in phyletic lineages might be explained by either higher levels of selective pressure due to crowding stress and competition as diverse subtropical faunas that formerly inhabited the whole North American continent were crowded onto a narrow isthmus in Central America, by great reductions in population size in the subtropical forms, or both. There is no evidence to suggest that the origin of modern mammalian orders during the Late Paleocene was accompanied by higher than normal rates of cladogenetic speciation, and high rates of cladogenesis would be unlikely during a time of contraction in the geographic ranges of the subtropical species. A very similar abrupt appearance of many modern mammalian families occurred in the Early Oligocene (Stehlin's "*Grande Coupure*", see Stehlin, 1909) following a major climatic deterioration, and a similar explanation may be offered for the abrupt appearance of new forms at that time.

Diversity through time

During the first two-thirds of their 200 m.y. history, mammals constituted a very small part of the terrestrial vertebrate fauna. By the end of the Cretaceous many previously important reptile groups were extinct, and mammals became the dominant terrestrial vertebrates. Their diversity has increased almost continuously since that time (Fig. 2). The total number of genera of mammals present in each successive subdivision of the Tertiary epochs appears to have increased at a nearly constant rate, but when this total number is corrected for the duration of each subdivision (i.e., when genera per million years is calculated), the increase in diversity of mammals through geological time approximates an exponential curve. The shape of this curve (Fig. 2) is undoubtedly influenced by the increasing probability of finding fossils in more recent strata, i.e., the fact that more recent fossil mammal faunas are more adequately sampled than older ones, but at the same time there is no denying a great increase in the diversity of mammals through the course of geological time.

Mammals were well established in insectivorous, herbivorous, and carnivorous adaptive zones by the Early Paleocene, and these continued to be important throughout the Tertiary and up to the present day. Fig. 3 shows that the relative importance of each of these three basic adaptive zones has been nearly constant since the Paleocene. As one would expect, genera of herbivorous mammals outnumbered terrestrial carnivorous genera by a fairly constant factor of about 3 or 4 to 1.

Marine mammals (Sirenia and Cetacea — sea cows, whales and porpoises) and volant mammals (Chiroptera — bats) first appear in the Early Eocene fossil record, and represent important invasions of new adaptive zones for mammals. Each subsequently underwent a major adaptive radiation. It is not clear

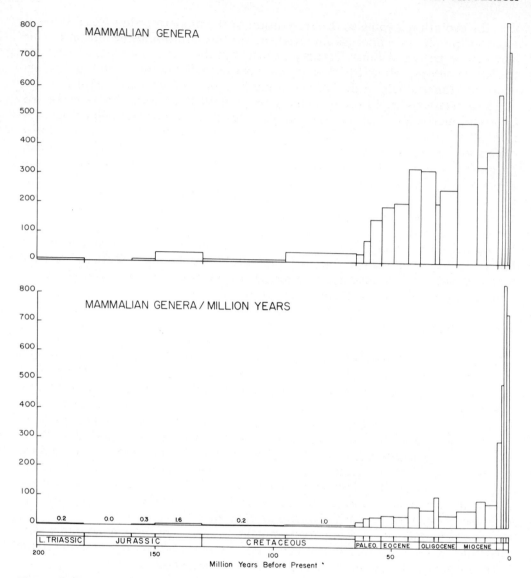

Fig. 2. Generic diversity in mammals through the Mesozoic and Cenozoic. Numbers of genera present in each subdivision of the geological time scale are shown in the upper figure. Numbers of genera per million years in each subdivision of the time scale are shown in the lower figure. Data from Romer (1966).

whether rodents, which also first appear in the Early Eocene, invaded a new adaptive zone, or one previously occupied by less efficient multituberculates and then archaic primates (Van Valen and Sloan, 1966; Hopson, 1967). With their wedge-shaped, self-sharpening, ever-growing incisors forming structural arches stressed by specialized and powerful masseteric musculature, rodents introduced a characteristic gnawing adaptation for ingesting food that has

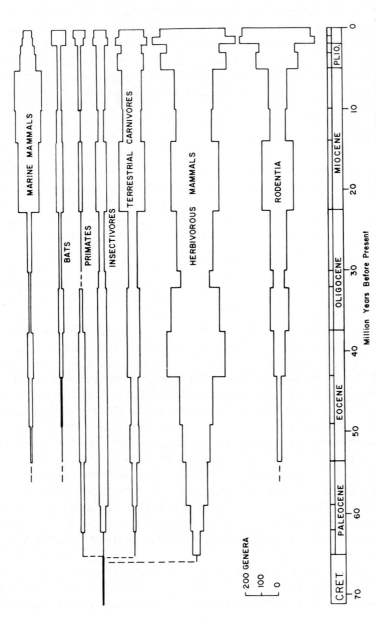

Fig. 3. Relative generic diversity within each major adaptive zone during the Cenozoic. Marine mammals include Sirenia, Desmostylia, Cetacea, and pinniped Carnivora. Terrestrial carnivores include Creodonta, fissiped Carnivora, and borhyaenid Marsupialia. Herbivorous mammals include Artiodactyla, Perissodactyla, Lagomorpha, Litopterna, Astrapotheria, Notoungulata, Embrithopoda, Hyracoidea, Proboscidea, Amblypoda, Condylarthra, Taeniodonta, and Tillodontia. Data from Romer (1966).

remained little modified during their 50 m.y. history. During this time rodents radiated rapidly to become very quickly the most important mammalian order in terms of generic diversity.

It has been mentioned that rodents may have replaced multituberculates and archaic primates in the small mammal herbivorous—omnivorous adaptive zone. Similarly, during the course of the Tertiary several specialized condylarth derivatives, the artiodactyls, perissodactyls, and others, replaced the remaining generalized ancestral condylarthran stock. Perissodactyls first underwent a broad radiation in the Eocene and Oligocene, only to be replaced in large part by a Miocene radiation of artiodactyls in the herbivorous adaptive zone. The carnivorous Creodonta were gradually replaced by true Carnivora during the course of the Tertiary. Thus, within each adaptive zone occurred important replacements of one taxonomic group by another group, the individual species of which were presumably better adapted in a variety of ways than the species they replaced.

Rates of origination and extinction

There are as yet very few groups of mammals in the fossil record that are sufficiently well known stratigraphically to permit tracing individual species lineages through time. The lineages of *Plesiadapis*, *Hyopsodus*, and *Pelycodus* discussed later in this chapter (Figs. 11—13) all show species durations of something on the order of one million years. In these examples, rate of origination and extinction of species in the fossil record is about one per million years in each lineage. Cladogenic branching tends to happen less frequently, varying from a rate of nearly one per million years in *Hyopsodus*, to one per three or four million years in the Plesiadapidae and in *Pelycodus*. Kurtén (1959) has calculated mean "species longevities" varying from 0.3 to 7.5 m.y. for various orders of Cenozoic mammals, and more recently Stanley (1976) has calculated mean species durations of 1.2 m.y. for Plio-Pleistocene mammals of Europe.

Since there are so few good examples of evolution at the species level in fossil mammals, considerations of rates of origination and extinction are generally based on analyses of the geological ranges of higher taxa. Simpson (1953, p. 38) has discussed the advantages and the limitations of such analyses, and his comments apply equally to the analysis presented here. The genus is the smallest taxonomic unit for which geological range data are readily available, and the genus is probably the unit most consistently defined by mammalian palaeontologists. Rates of origination and extinction have been calculated here for rodents, artiodactyls, terrestrial carnivores, and primates, and these rates are plotted in Fig. 4. Each point on the charts in Fig. 4 represents the number of genera making their first (or last, in the case of extinctions) appearance in each subdivision of the geological time scale, divided by the duration of that subdivision.

In Fig. 4, a general overall trend toward increasing rates of both origination and extinction is apparent since the Cretaceous, correlated with the general increase in generic diversity illustrated in Fig. 2. As noted above, this trend is

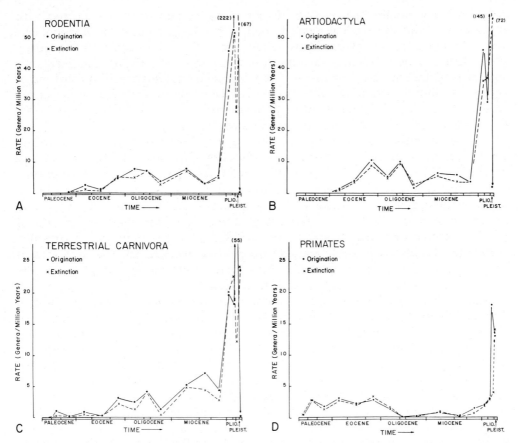

Fig. 4. Rates of origination and extinction at the generic level in Cenozoic mammals. Originations (solid circles) and extinctions (marked by x's) follow a similar pattern in small herbivores (Rodentia), large herbivores (Artiodactyla), carnivores (fissiped Carnivora), and Primates, with relatively high levels of faunal turnover in the Early Eocene, Early Oligocene, Early Miocene, and Plio-Pleistocene. Modern Primates, being largely confined to the tropics, are underrepresented in Neogene sediments. Note the close correlation of origination and extinction throughout the Cenozoic, and the very high rates of origination in the Early Pleistocene preceding high rates of extinction in the Late Pleistocene. Pseudo-originations and pseudo-extinctions (where one known genus evolved directly into another) may account for as much as 20% of rates shown here, although Van Valen (1973) has estimated that 5% is a more likely figure. Data from Romer (1966).

probably in part due to better sampling in the later epochs. Extinction follows origination very closely in each chart of Fig. 4, as one would expect from equilibrium theory and from the relative stability of generic diversity within each major adaptive group of mammals through the course of the Tertiary (Fig. 3). Thus the charts in Fig. 4 give a measure of faunal turnover during the Tertiary. Interestingly, the relatively high levels of generic turnover in the Early Eocene, Early to mid-Oligocene, and Early Miocene correspond to periods of major climatic change (cf. Wolfe and Hopkins, 1967). As was discussed above, a major influx of new genera into western North America and Europe occurred with expansion of the subtropical climatic belt at the beginning of the Eocene. The major faunal turnover of the Early Oligocene (*"Grande Coupure"*) was correlated with a major climatic warming (see Crochet et al., 1975). Similarly, the high rate of faunal turnover in the Early Miocene coincided with a third major period of climatic warming. Lillegraven (1972) has documented a similar correlation of high rates of faunal turnover at the ordinal and familial level in Cenozoic mammals during major periods of climatic warming.

The major extinction of mammals during the Pleistocene has justly received much attention in the literature on extinction (Axelrod, 1967; Martin and Wright, 1967; Webb, 1969; Van Valen, 1969). As Fig. 4 shows, rates of extinction at the generic level were very high in all groups (67 genera/m.y. in Rodentia, 72 genera/m.y. in Artiodactyla, 23 genera/m.y. in terrestrial Carnivora, and 13 genera/m.y. in Primates). In each example, the Late Pleistocene rate of extinction exceeded that of any other subdivision of the Cenozoic. Van Valen (1969) has tabulated possible causes proposed to account for the high rate of extinction in the Late Pleistocene. Most important among these are severe climatic deterioration and/or human intervention.

While the rates of extinction of mammalian genera were at their highest in the Late Pleistocene, these rates were far below the rates of origination of new genera in the Early Pleistocene. Considering the close correlation and general equilibrium of rates of origination and extinction shown in Fig. 4 (see also Webb, 1969), it is only to be expected that high extinction rates would follow the incredible rates of origination seen in the Early Pleistocene. *What requires explanation is not so much the high rate of Late Pleistocene extinctions, but rather the extraordinarily high rate of Early Pleistocene originations.* Late Pleistocene mammal extinctions can be explained as a simple return to faunal equilibrium following an extraordinary over-diversification in the Early Pleistocene. Early Pleistocene over-diversification of mammals was probably a result of abnormal spatial and temporal fragmentation of habitats due to Pleistocene climatic fluctuations and continental glaciations.

The Pleistocene extinction of mammals on many continents has sometimes been attributed to human interference (Martin, 1967). If Late Pleistocene extinctions were due to natural diversity equilibration, then the human contribution to Pleistocene extinctions was probably insignificant. As during the course of the Tertiary, climate more than anything else controlled the level of faunal diversity and the equilibrium level of rates of origination and extinction during the Pleistocene. Humans, rather than controlling mammal diversity in the Late Pleistocene, were apparently subjected to the same pattern of Plio-Pleistocene diversification as other mammals; there is good evidence of two distinct hominid lineages in the Early Pleistocene, but only one thereafter.

Survivorship

Having described the pattern of extinction through the course of the Cenozoic, we can now consider a related pattern — survivorship. In the previous section, the longevity of a mammalian species was given as something on the order of a million years (in the few cases where data are available). For genera and higher taxa there is information available for many more examples, and the following discussion will concentrate on generic-level longevity and survivorship.

Fig. 5 shows the distribution of generic longevity in rodents. When all rodent genera with a fossil record are considered, whether now living or extinct, the average length of life of a rodent genus is 5.85 m.y. As the figure clearly shows, 2 m.y. is the dominant modal longevity. The longevity distribution of rodent genera can be converted into survivorship by considering the total number of genera that survive for various intervals of time. All 558 genera survived at least 1 m.y., the minimum time interval considered. Of these, 521 survived for at least 2 m.y., 286 survived for at least 3 m.y., etc.

Plotting the number of genera surviving each interval of time (on the ordinate) versus the duration of that interval (on the abscissa) gives a survivorship curve. If a logarithmic scale is used on the ordinate, the resulting curve has the property that the probability of extinction at any given age is given by the slope of the curve at that age. Van Valen (1973) was the first to apply this property of survivorship curves in analyzing probabilities of extinction in mammals (and other groups of animals as well). He discovered that the survivorship curves of mammals are very nearly linear with constant slope, i.e., the probability of extinction of a genus is independent of the age of the genus. Raup (1975) has suggested pooling data on living and extinct taxa into a single survivorship curve, and such a curve is given in Fig. 5 for the genera of rodents. The survivorship curve given for rodent genera in Fig. 5, calculated independently

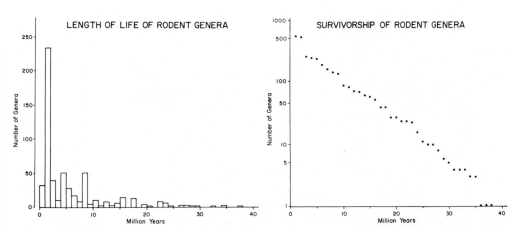

Fig. 5. Survivorship in Rodentia. Histogram (left) shows number of living and extinct genera with a fossil record known from various intervals of geological time. One genus lived for 38 m.y., and the average length of life of a genus is 5.85 m.y. Almost half of the genera known survived for only 2 m.y. Survivorship curve (at right) shows a cumulative plot of the same data, with number of genera on a logarithmic scale. This very nearly straight survivorship curve indicates that the probability of extinction of a rodent genus is nearly constant, regardless of its age (see Van Valen, 1973; Raup, 1975). Data from Romer (1966).

of that given by Van Valen (1973), demonstrates even more closely than his figures the linearity of the curve, and adds additional weight to Van Valen's law of the constancy of extinction. The fact that the probability of extinction of a genus is independent of its age means that extrinsic factors control the survival of the genus. Van Valen emphasizes stochastic deterioration of an animal's effective environment as a critical extrinsic factor.

Size increase — Cope's Rule

The earliest mammals were very small shrew-sized forms, with cheek teeth only 1—2 mm in length. By the Early Eocene, mammals of the size of hippopotami (*Coryphodon*, etc.) were present. The largest land mammal known, the rhinocerated *Baluchitherium*, comes from Oligocene and Early Miocene deposits in Asia. *Baluchitherium* stood nearly 5.5 m tall, and had a skull almost 1.2 m in length (Granger and Gregory, 1936).

In many groups of Cenozoic mammals, lineages can be traced in which there was a progressive trend toward larger size through time. The reasons for this general tendency toward larger body size ("Cope's Rule") have recently been discussed by Stanley (1973). Stanley presents data showing the number of North American rodent species with molars of a given size in the Eocene, Miocene, and Pliocene. These plots become progressively more right-skewed toward larger size through time, while the modal size category remains approximately constant near the small end of the range observed. This indicates that rodents began as small animals, and most remained small while some became larger and invaded niches requiring larger body sizes than those for which the group as a whole was adapted.

Thus, Cope's Rule as a generalization is not to be explained by the intrinsic advantages of large size. It is rather the tendency for new groups to arise at small size that accounts for the observed pattern of net size increase. Stanley (1973) has proposed that the specialized nature of large species, required by problems of similitude, renders these forms unlikely potential ancestors for major new descendant taxa, but it must be remembered that the simple fact that most mammals are small introduces a bias favouring origins from small size by chance alone. It has not yet been shown that more mammalian orders, for example, originated at small size than would be expected given the fact that the great majority of mammalian species at any given time were small. Van Valen (1975) has cited the repeated radiations of large mammals from smaller ones as an example of group selection favouring small mammals, but here again the tendency for small mammals to give rise to successive radiations of large mammals may be due simply to the fact that small mammal species have always been much more abundant than large ones.

Morphological diversity — dental complication and simplification

The broad radiation and diversification of different dental types in Cenozoic mammals has long been known, and as a result of recent work the outline of dental evolution in the Mesozoic is becoming clearer (see Kermack, 1967; Crompton, 1971, 1974; Clemens, 1968, 1970; Parrington, 1971; among others). Early work by Butler and by Mills on the significance of wear patterns on teeth has recently been considerably expanded by cineradiographic (Crompton and Hiiemae, 1970) and electromyographic (Kallen and Gans, 1972) studies of chewing in living mammals. The result is a fairly clear pattern of functional diversification including a general tendency toward the acquisition of progressively more complicated teeth, which was followed in some groups by specialization and secondary simplification of molar morphology.

The most generalized Late Triassic mammals have cheek teeth similar to those of *Eozostrodon*, illustrated in Fig. 6. A large apical cusp and a secondary smaller cusp behind it dominate the crown. Reptiles as a rule have simple pointed teeth that oppose each other in a point-to-point manner, but little precise occlusion is possible. The pointed teeth may function as the mandible is

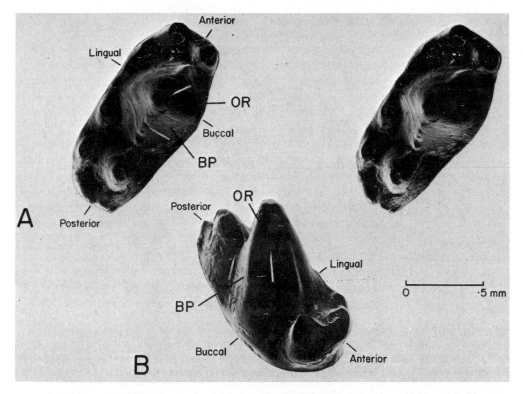

Fig. 6. Crown morphology of a lower molar of the Late Triassic mammal *Eozostrodon*. Scanning electron micrographs in stereoocclusal (A) and oblique (B) anterior views to show the pattern of wear facets. White arrows on facets indicate inferred directions of mandibular movement during function. OR facets with upward and backwardly oriented striations are associated with the apex of the major cusp. BP facets with upward and forwardly oriented striations are associated with the linear crest connecting the major and secondary cusps. Specimen is in the University of California Museum of Paleontology, Berkeley (UCMP 82771).

drawn up and backward by the adductor musculature, or up and forward by the pterygoid muscles. Points may function to hold or puncture food, but they cannot cut food into finer pieces. *Eozostrodon* has rather reptile-like molars with large puncturing cusps, but it also shows a clear advance over most reptiles in that the major cusps were connected by precisely occluding linear shearing edges, permitting food to be cut as well as punctured (see Fig. 6). Interestingly, these two functions appear (from study of minute occlusal wear facets on the teeth) to have been associated with different directions of mandibular motion during chewing; the puncturing cusps functioned as the mandible was drawn up and backward, whereas the shearing crests functioned as the mandible was drawn up and forward.

Several different molar types were derived from the primitive pattern seen in *Eozostrodon* (see Fig. 7). Multituberculates may have split off from the remain-

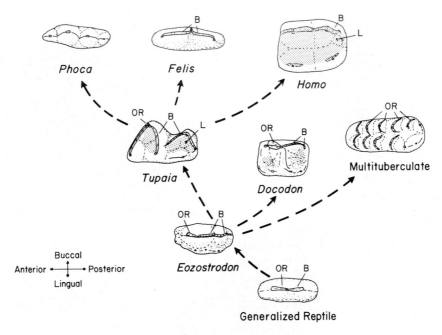

Fig. 7. Adaptive radiation of molar types in mammals. At the hypothetical generalized reptile stage, the teeth had a single puncturing cusp. This cusp functioned to puncture food with either an upward and backward (OR) movement or an upward and forward movement (B). In the Late Triassic mammal *Eozostrodon*, OR wear facets are associated with puncturing cusps, while B facets are associated with linear shearing crests (see also Fig. 6). *Docodon* and multituberculates (see Fig. 9) represent two evolutionary experiments in molar design. In therian mammals, representing a third experiment, the principal molar cusps became rotated to form interlocking triangles. Eventually, a heel or talonid was added to the back of the lower molar, as in *Tupaia*, providing a large planar grinding surface (L) for the upper molar protocone. Later mammals derived from a *Tupaia*-like ancestor specialized for puncturing only (*Phoca*), shearing only (*Felis*), for grinding (*Homo*), or for some combination of these three possibilities.

ing mammalian stock before upward and forwardly directed shearing like that present in *Eozostrodon* evolved. Instead multiple cusps were added to the upper and lower molar teeth and precise shearing occlusion evolved that was designed to function as the mandible was drawn upward and backward. As will be discussed below, this experiment represented a basically different evolutionary pathway to a microshearing dentition from that taken by most other mammals. The genus *Docodon*, like *Eozostrodon*, had both puncturing and shearing features well developed on its molars (Gingerich, 1973), and it is possible that *Docodon* represents an early experiment in the development of a crushing dentition as well.

Zero-dimensional points (cusps) and one-dimensional lines (crests) were well developed in *Eozostrodon*. Of particular interest and importance for the later evolution of mammals was the addition of two-dimensional planar areas (basins) in the therian mammals. These planar areas added a grinding capability to mammalian mastication. By the mid-Cretaceous both *Holoclemensia* and *Pappotherium* show the combination of geometrical points, lines, and planes correlated with puncturing, shearing, and grinding that is found in generalized living mammals such as the tree shrew *Tupaia*. The three functional features found in various combinations on the teeth of therian mammals are analogous to the corners, edges, and surfaces of a solid cube, and the therian molar pattern presumably represents the most efficient way of packing these features onto occluding tooth crowns. With the augmentation of shearing and the addition of grinding, generalized therians like *Tupaia* are able to triturate their food much more completely than their primitive precursors like *Eozostrodon*. The evolution of the mammalian molar from *Eozostrodon* to *Tupaia* provides a nice example of increasing geometrical complexity in the course of mammal evolution.

Most Late Cretaceous insectivores had molars functionally similar to those of *Tupaia*, and molar evolution in the Cenozoic can be seen as a series of trends towards specialization for one or a combination of the functional components present in *Tupaia* molars: puncturing, shearing, or grinding. Phocid seals provide an example of a group specialized for puncturing only, and felid carnivores illustrate molar specialization for shearing only. Human molars are specialized for grinding. These specializations provide examples of decreasing geometrical complexity in the course of mammal evolution.

Mammalian Adaptations

Adaptations of three general types are the ones most often studied in the fossil record: those having to do with the dentition, with the brain, or with locomotion. Evolution of adaptations can either be studied directly by determining morphological trends in established phylogenetic lineages, or indirectly by noting the independent acquisition of particular morphological features in closely related (acquisition via parallelism) or distantly related (acquisition via convergence) groups of animals. Where changes in several morphological features are studied simultaneously, it is often the case that different features change independently, providing examples of mosaic evolution.

Convergent evolution

Locomotor evolution provides a spectacular case of convergence in the independent elongation and reduction in number of the toes in North American horses and in South American litopterns during the Tertiary. Equally dramatic examples are seen in the dental adaptations of diverse mammals known from the fossil record (Fig. 8). The well-known Pleistocene sabre-toothed cat *Smilodon* was probably a scavenger, using its enlarged canine teeth to open and

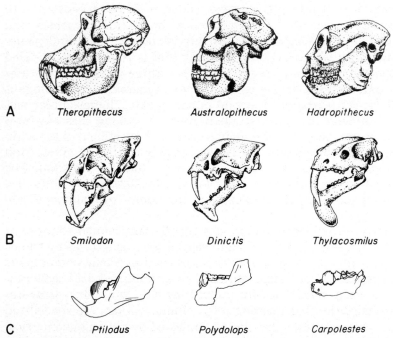

Fig. 8. Examples of convergence in the dental adaptations of mammals. A. *Theropithecus*-complex: the extant baboon *Theropithecus*, the Pleistocene hominid *Australopithecus*, and the subfossil lemur *Hadropithecus* all show a similar shortening of the facial region of the skull with a short, deep mandible in correlation with a diet of seeds, rhizomes, and other small tough objects (inferred in *Australopithecus* and *Hadropithecus*: see Jolly, 1970a,b). B. Sabre-toothed carnivores: the Pleistocene true sabre-toothed cat *Smilodon*, the Oligocene false sabre-toothed cat *Dinictis*, and the Pliocene borhyaenid marsupial *Thylacosmilus* all evolved enlarged shearing upper canine teeth independently (see Riggs, 1934; Miller, 1969). C. Plagiaulacoid herbivores of the Paleocene: the multituberculate *Ptilodus*, the marsupial *Polydolops*, and the primate *Carpolestes* all independently evolved an enlarged ribbed, blade-like tooth in the center of the lower dental series (see Simpson, 1933; Rose, 1975). Drawings are not to the same scale — primates and carnivores in A and B have skulls approximately the size of a human skull or slightly smaller, whereas the plagiaulacoids in C are much smaller and would have a skull approximately the size of a squirrel.

divide carcasses. An independent line of distantly related "false" sabre-toothed felids (including the Oligocene *Dinictis*) acquired a similar morphological adaptation, as did a very distantly related Pliocene South American borhyaenid marsupial *Thylacosmilus*. These forms too were presumably scavengers like *Smilodon*.

In the Paleocene of Europe and North America, the small herbivore niche was dominated by multituberculates, which were characterized by an enlarged, blade-like tooth in the center of the tooth row. The evolution of North American carpolestid primates can be traced in detail through the Middle and Late Paleocene, during which time they developed an enlarged blade-like tooth in the middle of the tooth row similar to that seen in multituberculates (Rose, 1975). Also, completely independently, a group of South American polydolopid marsupials acquired the same blade-like tooth in the center of the tooth row. This convergence of "plagiaulacoid" dental types has been described in greater detail by Simpson (1933).

More recently a very interesting "seed eating" model has been proposed by Jolly (1970a,b) to explain the origin of human dental and cranial morphology. Jolly's model is based on a baboon analogy: he noted that the morphological series from long-snouted mandrills to intermediate *Papio* baboons to short-snouted geladas is similar to the series from the great apes to humans. *Theropithecus* geladas have their mandibles tucked underneath the cranium like the condition in hominid primates. Geladas differ from other baboons in feeding on a greater proportion of small, tough seeds and rhizomes in more open savanna. Jolly proposed that a progressive ecological shift to open country seed eating might explain the morphological shift from long-jawed, forest living apes to short-jawed humans. Skulls of *Theropithecus* and the archaic hominid *Australopithecus* are illustrated in Fig. 8A, along with another remarkable primate showing a similar morphological pattern, *Hadropithecus*, a subfossil lemur from Madagascar that convergently evolved very hominid-like skull proportions.

Multiple evolutionary pathways

The examples of convergence discussed above all show similar morphological adaptations. Another example is known from the mammalian fossil record that shows a very similar functional adaptation acquired by very distantly related mammals: multituberculates and rats — animals whose last common ancestor lived some 200 m.y. ago. Both acquired a cheek tooth complex adapted for microshearing, but they acquired this from such different morphological backgrounds that the whole functional complex is oriented in opposite directions in the two (see Fig. 9).

The functional evolution of mammalian molars summarized in Fig. 7 shows that primitively upward and backward jaw movements powered by the mandibular adductor musculature were the most important. During the course of mammal evolution in the Mesozoic, upward and forward jaw movements became progressively more important for shearing and grinding. Correlated

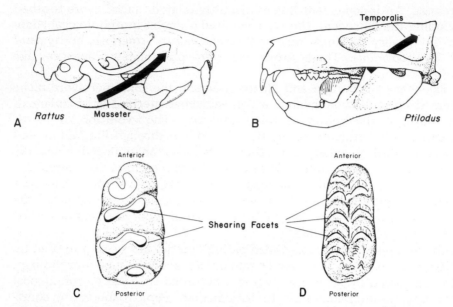

Fig. 9. Multiple evolutionary pathways to a microshearing dentition. The living rat *Rattus* (A) and the Paleocene multituberculate *Ptilodus* (B) were approximately the same size and had skulls that were fundamentally similar in having enlarged central incisors, a diastema, and a battery of multicusped cheek teeth, however the jaw musculature of *Rattus* is dominated by the masseter muscles, whereas multituberculates were temporalis dominated. Concave microshearing crests on *Rattus* molars (C) show that the molars functioned as the mandible was drawn upward and forward, whereas the concave microshearing crests are oriented to function as the mandible was drawn upward and backward in multituberculates (D). See also Fig. 7.

with this was progressive enlargement of the masseter musculature pulling the mandible upward and forward. Multituberculates developed a microshearing dentition by multiplying the number of cusps on upper and lower molars, and by pulling the lower jaw backwards during the power stroke of chewing with their enlarged adductor or temporalis musculature. The orientation of concave shearing facets on the upper and lower molars of multituberculates shows clearly that they functioned during the upward and backward power stroke. The mandibular mechanics of the earliest rodents, on the other hand, were already dominated by the masseter pulling the lower jaw upward and forward (an adaptation perhaps acquired in correlation with the gnawing incisors present in the earliest rodents). Thus, when rodents specialized for a microshearing cheek dentition, they did it by developing small concave shearing blades functioning when the jaw was drawn forward. Functionally the result was the same as in multituberculates but the two systems were oriented in opposite directions — an example of multiple evolutionary pathways to the same morphological adaptation.

Parallel evolution

Many examples of parallel evolution have been documented in the fossil record. The archaic family Plesiadapidae shows development of crenulated enamel, molarization and reduction in number of premolar teeth, increase in overall size, etc. in parallel but independent lineages during the Paleocene (Gingerich, 1976b). Numerous lineages of primates and other mammals independently evolved larger brain size through the course of the Tertiary (Jerison, 1973).

One of the most interesting cases of parallel evolution yet documented in the fossil record was described by Radinsky (1971). Radinsky's example shows the independent evolution of a cruciate sulcus at least four times in different families of modern Carnivora. Radinsky's data are presented in Fig. 10. Practically all modern carnivores have a cruciate sulcus dividing the frontal lobe of the brain. The earliest representatives of the Ursidae (bears), Procyonidae (racoons), and Mustelidae (weasels), known from the Late Eocene and Oligocene, all have a well-developed cruciate sulcus. However, the Felidae (cats), Viverridae (civets), and Canidae (dogs) can be traced back to distinctive Late Eocene and Oligocene genera that do not have a cruciate sulcus. Thus, this important morphological feature of the brain evolved at least four times: in Felidae, in Viverridae, in Canidae, and in the common ancestor of Ursidae, Procyonidae, and Mustelidae. Radinsky attributes the multiple independent

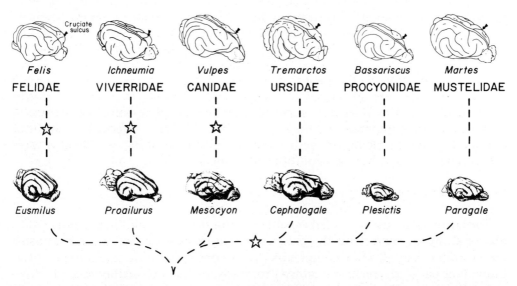

Fig. 10. Parallel evolution of the cruciate sulcus in fissiped Carnivora. Representatives of six modern carnivore families are shown at the top. Eocene or Oligocene ancestral forms of each of these families are shown at the bottom. Position of the cruciate sulcus is indicated by an arrow in the living genera. Stars show the independent evolution of a cruciate sulcus at least four times in Felidae, Viverridae, Canidae, and the common ancestor of Ursidae, Procyonidae, and Mustelidae. Data and inset figures from Radinsky (1971).

origins of a cruciate sulcus in carnivores to the functional requirements of folding the neocortex in association with expansion of the motor cortex. Since
these advances were presumably adaptive in all of the different families of carnivores, it is not surprising that all eventually acquired a similar cruciate sulcus.

The independent evolution of a mammal-like dentary-squamosal jaw articulation in different groups of advanced cynodont reptiles (Crompton and
Jenkins, 1973), and the independent origin of the mammalian middle-ear
mechanism in monotremes and therian mammals (Hopson, 1966) provide
additional examples of parallel evolution in complex anatomical systems.

Mosaic evolution

When intermediate stages connecting a primitive mammal with a more modern one are known, it is usually found that the characters by which the modern
form differs were not all acquired at the same time in its evolutionary history.
The appearance of different characteristics at different times is termed mosaic
evolution.

One of the most interesting examples of mosaic evolution in mammals has
been documented in the evolution of humans from more primitive ape-like
ancestors (as yet inadequately known). Humans differ from apes in two obvious ways — humans walk bipedally and have much larger brains. Lamarck,
Haeckel, and Darwin all postulated that human bipedality preceded the evolution of a large brain, but only over the past fifty years has evidence been collected and analyzed that demonstrates this to be so. The pelvis of *Australopithecus* is intermediate between modern apes and modern humans when its
total morphological pattern is analyzed. However, when the morphological features associated only with locomotion are studied, they show that the gait of
Australopithecus was that of a fully modern human biped (Lovejoy et al.,
1973). The features of the pelvis of *Australopithecus* which make it resemble
the pelvis of apes are all related to the small size of the birth canal, which is in
turn related to the relatively slight degree of encephalization of *Australopithecus* when compared to modern humans. The fossil record shows that
human bipedalism clearly preceded human encephalization, which illustrates
the mosaic nature of human evolution (McHenry, 1975).

Evolutionary reversals

Evolution, like history, is irreversible for the simple reason that time is unidirectional. Few would deny the progressive nature of evolution when organisms
are considered in all their complexity. As a consequence we might expect that
individual parts of organisms evolve progressively and irreversibly as well. However, this is not always the case. The occurrence of minor evolutionary reversals
in no way diminishes the irreversibility of evolution as a whole, but it does
again emphasize both the mosaic nature of the process and the importance of
adaptation.

Kurtén (1963) has described a most interesting case of minor evolutionary reversal in the reappearance in some specimens of *Felis lynx* of the lower second molar M_2 — a tooth unknown in the Felidae since the Miocene. A similar minor morphological reversal can be seen in the phylogenetic history of the primate family Plesiadapidae (Fig. 11), where a minute central cusp first appeared in the upper incisors in the species *Plesiadapis rex*. In one subsequent branch leading to *P. fodinatus* and *P. dubius* this cusp was reduced and lost, while in another derived branch leading to *P. tricuspidens* it was retained in well-developed form before being lost in the genus *Platychoerops*. After becoming fully developed, the cusp was reduced and lost in two independent lineages. To take another example, in the lineage from *Plesiadapis* to *Platychoerops* the paraconule cusp on the premolars was lost and then regained (Gingerich, 1976b). Thus it is not possible to infer from the general pattern of progressive evolution in Felidae or Plesiadapidae that individual characters will behave in the same progressive way.

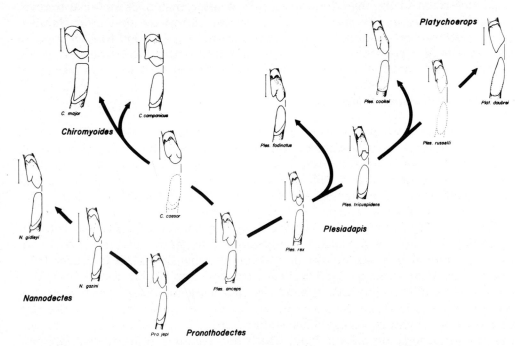

Fig. 11. Minor reversal in the evolution of a distinctive centroconule on the upper incisors of *Plesiadapis*. The primitive species of *Pronothodectes*, *Nannodectes*, *Chiromyoides*, and *Plesiadapis* all had a simple tricuspid apex on the upper incisors. In *Plesiadapis rex* a distinctive fourth cusp, the centroconule, was added in the center of the apex. In one derived North American lineage, leading to *Plesiadapis fodinatus*, this cusp was reduced. In another lineage derived from *Plesiadapis rex*, leading to the North American *Plesiadapis cookei*, the centroconule was lost completely. The centroconule may have been lost independently in Europe in the lineage leading from *Plesiadapis tricuspidens* to *Platychoerops daubrei*. Note also the gradual simplification of upper and lower incisors in the line leading to *Platychoerops*. Figures show left upper and lower incisors in posterior view. Scale bar = 5 mm. From Gingerich (1976b).

Oscillations in dental size and shape have been very nicely demonstrated and correlated with Pleistocene climatic fluctuations in European hamsters (*Cricetus*, see Kurtén, 1960) and in North American muskrats (*Ondatra*, see Nelson and Semken, 1970), and many other examples of this kind of reversal could probably be given. Another interesting example of evolutionary reversal in size occurred when large mammals became isolated on islands. Elephants, hippopotami, and deer, whose general evolutionary history was one of increasing size, are known from the fossil record to have become smaller when they were isolated on islands (Thaler, 1973). Finally, a third case of evolutionary reversal in size, due to character divergence, will be discussed and illustrated in the following section.

Origin of Species and Higher Taxa

As described in the introduction to this chapter, mammals are ideal for evolutionary studies at the species level. It is thus surprising that more detailed study has not been devoted to those mammals that have a good fossil record. In the remaining pages stratigraphically documented patterns of evolution at the species level, evolution at higher levels, and hominid evolution are discussed.

Evolution at the species level

For the past century, since 1876 when Marsh demonstrated his fossil collection to Huxley, the evolution of the horse has been the standard textbook example of mammalian evolution in the fossil record. Unfortunately, much of this remarkable sequence is still only understood at the level of the genus. For an understanding of evolution at the species level one must turn to smaller forms: condylarths, lagomorphs, rodents, and primates.

Simpson (1943) originally illustrated his "chronocline" concept with an example from the Early Eocene condylarths. The genus *Ectocion* shows continuous, gradual increase in size in a single evolving lineage as one goes from Clark Fork beds through Sand Coulee beds and into the overlying Gray Bull beds in Wyoming. Simpson's example is especially important because in presenting it he revised the diagnoses of the successive species to reflect the fact that they are stages in a single lineage. Previously, he and others had diagnosed the species strictly on morphology, which had resulted in an overlapping pattern of species ranges suggesting that multiple lineages were present and that the origin of one from another was an abrupt saltation. A similar example of the importance of using time planes to separate adjacent species in a single lineage is illustrated by Simpson's (1953, p. 387) modification of Trevisan's (1949) typological diagnosis of *Elephas meridionalis* from its ancestor *E. planifrons*.

Other chronoclines have been very nicely documented in the evolution of the mid-Tertiary lagomorph *Prolagus* (Hürzeler, 1962, see also Kurtén, 1965) and in the Middle Oligocene rodent *Theridomys* (Vianey-Liaud, 1972). In

Prolagus a very complete fossil record shows a remarkable but continuous and gradual reorganization of the premolar crown morphology in a single lineage. In two independent lineages of *Theridomys*, Vianey-Liaud has shown that "synclinide I" was present in an increasing percentage of specimens, and the molars became gradually more hypsodont in successive species. As more lineages are studied in detail it is virtually certain that many of the species already known will be found to intergrade continuously. Wood's statement (1954) that "whenever we do have positive palaeontological evidence, the picture is of the most extreme gradualism" was based on his stratigraphic studies of fossil rhinoceroses, and should probably not be lightly dismissed.

The above studies document the gradual evolution of one species into another within single lineages. In a series of more recent studies an attempt has been made to document the division and separation of lineages as well as evolutionary change within them. *Hyopsodus* is the most common fossil mammal in collections from Lower Eocene strata in North America. Several distinct biological species are present in single locality samples, as indicated by discrete gaps in the size variation of individual teeth (each species within the sample having the variation typical of a modern mammalian population). When all of the samples of *Hyopsodus* that can be placed in stratigraphic position are so ordered, the pattern of change in tooth size that emerges is one of continuous gradual change within lineages, with gradual divergence following the separation of new sister lineages (Gingerich, 1974). A more complete picture of *Hyopsodus* evolution based on additional collecting is presented in Fig. 12 (see Gingerich, 1976a, for discussion).

Other Early Eocene mammals can be studied in the same stratigraphic context, and Fig. 13 shows the pattern of change in dental size in the early primate *Pelycodus*. As in Simpson's *Ectocion* example, there is no evidence for more than a single evolving lineage of *Pelycodus* in Sand Coulee through Gray Bull strata. However, in the upper levels, during Lysite and Lost Cabin time, there is clear evidence that species of two lineages of *Pelycodus* were present. Tracing these two distinct species, *Pelycodus frugivorus* and *P. jarrovii* back in time, they converge with *Pelycodus abditus* in size, mesostyle development, and every other character available for study, and there can be little doubt that each was derived from that species. A similar pattern is seen in the North American Paleocene Plesiadapidae, stratigraphically the best known family of primates. *Nannodectes*, *Chiromyoides*, and two lineages of *Plesiadapis* can be traced back in the fossil record until each converges with a known species of known geological age (Gingerich, 1976b).

These examples are important for several reasons. First, they demonstrate that in some cases phylogenetic patterns, including branching sequences, can be determined empirically from the fossil record. Second, these examples illustrate again the importance of gradual phyletic evolution within single lineages and the importance of this mechanism (anagenesis) in the origin of new species. Finally, they provide the first palaeontological evidence on the geometry of cladogenic branching patterns in mammalian speciation.

Patterns such as these have sometimes been used to support the idea of sym-

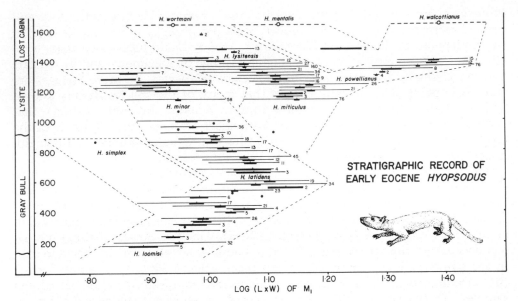

Fig. 12. Stratigraphic record of Early Eocene *Hyopsodus* in northwestern Wyoming. Figure shows variation and distribution of tooth size in samples from many levels spanning most of Early Eocene time (conventionally divided into Gray Bull, Lysite, and Lost Cabin intervals). Specimens come from localities at approximately 6-m intervals in or near a measured stratigraphic section totalling about 500 m in thickness. The small species *Hyopsodus loomisi* became larger gradually through time, until it differed sufficiently to be recognized as a different species *H. latidens*. *H. "simplex"* was another early species derived from *H. loomisi*. *H. latidens* apparently gave rise to both *H. minor* and *H. miticulus*, which in turn gave rise to *H. lysitensis* and *H. powellianus*. Note the regular pattern of divergence in tooth size (and by inference body size) in pairs of sympatric sister lineages. Vertical slash is sample mean, solid bar is standard error of mean, horizontal line is total range, and small number is sample size. From Gingerich (1976a).

patric speciation, but they do not, in fact, support a totally sympatric origin of new clades. Geographic disruption is probably essential in subdividing populations, some of which are or become genetically isolated from each other. However, the fact that size divergence in recently separated sister lineages is so pronounced suggests that character displacement is an important mechanism acting to make genetically separated populations into morphologically different species. The morphological features that distinguish two descendant species from each other and from their common ancestor are acquired gradually (albeit probably as rapidly as possible given the genetic basis underlying most morphological characteristics) only after the two descendants have become sympatric. Since in the above examples it is the close sympatric interaction of two sister species (which became genetically different allopatrically) that makes them different morphologically, I previously characterized speciation in these cases as "parapatric" to indicate that it was a form of speciation intermediate between strictly allopatric or sympatric speciation (Gingerich, 1976a). It now seems better just to recognize that in these examples the genetic separation stage of spe-

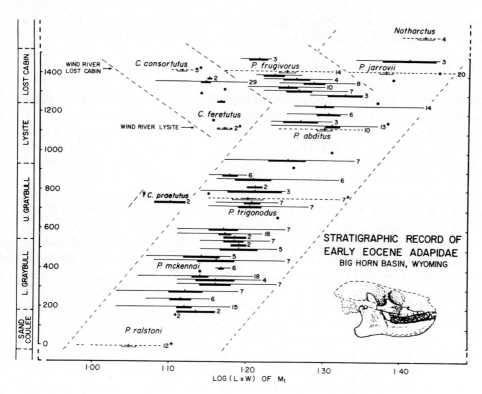

Fig. 13. Stratigraphic record of Early Eocene *Pelycodus* in northwestern Wyoming. Samples from same localities and stratigraphic section as Fig. 12, and the symbols are the same. Note again the continuous, gradual connection between successive species, and the reversed trend toward smaller tooth size in *Pelycodus frugivorus* with the appearance of *P. jarrovii*. Other characters available for study in this sequence, such as mesostyle development, show the same pattern of gradual evolutionary change, but mesostyle development continues progressively through the whole sequence and does not show the character divergence seen in tooth size. From Gingerich and Simons (1977). *C. praetutus*, *C. feretutus* and *C. consortutus* belong to a new genus related to *Pelycodus*.

ciation probably occurred allopatrically, while the subsequent stage of morphological differentiation occurred in large part sympatrically.

The examples discussed here certainly do not rule out the possibility that both genetic separation *and* morphological differentiation occur allopatrically in some cases, but they do suggest that the strictly allopatric view of speciation, such as that advanced by Eldredge and Gould (1972), may not be as important as it has been thought to be. Three common flaws in the fossil record and in palaeontological methodology would give a "punctuated" picture of phylogeny even where the actual case was one of gradual phyletic change. These potential sources of bias toward a pattern of "punctuated equilibria" (see Eldredge and Gould, 1972) have been discussed in detail elsewhere (Gingerich, 1976a,b) but are listed here as well: Gaps in the fossil record, coarse stratigraphic sampling of a continuous record, or typological analysis of fossils collected could all yield a pattern of "punctuated equilibria" as an artifact of methodology.

Evolution above the species level

Migration is another potentially important source of abrupt change in fossil sequences. At the beginning of this chapter the importance of climate was discussed in connection with the abrupt origin of many modern mammalian orders at the beginning of the Eocene. The first rodents appeared then, with their diagnostic gnawing incisors. Similarly, the first artiodactyls appeared abruptly at the beginning of the Eocene with their distinctive double pulley astragalus already fully developed (see Schaeffer, 1948). The earliest known bats appeared early in the Eocene with their wings fully adapted for flight (Jepsen, 1966). Saltational, or quantum, or punctuated evolution has been invoked to explain the sudden appearance of modern groups of mammals at different times in the fossil record. However, we do not yet know the fossil ancestors of these groups and, from the gradual generic level transitions seen, for example, in going from *Plesiadapis* to *Platychoerops* (Gingerich, 1976b), we might do well to wait until there is more evidence before hypothesizing special evolutionary mechanisms to explain sudden appearances in the fossil record.

One problem in understanding the origin of higher categories of mammals arises when authors assume that since the incisors, or wings, or astragali characteristic of a modern order of mammals were present in the Eocene, the mammals with such diagnostic characters differed at an "ordinal" level from other Eocene mammals. This is rarely true. Apart from the relatively minor specializations distinctive of each order, most "modern" Eocene mammals were as yet very little differentiated. The origin of higher categories of mammals may have involved higher than average rates of evolution, but there is as yet no reason to suppose that the basic mechanism was any different from that operating in normal speciation. To propose that natural selection operates too slowly to account for the major features of evolution (Stanley, 1975) probably underestimates potential rates of gradual evolutionary change within lineages, underestimates the control of natural selection on cladogenic speciation, and overestimates rates of morphological change during cladogenic speciation events. If nothing else, the patterns of phylogeny in *Hyopsodus* and *Pelycodus* presented in Figs. 12 and 13 indicate that new evidence on the origin of species and higher categories should be forthcoming from careful studies of fossil mammals with good stratigraphic records.

Hominid evolution

One of the most interesting patterns of evolution in the mammalian fossil record is that now emerging in the history of the human family Hominidae. The study of hominid evolution has suffered from the same methodological problems that have plagued study of the rest of primate and mammalian evolution. In the absence of a good fossil record, the study of comparative anatomy alone has often been used to "reconstruct" human evolution (in the manner that best suited each individual scholar). Typology and inadequate studies of variation have led to overly complex or overly simple patterns of human descent. Lack of fossil dating has led some workers to postulate that the human line has been distinct from other primates for many millions of years, while other workers have advocated a much shorter time. Fortunately, in the past few years concentrated efforts in the field have been rewarded by an abundance of new fossil hominid specimens, many of which can now be dated either absolutely or stratigraphically relative to other hominid fossils.

It is curious how little the dynamic nature of evolution is recognized in discussions of human evolution. Kortlandt (1974, p. 444) for example, is unable to understand how the profound changes making a hominid (*Ramapithecus*, dated at 10—14 m.y.) out of an ape (*Dryopithecus*, 18 m.y.) could have taken place in the space of a few million years, and he proposes that on the contrary, *Ramapithecus* must have diverged from apes several million years earlier, at about 20—25 m.y., if it was to be distinct by 14 m.y. ago. In reply to this one might note that first of all, *Ramapithecus* can hardly be distinguished from *Dryopithecus* morphologically. The differences which separate the two are barely sufficient to justify placing them in two genera, but they indicate that *Ramapithecus* was more like a hominid than *Dryopithecus* was. Since *Ramapithecus* is the only fossil form known before *Australopithecus* to show certain hominid dental features, it is grouped with *Australopithecus* and *Homo* in the family Hominidae, even though it differed only slightly from species of the ape *Dryopithecus*. This procedure is perfectly normal in mammalian palaeontology, and again illustrates the point made above about the earliest representatives of mammalian orders: the early representatives of new higher taxa usually differ only very slightly from each other.

The second point that should be noted in reply to Kortlandt is that species are dynamic — the time period from 18 m.y. to 14 m.y. is sufficient for a considerable amount of change to occur in a lineage of *Dryopithecus*, the amount necessary for it to become *Ramapithecus* being well within that occurring in 4 m.y. in other generic level transitions known in the fossil record. There is a strong predilection to regard species as fixed in time, with all change being concentrated in unknown transitions between fixed types (a predilection given some theoretical justification by Eldredge and Gould, 1972), but in practically all of the cases where samples are known from successive stratigraphic intervals, species are seen to be changing continuously.

Surprisingly little attempt has been made to look at patterns of variation in an explicitly stratigraphic or temporal context in the hominid fossil record. When Pilbeam and Zwell (1972) attempted this using dental measurements of all Plio-Pleistocene hominids, grouping them into 0.5-m.y. bands, they were able to show a fairly clear pattern of divergence of two hominid lineages at about 2—2.5 m.y. ago. To avoid problems of correlation between distant sites, Schoeninger (in Gingerich and Schoeninger, 1976) made a similar analysis of fossil hominids recently collected and described by R.E.F. Leakey from a restricted geographic area east of Lake Turkana in northern Kenya. The results

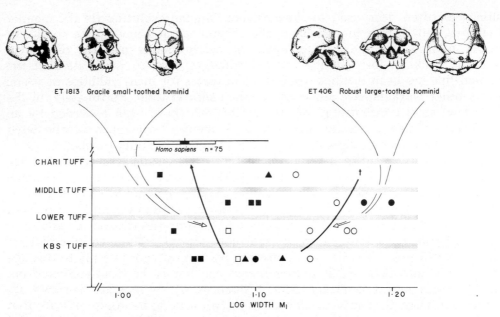

Fig. 14. Stratigraphic distribution of fossil hominids east of Lake Turkana in northern Kenya. Tooth size is plotted for four successive stratigraphic intervals separated by tuffs. The sample of a single population of modern *Homo sapiens* at the top gives a scale for interpreting the east Turkana fossil hominids at each level (horizontal line is range, vertical slash is mean, solid bar is standard error of the mean, open bar encloses one standard deviation from the mean, and the distribution is unimodal). In the east Turkana samples, squares represent *Homo* sp. of authors, circles represent *Australopithecus* sp., and triangles represent undetermined specimens. Open symbols are all estimated from other tooth dimensions for those specimens using regressions, and thus they are less reliable than the solid symbols. Note particularly the well-separated modes in the two middle intervals, suggesting the presence of two hominid lineages. Gracile, small-toothed skulls, some with relatively large endocranial volumes, are thought to be ancestral to modern *Homo sapiens*, while the robust, large-toothed skulls are a divergent extinct lineage. The KBS Tuff is dated at about 2.4 m.y., while the Middle Tuff is dated at about 1.5 m.y. Redrawn from Gingerich and Schoeninger (1976), with inset figures of skulls from Leakey (1976).

of her analysis are presented here in Fig. 14. Specimens discovered from below the KBS tuff (dated at about 2.4 m.y.) have teeth which do not differ greatly in range of distribution of variation from modern comparative populations, although two distinct hominoid lineages may already have been present. The stratigraphic intervals between the KBS and Middle tuffs (the latter dated at about 1.5 m.y.) have yielded samples which exceed the range of variation typical of modern human populations. More importantly, they show a bimodality in the distribution of size of the first molar that would be extremely unlikely in a single species, whether hominids, forest living gorillas, or savanna living baboons (with their extreme sexual dimorphism) were being sampled. Furthermore, these differences correlate with other cranial differences separating "gracile" and "robust" hominids. As variation in morphological characteristics such as endocranial volume, external cranial anatomy, and postcranial anatomy becomes better known it is to be hoped that these too will be studied in an explicitly stratigraphic context. The studies by Pilbeam and Zwell, and Schoeninger show clear evidence of two lineages of Plio-Pleistocene hominids, based on the available tooth measurements. The smaller toothed lineage appears to be the one leading to modern humans, and unless evidence to the contrary is forthcoming, it seems more reasonable to associate large endocranial volumes and human-like tali and ulnae with this dental lineage rather than postulating extension of a third dentally unknown *"Homo erectus*-like" lineage back to 5 m.y. as Oxnard (1975) has recently advocated. Whatever the true pattern of hominid evolution eventually found, it is now clear that hominid phylogeny was sufficiently complicated that only the discovery of well-dated fossil specimens will ever reveal its true course.

Conclusions

Several general trends are apparent in the evolutionary history of mammals. For the first two-thirds of their history, mammals were little diversified, of small body size, and basically insectivorous (multituberculates excluded). During the final one-third of mammalian history, spanning the past 65 m.y., mammalian diversity increased rapidly, mammals of large body size first appeared, and mammals invaded new herbivorous and carnivorous adaptive zones. Relative diversity within each major terrestrial adaptive zone has been fairly stable since the Paleocene. Flying mammals and swimming marine mammals are first known from the Early Eocene, and each underwent a separate major radiation during the Tertiary.

Rates of origination and extinction have remained in close equilibrium through the course of the Cenozoic, with major periods of faunal turnover in the Early Eocene, Early Oligocene, Early Miocene, and Early Pliocene. These periods of high faunal turnover correlate with major periods of climatic warming. The high rate of extinction of mammalian genera in the Late Pleistocene is seen to be less remarkable than the incredibly high rate of origination of mammalian genera in the Early Pleistocene. This high rate of origination requires explanation, whereas a high rate of extinction following such a high rate of origination is a predictable result, given the close equilibration of extinction with origination.

There has been a general trend toward increasing size and complexity during mammalian evolution, when the order Mammalia is considered as a whole, but many lineages became smaller, and many anatomical complexes became simpler through time. These facts, plus the great amount of parallelism, convergence, and minor evolutionary reversal all point to functional adaptation as the goal of morphological evolution. Species are dynamic, and natural selection acts continuously to maintain adaptations. Patterns of phylogenetic change are sometimes complex, and a detailed, dense, and continuous fossil record is usually required to decipher the course of evolution at the species level and within higher taxa.

Study of stratigraphically documented transitions from one species to another reveals that both anagenesis and cladogenesis are important in the origin of new species. Patterns of cladogenic speciation indicate that new lineages diverge morphologically when they occur sympatrically *after* genetic barriers have formed between populations during geographic separation. The genetic basis of cladogenic speciation is thus allopatric, but morphological divergence is probably due, in some cases at least, to character divergence during subsequent sympatry. Thus, it appears that cladogenic speciation normally involves both allopatric and sympatric phases associated with genetic and morphological differentiation, respectively.

New taxa of higher than species level sometimes appear abruptly in the fossil record. In one of the most important instances, the sudden appearance of many modern mammalian orders in the Early Eocene, the new ordinal level taxa appeared abruptly as immigrants (in the areas where fossil deposits are known) due to major climatic warming and expansion of their geographic ranges northward. The diagnostic differences between members of different orders sometimes receive disproportionate attention — Eocene mammals placed in different orders were not nearly as different as their modern counterparts are today. Generic level transitions documented in the fossil record of archaic plesiadapid primates show that higher categories may arise by gradual phyletic change in single lineages without any distinguishable periods of abrupt morphological reorganization. In other words, there is as yet no evidence that the origin of higher taxa involves any mode or process different from the origin of species.

Phylogenetic patterns documented in *Plesiadapis*, *Pelycodus*, and *Hyopsodus* give the impression that populations within a species are constantly being shuffled by vagaries of climate and geography, producing a constant supply of genetically separated populations partially isolated by reproductive barriers from sister populations. Sometimes, when a sufficient width of ecological niche is available, two genetically separated incipient sister species survive sympatrically, diverging in character to minimize competition. In other cases the two incipient species might recombine into one, or one might replace the other. Geographic shuffling produces a constant source of slightly different populations of a species, which is analogous to the production of slightly different individuals within a population by recombination and mutation. Natural selection limits this variability in populations through ecological competition between them, which is analogous to the way natural selection channels variation in individuals through differential survival within populations. In mammals like *Hyopsodus*, lineages of species which apparently differed only in body size were closely packed in narrow adjoining adaptive zones. Interestingly, the width of hominid niche available during the Late Pliocene and Early Pleistocene was sufficient to permit the coexistence of two lineages of early bipedal humans, but by the Late Pleistocene only one lineage remained — the gracile form *Homo* survived, while the robust one became extinct.

Acknowledgements

I thank Mrs. Gladys Newton for secretarial assistance in preparing the manuscript and Mr. Karoly Kutasi for photographic assistance with the figures. The photographs in Fig. 6 are by Kenneth Rose, and in addition, his comments have improved the manuscript considerably. Elizabeth Nof drafted Figs. 2—4, and drew the inset figures in Figs. 1, 8, and 14. Research reported here was supported in part by a Faculty Research Grant from the Rackham School of Graduate Studies, University of Michigan.

References

Alvesalo, L. and Tigerstedt, P.M.A., 1974. Heritabilities of human tooth dimensions. Hereditas, 77: 311—318.

Axelrod, D.I., 1967. Quaternary extinctions of large mammals. Univ. Calif. Publ. Geol. Sci., 74: 1—42.

Bader, R.S., 1965. Heritability of dental characters in the house mouse. Evolution, 19: 378—384.

Clemens, W.A., 1968. Origin and early evolution of marsupials. Evolution, 22: 1—18.

Clemens, W.A., 1970. Mesozoic mammalian evolution. Ann. Rev. Ecol. Syst., 1: 357—390.

Crochet, J.-Y., Hartenberger, J.-L., Sigé, B., Sudre, J. and Vianey-Liaud, M., 1975. Les nouveaux gisements du Quercy et la biochronologie du Paléogène d'europe. Réunion Annuelle des Sciences de la Terre, 3e, Montpellier, 1975: 114.

Crompton, A.W., 1971. The origin of the tribosphenic molar. In: D.M. Kermack and K.A. Kermack (Editors), Early Mammals. Academic Press, London, pp. 65—87.

Crompton, A.W., 1974. The dentitions and relationships of the southern African Triassic mammals, Erythrotherium parringtoni and Megazostrodon rudnerae. Bull. Br. Mus. (Nat. Hist.), Geol., 24: 397—437.

Crompton, A.W. and Hiiemae, K., 1970. Molar occlusion and mandibular movements during occlusion in the American opossum, Didelphis marsupialis L. Zool. J. Linn. Soc., Lond., 49: 21—47.

Crompton, A.W. and Jenkins, F.A., 1973. Mammals from reptiles: a review of mammalian origins. Ann. Rev. Earth Planet. Sci., 1: 131—154.

Eldredge, N. and Gould, S.J., 1972. Punctuated equilibria: an alternative to phyletic gradualism. In: T.J.M. Schopf (Editor), Models in Paleobiology. Freeman, San Francisco, Calif., pp. 82—115.

Gingerich, P.D., 1973. Molar occlusion and function in the Jurassic mammal Docodon. J. Mammal., 54: 1008—1013.

Gingerich, P.D., 1974. Stratigraphic record of early Eocene Hyopsodus and the geometry of mammalian phylogeny. Nature, 248: 107—109.

Gingerich, P.D., 1976a. Paleontology and phylogeny: patterns of evolution at the species level in early Tertiary mammals. Am. J. Sci., 276: 1—28.

Gingerich, P.D., 1976b. Cranial anatomy and evolution of early Tertiary Plesiadapidae (Mammalia, Primates). Univ. Mich. Pap. Paleontol., 15: 1—140.

Gingerich, P.D. and Schoeninger, M.J., 1976. The fossil record and primate phylogeny. J. Human Evol., in press.

Gingerich, P.D. and Simons, E.L., 1977. Systematic revision of the Early Eocene Adapidae (Mammalia, Primates) in North America. Contrib. Univ. Mich. Mus. Paleontol., 24, in press.

Granger, W. and Gregory, W.K., 1936. Further notes on the gigantic extinct rhinoceros, Baluchitherium, from the Oligocene of Mongolia. Bull. Am. Mus. Nat. Hist., 72: 1—73.

Hopson, J.A., 1966. The origin of the mammalian middle ear. Am. Zool., 6: 437—450.

Hopson, J.A., 1967. Comments on the competitive inferiority of the multituberculates. Syst. Zool., 16: 352—355.

Hürzeler, J., 1962. Kann die biologische Evolution, wie sie sich in der Vergangenheit abgespielt hat, exakt erfasst werden? Stud. Ber. Katholischen Akad. Bayern, 16: 15—36.

Jepsen, G.L., 1966. Early Eocene bat from Wyoming. Science, 154: 1333—1339.

Jerison, H.J., 1973. Evolution of the Brain and Intelligence. Academic Press, New York, 482 pp.

Jolly, C.J., 1970a. The seed-eaters: a new model of hominid differentiation based on a baboon analogy. Man, 5: 5—26.

Jolly, C.J., 1970b. Hadropithecus: a leumuroid small-object feeder. Man, 5: 619—626.

Kallen, F.C. and Gans, C., 1972. Mastication in the little brown bat. J. Morphol., 136: 385—420.

Kermack, K.A., 1967. The interrelations of early mammals. Zool. J. Linn. Soc., Lond., 47: 241—249.

Kortlandt, A., 1974. New perspectives on ape and human evolution. Curr. Anthropol., 15: 427—448.

Kurtén, B., 1959. On the longevity of mammalian species in the Tertiary. Soc. Sci. Fenn. Comment. Biol., 21: 4, 1—14.

Kurtén, B., 1960. Chronology and faunal evolution of the earlier European glaciations. Soc. Sci. Fenn. Comment. Biol., 21: 5, 1—62.

Kurtén, B., 1963. Return of a lost structure in the evolution of the felid dentition. Soc. Sci. Fenn. Comment. Biol., 26: 4, 1—12.

Kurtén, B., 1965. Evolution in geological time. In: J.A. Moore (Editor), Ideas in Modern Biology. Natural History Press, Garden City, N.Y., pp. 329—354.

Leakey, R.E.F., 1976. Hominids in Africa. Am. Sci., 64: 174—178.

Lillegraven, J.A., 1972. Ordinal and familial diversity of Cenozoic mammals. Taxon, 21: 261—274.

Lovejoy, C.O., Heiple, K.G. and Burstein, A.H., 1973. The gait of *Australopithecus*. Am. J. Phys. Anthropol., 38: 757—780.

Martin, P.S., 1967. Prehistoric overkill. In: P.S. Martin and H.E. Wright (Editors), Pleistocene Extinctions: the Search for a Cause. Yale University Press, New Haven, Conn., pp. 75—120.

Martin, P.S. and Wright, H.E. (Editors), 1967. Pleistocene Extinctions: the Search for a Cause. Yale University Press, New Haven, Conn., 453 pp.

McHenry, H.M., 1975. Fossils and the mosaic nature of human evolution. Science, 190: 425—431.

McKenna, M.C., 1975. Fossil mammals and early Eocene North Atlantic land continuity. Ann. Mo. Bot. Garden, 62: 335—353.

Miller, G.J., 1969. A new hypothesis to explain the method of food ingestion used by *Smilodon californicus* Bovard. Tebiwa, 12: 9—19.

Nelson, R.S. and Semken, H.A., 1970. Paleoecological and stratigraphic significance of the muskrat in Pleistocene deposits. Bull. Geol. Soc. Am., 81: 3733—3738.

Oxnard, C.E., 1975. The place of the australopithecines in human evolution: grounds for doubt? Nature, 258: 389—395.

Parrington, F.R., 1971. On the upper Triassic mammals. Philos. Trans. R. Soc., Lond., 261 (B): 231—272.

Pilbeam, D.R. and Zwell, M., 1972. The single species hypothesis, sexual dimorphism, and variability in early hominids. Yearb. Phys. Anthropol., 16: 69—79.

Radinsky, L., 1971. An example of parallelism in carnivore brain evolution. Evolution, 25: 518—522.

Raup, D.M., 1975. Taxonomic survivorship curves and Van Valen's law. Paleobiology, 1: 82—96.

Riggs, E.S., 1934. A new marsupial saber-tooth from the Pliocene of Argentina and its relationships to other South American predacious marsupials. Trans. Am. Philos. Soc., 24: 1—32.

Romer, A.S., 1966. Vertebrate Paleontology. University of Chicago Press, Chicago, Ill., 468 pp.

Rose, K.D., 1975. The Carpolestidae, early Tertiary primates from North America. Bull. Mus. Comp. Zool., 147: 1—74.

Schaeffer, B., 1948. The origin of a mammalian ordinal character. Evolution, 2: 164—175.

Simpson, G.G., 1933. The "plagiaulacoid" type of mammalian dentition, a study of convergence. J. Mammal., 14: 97—107.

Simpson, G.G., 1943. Criteria for genera, species, and subspecies in zoology and paleozoology. Ann. N.Y. Acad. Sci., 44: 145—178.

Simpson, G.G., 1953. The Major Features of Evolution, Simon and Schuster, New York, N.Y., 434 pp.

Sloan, R.E., 1969. Cretaceous and Paleocene terrestrial communities of western North America. Proc. North Am. Paleontol. Conv., Part E: 427—453.

Stanley, S.M., 1973. An explanation for Cope's rule. Evolution, 27: 1—26.

Stanley, S.M., 1975. A theory of evolution above the species level. Proc. Natl. Acad. Sci., U.S.A., 72: 646—650.

Stanley, S.M., 1976. Stability of species in geologic time. Science, 192: 267—269.

Stehlin, H.G., 1909. Remarques sur les faunules de mammifères des couches éocènes et oligocènes du Bassin de Paris. Bull. Soc. Géol. Fr., Sér., 4, 9: 488—520.

Thaler, L., 1973. Nanisme et gigantisme insulaires. Recherche, 4: 741—750.

Trevisan, L., 1949. Lineamenti dell'evoluzione del ceppo di elefanti eurasiatici nel Quaternario. Ric. Sci. (Suppl.), 19: 105—111.

Van Valen, L., 1969. Evolution of communities and late Pleistocene extinctions. Proc. North Am. Paleontol. Conv., Part E: 469—485.

Van Valen, L., 1973. A new evolutionary law. Evol. Theory, 1: 1—30.

Van Valen, L., 1975. Group selection, sex, and fossils. Evolution, 29: 87—94.

Van Valen, L. and Sloan, R.E., 1966. The extinction of the multituberculates. Syst. Zool., 15: 261—278.

Vianey-Liaud, M., 1972. L'évolution du genre *Theridomys* a l'Oligocène moyen. Bull. Mus. Nat. Hist. Nat., Sci. Terre, 18: 295—372.

Webb, S.D., 1969. Extinction—origination equilibria in late Cenozoic land mammals of North America. Evolution, 23: 688—702.

Wolfe, J.A. and Hopkins, D.M., 1967. Climatic changes recorded by Tertiary land floras in northwestern North America. Symp. Pac. Sci. Congr., 25: 67—76.

Wood, H.E., 1954. Patterns of evolution. Trans. N.Y. Acad. Sci., 16: 324—336.

CHAPTER 16

PATTERNS OF EVOLUTION IN EARLY ANGIOSPERMS

JAMES A. DOYLE

Introduction

Although micro-evolutionary studies of Recent plants are numerous and have made important contributions to modern evolutionary theory (cf. Stebbins, 1950, 1974; Grant, 1963), studies of major patterns of evolution based on the fossil record of plants have been relatively few and have had little general impact compared to those based on fossil animals. In further contrast to the situation in palaeozoology, where studies on fossil mammals have played a central role in evolutionary thinking, the fossil record of the angiosperms, the most recent, successful, and intensively studied group of living plants, is even less well understood from an evolutionary point of view than is the record of more ancient, less familiar groups. Thus palaeobotany has provided critical evidence on the course and tempo of the initial Silurian—Devonian radiation of land plants (cf. Chaloner, 1967, 1970; Banks, 1968, 1970) and the origin and relationships of the first Permo-Carboniferous conifers (Florin, 1951), but the problem of the origin and evolution of the angiosperms seemed even to quite recent palaeobotanical investigators as intractable as when Darwin labelled it "an abominable mystery" nearly 100 years ago. Since the first unquestionable angiosperm remains (mostly leaves) from the mid-Cretaceous appeared identifiable with diverse and advanced modern taxa and showed no obvious links to earlier groups, suggesting a long period of prior diversification in some region unrepresented in the fossil record (such as the tropical uplands: cf. Axelrod, 1952, 1960, 1970; Takhtajan, 1969), most botanists have considered the fossil record useless as a source of evidence on early angiosperm evolution. They have therefore either attempted to reconstruct angiosperm phylogeny entirely on the basis of comparative morphological studies of extant forms, or else despaired of ever establishing the course of angiosperm evolution (e.g., Davis and Heywood, 1963).

This situation has changed significantly in the past two decades, thanks in large part to the growth of palynology — the study of spores and pollen grains, or more specifically their acid-resistant walls, or exines. Because of the extreme chemical inertness and transportability of exines, the palynological record is generally more representative of the regional flora, including elements removed from the actual site of deposition, than is the megafossil record, and it is of corresponding value in both stratigraphic and evolutionary studies (cf. Kuyl et al., 1955; Muller, 1959, 1970; Doyle, 1976). Hence the consistent failure of palynologists to find distinctively angiospermous pollen types in Triassic, Jurassic, and earliest Cretaceous strata and the generalized morphology of the first mid-Cretaceous angiosperm pollen grains were quickly recognized as evidence against the concept of extensive pre-Cretaceous diversification of the angiosperms and in favour of the alternative concept of a Cretaceous radiation (Scott et al., 1960; Hughes, 1961; Pierce, 1961; Pacltová, 1961; Brenner, 1963; Kemp, 1968; etc.). This interpretation has subsequently been strengthened by more detailed morphological and stratigraphic investigations of mid-Cretaceous pollen floras, which reveal a pattern of appearance and diversification of suc-cessively more differentiated angiosperm pollen types, as would be expected in a genuine evolutionary radiation (Doyle, 1969, 1973; Muller, 1970; Pacltová, 1971; Dettmann, 1973; Jarzen and Norris, 1975; Laing, 1975, 1976; Wolfe et al., 1975; and below). These results, plus the success of similar studies in eluci-dating the evolutionary significance of the Silurian—Devonian record of early land plants (Chaloner, 1967, 1970; Banks, 1968), have stimulated a critical re-evaluation of the early angiosperm leaf record using a comparable stratig-raphic—morphological approach, independent of supposed affinities with modern taxa (Pacltová, 1961; Wolfe, 1972a; Doyle and Hickey, 1972, 1976; Wolfe et al., 1975). When viewed in this way, the leaf record too reveals a diversification pattern consistent with a Cretaceous radiation, rather than the previously inferred sudden appearance of modern forms (cf. below). Since these results have clear implications on directions of evolution in early angio-sperm pollen and leaf characters, and hence on the relative advancement and relationships of extant high-rank angiosperm taxa, they represent the first posi-tive indication that the fossil record may have something to say about angio-sperm phylogeny after all (cf. Doyle, 1969, 1973; Muller, 1970; Wolfe et al., 1975; Hughes, 1976; Doyle and Hickey, 1976).

The main purpose of the present paper is to summarize mid-Cretaceous palaeontological evidence on the timing and pattern of early angiosperm evolu-tion, stressing general questions relating to the origin and evolution of major groups. Phylogenetic and systematic implications of more specifically botanical interest have been discussed sufficiently elsewhere (Doyle, 1969, 1973; Muller, 1970; Wolfe et al., 1975; Doyle and Hickey, 1976) and will be mentioned only when relevant to more general problems. Much of this analysis will be fairly qualitative: because of the highly questionable, non-evolutionary assumptions of most early angiosperm palaebotany (cf. Wolfe, 1973; Dilcher, 1974; Wolfe et al., 1975; and below), compounded by the overwhelming diversity of angio-sperms from the Late Cretaceous onward, there are simply not yet enough crit-ically and uniformly evaluated basic taxonomic and stratigraphic data for the sort of across-the-board quantitative analysis of evolutionary rates that is possi-ble in many animal groups. Any attempt to use such frankly uncritical com-pendia as the list by Chesters et al. (1967) of first appearances of Recent fami-lies would involve so many demonstrably false identifications (e.g., modern genera used as form genera), especially in the crucial Cretaceous phases of the record, that it would almost surely do more harm than good. Likewise, the fact that the fossil record of angiosperms consists largely of dispersed pollen grains and leaves makes it more difficult to draw meaningful conclusions on the ecol-ogy and adaptations of whole organisms from rates of morphological evolution (e.g., size trends) than with invertebrate shells or even mammalian teeth. However, for the early phases of the angiosperm record I will present a semiquantitative analysis of taxonomic frequency changes and morphological evolution, using a composite advancement index incorporating largely qualitative characters, based on my own studies of angiosperm pollen from the Potomac Group of the eastern U.S.A. (Doyle, 1969, 1973; Doyle and Hickey, 1976; Doyle and Robbins, 1977). Although some of the pollen types involved may be relatable to high-rank Recent taxa, it should be noted that this analysis itself is independent of any such relationships. Finally, I will attempt to draw inferences on the adaptive bases of evolutionary phenomena seen in the mid-Cretaceous record from a consideration of the geographic and facies distribution and functional morphology of early angiosperm pollen and leaves. This discussion, largely the result of joint work by L.J. Hickey and myself (Doyle and Hickey, 1976), may serve as a basis for some speculations on the possible role of biological innovations and external environmental change (climatic, tectonic, biotic) in the origin, diversification, and phenomenal success of the angiosperms, as well as possible effects of the rise of angiosperms on the evolution of other members of the terrestrial biota.

Although I will consider only the Early Cretaceous and Cenomanian record in detail, it is important to realize that angiosperm diversification continued unabated throughout the later Cretaceous and Tertiary. Some of the more important evolutionary phenomena and problems seen in the post-Cenomanian angiosperm record and recent conceptual and methodological advances in dealing with them are discussed by Góczán et al. (1967), Germeraad et al. (1968), Muller (1970), Tschudy (1970), Wolfe (1971, 1972b, 1973), Leopold and MacGinitie (1972), Dilcher (1974), Hickey and Wolfe (1975), and Wolfe et al. (1975).

Systematic Background

Before discussing patterns of evolution seen in the fossil record of angiosperms, it is useful to provide some general background on the place of angiosperms in the evolutionary history of land plants and on major subgroups of angiosperms and their presumed interrelationships (Fig. 1).

The evolution of vascular land plants (i.e., excluding the bryophytes, or mosses and liverworts) can be conveniently treated in terms of three main grades of evolution — pteridophytes, gymnosperms, and angiosperms — corresponding to progressively more advanced versions of the same basic life cycle: an alternation of the familiar diploid sporophyte, which produces haploid spores by meiosis, and the less conspicuous haploid gametophyte, which produces gametes and hence the next sporophyte generation. In the pteridophytes, which include the first land plants of the Late Silurian and Devonian (Psilophytopsida), club mosses (Lycopsida), horsetails (Sphenopsida), and ferns (Pteropsida or Filicopsida), the spores are shed freely from the sporangia, and at least a thin film of water is required for fertilization. The Devonian record shows good evidence of the trend from homospory, where spores are of one

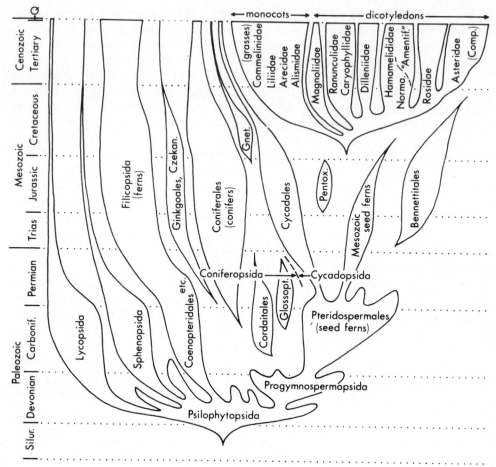

Fig. 1. Stratigraphic distribution, changes in abundance (generalized), and presumed phylogenetic relationships of major groups of vascular land plants. Widths of bars not to scale in different groups. Names ending in -opsida are classes; -idae, subclasses; -ales, orders (Czekan. = Czekanowskiales, Glossopt. = Glossopteridales, Gnet. = Gnetales, Pentox. = Pentoxylales, Norma. = Normapolles pollen group, "Amentif." = porate "Amentiferae", Comp. = family Compositae).

size and produce free-living, bisexual gametophytes, to heterospory, where spores are differentiated into microspores and megaspores, producing male and female gametophytes respectively, which undergo their development within the spore wall (cf. Chaloner, 1967, 1970). The lycopsids, sphenopsids, and ferns differentiated from the basic psilophyte complex during the Devonian radiation of land plants and reached their zenith in the Late Carboniferous, when they included dominant tree forms and other bizarre extinct types with no close analogues in the modern flora.

The gymnosperms, often divided into the two classes Cycadopsida and Coniferopsida, have a heterosporous life cycle, but with special advances as important in liberating land plant life from its aquatic heritage as evolution of the amniote egg was for the vertebrates. The megaspore and enclosed female gametophyte are not shed, but retained inside the megasporangium, which together with a protective integument of sporophytic tissue constitutes the ovule (at maturity, the seed). Fertilization occurs inside the ovule, following wind transfer of the microspores and enclosed male gametophytes, now referred to as pollen grains, to the opening or micropyle of the ovule. Seeds are first known from the latest Devonian (Pettitt and Beck, 1968), but many of the anatomical specializations of modern gymnosperms, particularly the capacity for secondary increase in diameter of the stem by production of secondary xylem (wood), phloem, and periderm, had already originated in their presumed Middle and Late Devonian ancestors, the progymnosperms — a good example of mosaic evolution (Beck, 1960, 1970; Scheckler and Banks, 1971). Curiously, although primitive representatives of both cycadopsid and coniferopsid gymnosperms are well represented in the Late Carboniferous (seed ferns and Cordaitales), the definitive rise to dominance of gymnosperms did not occur until the Permian, the so-called Palaeophytic—Mesophytic transition. In the most familiar, then-equatorial Euramerican province, this transition corresponds to the extinction of the tree lycopsids and sphenopsids and primitive gymnosperms and the dramatic rise of conifers, which are known from the Late Carboniferous but rarely occurred in the coal basins themselves. Considering both geological evidence and the reduced leaves, thick cuticles, and other xeromorphic (drought-adapted) features of the conifers and their associates, these changes seem clearly related to the disappearance of coal swamp conditions and increasing aridity (cf. Frederiksen, 1972). Whether or not the Palaeophytic—Mesophytic transition is a valid concept on a world scale (it is less marked or characterized by different changes in provinces then to the north and south: cf. Chaloner and Lacey, 1973), conifers, a variety of other "higher" gymnosperms (Ginkgoales, Czekanowskiales, Cycadales, Bennettitales, and Mesozoic seed ferns), and ferns remained dominant until the rise of angiosperms in the mid-Cretaceous, the so-called Mesophytic—Cenophytic transition.

Anatomically and in their basic life cycle, the angiosperms are clearly seed plants, but they differ from gymnosperms in having several important advances of their own. As the term "flowering plant" implies, one of these advances is the flower, which in its most complete form consists of protective and/or showy perianth parts (sepals and petals) and pollen- and ovule-bearing structures (stamens and carpels) grouped in that order on a shortened axis. Flowers in this loose, functional sense are not unique to angiosperms, but are also conspicuously present in the apparently unrelated Mesozoic gymnosperm order Bennettitales, where they presumably played the same role in ensuring cross-pollination by insects or other animals rather than the wind (cf. Grant, 1950; Takhtajan, 1969). The more definitively angiospermous features are in the detailed morphology of the fertile appendages: with rare and apparently insignificant exceptions, the stamen is a simple structure with four pollen sacs, and the carpel is a closed structure, most diagrammatically illustrated by a pea pod, bearing ovules on its inner surface (hence the term angiosperm, = vessel seed, vs. gymnosperm, = naked seed). Functionally, the closed character of the carpel means that the pollen grains germinate not in the micropyle or pollen chamber of the ovule, but rather on a sticky surface (stigma) of the carpel, and that pollination and fertilization can occur when the ovules are still at a very early stage of their development and can be discarded with a minimum of waste if fertilization does not occur (cf. Takhtajan, 1969; Stebbins, 1974; and below). Correlated with these features are embryological specializations: drastic reduction of the haploid generation, beyond the point seen in any gymnosperm (three nuclei in the male gametophyte, usually eight nuclei in the female gametophyte); and the unique process of double fertilization, where one of the two sperm nuclei produced by the pollen grain fuses with two (rarely one, three, or more) other nuclei of the female gametophyte, triggering development of the endosperm tissue which serves to nourish the developing embryo.

The angiosperms have long been divided into two classes (or subclasses), the monocotyledons and dicotyledons, distinguished on the number of embryonic leaves or cotyledons in the seed, and correlated characters such as "parallel" vs. "reticulate" leaf venation (cf. below). With the usual allowances for lumping and splitting, grouping of the quarter of a million species of Recent angiosperms into genera and some 300—500 families is also fairly uniform, but arrangements between the family and class level are more fluid and controversial, reflecting confusing, often repetitive patterns of variation and uncertainties on criteria for phylogenetic relationships. In the present paper I will follow the system of Takhtajan (1969), which in most cases agrees with that of Cronquist (1968). Takhtajan divides the monocots (Liliatae) into four subclasses: Alismidae (largely aquatic herbs, with many presumed primitive floral characters), Liliidae (lilies, orchids, etc.), Commelinidae (including the wind-pollinated and unquestionably specialized grasses and sedges), and Arecidae (palms, aroids, etc.). The dicots (Magnoliatae) are divided into seven subclasses: Magnoliidae (*Magnolia*, waterlilies, etc.), Ranunculidae (buttercups, etc., retained in Magnoliidae by Cronquist, 1968), Hamamelididae (including most of the wind-pollinated, catkin-bearing "Amentiferae" of the north temperate zone), Caryophyllidae (Centrospermae of many systems), Dilleniidae, Rosidae, and Asteridae (including the highly successful family Compositae).

The Takhtajan (1969) system is based on the "classical" or "ranalian" theory of angiosperm evolution: namely, that the angiosperms are a monophyletic group whose earliest members were woody and had bisexual flowers, free carpels, monosulcate pollen (cf. below), and other features seen in least modified form in extant Magnoliidae, and that groups with no secondary growth (e.g., monocots and many herbaceous dicots), unisexual flowers (e.g., Amentiferae), and fused floral parts (e.g., Asteridae) are derived. This theory is far from universally accepted; for instance, the view that the Amentiferae, with simple, unisexual flowers grouped into superficially gymnosperm-like catkins, are primitive was widely accepted earlier in the century and is still incorporated into many polyphyletic theories (e.g., Meeuse, 1965). Meeuse in fact postulates that the "carpels" of some angiosperm groups are homologous with the "ovules" of others, and denies that the flower is a valid morphological concept.

"Ranalian" phylogenists cite the universal presence of double fertilization and other characters as evidence that the angiosperms are monophyletic, and use correlations with presumed advancement trends in other characters (e.g., wood anatomy: Bailey, 1944) and the existence of morphologically intermediate groups in the modern flora as evidence that marked deviations from the magnoliid prototype (e.g., Amentiferae) represent specializations which occurred within the angiosperms (cf. Takhtajan, 1969). However, it is important to realize that until recently there was practically no direct fossil evidence in favour of either interpretation (cf. Axelrod, 1952, p. 29). Even today, most of the floral and other reproductive characters which have played such an important role in defining major groups and trends within the angiosperms and in separating them from gymnosperms are rarely if ever visible in the fossil record, which consists largely of pollen, leaves, wood, and some mature reproductive structures (fruits and seeds), rather than flowers. This has contributed to the widespread belief that the fossil record is "wholly inadequate" (Walker, 1976) as a source of evidence on angiosperm phylogeny. However, the situation is not intrinsically different from that in the vertebrates, where many of the features used to define classes and orders are reproductive, physiological, and soft-anatomical. Just as progress in unraveling vertebrate phylogeny has resulted from coordinated comparative, functional—morphological, and stratigraphic studies of bones and teeth of living and fossil vertebrates, similar studies of Recent and Cretaceous pollen and leaves are beginning to make evolutionary sense of the early angiosperm record.

Evolutionary Patterns in the Mid-Cretaceous Angiosperm Record

Previous interpretations

Although recognition of Early Cretaceous floras with low proportions of angiosperms has led most recent authors to reject the concept that angiosperms rose to dominance with a suddenness unparallelled in other groups (Darwin's "abominable mystery"), controversy continues on the correct evolutionary interpretation of the early angiosperm record. As mentioned earlier, some have attempted to reconcile the apparent systematic diversity of mid-Cretaceous angiosperm leaves with "normal" rates of evolution by postulating that the angiosperms had been diversifying in upland areas for two or more geological periods before the Cretaceous (e.g., Axelrod, 1952, 1960, 1970; Takhtajan, 1969). Others, impressed by the low diversity of mid-Cretaceous angiosperm pollen and skeptical of the degree of modernity inferred from leaf identifications, have proposed that the angiosperms were diversifying at rapid but not abnormal rates during their rise to dominance and may have originated not long before (e.g., Scott et al., 1960; Hughes, 1961, 1976; Pierce, 1961; Pacltová, 1961; Brenner, 1963; Kemp, 1968; Doyle, 1969, 1973; Muller, 1970; Wolfe et al., 1975; Doyle and Hickey, 1976). Of the latter, some have argued that the pollen diversification pattern is consistent with radiation of a monophyletic group (Doyle, 1969; Muller, 1970), but others have adopted or inclined toward the view that the angiosperms originated polyphyletically from distantly related lines of Late Jurassic—Early Cretaceous gymnosperms (Meeuse, 1965; Krassilov, 1973, 1975; Hughes, 1976; Krassilov combines this with the unorthodox concept of transfer of genes for key angiosperm features from one line to another by viral transduction). Many of these concepts and arguments are of course familiar from palaeontological discussions of the origins of other major groups (cf. Simpson, 1953, chapter 11).

Of previous interpretations of early angiosperm history, Axelrod's theory (1952, 1959, 1960, 1970) of a Permo-Triassic origin of the angiosperms in the tropical uplands deserves special attention, both as the most coherent exposition of the prior diversification concept, and as one of the few attempts to interpret the early fossil record of angiosperms in terms of modern ideas on the origin and early evolution of higher taxa (e.g., Simpson, 1953). Axelrod's principal reasons for postulating a long pre-Cretaceous history are: (1) the apparent high level of advancement and systematic differentiation of mid-Cretaceous megafossil remains, particularly leaf identifications implying that "modern genera had become established within both primitive and derived groups" (Axelrod, 1952, p. 31); (2) rare finds of supposed angiosperm fossils from Upper Triassic and Jurassic rocks; and (3) pantropical and African—South American distributions of Recent angiosperm families and genera, which he has recently argued (Axelrod, 1970) must date at least from the Albian—Cenomanian separation of Africa and South America (see, however, Raven and Axelrod, 1974; and below). His reasons for choosing the tropical uplands as the site of pre-Cretaceous angiosperm evolution include: (1) the fact that the fossil record is strongly biased toward lowland basins of deposition, so that upland forms may be totally unrepresented; (2) analogies with other cases (cf. below) where advanced plant groups seem to have existed in upland areas before migrating into the lowlands in response to climatic change; (3) the concept that the diverse and fragmented environments of the uplands would produce population structures and selective factors ideal for speciation, rapid evolution, and origin of new adaptive types (cf. Simpson, 1953); (4) Recent distribution patterns with greatest diversity and primitive elements centred in the tropics (cf. Bews, 1927; and below), especially the tropical—subtropical wet upland distribution of the putatively primitive order Magnoliales (cf. Takhtajan, 1969; Axelrod, 1970); and (5) the progressively younger first stratigraphic appearance of Cretaceous angiosperms at higher latitudes (Axelrod, 1959). Most recently, Axelrod (1970) suggests that the factor which caused the angiosperms to invade the lowlands in the Cretaceous rather than long before (previously an unsatisfactory aspect of his theory) was a world-wide increase in climatic equability (i.e., decrease in seasonality) associated with the mid-Cretaceous breakup of Gondwana.

Although I believe that the most cogent evidence concerning the merits of Axelrod's theory comes from a consideration of patterns in the Cretaceous record, it should be noted that many of the assumptions behind his arguments have been questioned at one time or another on other grounds. For instance,

Stebbins (1965, 1974) has argued that wet tropical environments would not favour origin of the sort of reproductive specializations which unify the angiosperms, and that living "primitive" angiosperms of the wet tropical uplands are hence ecologically specialized (cf. below). In the latest formulation of his theory, Axelrod (1970) himself tends toward Stebbins' view that the angiosperms originated in seasonally dry rather than wet tropical upland environments (cf. also Raven and Axelrod, 1974). Similarly, Raven and Axelrod (1974) argue that significant trans-Atlantic floristic interchange might still have been possible up to the beginning of the Tertiary, hence blurring Axelrod's (1970) earlier argument based on African—South American distributions.

Recent investigations of other major events in the fossil record of plants have also weakened Axelrod's analogical argument that new groups tend to originate in the uplands and later migrate into the lowlands. Here the fact that upland groups are represented in the Recent lowland and near-shore marine pollen record (Muller, 1959) is especially significant (cf. Scott et al., 1960). Palynological and improved stratigraphic data have supported the concept that the Late Silurian—Devonian rise of early land plants reflects their initial adaptive radiation (Chaloner, 1967, 1970; Banks, 1968, 1970), and the Middle and Late Devonian trunks taken by Axelrod (1952) as evidence of more advanced gymnosperms in the uplands have turned out to be remains of the dominant lowland group, the progymnosperms (Beck, 1960, 1970). Stratigraphic and morphological analyses of Tertiary leaf floras have also contradicted the hypothesis that temperate deciduous forests existed in the Arctic and upland regions in the Early Tertiary and simply migrated into the middle-latitude lowlands in response to Late Tertiary climatic deterioration (Wolfe, 1971, 1972b; Leopold and MacGinitie, 1972). Even the best example of an invasion of the lowlands by upland plants, the Permian replacement of the Palaeophytic coal swamp vegetation by conifers, taeniopterids, and other drought-adapted Mesophytic elements, is not strictly analogous to the process postulated by Axelrod. Conifers, taeniopterids, etc., are occasionally seen in the Late Carboniferous megafossil record and are well represented in the pollen record (cf. Frederiksen, 1972); and their first representatives are not nearly as diversified or as removed systematically from contemporaneous lowland groups (Cordaitales, neuropterids) as Early Cretaceous angiosperms are supposed to have been.

Critical examination of supposed pre-Cretaceous (and even earliest Cretaceous) angiosperm remains has removed the most tangible evidence for the prior diversification theory by showing that most if not all of the fossils in question are either: (1) stratigraphically misplaced (e.g., the "Jurassic" palm stems and roots of Tidwell et al., 1970a, b, shown by Scott et al., 1972, to be of Tertiary age); (2) members of extinct gymnosperm groups poorly known at the time (e.g., *Eucommiidites*, *Classopollis*, *Sahnioxylon*, and supposed nymphaeaceous pollen from the Jurassic of Scotland: Scott et al., 1960; Hughes, 1961, 1976; Doyle et al., 1975); or (3) not definitely assignable to any known gymnosperm group, but lacking sufficiently diagnostic characters to be assignable to the angiosperms (e.g., *Sanmiguelia* and *Furcula* from the Upper Triassic, *Propalmophyllum* from the Jurassic, and *Onoana* from the Lower Cretaceous:

Scott et al., 1960; Hughes, 1961, 1976; Read and Hickey, 1972; Doyle, 1973; Wolfe et al., 1975). Clearly, some members of the last group may deserve additional study as potential angiosperm relatives, but they cannot be considered evidence of full-fledged angiosperms in pre-Cretaceous times. Finally, many authors have pointed out that the early leaf identification methods which form the basis for the concept of a modernized mid-Cretaceous angiosperm flora — essentially searching for the closest Recent "match" for each fossil leaf type — tend because of their non-evolutionary assumptions to obscure evidence for evolution even when it is found (cf. Stebbins, 1950; Pacltová, 1961; Wolfe, 1973; Dilcher, 1974; Wolfe et al., 1975; Hughes, 1976).

In the following sections, I will argue that recent studies of the Cretaceous angiosperm record provide the first strong positive evidence for the concept of adaptive radiation of the angiosperms within the Cretaceous, and place severe limits on the amount of pre-Cretaceous diversification or polyphylesis which can be postulated without excessive multiplication of *ad hoc* hypotheses. Although these considerations greatly alter Axelrod's time table for early angiosperm evolution, it is interesting to note that the pollen and leaf records do support certain aspects of his theory which can be divorced from the assumption of a pre-Cretaceous upland origin. Such aspects include his conclusion (Axelrod, 1959) that angiosperms appeared later at higher latitudes (cf. Brenner, 1976), and his (Axelrod, 1970) and Stebbins' (1965, 1974) largely theoretical arguments for the role of seasonal aridity and unstable, heterogeneous environments in the origin and early adaptive radiation of angiosperms (cf. Raven and Axelrod, 1974; Brenner, 1976; Doyle and Hickey, 1976).

The Potomac—Raritan angiosperm pollen sequence

As my main source of detailed data on patterns of evolution in early angiosperms, I will concentrate on fossil pollen from the Potomac Group and Raritan Formation, the basal fluvial-deltaic sedimentary units of the Atlantic Coastal Plain of Virginia, Maryland, Delaware, and New Jersey, U.S.A. (Figs. 2, 3). The Potomac—Raritan sequence is a suitable choice for such purposes, since it is relatively continuous, rich in fossil pollen and leaves, and unusually intensively studied from both palynological and macropalaeobotanical points of view, besides being the section with which I am personally most familiar. The greatest disadvantage of this sequence is that it is not independently dated by marine animal fossils (except for the Woodbridge Clay Member of the Raritan Formation: cf. Wolfe and Pakiser, 1971). However, thanks to the availability of closely sampled well sections and abundant outcrop material, vertical changes in the pollen and spore flora are firmly established, leading to a biostratigraphic zonation of considerable practical value in regional stratigraphy (Fig. 2, middle; cf. Brenner, 1963, 1967; Doyle, 1969, 1973, 1976; Wolfe and Pakiser, 1971; Doyle and Hickey, 1976; Doyle and Robbins, 1977). Recognition of closely comparable pollen and spore successions in other areas where there is better independent dating (e.g., England and France: Kemp, 1968, 1970; Laing, 1975, 1976; Czechoslovakia: Pacltová, 1971; U.S. Western Interior and Gulf Coastal

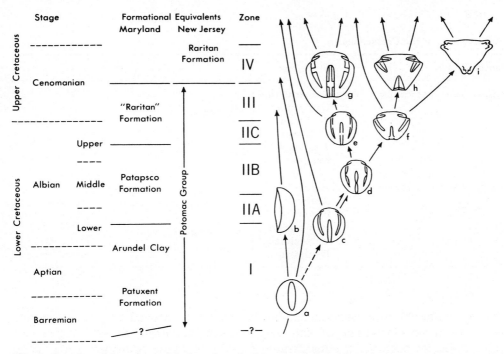

Fig. 2. Stratigraphic units and inferred evolutionary relationships of major angiosperm pollen types in the Potomac Group and lower Raritan Formation of the Atlantic Coastal Plain of the U.S.A. (modified from Doyle, 1973). Left: proposed correlations with the standard European stage sequence (cf. Table I); middle: informal palynostratigraphic zonation of Brenner (1963), Doyle (1973), Doyle and Hickey (1976), and Doyle and Robbins (1977). Pollen types indicated: *a*, generalized tectate-columellar monosulcates (*Clavatipollenites, Retimonocolpites, Stellatopollis*); *b*, monocotyledonoid reticulate monosulcates, with finer sculpture at the ends of the grain and/or sulcus margins (*Liliacidites*); *c*, reticulate to tectate tricolpates (*Tricolpites*); *d*, reticulate to tectate tricolporoidates (*Tricolpites, Tricolporoidites*); *e*, small, usually smooth, prolate tricolporoidates (*Tricolporoidites*); *f*, small, usually smooth, oblate-triangular tricolporoidates (*Tricolporoidites*); *g*, larger, smooth to reticulate, prolate tricolpor(oid)ates (*Tricolporopollenites*); *h*, larger, usually smooth, oblate-triangular tricolpor(oid)ates (*Tricolporopollenites*); *i*, early members of the triangular triporate Normapolles group (*Complexiopollis, Atlantopollis*). Arrows indicate modes of origin, not necessarily single evolutionary lines. The dashed arrow indicates that the inferred transition from monosulcate to tricolpate is not directly documented in the Potomac sequence; the double arrow emphasizes the multiple origin of tricolporoidates.

Plain: Pierce, 1961; Hedlund and Norris, 1968; western Canada: Norris, 1967; Singh, 1971; Jarzen and Norris, 1975) has in turn permitted rather precise palynological correlations with the standard stage sequence (Table I; Fig. 2, left; cf. references listed above). Some such comparisons will be discussed below in connection with geographic patterns and the role of migration vs. evolution in the early angiosperm record.

The angiosperm pollen flora shows a general stratigraphic increase in abundance, number of species, and total range of morphological types throughout

TABLE I

Subdivisions of the Cretaceous

Series/Epoch	Stage/Age	
Upper/Late Cretaceous	Maestrichtian Campanian Santonian Coniacan Turonian Cenomanian	"Senonian"
Lower/Early Cretaceous	Albian Aptian Barremian Hauterivian Valanginian Berriasian	"Neocomian"

the Potomac—Raritan sequence, and it has hence proved the most valuable element in establishment of the palynological zonation. Several angiosperm pollen species showing a considerable variety of exine sculpture patterns are already present at the base of the Potomac Group (lower Zone I of Brenner, 1963: Barremian—Aptian?), but they are markedly subordinate to the diverse fern spores and gymnosperm (especially conifer) pollen grains which dominate the flora, and they are exclusively monosulcate, with a single distal furrow (sulcus) for pollen tube germination (Plate I, 1—6). In the Recent flora, monosulcate pollen is produced only by gymnosperms (Cycadales and Ginkgoales, plus Bennettitales and other extinct groups), many monocots, and certain members of the dicot subclass Magnoliidae (*sensu* Takhtajan, 1969), considered primitive under "ranalian" theories of angiosperm evolution. The principal criteria for distinguishing these grains from the more abundant monosulcate pollen of gymnosperms are technical features of their exine structure (cf. Van Campo, 1971; Doyle et al., 1975): they are columellar, with a perforated to coarsely reticulate outer exine layer or **tectum** connected to the inner exine layer or **nexine** by radial rods or **columellae**, in contrast to the honeycomb-like or alveolar structure of the superficially similar pollen of cycads, or they belong to an apparently related type with a coarse tectal reticulum but no columellae (e.g., *Retimonocolpites peroreticulatus*: Plate I, 3, 4). In addition, the two species investigated with transmission electron microscopy (TEM) (*Clavatipollenites* cf. *hughesii, R. peroreticulatus*), plus a Zone II monosulcate species (*Stellatopollis barghoornii*), lack the laminated endexine (chemically differentiated inner layer of the nexine) characteristic of all Recent and fossil gymnosperms so far investigated (Doyle et al., 1975). Interestingly, a few Zone I monosulcates (*Liliacidites* sp. A: Plate I, 6) show exine sculpture patterns (coarsely reticulate on most of the grain surface, but fine at the ends of the grain and sulcus margins) common today among monocots but unknown in magnoliid dicots, suggesting that the basic split between monocots and dicots had already occurred (Doyle, 1973). Previously, the fact that only the less extreme of these types had been reported from dated Barremian rocks (*C. hughesii* from the upper Wealden of England: Couper, 1958; Kemp, 1968) seemed to suggest that Zone I might be Aptian or younger (cf. Wolfe and Pakiser, 1971; Doyle, 1973; Doyle and Hickey, 1976), but recently a suite of very similar monosulcate types has been reported from the Barremian of England (Hughes and Laing, unpublished data presented at the XII International Botanical Congress, Leningrad, 1975) and from possibly correlative rocks in equatorial Africa (Doyle, Biens, Doerenkamp and Jardiné, unpublished, and below).

These observations have certain broad implications on the extent of pre-Barremian angiosperm diversification. On the one hand, some prior evolution was clearly necessary to produce the variation in size, shape, and exine sculpture seen in basal Potomac angiosperm pollen. In addition, as several authors have pointed out (Doyle, 1969; Muller, 1970: Walker and Skvarla, 1975; Doyle et al., 1975; Walker, 1976), there are extant magnoliid dicots with non-columellar monosulcate pollen distinguishable from that of gymnosperms only by TEM studies of endexine structure which may represent relicts of a still earlier, pre-columellar phase of angiosperm evolution not yet recognized (though not necessarily lacking!) in the fossil record. On the other hand, basal Potomac angiosperm pollen grains cover only a small fraction of the vast morphological spectrum seen in Recent angiosperm pollen: in their monosulcate aperture condition and exine structure alone, they are more primitive than all six non-magnoliid dicot subclasses, many monocots, and even many Magnoliidae (e.g., the inaperturates, dicolpates, and forms with verrucate and spiny sculpture figured by Walker, 1976). Furthermore, the protean and taxonomically frustrating patterns of morphological variation in Zone I monosulcates (numerous intermediate and overlapping generalized types, characters recurring in peculiar combinations) correspond well to Simpson's (1953, pp. 228—229) description of a group undergoing active adaptive radiation, and hence caution against supposing that the plants that produced them were as specialized in non-palynological characters as particular highly relict modern monocots and Magnoliidae which happen to have retained similar, relatively primitive aperture and sculpture conditions (as often implied in comparisons between *Clavatipollenites* and *Ascarina* of the extant Chloranthaceae: e.g., Walker, 1976, p. 291). Finally, it may be noted that some of the more extreme Potomac monosulcates (e.g., *R. peroreticulatus*: Plate I, 3, 4) lack modern analogues and are reasonably interpreted as extinct "experimental" lines of the sort frequently observed in the early stages of adaptive radiations (cf. Doyle et al., 1975). The possible functional significance of the reticulate sculpture of Zone I monosulcates as an adaptation for insect pollination and/or stigmatic germination is discussed further below.

Near the top of Zone I (Arundel Clay and equivalents: Aptian—early Albian?), monosulcate angiosperms are joined by the first representatives of the tricolpate pollen class, with radial rather than bilateral symmetry and three

PLATE I

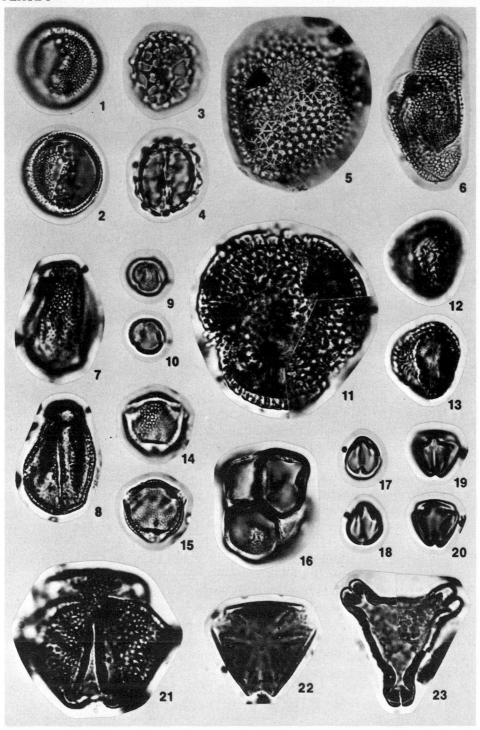

germination furrows or colpi arranged along lines of longitude (Plate I, 7, 8; Wolfe et al., 1975; Doyle and Hickey, 1976; Doyle and Robbins, 1977). Tricolpate pollen today is restricted to and apparently primitive in the six dicot subclasses exclusive of Magnoliidae (*sensu* Takhtajan, 1969; cf. Walker and

---

PLATE I

Selected angiosperm pollen types from the Potomac Group and Raritan Formation. All figures are light micrographs at 1000×.

1,2.   *Clavatipollenites* cf. *hughesii* Couper, lower Zone I (Barremian—Aptian?), tectate-columellar monosulcate, surface view and optical section (Aq 27-11, Baltimore—Susquehanna Aqueduct, Md.).

3,4.   *Retimonocolpites peroreticulatus* (Brenner) Doyle, upper Zone I (Aptian—early Albian?), coarsely reticulate monosulcate apparently lacking typical columellae, surface view and optical section (69-21-27, United Clay Mine near Poplar, Md.).

5.   *Stellatopollis* sp., lower Zone I (Barremian—Aptian?), monosulcate (zonasulculate?) grain with "crotonoid" or "stellate" sculpture pattern (71-15-1a, Dutch Gap Canal, Va.).

6.   *Liliacidites* sp. A (=*Retimonocolpites* sp. C of Doyle, 1973), lower Zone I (Barremian—Aptian?), monocotyledonoid monosulcate with finer sculpture at the ends of the grain, two grains (71-8-1d, Trent's Reach, Va.).

7,8.   aff. *Tricolpites crassimurus* (Groot and Penny) Singh, upper Zone I (Aptian—early Albian?), early reticulate tricolpate, surface view and optical section (69-21-1a, United Clay Mine near Poplar, Md.).

9,10. *Tricolpites minutus* (Brenner) Dettmann, upper Subzone II-B (late Albian?), very small, finely reticulate tricolporoidate, surface view and optical section (D13-555-1b, Delaware City well D13).

11.   "*Retitricolpites*" *geranioides* (Couper) Brenner, upper Subzone II-B (late Albian?), large, coarsely reticulate tricolpate, optical section with inset of surface view (D13-540-1b, Delaware City well D13).

12,13. "*Retitricolpites*" *vermimurus* Brenner, middle Subzone II-B (late middle Albian?), rugulate—reticulate tricolpate, high surface view and composite of lower surface view and optical section (D12-515-1c, Delaware City well D12).

14,15. aff. *Tricolpites micromunus* (Groot and Penny) Burger, upper Subzone II-B (late Albian?), oblate, finely reticulate tricolporoidate, surface view and optical section (65-2a-1j, West Bros. Brick Co., Md.).

16.   aff. *Ajatipollis* sp. A, middle Subzone II-B (late middle Albian?), permanent tetrad of grains with irregular poroid apertures (D12-515-1a, Delaware City well D12).

17,18. cf. *Tricolporoidites subtilis* Pacltová, Subzone II-C (latest Albian?), very small, smooth, prolate tricolporoidate, surface view and optical section (D13-420-1d, Delaware City well D13).

19,20. cf. "*Tricolporopollenites*" *triangulus* Groot, Penny and Groot, lower Zone III (early Cenomanian?), very small, smooth, triangular tricolporoidate, surface view and optical section (D13-370-1c, Delaware City well D13).

21.   *Tricolporopollenites* sp. D, Zone IV (middle—late Cenomanian?), large, reticulate tricolporate (68-8-1a, Sayreville, N.J.).

22.   *Tricolporopollenites* sp. C, Zone IV (middle—late Cenomanian?), medium-sized triangular tricolporate (69-13-1d, near Mill Brook, N.J.).

23.   *Complexiopollis* sp. A, Zone IV (middle—late Cenomanian?), early blotchy-verrucate triangular triporate of the Normapolles group, optical section with inset of surface view (NJ 2-1b, Woodbridge, N.J.).

See Doyle (1969), Wolfe et al. (1975), Doyle et al. (1975), Doyle and Hickey (1976), and Doyle and Robbins (1977) for detailed locality data and discussion of stratigraphy.

Doyle, 1975; Wolfe et al., 1975). Like the first tricolpates of the lower Albian of England (Kemp, 1968; Laing, 1975, 1976) and the Aptian of Africa—South America (Brenner, 1976; Doyle, Biens, Doerenkamp and Jardiné, unpublished; cf. below), the oldest Potomac tricolpates are small- to medium-sized (20—35 µm), prolate in shape, and reticulate in exine sculpture, consistent with their postulated origin from reticulate monosulcates of the *Clavatipollenites—Reti-monocolpites* complex (cf. Doyle, 1969; Wolfe et al., 1975). However, the Potomac record gives no hint on the exact mode of origin of the tricolpates; possible reasons for this gap, one of the most bothersome in the Potomac—Raritan record, are discussed further below in connection with geographic patterns in the early angiosperm record.

The abundance and morphological diversity of both tricolpates and continuing monosulcates increases stratigraphically in Zone II (Patapsco Formation and basal beds of a Maryland and Delaware unit formerly incorrectly identified with the Raritan Formation of New Jersey: cf. Wolfe and Pakiser, 1971; Doyle and Hickey, 1976, etc.), as in middle and late Albian sequences elsewhere in the world (cf. Hedlund and Norris, 1968; Singh, 1971; Dettmann, 1973; Laing, 1975, 1976). Very small, finely reticulate to smooth tricolpates appear in Subzone II-A; by the upper part of Subzone II-B the spectrum of tricolpate types varies from extremely small and finely reticulate (Plate I, 9, 10) to large and coarsely sculptured (Plate I, 11), smooth and tectate, striate, rugulate (Plate I, 12, 13), reticulate but differentiated into coarse and fine areas, oblate rather than prolate in shape (Plate I, 14, 15), and even one species with grains shed in permanent tetrads (Plate I, 16). A few upper Subzone II-B samples have been found in which angiosperms constitute more than 50% of the pollen and spore flora for the first time, but ferns and conifers are still unquestionably dominant on a regional scale (cf. Brenner, 1963; Doyle, 1969; Doyle and Hickey, 1976). In Subzones II-B and II-C, several distinct groups of tricolpates as defined on size, shape, and sculpture begin to show rudimentary thin areas at the centers of their colpi (tricolporoidate: Plate I, 9, 10, 14, 15, 17, 18), suggesting the tricolporate condition which is stabilized in Zone III and characteristic of most "average" modern dicot groups; this constitutes a good example of parallel evolution, or more specifically program evolution, since the change occurs at roughly the same time in related lines (cf. Simpson, 1953; Wolfe et al., 1975). Perhaps even more than Zone I monosulcates, Zone II tricolpates show the sort of exuberant but repetitive and overlapping morphological variation characteristic of groups undergoing adaptive radiation (Simpson, 1953).

Small, smooth tricolporoidate grains (Plate I, 17—20) become more common going from Subzone II-C into Zone III (i.e., within the "Maryland Raritan": cf. Wolfe and Pakiser, 1971; Doyle and Hickey, 1976), as near the Albian—Cenomanian boundary in other areas (cf. Pacltová, 1971; Singh, 1971; Jarzen and Norris, 1975; Laing, 1975, 1976). These include oblate-triangular forms with flat sides (Plate I, 19, 20), believed to be the result of continuation of the trend from prolate to oblate shape (cf. Doyle, 1969; Wolfe et al., 1975). Although most Subzone II-C and lower Zone III tricolpor(oid)ates are smaller, smoother, and thinner walled than their presumed tricolpate ancestors, apparently related tricolporates which enter in upper Zone III and Zone IV (lower Raritan Formation of New Jersey, including the Woodbridge Clay Member: middle and late Cenomanian?) are larger and thickerwalled and often have reticulate or low, blotchy sculpture, suggesting reversal of earlier evolutionary trends in these characters (Plate I, 21, 22; cf. Wolfe et al., 1975). Finally, Zone IV yields the first triporate pollen grains of the Normapolles complex (Plate I, 23), a group which diversifies throughout the rest of the Late Cretaceous of Europe and eastern North America (Góczán et al., 1967; Doyle, 1969; Wolfe and Pakiser, 1971). Because of their extreme, concave-sided shape and smooth to blotchy exine sculpture, these first Normapolles have been interpreted as the culmination of the trend toward oblate shape and shortening of the apertures (Doyle, 1969; Muller, 1970; Wolfe et al., 1975).

The fact that the major angiosperm pollen types seen in the Potomac—Raritan sequence can be fitted into a morphologically and stratigraphically consistent "pollen phylogeny" (Fig. 2, right), in which the first representatives of each new pollen class differ least from members of pre-existing classes, has been taken as evidence that the mid-Cretaceous rise of angiosperm pollen reflects fairly closely the evolutionary diversification of a natural, monophyletic group, rather than immigration of already highly differentiated taxa, or polyphyletic origin of angiosperms from unrelated gymnosperm groups (cf. Doyle, 1969, 1973; Muller, 1970; Wolfe et al., 1975; Doyle and Hickey, 1976). Interestingly, this pollen phylogeny is essentially the same as that postulated on the basis of comparative studies of living angiosperms, particularly assuming "ranalian" theories of angiosperm evolution (e.g., Wodehouse, 1936; Takhtajan, 1959, 1969; cf. Doyle, 1969; Muller, 1970; Wolfe et al., 1975; Walker and Doyle, 1975). However, it is desirable to test this somewhat subjective, qualitative interpretation with a semiquantitative analysis of stratigraphic changes in taxonomic diversity and morphological advancement in the Potomac sequence.

The basic data for my analysis of taxonomic frequency changes in the early angiosperm record (Fig. 3) consist of the stratigraphic ranges of pollen "species" in a composite section based on core samples from two wells drilled through the Potomac Group near Delaware City, Delaware (Doyle and Robbins, 1977; treated in less detail by Brenner, 1967; Doyle and Hickey, 1976; Doyle, 1976). Solid portions of ranges indicate forms closely comparable with the named species; dashed portions, forms differing consistently from but apparently closely related to the named species ("aff."). Solid and dashed portions of the same line are lumped in the present analysis, as they would in fact be lumped taxonomically by many palynologists. It is often suspected that such broadly defined palaeopalynological "species" would correspond more to genera than to species in a Recent whole-plant classification (cf. Muller, 1970, p. 418).

For the purposes of this analysis, the Delaware well sequence has first been divided into six stratigraphic units of roughly equal sediment thickness (ca. 125 ft. or 38 m), corresponding to Zone I, Subzone II-A plus lower Subzone II-B, middle and upper Subzone II-B, Subzone II-C, lower Zone III, and upper Zone III. In order to clarify the especially great changes which occur in the

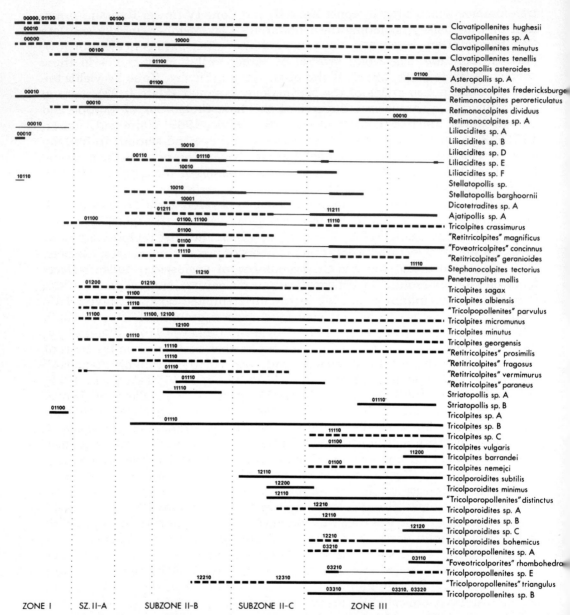

Fig. 3. Stratigraphic ranges and advancement indices of angiosperm pollen types in Delaware City wells D12 (Zone I) and D13 (Zones II and III), used in analyses of taxonomic frequency changes (Fig. 4) and morphological evolution (Fig. 5) (modified from Doyle and Robbins, 1977). Roman numerals refer to the palynostratigraphic units in Fig. 2; dotted horizontal lines delimit intervals of approximately equal stratigraphic thickness (125 ft. or 38 m), except for Subzone II-A and lower Subzone II-B. Dashed portions of ranges indicate forms differing significantly from but apparently related to the named species ("aff."); fine lines correspond to intervals where species have not been observed in more than two successive well samples, or are known only from putatively correlative outcrop samples. The advancement index for each species is the sum of the five digits next to its range (see Table II and text for discussion).

second of these units, it has been subdivided into two, Subzone II-A and lower
Subzone II-B. It is assumed as a first approximation that equal sediment thicknesses represent equal lengths of time; the fact that Zone I is much thinner in
Delaware than in Maryland and Virginia may introduce some error here, but
there is otherwise no evidence of significant hiatuses or variations in rate of
deposition. Because of the lack of reliable data on absolute lengths of the
twelve stages of the Cretaceous (cf. Casey, 1964) and uncertainties in correlation of the Potomac, it would be premature to translate these intervals into millions of years. However, some feeling for the amount of time involved may be
gained by assuming that rocks of (Barremian?—) Aptian, Albian, and early
Cenomanian age are represented, and that the average Cretaceous stage was
6 m.y. long (cf. Casey, 1964). The resulting estimate of 2—3 m.y. per basic
38-m interval may be an underestimate, considering that the Albian appears to
be one of the longest stages of the Cretaceous (cf. Kauffman, cited in Doyle
and Hickey, 1976).

Fig. 4 presents curves for total frequency of angiosperm pollen "species,"
constructed by counting all "species" present in each interval in Fig. 3, and
three taxonomic frequency rates (cf. Simpson, 1953, chapter 2): rate of change

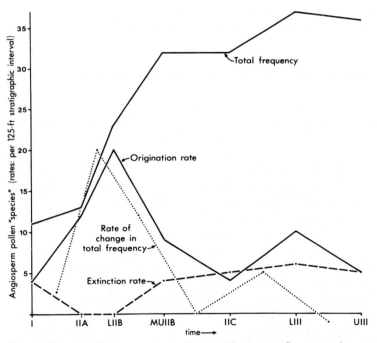

Fig. 4. Taxonomic frequency analysis of Potomac Group angiosperm pollen, based on data
in Fig. 3. The total frequency curve is in terms of the total number of palynological "species" at each of the stratigraphic intervals recognized in Fig. 3 (solid and dashed portions of
ranges are lumped). The rate curves are in terms of first occurrences (origination rate), last
occurrences (extinction rate), and rate of change in total number of "species" per 125-ft.
(38-m) stratigraphic thickness interval. See text for discussion.

in total frequency (approximated as between midpoints of successive intervals), origination rate (first appearances, except for those at the base of Zone I), and extinction rate (last appearances, except at the top of Zone III). All rates are calculated in terms of "species" per standard 38-m interval, i.e., correcting for the shorter lengths of Subzone II-A and lower Subzone II-B.

Fig. 4 shows that the total frequency of angiosperm pollen "species" generally rises throughout the Delaware City Potomac section, but both the rate of increase and the origination rate are conspicuously higher earlier in the sequence, with highest rates in both around the base of Subzone II-B. Falling origination rates and rising extinction rates result in a temporary levelling-off of the total frequency curve in Subzone II-C and a slight drop in upper Zone III, though there is a moderate secondary peak in origination in lower Zone III. As Simpson (1953, chapter 7) notes, such patterns of an early "explosive" rise in diversity reflecting high origination rates somewhat after the appearance of a group, followed by a more gradual levelling-off as origination rates decline, are common though by no means universal in the fossil record. He interprets them in terms of an initial phase of rapid adaptive radiation characterized by centrifugal selection and release of variability as the group enters a new adaptive zone, and a subsequent phase of less rapid intrazonal evolution characterized by centripetal selection and "weeding out" of earlier experimental lines as the group adjusts to a limited number of stable, non-overlapping adaptive subzones. Thus it is tempting to conclude that the main peak in origination rates in lower Subzone II-B corresponds to the climax of a Barremian—middle Albian phase of rapid adaptive radiation of the angiosperms, while the declining origination rates and rising extinction rates of upper Subzone II-B and Subzone II-C mark the gradual transition to a late Albian—Cenomanian phase of intrazonal evolution and weeding out of intermediate types. However, evidence presented below that the middle Albian peak may be somewhat exaggerated and too late because of immigration of tricolpate groups which had originated somewhat earlier (late Aptian?) south of the Tethys, and such anomalies as the zero extinction rates in Subzone II-A and lower Subzone II-B (Simpson predicts that extinction rates should be high during adaptive radiation as well as intrazonal evolution) caution against interpreting detailed features of the curves in Fig. 4 as exact reflections of evolutionary phenomena.

There is reason to suspect that the temporary levelling-off of the total frequency curve in Subzone II-C reflects some sort of external environmental fluctuation, particularly because it corresponds to a time of "local extinctions" (not counted in constructing Fig. 4, but seen as fine lines in Fig. 3), i.e., cases where "species" disappear in Subzone II-C and reappear in Zone III. This phenomenon is apparently not simply a local facies effect, since similar changes occur not only in the Maryland outcrop area at the transition from Patapsco variegated clays to "Raritan" sands, but also in the Delaware City wells in the middle of a thick clay unit (cf. Doyle and Hickey, 1976; Doyle, 1976). Since a general influx of new, typically Late Cretaceous conifers (e.g., *Rugubivesiculites*) is also occurring at about this point, it is tempting to associate the Subzone II-C drop in angiosperm diversity with climatic deterioration from the

Albian high to the Cenomanian low inferred from oxygen isotope studies (Bowen, 1961; Lowenstam, 1964). However, the changes in the conifer element begin earlier in the Potomac sequence (cf. Brenner, 1963), and it is difficult to say whether they reflect climatic or biotic changes (cf. Zherikhin, discussed below).

Following Simpson (1953, chapter 7), one would expect that any group as diverse as the angiosperms are today must be the product of a series of adaptive radiations, corresponding to attainment of new adaptive levels by various subgroups, rather than a single radiation. The secondary peak in origination in lower Zone III might be interpreted as an example of such a secondary radiation at the tricolporate level. More striking examples are the proliferation of the triporate Normapolles group later in the Cretaceous (Góczán et al., 1967), and of Compositae and other herbaceous groups in the Late Tertiary (cf. Germeraad et al., 1968; Muller, 1970; Leopold and MacGinitie, 1972).

While the study of taxonomic diversity changes summarized in Fig. 4 shows rate patterns consistent with an adaptive radiation, a quantitative analysis of changes in morphological characters of the pollen grains involved might provide a more direct, critical test of the concept that the general increase in number of species reflects evolutionary diversification rather than immigration, and reveal new phenomena of evolutionary interest. A previous palynological example of such an approach is Chaloner's (1967) study of size trends in Silurian—Devonian spores: histograms of mean spore size at successive horizons showed a progressive increase in maximum and mean spore size throughout the Silurian—Devonian interval, which Chaloner argued documents the transition from homospory to heterospory and supports an evolutionary interpretation of the early land plant record. I suspect that a size analysis of mid-Cretaceous angiosperm pollen would also show significant trends: the earliest Zone I angiosperm pollen is small- to medium-sized, and very small and large grains appear later. However, since so many of the more striking variations in Cretaceous angiosperm pollen morphology involve characters other than size, including many difficult-to-quantify characters such as symmetry, aperture type, sculpture, etc., a more profitable approach may be to use a semiquantitative measure of morphological evolution (advancement index) derived by adding scores (0, 1, 2, ...) for a series of qualitative characters, as in Westoll's (1949) analysis of morphological evolution in lungfish (reformulated somewhat more lucidly by Simpson, 1953, pp. 22—25). Westoll's lungfish advancement indices fall near a single smooth curve when plotted against time, rising most rapidly soon after appearance of the group, then more and more slowly as the modern condition is approached, reflecting progressive modernization in a limited number of phyletic lineages. Since the angiosperms (like early land plants) consist of so many divergent lines, including some which retain relatively primitive pollen types to the present day, the angiosperm advancement data are more appropriately presented as a series of histograms for the same stratigraphic intervals as recognized in Figs. 3 and 4, and curves for the mean advancement of the whole flora and rate of change in mean advancement per 125-ft. (38-m) stratigraphic interval (Fig. 5).

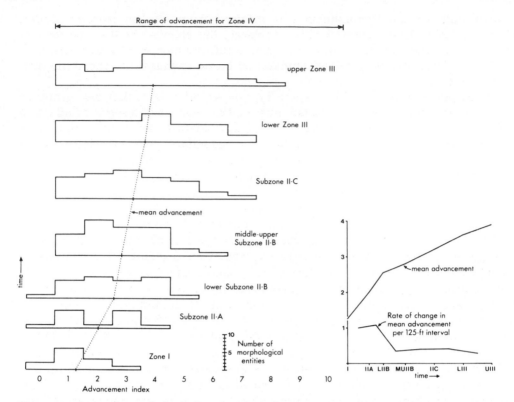

Fig. 5. Analysis of morphological evolution in Potomac Group angiosperm pollen, based on data in Fig. 3. Histograms show numbers of morphological entities with particular advancement indices (Table II, Fig. 3) at each of the stratigraphic intervals recognized in Fig. 3. Morphological entities generally correspond to "species" in Figs. 3 and 4, except that forms compared with the named species which have different advancement indices are distinguished rather than lumped. Curves in the graph to the right show mean advancement of the entire angiosperm flora at each stratigraphic interval (dotted line in histograms) and rate of change in mean advancement per 125-ft. (38-m) stratigraphic thickness interval. See text for discussion.

The bases for assigning advancement scores in early angiosperm pollen are summarized in Table II. The ancestral condition — zero advancement in all five characters considered (size, aperture condition, shape, exine sculpture, and pollen unit) — is assumed to be medium-sized (18–35 μm), monosulcate, bilaterally symmetrical, finely reticulate (tectate to semitectate), and shed as single grains, a sort of "average" member of the *Clavatipollenites—Retimonocolpites* complex of lower Zone I. Deviations from these conditions in a variety of directions are given scores of 1 or more, depending on whether or not several stages of a transformation series can be recognized on the basis of present data (i.e., morphological intermediates, stratigraphic sequence, etc., as summarized briefly above; cf. Doyle, 1969, 1973; Wolfe et al., 1975). For example, since there is evidence that triporate grains of the type seen in Zone IV passed

TABLE II

Bases for assignment of advancement scores for analysis of morphological evolution in Potomac—Raritan angiosperm pollen (Figs. 3, 5; see text for discussion)

Characters	Score				
	0	1	2	3	4
Mean size	18–35 μm	<18 μm >35 μm			
Aperture condition	monosulcate	branched sulcus polycolpoidate tricolpate	tricolporoidate triporoidate	tricolporate	triporate
Shape	bilateral, length > width	width ≥ length (uniaperturates) prolate (triaperturates)	oblate, speroidal	oblate, triangular	oblate, concave sides
Exine sculpture	fine reticulate (meshes <2 μm)	coarse reticulate (meshes ≥2 μm) heterogeneous striate, rugulate tectate, smooth	"blotchy", verrucate		
Pollen unit	single grains	tetrads			

through monosulcate, tricolpate, tricolporoidate, tricolporate, and triporate stages, these conditions are scored 0, 1, 2, 3, 4; but because coarsely reticulate, striate, rugulate, smooth, and other deviations from finely reticulate sculpture appear to have arisen independently or in uncertain sequence, they are all (except blotchy-verrucate, clearly derived from smooth) assigned scores of 1. Clearly, addition of other characters and improved evidence on species-level phylogenetic relationships and the nature of the ancestral type might permit considerable refinement of this system. The advancement index for each pollen type is the sum of the five digits (scores for each of the five characters) recorded next to its range in Fig. 3; note that when forms from two portions of the range of one pollen "species" or forms lumped at the same horizon have different advancement indices, they are treated separately, so the number of entities in this analysis is not precisely the same as in the taxonomic frequency analysis (Fig. 4).

Fig. 5 provides striking graphical confirmation of the concept of regularly increasing morphological differentiation in Potomac angiosperm pollen: both the mean and maximum advancement indices move continuously to the right, with no sudden appearances of types more than one advancement point beyond those existing at the previous level. The largest "jump" is the appearance of a secondary peak at advancement 3 in Subzone II-A, reflecting largely the variety of small tricolpates entering at that level; this re-emphasizes the point made earlier that there are no transitional forms between tricolpates and older types in the Potomac record. The persistence through Zone III of peaks corresponding to advancements 1 and 4 and the lowering of the intervening bars is suggestive of the weeding-out process associated with the shift from rapid adaptive radiation to intrazonal evolution, inferred above to be occurring at similar levels. No attempt has been made to construct advancement histograms for later stages of the angiosperm record, but as is indicated by the double-headed arrow at the top of Fig. 5, most of the less advanced types seen in Zone III continue into Zone IV, but entry of triporate, concave-sided, smooth to verrucate Normapolles (advancement indices 04410, 04420) raises the maximum advancement to 10. In terms of the pollen characters considered here, many "average" or intermediate-level extant dicots — e.g., members of less specialized families of Rosidae and Dilleniidae with small to medium-sized, tricolporate, sometimes triangular pollen (cf. Walker and Doyle, 1975) — would have advancement indices in the range of 4—7, but many other common modern types, such as secondarily spheroidal triporate "Amentiferae" (cf. Doyle, 1969; Wolfe, 1973) and polyaperturate types, would require additional grades of advancement beyond the scale used here. Hence by the end of the Potomac sequence (mid-Cenomanian?), some of the more advanced angiosperms had reached an evolutionary grade in pollen morphology comparable to that of intermediate-level modern dicots, but nowhere near the grade of other modern types, and the flora as a whole was much less advanced than today (cf. Scott et al., 1960; Pierce, 1961; Doyle, 1969; Muller, 1970).

The curve for rate of change in average advancement (Fig. 5, right) shows markedly higher rates early in the Potomac sequence, with the peak between

Subzone II-A and Subzone II-B corresponding closely to the peaks in origination rate and rate of change in total frequency seen in Fig. 4, and a decline to lower rates later in the sequence. This curve is remarkably similar to Simpson's curve (1953, p. 23) for rate of change in advancement based on Westoll's (1949) lungfish data, except that in the Potomac angiosperm record there is only a slight indication of an initial phase of moderate rates immediately preceding the peak rate of change, consistent with the inference that the Potomac record begins somewhat after the very earliest stages of angiosperm diversification (cf. above). Westoll (1949) and Simpson (1953) identify the peak rates early in lungfish history with the phase of most rapid adaptive radiation; this complements well the similar interpretation of the peak in origination rate at the corresponding point in the Potomac angiosperm record. Interestingly, the rate curve in Fig. 5 shows no slowing of increase in mean advancement in Subzone II-C, despite the temporary decline in regional species diversity at this horizon; this supports the concept that morphological trends at this point in the Potomac record reflect true progressive evolution, rather than simply changes in the total number of species resulting from environmental fluctuations.

It is difficult to imagine how the patterns in Figs. 4 and 5 could be obtained under extreme prior diversification or polyphyletic theories, or if the ancestral conditions assumed for the advancement analysis were seriously in error. On the other hand, these data do not rule out smaller-scale prior origination and immigration of particular types, nor the possibility that *Clavatipollenites* and its relatives were preceded by angiosperm pollen types more difficult to distinguish from gymnosperm pollen (perhaps though not necessarily like the large, smooth, monosulcate grains of certain Magnoliales: cf. Doyle, 1969; Muller, 1970; Walker and Skvarla, 1975; Doyle et al., 1975; Walker, 1976), or that the angiosperms originated by parallel evolution from a complex of closely related gymnosperm lines with generalized monosulcate pollen (cf. Doyle, 1969).

Speciation models

In theory, the abundance of pollen and spores and the possibility of close stratigraphic sampling might make palynology ideal for testing alternative models of finer-scale patterns of evolution, such as the "phyletic gradualism" and "punctuated equilibria" models of speciation contrasted by Eldredge and Gould (1972), Gingerich (1974; this volume), and others. However, the relatively small number of characters extractable from pollen grains and the resulting practical difficulties in recognizing species limits and phyletic lineages (cf. above) preclude such tests using the sort of data presently available. Many appearances of new "species" in the Potomac sequence are abrupt, as is generally true in palaeontology and as is predicted by allopatric speciation models in general, and others persist with no appreciable morphological change through almost the whole section (e.g., *Retimonocolpites peroreticulatus*), as predicted by the punctuated equilibria model in particular (Eldredge and Gould, 1972). However, the morphological resolution is not good enough to say whether the "gradual" series of intermediates between types alluded to above (e.g., between tricolpates and tricolporates), or the changes between dashed and solid portions of "species" ranges in Fig. 3, represent gradual phyletic transformation series between speciation events, or successions of slightly differentiated, allopatrically derived species (i.e., Ahnenreihen or Stufenreihen: cf. Simpson, 1953). I suspect that some of these trends, such as the accentuation of sculpture patterns and increasing proportions of tricolporoidate specimens among Zone II tricolpates, reflect phyletic evolution, e.g., readjustment to new physical and biotic factors (including competitive displacement: cf. Gingerich, 1974) following entry of former peripheral isolates into new environments. I would further speculate that such cases should be more common in groups actively radiating into new adaptive zones (such as Early Tertiary mammals or Early Cretaceous angiosperms), where more "ecological space" should be available for competitive divergence and coexistence of sister species after re-establishment of sympatry, than in groups already occupying stable or even contracting adaptive zones (such as Devonian trilobites), where because of the more closed ecological situation extinction of one lineage might be the more likely outcome of speciation. However, further quantitative studies would be necessary to raise this suggestion above the level of speculation.

Geographic variations in the pollen record

Comparisons among mid-Cretaceous angiosperm pollen sequences from different geographic areas are desirable both as an independent test of the extent to which the diversification pattern seen in the Potomac—Raritan record (Figs. 2—5) reflects evolutionary diversification rather than migration, and for possible evidence on climatic preferences, directions of dispersal, and ecological evolution of early angiosperms. Clearly, some migrational effects are to be expected in the fossil record of any adaptive radiation, if only because of the unlikelihood of sympatric speciation and the fact that any new adaptive type is likely to originate in a particular area and environment and become adapted to a wider range of ecological conditions as it diversifies. To detect such effects of course requires independent dating by fossils other than angiosperms, so sections such as the Potomac Group can be used only in a very general, indirect way if circular arguments are to be avoided. The most comprehensive study along these lines is Brenner's (1976) synthesis based on a series of faunally dated localities from Arctic North America to South America; to his data may be added recent information from England and France (Laing, 1975, 1976; Hughes and Laing, unpublished), western Canada (Jarzen and Norris, 1975), Brazil (Herngreen, 1973, 1974), equatorial Africa (Doyle, Biens, Doerenkamp and Jardiné, unpublished), and Australia (Dettmann, 1973; Burger, 1973).

Brenner (1976) recognizes four palynofloristic provinces in the Early Cretaceous, distinguished largely on differences in the non-angiospermous dominants, and showing clear relationships to the positions of the continents and latitudinal belts at the time. The Potomac flora belongs to Brenner's Southern Laurasian province, represented also by floras from Europe, Central Asia, other parts of the U.S.A., and (with some features transitional to the Northern Laurasian province) western Canada. These floras are characterized by a high diversity of conifer pollen, including many bisaccates, and fern spores, including abundant representatives of the family Schizaeaceae. Monosulcate angiosperm pollen of the sort seen in lower Zone I is first seen elsewhere in the Southern Laurasian province in the Barremian of England (Couper, 1958; Kemp, 1968; Hughes and Laing, unpublished data, XII International Botanical Congress, Leningrad, 1975). Tricolpates first appear in strata of early or middle Albian age and attain much the same diversity seen in middle—upper Subzone II-B in the middle and late Albian (Kemp, 1968; Hedlund and Norris, 1968; Laing, 1975, 1976). Tricolporoidates appear in the late Albian (cf. Pacltová, 1971); very small, smooth, and triangular forms like those of Subzone II-C and Zone III near the Albian—Cenomanian boundary (Norris, 1967; Singh, 1971; Jarzen and Norris, 1975; Laing, 1975, 1976); and triporates of the Normapolles group (as in Zone IV) in the middle Cenomanian (Góczán et al., 1967; Médus and Pons, 1967; Pacltová, 1971; Laing, 1975, 1976).

Despite the sometimes wide margins of error in correlation, significant deviations from this pattern can be detected both to the north and the south. In the transitional area of western Canada, a Barremian—early Albian phase with monosulcate angiosperm pollen has not been recognized, except possibly in Saskatchewan (Playford, 1971); elsewhere, tricolpates and angiospermous monosulcates appear simultaneously in the middle or even late Albian, after which the sequence is essentially the same as in the Potomac—Raritan section (Norris, 1967; Singh, 1971; Jarzen and Norris, 1975). In Brenner's (1976) Northern Laurasian province proper (Arctic North America and Siberia), where schizaeaceous fern spores are much less diverse and conifer pollen and *Sphagnum*-type moss spores proportionally more common than in Southern Laurasia, angiosperm pollen does not appear until much later, in the Cenomanian. A pattern similar to that seen in western Canada may also exist in Australia, part of Brenner's (1976) Southern Gondwana province (including also Patagonia and South Africa): tricolpates and angiospermous monosulcates both enter in the middle or late Albian [in fact, Dettmann (1973) reports that *Clavatipollenites* enters later than the tricolpates], but the order of appearance of later types (tricolporate, triporate) is similar to that in Southern Laurasia (Burger, 1970, 1973; Dettmann, 1973). These exceptions do not contradict the evolutionary trends inferred from the Potomac—Raritan sequence, since all the major types in question already exist at correlative horizons at lower latitudes (cf. Doyle, 1976), but they do show a significant lag in migration of angiosperms into higher latitudes, presumably because time was required for evolution of physiological adaptations to cooler climates (cf. Brenner, 1976).

In contrast, there is evidence of greater diversity of angiosperms and significantly earlier appearances of some pollen types in Brenner's (1976) Northern Gondwana province (South America except Patagonia, Africa except South Africa, Israel). In coastal basins of Brazil (Müller, 1966; Brenner, 1976) and equatorial Africa (Doyle, Biens, Doerenkamp and Jardiné, unpublished), reticulate tricolpate pollen appears below salt deposits marking the first incursion of marine waters into the South Atlantic. Despite the lack of reliable faunal dating below the salt, the fact that the salt is overlain by marine rocks containing latest Aptian and early Albian ammonites (Reyment and Tait, 1972) securely brackets the age of these tricolpates as older than the first well-dated early and middle Albian occurrences of tricolpates in Europe (Kemp, 1968; Laing, 1975, 1976). Brenner (1976) also reports tricolpates from pre-late Aptian beds in Israel. In addition, several authors have documented the greater diversity and abundance of angiosperm pollen in younger Albian and Cenomanian beds (Jardiné and Magloire, 1965; Boltenhagen, 1965; Müller, 1966; Brenner, 1968, 1976; Herngreen, 1973, 1974): tricolpates are joined by polyporates (unknown in the Lower Cretaceous of Laurasia) early in the Albian, and by quite endemic tricolporoidates (with two ora in each of the three colpi) later in the Albian. However, the order of appearance of major triaperturate types is the same as in Laurasia: tricolpate, tricolpor(oid)ate (middle or late Albian), triporate (middle? Cenomanian, but a different group from the Normapolles); and recently monosulcate angiosperm pollen covering much the same range of exine sculpture conditions seen in the basal Potomac Group and the Barremian of England has been found in beds predating the first appearance of tricolpates in Gabon and the Congo (Doyle, Biens, Doerenkamp and Jardiné, unpublished).

These data suggest that lags in migration across the Tethys were not great enough to distort seriously the evolutionary sequence of pollen types in Laurasia, but that some types (particularly the tricolpates) originated first in Northern Gondwana, so that their first appearance in Laurasia may represent immigration rather than evolution in place. This would explain the lack of obvious morphological intermediates between monosulcates and tricolpates in the Potomac sequence. Furthermore, although it would be premature to conclude that the angiosperms originated in Northern Gondwana (within the uncertainties of correlation, the first monosulcate angiosperms in Africa, England, and the Potomac could all be the same age), the greater diversity of types suggests that ecological conditions in Northern Gondwana were generally more favourable for early angiosperm diversification. As Brenner (1976) notes, palaeoclimatic indicators such as the presence of salt deposits, the dominance of *Classopollis* (known to have been produced by xeromorphic conifers such as *Brachyphyllum*) and of *Ephedra*-like pollen, and the rarity of fern spores (cf. also Kuyl et al., 1955; Jardiné et al., 1974) suggest that these conditions were not hot and wet, as usually postulated for angiosperm origins (e.g., Bews, 1927; Axelrod, 1952; Takhtajan, 1969), but rather hot and at least seasonally dry, as recently proposed by Stebbins (1965, 1974), Axelrod (1970), and Raven and Axelrod (1974).

As Brenner (1976) points out, these observations support in a general way Axelrod's (1959) concept of a tropical origin and poleward spread of the early angiosperms, especially his more recent formulation (Axelrod, 1970) stressing the role of seasonal aridity in their origin and diversification, but not his concept of a long period of pre-Cretaceous diversification. Rather, spread and adaptive radiation of the angiosperms were simultaneous, and except for high-latitude areas, most stages of the radiation were essentially world-wide and distorted by migrational effects only in a relatively minor way (cf. Doyle, 1976). Recalling the quantitative analyses of diversity and advancement trends in the Potomac record (Figs. 4 and 5), removal of the migrational distortions that do exist would probably tend to spread out the peaks in origination rate and increase in advancement (both in part reflecting increasing numbers of tricolpates) to the left, but I doubt if it would change the shapes of the curves radically.

Evolutionary patterns in the early angiosperm leaf record

Re-awakening of interest in Cretaceous angiosperm leaves, nearly dormant for over fifty years except in the U.S.S.R. (cf. Vachrameev, 1952, 1973; Samylina, 1960, 1968; Krassilov, 1967), has been stimulated both by recognition of the contrast between the apparently evolutionary patterns revealed by palynology (cf. above) and the concept of a highly diversified mid-Cretaceous angiosperm flora based largely on early leaf identifications, and by the hope that re-examination of the leaf record from a new perspective might also reveal patterns of evolutionary interest. Some authors (e.g., Axelrod, 1970, pp. 287—290; Stebbins, 1974, pp. 207—208) have attempted to resolve the apparent conflict between the pollen and leaf records by postulating systematic biases in the pollen record, such as lagging of pollen evolution behind systematic diversification, fragility and selective non-preservation of early angiosperm pollen, obligate insect pollination, etc.

It should be noted, however, that Axelrod exaggerates the conflict between the pollen and megafossil records to some extent by following the earlier assumption of some palynologists (e.g., Hughes, 1961; Brenner, 1963) that the appearance of tricolpate pollen near the Aptian—Albian boundary represents the first definite palynological evidence of angiosperms: such occurrences as angiosperm woods in the Lower Greensand (Aptian) of England, angiosperm leaves in the lower Potomac Group, etc., become less anomalous with recognition of diverse monosulcate angiosperm pollen (*Clavatipollenites, Retimonocolpites, Liliacidites, Stellatopollis*) in Barremian—Aptian deposits (cf. Kemp, 1968; Wolfe et al., 1975; Doyle et al., 1975; and above). Furthermore, it is difficult to imagine such severe biases in the pollen record as Axelrod and Stebbins suggest, in view of the systematic consistency and usefulness of pollen morphology in modern angiosperm taxonomy (cf. Erdtman, 1952; Doyle, 1969; Muller, 1970; Walker and Doyle, 1975), the integral role of pollen in floral biology, the diverse and readily preserved pollen types produced by Recent members of genera and families supposedly present in the mid-Cretaceous (cf. Pierce, 1961; Brenner, 1963), and the good representation of insect-pollinated forms in the Tertiary tropical pollen record (cf. Germeraad et al., 1968; Muller, 1970).

These considerations, together with the doubts mentioned above on the validity of the identification methods used in older studies of angiosperm leaves, raise the alternative possibility that the apparent high diversity of mid-Cretaceous angiosperm leaves is a methodological artifact (cf. Stebbins, 1950;

Pacltová, 1961; Doyle, 1969; Wolfe, 1972a, 1973; Dilcher, 1974; Wolfe et al., 1975; Hughes, 1976). This suggests that the best test of whether the evolutionary implications of the pollen and leaf records are really inconsistent is a critical re-examination of the leaf record itself using a morphological—stratigraphic approach initially independent of possible affinities with Recent taxa (cf. Wolfe et al., 1975; Doyle and Hickey, 1976; Hughes, 1976).

In the following sections, I will summarize results of recent studies along these lines, again emphasizing the intensively studied Potomac Group section of the eastern U.S.A. (Fontaine, 1889; Ward, 1895, 1905; Berry, 1911; Wolfe, 1972a; Doyle and Hickey, 1972, 1976; Wolfe et al., 1975). However, although geographic sampling is less extensive than with palynology, it should be noted that many of the same patterns can be discerned in Cretaceous leaf sequences of other parts of Laurasia (cf. Vachrameev, 1952, 1973; Samylina, 1968; Doyle and Hickey, 1976). I will not consider angiosperm megafossil remains other than leaves (wood, fruits, etc.), despite their potentially great botanical interest; they are still too rare and inadequately studied to provide much evidence on general patterns of evolution (see however Samylina, 1960, 1968; Wolfe et al., 1975; Dilcher et al., 1976).

Understanding of the stratigraphic sequence of angiosperm leaf types in the Potomac Group (Fig. 6) has been greatly improved by palynological correlations of megafossil localities with the Delaware City well sections discussed above (Doyle and Hickey, 1972, 1976). As already mentioned, the sporadic distribution and facies restriction of leaves make it more difficult to gain a complete picture of the regional leaf flora at successive horizons and hence to analyze evolutionary patterns in the leaf record in such detail as was possible with the pollen record. However, palynological correlations of leaf floras from different lithofacies at three horizons — Zone I, middle Subzone II-B, and upper Subzone II-B — have helped sort out which changes are of regional significance and which due simply to local facies shifts. These correlations have the side-benefit of permitting tentative inferences on the habitat preferences and ecological evolution of early angiosperms (Doyle and Hickey, 1976; discussed further below in connection with adaptive aspects of the early angiosperm record).

Like Zone I angiosperm pollen, Zone I angiosperm leaves are rare and subordinate to fern and gymnosperm remains and cover only a limited portion of the spectrum of morphological types seen in the Recent flora (Fig. 6, *g*—*m*). In shape, they vary from narrowly elliptical (Fig. 6, *j*) to reniform (Fig. 6, *h*), and some have marginal lobes or serrations (Fig. 6, *i*, *l*), but all (except perhaps *Vitiphyllum*: Fig. 6, *l*) have basically pinnate venation, with secondary veins arranged along a single primary vein or midrib. Although it is impossible to be sure whether such remains are simple leaves or leaflets of compound leaves without actually finding them attached to stems or rachises, there is no positive evidence that any Zone I angiosperm leaves were compound, in contrast to the situation in Subzone II-B (cf. Fig. 6, *q*, *s*). The most striking feature of Zone I leaves is their ''first rank'' leaf architecture: i.e., poor definition of vein orders, irregularity of spacing, angle of departure, course, and branching pat-

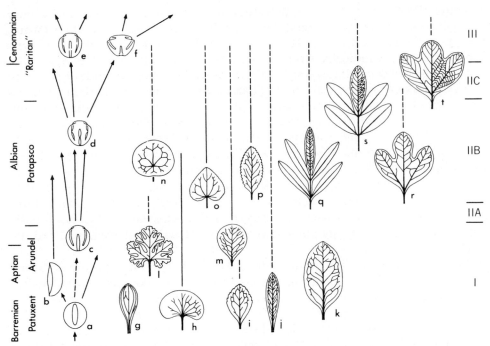

Fig. 6. Principal angiosperm leaf types in the Potomac Group, plotted against the pollen sequence and stratigraphic units recognized in Fig. 2 (modified from Doyle and Hickey, 1976). Dashed lines indicate range extensions inferred from other geologic sections (e.g., Kansas Dakota Group, Portugal, Kazakhstan). Pollen types indicated are as in Fig. 2. Leaf types: *g*, narrowly obovate, monocotyledonoid, with apically fusing secondary veins (*Acaciaephyllum*); *h*, pinnately veined, reniform (*Proteaephyllum reniforme*); *i*, pinnately veined, serrate (*Quercophyllum*); *j*, pinnately veined, narrowly obovate (*Rogersia*); *k*, pinnately veined, broadly elliptical (*Ficophyllum*); *l*, palmately veined, reniform, lobate (*Vitiphyllum*); *m*, pinnately veined, obovate (*Celastrophyllum*); *n*, palmately veined, peltate (*Nelumbites, Menispermites*); *o*, palmately veined, ovate-cordate ("*Populus*" *potomacensis, Populophyllum reniforme*); *p*, pinnately veined, serrate (*Celastrophyllum*); *q*, pinnately lobed (*Sapindopsis variabilis*); *r*, palmately lobed, with irregular tertiary venation (*Araliaephyllum*); *s*, pinnately compound, with tendency for regular tertiary venation (*Sapindopsis* sp.?); *t*, palmately lobed, with regular tertiary venation ("platanoids": *Araliopsoides*, "*Sassafras*", etc.).

terns of secondary and higher-order veins, and incomplete differentiation of blade and petiole, a syndrome of characters originally postulated to be primitive in dicots on the basis of comparative studies of Recent forms (Hickey, 1971). In the extant flora, first rank leaves are characteristic of (though not restricted to) certain members of the Magnoliidae, notably the Winteraceae (cf. Wolfe, 1972a) and Canellaceae, but as in the case of the pollen record, some Zone I leaf types have no close analogues in any extant groups: e.g., the reniforms (Fig. 6, *h*), the lobate *Vitiphyllum* (Fig. 6, *l*), and *Eucalyptophyllum* (discussed in Wolfe et al., 1975). Ironically, although most recent authors have tended to claim that early angiosperm leaves "do not have any obviously primitive features" (Axelrod, 1970, p. 289), many of the peculiar features of Zone I leaves were recognized and termed "archaic" by their original describers (Ward, 1888, pp. 129—130; Fontaine, 1889, pp. 291, 347). The one Zone I leaf type which departs most radically from the others, *Acaciaephyllum* (Fig. 6, *g*), is of special interest because its upward-arching, apically fusing secondary venation and chevron-like cross-veins are more characteristic of monocots than of dicots today, suggesting again that the basic split of angiosperms into monocots and dicots had already occurred by earliest Potomac times (Barremian—Aptian? cf. Doyle, 1973; Doyle and Hickey, 1976), contrary to Hughes (1976) but in agreement with Samylina (1960, 1968). Finally, it should be noted that Zone I angiosperm leaves are even more sporadically distributed than angiosperm pollen and show a characteristic facies restriction: they are sometimes present in coarser-grained sediments interpreted as high-energy point-bar and levee deposits, but they are absent in fine-grained back-swamp and flood-basin deposits rich in fern and gymnosperm remains (Doyle and Hickey, 1976; cf. below).

By middle Subzone II-B (late middle Albian?), the next Potomac Group horizon with megafossil localities, angiosperm leaves, like angiosperm pollen, are locally abundant and include several important new morphological complexes (Doyle and Hickey, 1972, 1976; Wolfe et al., 1975). One new complex consists of a series of forms which show some relations to the reniforms of Zone I (Fig. 6, *h*) but which differ in their cordate (Fig. 6, *o*) to peltate shape (Fig. 6, *n*, with the petiole attached to one surface of the blade) and their truly palmate venation. Because of certain leaf architectural features (dichotomy and looping of the primary veins well in from the margin; long, thin petioles, etc.) and sedimentary relationships (occurrence in apparently *in situ* clumps in parallel-laminated pond facies), the cordates and especially the peltates have been interpreted as floating- or emergent-leafed aquatics (Doyle and Hickey, 1972, 1976; cf. Samylina, 1968). Another new complex, sometimes found in great abundance but usually in different facies from the cordate-peltates, consists of nearly compound, pinnately lobed leaves (*Sapindopsis*: Fig. 6, *q*). In upper Subzone II-B (late Albian?), these are largely replaced by leaves which are apparently related on cuticle structure, fine venation, and other features (Doyle and Hickey, 1972, 1976; Mersky, 1973), but which are truly pinnately compound and show a tendency for more regular, rigid orientation of the tertiary veins perpendicular to the leaflet midrib ("third rank" venation of Hickey, 1971: Fig. 6, *s*). Finally, large, palmately lobed "platanoid" leaves also appear in middle Subzone II-B, and in Subzone II-B and especially Subzone II-C (latest Albian?) they too show a trend to more rigid, third-rank venation, but in this case with the tertiary veins perpendicular to the secondaries rather than the midrib (Fig. 6, *r*, *t*). Most of the Subzone II-C flora in fact consists of platanoids, but this is probably a reflection of the predominance in the "Maryland Raritan" of the fluvial sand facies with which they are typically associated (cf. Doyle and Hickey, 1976).

Few angiosperm leaves of Zone I age (Barremian—early Albian?) have been reported from other parts of the world. Exceptions include a crenulate-margined reniform leaf from the probable Aptian of Portugal (Teixeira, 1948), and, especially significant because of its possible pre-Barremian age, a very small, pinnately veined, elliptical leaf from Transbaikalian Siberia (Vachrameev, 1973). On the other hand, angiosperm leaf floras comparable to those of Sub-zone II-B are well known from sediments of middle and late Albian age from such diverse parts of Laurasia as Kansas (Cheyenne Sandstone: Berry, 1922), the Black Hills (Ward, 1899), western Canada (Bell, 1956), Portugal (Buarcos, etc.: Teixeira, 1948), Kazakhstan (Vachrameev, 1952), and eastern Siberia (Samylina, 1960, 1968; Krassilov, 1967). Platanoids of the sort seen in Sub-zone II-C and a variety of other angiosperm leaf types are abundant in latest Albian and Cenomanian rocks of Kansas (Dakota Group: Lesquereux, 1892), Kazakhstan (Vachrameev, 1952), and other areas (see Doyle and Hickey, 1976).

In summary, re-examination of the mid-Cretaceous angiosperm leaf record reveals morphological and stratigraphic patterns analogous to those seen in the pollen record and consistent with the concept of a Cretaceous adaptive radiation. Several rare angiosperm leaf types, including forms with both monocot- and dicot-like venation patterns, are present at the base of the Potomac Group (Barremian—Aptian?), but contrary to previous interpretations they include "experimental" extinct types and show a limited range of morphological variation on themes which are conspicuously primitive when compared with the Recent flora. Higher in the Potomac section (middle—late Albian?), angiosperm leaves become more abundant and morphologically diverse. Divergent trends can be inferred from simple leaves with pinnate venation to cordate and peltate leaves with palmate venation in one direction, and to pinnately and palmately lobed and eventually pinnately compound leaves in another. Parallel trends for regularization and more rigid orientation of both lower and higher vein orders can be inferred especially in the lobate and compound groups. It is true that initial examination suggests that upper Potomac (late Albian?) angiosperms cover a greater part of the morphological spectrum of extant types than does the pollen (cf. Axelrod, 1970; Vachrameev, 1973). However, most angiosperm leaves (except the platanoids and some of the compound forms) have still not attained the third-rank venation characteristic of most "average" Recent dicots (cf. Hickey, 1971), and none of the leaf types present are more advanced (and many are more primitive) than leaves found today in some groups which retain pollen of the evolutionary grade represented at the same horizons (cf. Doyle and Hickey, 1976, pp. 181—183; Wolfe et al., 1975). Hence there is no need to postulate special complicating factors such as lagging of pollen evolution behind systematic diversification, non-preservation of early angiosperm pollen, etc. in order to reconcile the pollen and leaf records.

Adaptive Aspects of the Early Angiosperm Record

Today, angiosperms have replaced older groups in most land plant adaptive zones, with notable exceptions such as the coniferous trees which dominate boreal and equable, wet temperate forests (e.g., northwestern U.S.A., New Zealand) and the ferns which occupy a variety of subordinate forest niches, and they even defy characterization as land plants by having secondarily invaded fresh-water and even marine environments. The patterns seen in the fossil record as summarized above suggest that occupation of this vast array of adaptive zones involved a process of adaptive radiation analogous to that seen in other groups. However, as Muller (1970, p. 435) points out, it is necessary to understand the adaptive significance of these patterns in order to relate them to specific models of the adaptive evolution of the angiosperms and to general theoretical questions such as the relative role of biological innovations, external environmental changes, and biotic interactions in the origin, diversification, and success of major groups (cf. Simpson, 1953, chapter 11).

In the following discussion (drawing heavily on Doyle and Hickey, 1976), I will approach these problems by analyzing the functional morphology and facies distribution of mid-Cretaceous pollen and leaves for evidence on the adaptations and ecological evolution of early angiosperms. Since the factors determining the distribution of plant fossils in sediments and the functional significance of most aspects of pollen and leaf morphology are still imperfectly understood, it would be premature to attempt to construct a theory of adaptive evolution of the angiosperms on fossil evidence alone, or to hope to make more than a few heuristic observations on theoretical questions related to the origin of major groups. Hence I will concentrate on the bearing of the mid-Cretaceous record on the relative merits of two contrasting, largely neobotanical models of the ecological evolution of the angiosperms.

The first is Bews' hypothesis (Bews, 1927; cf. also Axelrod, 1952; Takhtajan, 1969) that the angiosperms originated as trees of wet tropical forest environments, and subsequently radiated into cooler and drier areas and underwent reduction to shrubby, lianous, and herbaceous growth habits. As mentioned above, this concept is based largely on the great diversity of angiosperms in tropical forests, correlation of presumed levels of advancement in morphological features (e.g., floral structure) with ecology, and the predominantly wet tropical to warm temperate aspect of Late Cretaceous and Early Tertiary fossil floras. The second model (Stebbins, 1965, 1974; cf. also Axelrod, 1970; and above) postulates that the first angiosperms were weedy, fast-growing, pioneering shrubs or subshrubs of disturbed habitats in seasonally dry tropical or subtropical climates, and subsequently radiated into wetter, drier, and cooler environments and underwent divergence toward both tree and herbaceous habits. Stebbins bases this concept largely on arguments that the varied selective factors and seasonality of such environments would be more likely to favour adaptive diversification in general and origin of the syndrome of rapid reproduction and vegetative flexibility which unites angiosperms and separates them from gymnosperms in particular than would selective factors operating in wet, nonseasonal tropical forests (cf. above). According to Stebbins (1974), the diversity of primitive (and advanced!) types in wet tropical forests reflects the fact that such environments have served as "museums" where divergent early lines were preserved because of lower extinction rates than in the geologically less stable semi-arid areas where they originated.

One more general question to be considered, not necessarily related to the contrast between the two models above, is whether particular defining features of modern angiosperms arose before or during the observed mid-Cretaceous radiation. Simpson (1953, chapter 11) notes that in some cases, especially where groups are united by rather specific adaptive characters (e.g., bats with wings, rodents with persistently growing incisors), the fossil record suggests that these characters arose before the group became distinct and may in fact be interpreted as key innovations which enabled it to enter a new adaptive zone. Bock (1965) emphasizes that such innovations are the result of a step-wise process of acquisition of what are seen in retrospect as pre-adaptations to the new zone under the different selective pressures of a transitional adaptive zone; they may provide hints on the conditions under which a group originated. In other cases, particularly where groups are united by rather broad adaptive characters (e.g., carnivores and ungulates), Simpson (1953) shows that their distinctive features arose in the course of their radiation (in this specific example, co-evolutionary predator—prey interactions are clearly involved). Comparative evidence indicates that some of the most striking functional advances of the angiosperms also arose during their radiation — e.g., vessels in the wood, as inferred from the persistence of extant vesselless forms such as Winteraceae (cf. Bailey, 1944; Takhtajan, 1969) — but there is controversy over whether or not this is true of their unique reproductive features (e.g., flowers, double fertilization, closed carpels: cf. above).

In the near-absence of direct evidence from fossil flowers, interpretations of functional links between morphological features of fossil pollen and aspects of floral and pollination biology perhaps offer the most promise in determining when reproductive innovations of the angiosperms arose. At least two functions, not necessarily mutually exclusive, have been proposed for reticulate—columellar exine structure of the sort already seen in Barremian angiospermous monosulcates (e.g., Plate I, 1—6; cf. Doyle and Hickey, 1976; Hughes, 1976). First, the surface irregularities of a reticulate exine enable pollen grains to stick better together and to other objects and are hence advantageous in insect vs. wind pollination (indeed, exine sculpture belongs to the syndrome of characters associated with insect pollination by Faegri and Van der Pijl, 1966; cf. Heslop-Harrison, 1971). Second, Heslop-Harrison et al. (1973) have shown that the intercolumellar spaces and tectal perforations in living angiosperm pollen grains are the storage sites of recognition proteins involved in controlling germination on the appropriate stigma.

Although this leaves ambiguity as to the primary factor, it may be significant than both functions are closely tied to basic angiosperm reproductive innovations: insect pollination has been postulated as the principal selective factor in origin of the flower and the closed carpel (Grant, 1950; Takhtajan, 1969), while germination on a stigma is functionally almost synonymous with carpel closure, i.e., with angiospermy itself (cf. above). The appearance of the tricolpate condition is more unambiguous evidence that closed carpels had evolved by Aptian times (if not well before): whereas in gymnosperms the pollen grains usually germinate in the moist pollen chamber or micropyle of the ovule where there is no special advantage for more than one aperture, pollen grains with several apertures for pollen tube germination should have a definite advantage over those with only one on the exposed surface of a stigma, where a mono-sulcate grain would often land with its single aperture pointing out into the air (cf. Hughes, 1961, 1976; Walker, 1971; Doyle and Hickey, 1976). Finally, fructifications with closed carpels are well known from the Albian and Cenomanian (Samylina, 1960; 1968; Dilcher et al., 1976).

These considerations support the concept that the angiospermous syndrome of enclosure and reproductive efficiency (cf. Stebbins, 1974; and above) was already present at or near the beginning of the observed radiation of angiosperms and may have been a key factor allowing them to enter their present adaptive zone, contrary to Hughes' suggestion (1976) that angiospermous features arose in many separate mid-Cretaceous lines under similar selective pressures, particularly progressive ecological integration with the animal world and gradually warming climates. [It should be noted in this context that Hughes' concept of rising temperatures culminating in a "Radmax" at the end of the Cretaceous and followed by gradual cooling throughout the Tertiary is contradicted by palaeobotanical and oxygen isotope evidence for important temperature fluctuations in both periods: cf. Bowen (1961), Lowenstam (1964), Wolfe (1971), Savin et al. (1975).] Co-evolutionary and climatic factors may of course have been involved in origin of angiosperm features in their hypothetical transitional adaptive zone — Takhtajan (1969) emphasizes the role of insect pollination in origin of the flower and closed carpels, Stebbins (1974) the role of seasonal drought in favouring the various contractions in the angiosperm life cycle — and adaptive radiation on pollination ecological themes is undoubtedly reflected (though in poorly understood ways) in the continuing diversification in size, shape, exine structure, and aperture conditions seen in the Albian—Cenomanian pollen record (cf. Muller, 1970).

Whereas morphological trends in the pollen record should have the most bearing on reproductive adaptations of early angiosperms, analysis of the facies distribution and functional morphology of their leaves should permit inferences on their habitat preferences and vegetative adaptations. Interestingly, the diverse trends seen in the leaf record seem generally more easily reconciled with Stebbins' concept (1965, 1974) that the ancestral angiosperms were semixeric, weedy shrubs than with the more conventional postulate (e.g., Bews, 1927; Takhtajan, 1969) that they were mesic trees (adapted to average moisture regimes) comparable to extant Magnoliales (cf. Doyle and Hickey, 1976). As already noted above, palynological evidence (Brenner, 1976) for more active diversification of early angiosperms in Africa—South America, associated with indications of aridity, is also consistent with Stebbins' view, but since the first phases of the angiosperm pollen record are not demonstrably older in Africa—South America than in Laurasia (cf. above), and since the oldest angiosperm leaves so far discovered are from Siberia (Vachrameev, 1973), this evidence cannot be considered conclusive.

Upon entering an area of mesic climates such as Southern Laurasia (cf.

Brenner, 1976; and above), one might expect that Stebbins' opportunistic, semixerophytic shrubs would be preadapted as "weeds" to disturbed, unstable sites, where their rapid reproduction and vegetative growth would give them an advantage over gymnosperms (with more sluggish growth and reproduction) and ferns (with special requirements for gametophyte growth). This prediction agrees well with the observed restriction of the earliest Potomac angiosperm leaves to coarser point-bar and stream-levee facies (stream margins are among the principal disturbed habitats in a fluvial-deltaic environment) and their absence from fern- and gymnosperm-dominated back-swamp deposits (cf. Doyle and Hickey, 1976; and above). Likewise, Vachrameev (1952) has called attention to the small size of Albian angiosperm leaves of Kazakhstan as possible evidence of a xeromorphic ancestry; however, this size effect is less striking in the Potomac Group. Finally, the evidence from both pollen and leaf records of monocot-like forms from the beginning of the Potomac record (Plate I, 6; Fig. 6, g), and the early trend toward apparent aquatic adaptations, culminating in the peltates of Subzone II-B pond facies (Fig. 6, n), are easier to explain by assuming a weedy, semiherbaceous ancestral condition than an arborescent one. Stream margin plants in particular would often be subjected to selective pressures for aquatic adaptations, and the aquatic adaptive zone would have been relatively unoccupied by other groups of vascular plants (except perhaps reedy *Equisetites*, Isoetales, and early heterosporous ferns: cf. Doyle and Hickey, 1976).

Other aspects of the mid-Cretaceous leaf record are consistent with the concept of an initially shrubby habit and subsequent increase in stature, resulting in trees capable of competing directly with the conifers only near the beginning of the Late Cretaceous. One forest ecological niche which shrubby early angiosperms might be able to occupy with relatively few changes, and one where their ability to produce broad, shade-tolerant leaves might put them at an advantage over most gymnosperms, is the shaded forest understory. Possible representatives of such a trend are the largest Zone I angiosperm leaves (*Ficophyllum*: Fig. 6, k), whose broad, elliptical shape is typical of the "monolayer" light-gathering strategy which Horn (1971) argues is optimal for intercepting low-intensity light. This interpretation is consistent with their subordinate role relative to fern and conifer remains, and with their loose first-rank venation and poor petiole differentiation, features poorly adapted to the mechanical stresses of more exposed habitats; in fact, extant forms with large first-rank leaves (e.g., Winteraceae) are largely restricted to understory habitats (cf. Takhtajan, 1969; Carlquist, 1975; Doyle and Hickey, 1976). In contrast, the younger pinnately lobed and compound leaves of Subzone II-B (*Sapindopsis*: Fig. 6, q, s) are locally dominant and show a trend toward more rigid, third-rank venation, together suggesting dense stands growing in exposed conditions (cf. Doyle and Hickey, 1976).

Interestingly, two different ecological arguments have been presented which conclude that deeply lobed and compound leaves are more advantageous for fast-growing early successional trees than for climax forest trees: Horn (1971) argues that leaf dissection, by allowing lower leaves to be stacked in the partial shade of higher leaves in a "multilayer" configuration, allows early successional trees of exposed situations to make most effective use of high-intensity light, while Givnish (1977) points out that a compound leaf with a "throw-away" rachis represents a more efficient use of energy for rapid upward growth in such conditions than a permanent branch with small leaves. The first mid-Cretaceous leaves with a morphology typical of modern broad-leafed forest trees are perhaps the palmately lobed, third-rank "platanoids" of upper Subzone II-B and Subzone II-C, the Dakota of Kansas, etc. (late Albian—early Cenomanian?, Fig. 6, *t*), but even these are most characteristic of sandy facies, suggesting riparian habitats (like modern *Platanus*!).

These data all favour the concept that angiosperms first infiltrated Mesophytic communities as opportunistic, weedy plants capable of occupying poorly exploited, disturbed habitats, and then radiated into also underexploited aquatic, forest understory, and fast-growing early successional tree adaptive zones, rather than directly replacing the dominant conifers and ferns by "frontal assaults by entire communities" (Takhtajan, 1969, p. 134; as Takhtajan recognizes, this would require extensive prior diversification and invasion). In fact, palynological evidence suggests that conifers were still the regional vegetational dominants until into the early Late Cretaceous (Zone III or IV?), as Pierce (1961) noted in his study of pollen from the Cenomanian Dakota Formation of Minnesota (cf. also Muller, 1970; Doyle and Hickey, 1976; this incidentally explains some of the differences in the proportions of angiosperms in contemporaneous pollen and leaf floras cited by Axelrod, 1970, p. 287). Pierce also argued that the localized, shifting, early successional habitats of lowland fluvial-deltaic environments would tend to favour fragmented population structures and rapid evolutionary diversification, in much the same way as the upland and semi-arid environments postulated by Axelrod (1952, 1960, 1970), Takhtajan (1969), and Stebbins (1974). Interestingly, it is not until Zone IV (middle to late Cenomanian?) that we see combinations of pollen morphological features (the reduced sculpture and porate apertures of the Normapolles complex: cf. Doyle and Hickey, 1976) suggestive of wind pollination, an adaptive strategy most advantageous in dense stands of usually deciduous trees (cf. Whitehead, 1969). It may be significant that conifers remain dominant today as trees in equable temperate rainforests most unlike the environments which this model postulates for angiosperm origins (cf. Stebbins, 1974), and that it is primarily in temperate areas that angiosperms have sacrificed one of their most distinctive reproductive specializations by reverting to wind pollination (cf. Doyle and Hickey, 1976).

The concept that angiosperm evolution began with opportunistic, pioneer species and subsequently "ran up" ecological succession by occupation of more specialized niches of more mature communities agrees well with the prediction of Margalef (1968) that evolutionary trends should tend to parallel trends in succession. Since Margalef also suggests that neoteny may be involved in permitting groups to "drop back" to earlier successional stages, it is also intriguing that Takhtajan (1969, 1976) argues that many of the basic peculiarities of angiosperms, such as the numerous contractions of the life cycle mentioned above, the closed carpel, and the simple, first-rank leaves of Early Cretaceous angiosperms (cf. Doyle and Hickey, 1976), can be partially explained as juvenile characters retained by neoteny.

Generalizing further from this model, I would speculate that the factor triggering the radiation of the angiosperms was not any large-scale external climatic or tectonic change producing new environments, but rather acquisition of key biological innovations (rapid, efficient reproduction; insect pollination and capacity for symbiotic interaction with the animal world; rapid and flexible vegetative growth and capacity for producing broad, reticulate-veined leaves), perhaps evolved under the selective pressures of small-scale environmental fluctuations of a semi-arid climate, which preadapted the angiosperms to exploit existing environments more effectively than other plant groups (cf. Von Wahlert, 1965). In this respect, the angiosperm radiation might be considered more analogous to the initial Silurian—Devonian invasion of the land, an environment which existed earlier but was unoccupied because plants had not yet evolved the necessary preadaptations (e.g., desiccation-resistant spores, cuticles, three-dimensional tissue structure, etc., themselves explainable as adaptations to a fluctuating fresh-water environment: cf. Jeffrey, 1962), than to the Permian radiation of conifers and other xeromorphic gymnosperms, associated with geological evidence for a shift from coal swamp to semi-arid red bed conditions (cf. Frederiksen, 1972; and above). [Note, however, that some authors postulate that earlier invasion of the land was prevented by lack of an effective ozone screen against UV radiation: Banks (1970), Smart and Hughes (1972).]

This is not to say that large-scale climatic, geographic, and biotic changes had no effect on rates and patterns of early angiosperm evolution. For example, regional aridization of the sort which Jardiné et al. (1974) argue was occurring in the Early Cretaceous of Africa—South America may have increased the areas with climates favourable for origin of angiosperm innovations (cf. above), while the increased environmental diversity, marine transgressions, fluvial-deltaic sedimentation, and climatic amelioration associated with the Albian breakup of Gondwana (cf. Axelrod, 1970; Raven and Axelrod, 1974), and overgrazing by dinosaurs (Bakker, Ch. 14) may have stimulated the radiation of angiosperms into other areas. Later in the Cretaceous, the spread of north—south trending epicontinental seas was clearly responsible for increased floral provincialism and concomitant multiplication of the number of angiosperm taxa (Góczán et al., 1967; Tschudy, 1970; Muller, 1970; Wolfe and Pakiser, 1971). Likewise, it may be suggested that the temporary leveling-off of angiosperm pollen "species" diversity in Subzone II-C of the Potomac Group sequence (Fig. 4) and the expansion of lobate, third rank "platanoid" leaves (deciduous? cf. Doyle and Hickey, 1976, p. 175) at corresponding horizons throughout Laurasia are both effects of an Albian—Cenomanian climatic deterioration (cf. above). During the Tertiary, climatic changes had clearer effects on patterns of angiosperm evolution, the most striking being the spread of grasslands, deserts, and deciduous forests and the radiation of grasses, Compositae, Acanthaceae, and other herbaceous groups adapted to the cooler, drier, and more seasonal climates which followed the Oligocene deterioration (Germeraad et al., 1968; Muller, 1970;

Wolfe, 1971, 1972b; Leopold and MacGinitie, 1972).

Whatever its causes, the radiation of the angiosperms and the resulting changes in vegetational dominants, community structure, soil chemistry, etc., must have had major repercussions on the rest of the terrestrial biota, the clearest of these being on the insect world (Rodendorf and Zherikhin, 1974). The important changes in fossil insect faunas between the Jurassic and the Early Tertiary (appearance of termites, Lepidoptera, aculeate Hymenoptera, higher Diptera) have long been considered related in some way to the rise of the angiosperms (cf. Carpenter, 1953; Smart and Hughes, 1972; Hughes, 1976), but because of the rarity of Cretaceous fossil insects there was previously little concrete palaeontological basis for more detailed speculation. However, recent studies of new insect faunas from the Cretaceous of the U.S.S.R. (Zherikhin and Sukacheva, 1973; Rodendorf and Zherikhin, 1974) show that changes in the insect fauna did coincide closely with the rise of angiosperms: extinction of nearly a third of then-existing insect families occurred within the earlier stages of the Late Cretaceous, accompanied by a more gradual rise of new groups, some though not all of them now ecologically associated with angiosperms (e.g., Lepidoptera, aculeate Hymenoptera). By comparison, changes in the insect fauna at the Cretaceous—Tertiary transition were minor. Interestingly, in order to explain why the insect extinctions occurred even in more northern regions where the angiosperms had not yet reached dominance (Zherikhin and Sukacheva, 1973), V.V. Zherikhin (personal communication, 1975) speculates that the entry of angiosperms might have had amplified ecological effects if they were initially early successional forms, by unbalancing the course of later stages in ecological succession and causing changes in the composition of the gymnosperm-dominated climax forests. In fact, mid-Cretaceous sequences show gradual changes in the conifer element from a "Jurassic" to a "Late Cretaceous" aspect (exemplified by the decline of *Classopollis* and *Exesipollenites* and the rise of *Rugubivesiculites*, *Phyllocladidites*, *Araucariacites*, and "modern" Pinaceae and Taxodiaceae in Subzones II-B and II-C of the Potomac sequence: cf. Brenner, 1963; Doyle and Robbins, 1977) which might be interpreted as evidence of such an effect. However, more data on the timing of these changes and their relation to diversity changes in the fern and angiosperm elements are needed to decide whether they represent biotic or climatic effects, as discussed above.

Summary and Conclusions

Although earlier attempts at systematic identification of Cretaceous angiosperm megafossils suggested that the angiosperms had originated and diversified long before their first general appearance in the mid-Early Cretaceous, recent critical reappraisal of supposed pre-Cretaceous angiosperm remains and more detailed examination of the mid-Cretaceous pollen and leaf records from a stratigraphic and morphological point of view have revealed patterns of taxonomic and morphological diversification comparable with those seen in adaptive radiations of other groups. The oldest known angiosperm pollen and leaves of the Barremian—Aptian already show some morphological diversity, including both monocot- and dicot-like types, but they occupy only the most primitive portion of the morphological spectrum of Recent angiosperms and include "experimental" types with no analogues in the living flora. Rates of taxonomic origination and increase in morphological advancement are high from the Barremian to the middle Albian, with highest rates in the early—middle Albian, interpreted as the phase of most rapid adaptive radiation. These rates begin to decline in the late Albian and Cenomanian, suggesting a transition to intrazonal evolution as the most advanced lines reached the evolutionary level of more primitive members of some intermediate modern orders, but there are numerous secondary radiations of major angiosperm subgroups in the later Cretaceous and Tertiary. There is evidence of more active diversification and earlier origin of some types (Aptian tricolpates) in dry tropical regions of Africa—South America, and some lag in invasion of higher latitudes, but in general the initial angiosperm radiation seems to have been essentially world-wide, relatively uncomplicated by large-scale migrational effects. While the exact time of origin and ancestors of the angiosperms are still unknown, the pattern of regular increase in morphological diversity and advancement from the oldest known monosulcate types greatly limits the amount of pre-Cretaceous diversification or polyphyletic origin which can be seriously considered.

Analysis of the functional morphology and facies distribution of Cretaceous pollen and leaves appears consistent with a model of adaptive evolution in which many of the distinctive reproductive and vegetative features of the angiosperms (closed carpels, insect pollination, reticulate-veined leaves) had originated by their appearance in the Barremian—Aptian, perhaps under selective pressures for rapid reproduction and growth in a seasonally arid climate (cf. Stebbins, 1974). They entered wetter subtropical areas first as "weeds" capable of occupying disturbed habitats (e.g., stream margins) more effectively than the dominant gymnosperms and ferns, then radiated into also poorly exploited aquatic, forest understory, and early successional tree niches, and finally evolved into tree forms capable of competing directly with conifers as climax forest dominants only in the early Late Cretaceous. This suggests that the factor which triggered the radiation of the angiosperms was evolution of their key biological innovations (efficient reproduction, capacity for symbiotic interactions with the animal world, vegetative flexibility), rather than large-scale changes in the external physical or biotic environment. However, climatic and geographic factors had important influences on the pattern of angiosperm evolution: e.g., Late Cretaceous floral provincialism and multiplication of taxa related to the spread of epicontinental seas, and the expansion of grasslands, herbaceous groups, and deciduous forests related to the Oligocene climatic deterioration.

These considerations should help dispel the belief that angiosperm evolution was any more rapid or mysterious than that of other groups of organisms, or that the fossil record is useless as a source of evidence on angiosperm phylogeny (cf. Hughes, 1976). The interval from appearance (Barremian) through most rapid adaptive diversification (early—middle Albian) to dominance (Cenomanian—Turonian) — perhaps 25—30 m.y. — is shorter than the comparable interval in some groups (e.g., lungfishes: Westoll, 1949), but comparable with the initial Late Silurian to early Middle Devonian phase of land plant evolution (from the first delicate psilophytes to the first progymnosperm trees) or the Paleocene—Eocene radiation of mammals. While I would not accept the extreme viewpoint of Hughes (1976) that angiosperm phylogeny can and should now be approached on the basis of purely palaeontological data, further studies of the Cretaceous and Tertiary record emphasizing more refined analysis of the morphology and stratigraphic, facies, and geographic distribution of all available fossil remains, coordinated with but not prejudiced by comparative, functional morphological, and depositional studies of living plants, should continue to yield novel insights on patterns of angiosperm evolution which could not have been derived from neontological evidence alone. Because of the central role of angiosperms in the structure and dynamics of modern terrestrial ecosystems, such studies should have important implications for discussions of the evolution of other members of the terrestrial biota, not only insects, but also fungi, mammals, and birds.

Acknowledgements

I am indebted to Philip D. Gingerich, Jan Muller, G. Ledyard Stebbins, Andrew G. Stephenson, Louis Thaler, James W. Walker, Vladimir V. Zherikhin, and especially Leo J. Hickey for discussions which stimulated development of many of the ideas and arguments expressed in this paper. I wish also to thank M.C. Rudy and Garland R. Upchurch for editorial assistance, Gladys Newton for typing, and Carlton E. Brett for help in preparing the figures.

References

Axelrod, D.I., 1952. A theory of angiosperm evolution. Evolution, 6: 29—60.
Axelrod, D.I., 1959. Poleward migration of early angiosperm flora. Science, 130: 203—207.
Axelrod, D.I., 1960. The evolution of flowering plants. In: S. Tax (Editor), The Evolution of Life. University of Chicago Press, Chicago, Ill., pp. 227—305.
Axelrod, D.I., 1970. Mesozoic paleogeography and early angiosperm history. Bot. Rev., 36(3): 277—319.
Bailey, I.W., 1944. The development of vessels in angiosperms and its significance in morphological research. Am. J. Bot., 31(7): 421—428.
Banks, H.P., 1968. The early history of land plants. In: E.T. Drake (Editor), Evolution and Environment. Yale University Press, New Haven, Conn., pp. 73—107.
Banks, H.P., 1970. Evolution and Plants of the Past. Wadsworth Publishing Company, Belmont, Calif., 170 pp.
Beck, C.B., 1960. The identity of *Archaeopteris* and *Callixylon*. Brittonia, 12: 351—368.
Beck, C.B., 1970. The appearance of gymnospermous structure. Biol. Rev., Cambridge Philos. Soc., 45(3): 379—400.
Bell, W.A., 1956. Lower Cretaceous floras of Western Canada. Geol. Surv. Can. Mem., 285: 1—331.
Berry, E.W., 1911. Systematic paleontology, Lower Cretaceous (Pteridophyta—Dicotyledonae). In: W.B. Clark (Editor), Lower Cretaceous. Maryland Geological Survey, Johns Hopkins Press, Baltimore, Md., pp. 214—508.
Berry, E.W., 1922. The flora of the Cheyenne Sandstone of Kansas. U.S. Geol. Surv. Prof. Pap., 127I: 199—225.

Bews, J.W., 1927. Studies in the ecological evolution of angiosperms. New Phytol., 26: 1—21, 65—84, 129—148, 209—248, 273—294.

Bock, W.J., 1965. The role of adaptive mechanisms in the origin of higher levels of organization. Syst. Zool., 14(4): 272—287.

Boltenhagen, E., 1965. Introduction à la palynologie stratigraphique du bassin sédimentaire de l'Afrique équatoriale. Mém. Bur. Rech. Géol. Min., 32: 305—326.

Bowen, R., 1961. Paleotemperature analyses of Mesozoic Belemnoidea from Germany and Poland. J. Geol., 69: 75—83.

Brenner, G.J., 1963. The spores and pollen of the Potomac Group of Maryland. Md. Dep. Geol., Mines Water Resour. Bull., 27: 1—215.

Brenner, G.J., 1967. Early angiosperm pollen differentiation in the Albian to Cenomanian deposits of Delaware (U.S.A.). Rev. Palaeobot. Palynol., 1: 219—227.

Brenner, G.J., 1968. Middle Cretaceous spores and pollen from northeastern Peru. Pollen Spores, 10(2): 341—383.

Brenner, G.J., 1976. Middle Cretaceous floral provinces and early migrations of angiosperms. In: C.B. Beck (Editor), Origin and Early Evolution of Angiosperms. Columbia University Press, New York, N.Y., pp. 23—47.

Burger, D., 1970. Early Cretaceous angiospermous pollen grains from Queensland. Bur. Min. Resour., Geol. Geophys., Canberra, Bull., 116: 1—10.

Burger, D., 1973. Palynological observations in the Carpentaria Basin, Queensland. Bur. Min. Resour., Geol. Geophys., Canberra, Bull., 140: 27—44.

Carlquist, S., 1975. Ecological Strategies of Xylem Evolution. University of California Press, Berkeley, Los Angeles, Calif., 259 pp.

Carpenter, F.M., 1953. The geological history and evolution of insects. Am. Sci., 41(2): 256—270.

Casey, R., 1964. The Cretaceous Period. In: W.B. Harland (Editor), The Phanerozoic Time-Scale. Q. J. Geol. Soc. Lond., 120s: 193—202.

Chaloner, W.G., 1967. Spores and land-plant evolution. Rev. Palaeobot. Palynol., 1: 83—93.

Chaloner, W.G., 1970. The rise of the first land plants. Biol. Rev., Cambridge Philos. Soc., 45(3): 353—377.

Chaloner, W.G. and Lacey, W.S., 1973. The distribution of Late Palaeozoic floras. In: N.F. Hughes (Editor), Organisms and Continents through Time. Palaeontol. Assoc. Lond., Spec. Pap. Palaeontol., 12: 271—289.

Chesters, K.I.M., Gnauck, F.R. and Hughes, N.F., 1967. Angiospermae. In: W.B. Harland et al. (Editors), The Fossil Record. Geological Society, London, pp. 269—288.

Couper, R.A., 1958. British Mesozoic microspores and pollen grains. Palaeontographica, Abt. B, 103: 75—179.

Cronquist, A., 1968. The Evolution and Classification of Flowering Plants. Houghton Mifflin Co., Boston, Mass., 396 pp.

Davis, P.H. and Heywood, V.H., 1963. Principles of Angiosperm Taxonomy. Oliver and Boyd, Edinburgh, 556 pp.

Dettmann, M.E., 1973. Angiospermous pollen from Albian to Turonian sediments of eastern Australia. Geol. Soc. Aust., Spec. Publ., 4: 3—34.

Dilcher, D.L., 1974. Approaches to the identification of angiosperm leaf remains. Bot. Rev., 4: 1—157.

Dilcher, D.L., Crepet, W.L., Beeker, C.D. and Reynolds, H.C., 1976. Reproductive and vegetative morphology of a Cretaceous angiosperm. Science, 191: 854—856.

Doyle, J.A., 1969. Cretaceous angiosperm pollen of the Atlantic Coastal Plain and its evolutionary significance. J. Arnold Arbor., 50: 1—35.

Doyle, J.A., 1973. Fossil evidence on early evolution of the monocotyledons. Q. Rev. Biol., 48(3): 399—413.

Doyle, J.A., 1976. Spores and pollen: The Potomac Group (Cretaceous) angiosperm sequence. In: J.E. Hazel and E.G. Kauffman (Editors), Concepts and Methods in Biostratigraphy. Dowden, Hutchinson and Ross, Stroudsburg, Penn.

Doyle, J.A. and Hickey, L.J., 1972. Coordinated evolution in Potomac Group angiosperm pollen and leaves. Am. J. Bot., 59(6, part 2): 660 (abstract).

Doyle, J.A. and Hickey, L.J., 1976. Pollen and leaves from the mid-Cretaceous Potomac Group and their bearing on early angiosperm evolution. In: C.B. Beck (Editor), Origin and Early Evolution of Angiosperms. Columbia University Press, New York, N.Y., pp. 139—206.

Doyle, J.A. and Robbins, E.I., 1977. Angiosperm pollen zonation of the continental Cretaceous of the Atlantic Coastal Plain and its application to deep wells in the Salisbury Embayment. Am. Assoc. Stratigr. Palynol. Proc.

Doyle, J.A., Van Campo, M. and Lugardon, B., 1975. Observations on exine structure of *Eucommiidites* and Lower Cretaceous angiosperm pollen. Pollen Spores, 17(3): 429—486.

Eldredge, N. and Gould, S.J., 1972. Punctuated equilibria: An alternative to phyletic gradualism. In: T.J.M. Schopf (Editor), Models in Paleobiology. Freeman, San Francisco, Calif., pp. 82—115.

Erdtman, G., 1952. Pollen Morphology and Plant Taxonomy. Part I. Angiosperms. Chronica Botanica Co., Waltham, Mass., 539 pp.

Faegri, K. and Van der Pijl, L., 1966. The Principles of Pollination Ecology. Pergamon Press, Oxford, 248 pp.

Florin, R., 1951. Evolution in cordaites and conifers. Acta Horti Bergiani, 15(11): 285—388.

Fontaine, W.M., 1889. The Potomac or Younger Mesozoic flora. U.S. Geol. Surv. Monogr., 15: 1—375.

Frederiksen, N.O., 1972. The rise of the Mesophytic flora. Geosci. Man, 4: 17—28.

Germeraad, J.H., Hopping, C.A. and Muller, J., 1968. Palynology of Tertiary sediments from tropical areas. Rev. Palaeobot. Palynol., 6: 189—348.

Gingerich, P.D., 1974. Stratigraphic record of Early Eocene *Hyopsodus* and the geometry of mammalian phylogeny. Nature, 248: 107—109.

Givnish, T.J., 1977. The adaptive significance of compound leaves, with particular reference to tropical trees. In: P.B. Tomlinson and M.H. Zimmermann (Editors), Tropical Trees as Living Systems. Cambridge University Press, Cambridge.

Góczán, F., Groot, J.J., Krutzsch, W. and Pacltová, B., 1967. Die Gattungen des "Stemma Normapolles Pflug 1953b" (Angiospermae). Paläontol. Abh., Abt. B., 2(3): 429—539.

Grant, V., 1950. The pollination of *Calycanthus occidentalis*. Am. J. Bot., 37(4): 294—297.

Grant, V., 1963. The Origin of Adaptations. Columbia University Press, New York, N.Y., 606 pp.

Hedlund, R.W. and Norris, G., 1968. Spores and pollen grains from Fredericksburgian (Albian) strata, Marshall County, Oklahoma. Pollen Spores, 10: 129—159.

Herngreen, G.F.W., 1973. Palynology of Albian—Cenomanian strata of borehole 1-QS-1-MA, State of Maranhao, Brazil. Pollen Spores, 15(3—4): 515—555.

Herngreen, G.F.W., 1974. Middle Cretaceous palynomorphs from northeastern Brazil. Sci. Géol., Bull., Strasbourg, 27(1—2): 101—116.

Heslop-Harrison, J., 1971. Sporopollenin in the biological context. In: J. Brooks, P.R. Grant, M. Muir, P. van Gijzel and G. Shaw (Editors), Sporopollenin. Academic Press, London, pp. 1—30.

Heslop-Harrison, J., Heslop-Harrison, Y., Knox, R.B. and Howlett, B., 1973. Pollen-wall proteins: "Gametophytic" and "sporophytic" fractions in the pollen walls of the Malvaceae. Ann. Bot., 37: 403—412.

Hickey, L.J., 1971. Evolutionary significance of leaf architectural features in the woody dicots. Am. J. Bot., 58(5, part 2): 469 (abstract).

Hickey, L.J. and Wolfe, J.A., 1975. The bases of angiosperm phylogeny: Vegetative morphology. Ann. Mo. Bot. Garden, 62(3): 538—589.

Horn, H.S., 1971. The Adaptive Geometry of Trees. Princeton University Press, Princeton, N.J., 144 pp.

Hughes, N.F., 1961. Fossil evidence and angiosperm ancestry. Sci. Progr., 49: 84—102.

Hughes, N.F., 1976. Palaeobiology of Angiosperm Origins. Cambridge University Press, Cambridge, 242 pp.

Jardiné, S. and Magloire, L., 1965. Palynologie et stratigraphie du Crétacé des bassins du Sénégal et de Côte d'Ivoire. Mém. Bur. Rech. Géol. Min., 32: 187—245.

Jardiné, S., Doerenkamp, A. and Biens, P., 1974. *Dicheiropollis etruscus*, un pollen caractéristique du Crétacé inférieur afro-sudaméricain. Conséquences pour l'évaluation des unités climatiques et implications dans la dérive des continents. Sci. Géol., Bull., Strasbourg, 27(1—2): 87—100.

Jarzen, D.M. and Norris, G., 1975. Evolutionary significance and botanical relationships of Cretaceous angiosperm pollen in the western Canadian interior. Geosci. Man, 11: 47—60.

Jeffrey, C., 1962. The origin and differentiation of the archegoniate land-plants. Bot. Not., 115(4): 446—454.

Kemp, E.M., 1968. Probable angiosperm pollen from British Barremian to Albian strata. Palaeontology, 11(3): 421—434.

Kemp, E.M., 1970. Aptian and Albian miospores from southern England. Palaeontographica, Abt. B, 131: 73—143.

Krassilov, V.A., 1967. Rannemelovaya Flora Yuzhnogo Primorya i Yeye Znachenie dlya Stratigrafii. Nauka, Moscow, 264 pp.

Krassilov, V.A., 1973. Mesozoic plants and the problem of angiosperm ancestry. Lethaia, 6: 163—178.

Krassilov, V.A., 1975. Dirhopalostachyaceae — a new family of proangiosperms and its bearing on the problem of angiosperm ancestry. Palaeontographica, Abt. B, 153: 100—110.

Kuyl, O.S., Muller, J. and Waterbolk, H.T., 1955. The application of palynology to oil geology, with special reference to western Venezuela. Geol. Mijnbouw, N.S., 17(3): 49—76.

Laing, J.F., 1975. Mid-Cretaceous angiosperm pollen from southern England and northern France. Palaeontology, 18(4): 775—808.

Laing, J.F., 1976. The stratigraphic setting of early angiosperm pollen. In: I.K. Ferguson and J. Muller (Editors), The Evolutionary Significance of the Exine. Linn. Soc. Symp. Ser., 1: 15—26.

Leopold, E.B. and MacGinitie, H.D., 1972. Development and affinities of Tertiary floras in the Rocky Mountains. In: A. Graham (Editor), Floristics and Paleofloristics of Asia and Eastern North America. Elsevier, Amsterdam, pp. 147—200.

Lesquereux, L., 1892. The flora of the Dakota Group. U.S. Geol. Surv. Monogr., 17: 1—400.

Lowenstam, H.A., 1964. Palaeotemperatures of the Permian and Cretaceous periods. In: A.E.M. Nairn (Editor), Problems in Palaeoclimatology. Interscience Publishers, London, pp. 227—252.

Margalef, R., 1968. Perspectives in Ecological Theory. University of Chicago Press, Chicago, Ill., 111 pp.

Médus, J. and Pons, A., 1967. Etude palynologique du Crétacé Pyrénéo-Provençal. Rev. Palaeobot. Palynol., 2: 111—117.

Meeuse, A.D.J., 1965. Angiosperms — Past and Present. Advancing Frontiers of Plant Sciences, New Delhi, 11: 228 pp.

Mersky, M.L., 1973. Lower Cretaceous (Potomac Group) angiosperm cuticles. Am. J. Bot., 60(4, suppl.): 17—18 (abstract).

Müller, H., 1966. Palynological investigations of Cretaceous sediments in northeastern Brazil. In: J.E. van Hinte (Editor), Proceedings of the Second West African Micropaleontological Colloquium (Ibadan, 1965). Brill, Leiden, pp. 123—136.

Muller, J., 1959. Palynology of Recent Orinoco delta and shelf sediments. Micropaleontology, 5: 1—32.

Muller, J., 1970. Palynological evidence on early differentiation of angiosperms. Biol. Rev., Cambridge Philos. Soc., 45(3): 417—450.

Norris, G., 1967. Spores and pollen from the Lower Colorado Group (Albian—?Cenomanian) of central Alberta. Palaeontographica, Abt. B, 120: 72—115.

Pacltová, B., 1961. Zur Frage der Gattung *Eucalyptus* in der böhmischen Kreideformation. Preslia, 33: 113—129.

Pacltová, B., 1971. Palynological study of Angiospermae from the Peruc Formation (?Albian—Lower Cenomanian) of Bohemia. Ústřed. Ústav Geol., Sb. Geol. Věd, Paleontol., Řada P, 13: 105—141.

Pettitt, J.M. and Beck, C.B., 1968. *Archaeosperma arnoldii* — a cupulate seed from the Upper Devonian of North America. Contrib. Mus. Paleontol., Univ. Mich., 22(10): 139—154.

Pierce, R.L., 1961. Lower Upper Cretaceous plant microfossils from Minnesota. Minn. Geol. Surv. Bull., 42: 1—86.

Playford, G., 1971. Palynology of Lower Cretaceous (Swan River) strata of Saskatchewan and Manitoba. Palaeontology, 14(4): 533—565.

Raven, P.H. and Axelrod, D.I., 1974. Angiosperm biogeography and past continental movements. Ann. Mo. Bot. Garden, 61(3): 539—673.

Read, R.W. and Hickey, L.J., 1972. A revised classification of fossil palm and palm-like leaves. Taxon, 21: 129—137.

Reyment, R.A. and Tait, E.A., 1972. Biostratigraphical dating of the early history of the South Atlantic Ocean. R. Soc. Lond. Philos. Trans., Ser. B, 264: 55—95.

Rodendorf, B.B. and Zherikhin, V.V., 1974. Paleontologiya i okhrana prirody. Priroda, 1974(5): 82—91.

Samylina, V.A., 1960. Pokrytosemennye rasteniya iz nizhnemelovykh otlozheniy Kolymy. Bot. Zh., 45(3): 335—352.

Samylina, V.A., 1968. Early Cretaceous angiosperms of the Soviet Union based on leaf and fruit remains. J. Linn. Soc. (Bot.), 61: 207—218.

Savin, S.M., Douglas, R.G. and Stehli, F.G., 1975. Tertiary marine paleotemperatures. Geol. Soc. Am. Bull., 86(11): 1499—1510.

Scheckler, S.E. and Banks, H.P., 1971. Anatomy and relationships of some Devonian progymnosperms from New York. Am. J. Bot., 58(8): 737—751.

Scott, R.A., Barghoorn, E.S. and Leopold, E.B., 1960. How old are the angiosperms? Am. J. Sci., 258-A (Bradley Volume): 284—299.

Scott, R.A., Williams, P.L., Craig, L.C., Barghoorn, E.S., Hickey, L.J. and MacGinitie, H.D., 1972. "Pre-Cretaceous" angiosperms from Utah: Evidence for Tertiary age of the palm woods and roots. Am. J. Bot., 59(9): 886—896.

Simpson, G.G., 1953. The Major Features of Evolution. Columbia University Press, New York, N.Y., 434 pp.

Singh, C., 1971. Lower Cretaceous microfloras of the Peace River area, northwestern Alberta. Bull. Res. Counc. Alta., 28: 1—310.

Smart, J. and Hughes, N.F., 1972. The insect and the plant: Progressive palaeoecological integration. In: H.F. van Emden (Editor), Insect/Plant Relationships. Symp. R. Entomol. Soc. Lond., 6: 143—155.

Stebbins, G.L., 1950. Variation and Evolution in Plants. Columbia University Press, New York, N.Y., 643 pp.

Stebbins, G.L., 1965. The probable growth habit of the earliest flowering plants. Ann. Mo. Bot. Garden, 52(3): 457—468.

Stebbins, G.L., 1974. Flowering Plants: Evolution above the Species Level. Harvard University Press, Cambridge, Mass., 399 pp.

Tahktajan, A.L., 1959. Der Evolution der Angiospermen. VEB Gustav Fischer Verlag, Jena, 344 pp.

Takhtajan, A.L., 1969. Flowering Plants: Origin and Dispersal. Smithsonian Institution, Washington, D.C., 310 pp.

Takhtajan, A.L., 1976. Neoteny and the origin of flowering plants. In: C.B. Beck (Editor), Origin and Early Evolution of Angiosperms. Columbia University Press, New York, N.Y., pp. 207—219.

Teixeira, C., 1948. Flora Mesozóica Portuguesa. Serviços Geológicos de Portugal, Lisbon, 119 pp.

Tidwell, W.D., Rushforth, S.R., Reveal, J.L. and Behunin, H., 1970a. *Palmoxylon simperi* and *Palmoxylon pristina*: Two pre-Cretaceous angiosperms from Utah. Science, 168: 835—840.

Tidwell, W.D., Rushforth, S.R. and Simper, A.D., 1970b. Pre-Cretaceous flowering plants: Further evidence from Utah. Science, 170: 547—548.

Tschudy, R.H., 1970. Palynology of the Cretaceous—Tertiary boundary in the northern Rocky Mountain and Mississippi Embayment regions. In: R.M. Kosanke and A.T. Cross (Editors), Symposium on Palynology of the Late Cretaceous and Early Tertiary. Geol. Soc. Am. Spec. Pap., 127: 65—111.

Vachrameev, V.A., 1952. Stratigrafiya i iskopayemaya flora melovykh otlozheniy Zapadnogo Kazakhstana. Reg. Stratigraf. S.S.S.R., 1: 1—340.

Vachrameev, V.A., 1973. Pokrytosemennye i granitsa nizhnego i verkhnego mela. In: A.F. Chlonova (Editor), Palinologiya Mezofita (Trudy III Mezhdunarodnoy Palinologicheskoy Konferentsii). Nauka, Moscow, pp. 131—135.

Van Campo, M., 1971. Précisions nouvelles sur les structures comparées des pollens de Gymnospermes et d'Angiospermes. C. R. Acad. Sci., Paris, Sér. D, 272(16): 2071—2074.

Von Wahlert, G., 1965. The role of ecological factors in the origin of higher levels of organization. Syst. Zool., 14(4): 288—300.

Walker, J.W., 1971. Pollen morphology, phytogeography, and phylogeny of the Annonaceae. Contrib. Gray Herb., 202: 1—131.

Walker, J.W., 1976. Comparative pollen morphology and phylogeny of the ranalean complex. In: C.B. Beck (Editor), Origin and Early Evolution of Angiosperms. Columbia University Press, New York, N.Y., pp. 241—299.

Walker, J.W. and Doyle, J.A., 1975. The bases of angiosperm phylogeny: Palynology. Ann. Mo. Bot. Garden, 62(3): 664—723.

Walker, J.W. and Skvarla, J.J., 1975. Primitively columellaless pollen: A new concept in the evolutionary morphology of angiosperms. Science, 187: 445—447.

Ward, L.F., 1888. Evidence of the fossil plants as to the age of the Potomac Formation. Am. J. Sci., 36: 119—131.

Ward, L.F., 1895. The Potomac Formation. U.S. Geol. Surv., 15th Annu. Rep., pp. 307—397.

Ward, L.F., 1899. The Cretaceous formation of the Black Hills as indicated by the fossil plants. U.S. Geol. Surv., 19th Annu. Rep., Part 2, pp. 521—958.

Ward, L.F., 1905. Status of the Mesozoic floras of the United States. U.S. Geol. Surv. Monogr., 48: 1—616.

Westoll, T.S., 1949. On the evolution of the Dipnoi. In: G.L. Jepsen, E. Mayr and G.G. Simpson (Editors), Genetics, Paleontology, and Evolution. Princeton University Press, Princeton, N.J., pp. 121—184.

Whitehead, D.R., 1969. Wind pollination in the angiosperms: Evolutionary and environmental considerations. Evolution, 23: 28—35.

Wodehouse, R.P., 1936. Evolution of pollen grains. Bot. Rev., 2(2): 67—84.

Wolfe, J.A., 1971. Tertiary climatic fluctuations and methods of analysis of Tertiary floras. Palaeogeogr., Palaeoclimatol., Palaeoecol., 9: 27—57.

Wolfe, J.A., 1972a. Phyletic significance of Lower Cretaceous dicotyledonous leaves from the Patuxent Formation, Virginia. Am. J. Bot., 59(6, part 2): 664 (abstract).

Wolfe, J.A., 1972b. An interpretation of Alaskan Tertiary floras. In: A. Graham (Editor), Floristics and Paleofloristics of Asia and Eastern North America. Elsevier, Amsterdam, pp. 201—233.

Wolfe, J.A., 1973. Fossil forms of Amentiferae. Brittonia, 25(4): 334—355.

Wolfe, J.A. and Pakiser, H.M., 1971. Stratigraphic interpretations of some Cretaceous microfossil floras of the Middle Atlantic States. U.S. Geol. Surv. Prof. Pap., 750B: B35—B47.

Wolfe, J.A., Doyle, J.A. and Page, V.M., 1975. The bases of angiosperm phylogeny: Paleobotany. Ann. Mo. Bot. Garden, 62(3): 801—824.

Zherikhin, V.V. and Sukacheva, I.D., 1973. O melovykh nasekomonosnykh "yantaryakh" (retinitakh) severa Sibiri. In: Voprosy Paleontologii Nasekomykh. Nauka, Leningrad, pp. 3—48.

CHAPTER 17

PATTERNS OF EVOLUTION: A SUMMARY AND DISCUSSION

THOMAS J.M. SCHOPF

Introduction

This volume clearly reflects the wide spectrum of opinion on what palaeontologists think they should be doing in evolutionary palaeontology. Issues which seem crucial to one author are not even mentioned by another. The degree of explicit interpretation, implicit interpretation and absence of interpretation varies considerably. And where views are clearly stated, there can be enormous contradictions without any indication that an opposing opinion exists.

This diversity of interests can be highlighted by the ways in which the authors deal with three themes of historical and current interest: (1) directionality versus steady state (in diversity and in morphology); (2) environmental controls versus biological limitations on diversity; and (3) a punctuational versus gradual method and rate of change of diversity.

The summary needs two disclaimers. I have not emphasized for their sake alone many time-honoured but no longer highly controversial topics even though these are dealt with in some of the chapters, as is befitting the state of the art of the study in different taxa. Also reluctantly omitted are synopses of phylogenetic findings whether they refer to metazoan patterns (most of Valentine's chapter) or to specialist groups. To have included these would have made a summary unduly exhaustive.

Three Themes

In the initial chapter, Gould proposes that "the basic questions palaeontologists have asked about the history of life are three in number... The major contemporary issues in palaeobiology represent the latest reclothing of the ancient questions." We can use his organizational framework to ask what are the ideas of this group of authors with regard to these same questions to see what changes (if any) have occurred over the last century.

(1) Directionality versus steady state

"Does the history of life have definite directions?", Gould asks. We will consider two contexts in which this topic has been debated: diversity and morphology.

TABLE I

Examples of groups which may have a steady-state taxonomic diversity through a significant portion of geological time, aside from changing numbers of faunal provinces

Group	Comments	Author
Metazoans	"Peaks of diversification, in the Silurian, Devonian and Carboniferous . . . seem to . . . indicate the appearance of new provincial regions and regimes. The important differences between Permian and Recent marine diversities are accommodated chiefly on the provincial level."	Valentine
Bryozoans	Follow Species—Area equilibrium.	Schopf
Ammonites	The ceratites and ammonites are two major radiations, the products of one filling the ecological vacuum left by the demise of the other.	Kennedy
Trilobites	Infers a fixed number of niches for *Phacops* distribution.	Eldredge
Graptolites	Displacement of species by competition.	Rickards
Fishes	"At the end of the Devonian or the beginning of the Carboniferous, fishes had come to occupy all available carnivorous niches in the aquatic environment. Thereafter, fishes remained at (declined to) an *apparent* equilibrium in diversity demonstrated as an effective upper limit on diversity (at whatever rank) which they could not/did not exceed. Although they can be seen to have undergone significant morphological advancement . . . these improvements turned out only to be sufficient to allow the fishes to keep their heads above water . . . in competition in evolving aquatic ecosystems."	Thomson
Amphibians	Considers it unlikely that adaptive zones would have been empty.	Carroll
Reptiles and mammals	Considers that the fossil record shows many cases of a clade disappearing and being replaced by a new clade of similar body size, habitat and trophic guild.	Bakker
Mammals	Rates of origination and extinction have remained in close equilibrium through the course of the Cenozoic.	Gingerich
Mammals	Since the Oligocene, macro-evolution has been largely a process of refinement of adaptation within what seems to have been a generally saturated and only slightly expanding adaptive zone.	Stanley

(a) Is diversity on a continuing "upward spiral", or is it in a steady state? Most surprisingly, nearly all authors now seem to think that a steady state has prevailed for their group at the species level over a considerable part of its geologic time range. Tables I and II summarize these opinions. Since the sum of diversity at any given time is simply the sum of individual components, we reach the (historically) new conclusion that many palaeontologists seem to regard total diversity as being close to a steady state. This is simply to say that for most authors the mechanism of controlling the carrying capacity of the environment is not noticeably changing through time. Of course diversity will change as different geographic settings allow for greater or lesser degrees of endemism — it could hardly be different — but increasing diversity in and of itself by finer and finer partitioning of resources, so that an increasing number of species is packed into the biosphere of each faunal province, is a minority view.

To examine the anticipated properties of a long-term "steady state", Raup outlines the essence of the MBL program, which "may be thought it as a giant null hypothesis to describe phylogeny". The MBL program has now been used in several papers concerning clade trends (Raup et al., 1973), morphologic gradients (Raup and Gould, 1974; Schopf et al., 1975), and biogeographic controls on diversity (Simberloff, 1974). Raup discusses this Monte Carlo approach and is emphatic that "trends and patterns [in clade shape or morphological gra-

TABLE II

Groups which are said to be characterized as having an increasing taxonomic diversity through geologic time specifically owing to a factor other than increased Tertiary endemism

Group	Comments	Author
Ichnofossils	Uniform diversity in shallow-water deposits, but a three-fold increase in diversity in flysch deposits, the result of a gradual build-up in a very stable environment.	Seilacher
Bivalves	Both number of bivalve families and rates of origin of families have continued to increase into the Cenozoic. Adaptive breakthroughs of all kinds have occurred again and again, in one group after another.	Stanley
Reptiles and mammals	Believes the fossil record suggests that evolution can continue to increase the number of species and families of large tetrapods until a mass extinction empties the habitat.	Bakker
Angiosperms	Diversity increased through the Cretaceous and Tertiary, with early opportunistic species being replaced by more mature ecological communities as more specialized niches were progressively occupied.	Doyle

dients] cannot be used *by themselves* as evidence for Darwinian theory... In fact, in the absence of independent evidence, we have no reason to reject the hypothesis that some morphological traits are adaptively neutral and have evolved in a strictly random walk fashion. The MBL simulation therefore opens the way for interpreting some aspects of morphology as being adaptively neutral even though they show considerable regularity and order in their phylogeny.''

Three authors (Paul, Schopf and Thomson) set out to test for their groups of organisms (echinoderms, bryozoa, fishes) whether diversity is in a steady state or whether it is constantly expanding.

Paul compares echinoderm clade shapes with random clades previously generated by computer by Raup et al. (1973; given in Schopf's article, figure 19), and draws attention to several differences. Specifically: (1 and 2) real echinoderm clades are considered to have too many times of high simultaneous extinctions and expansions; (3) real clades have a "coelacanth" effect in the echinoderm somasteroids — a distinctive morphology of very great persistence and very low taxonomic diversity; (4 and 5) real clades have too many short-lived and small taxa of high rank (classes), thus yielding a "genuinely explosive" evolution that "requires explanation". Differences 3, 4 and 5 rest upon the use of morphology. Morphology was specifically omitted by Raup et al. and its inclusion might well yield clades more similar to those found in the real world. Paul's analysis is precisely the sort of effort that the paper of Raup et al. was designed to stimulate — and which is discussed at length by Raup in this chapter — the stochastic pattern as a criterion of subtraction.

Schopf cites literature (chiefly deterministic) explanations for changes in clade shape, in morphological trends (as represented in r to K selection), and in comparing presumed slow versus rapid rates of evolution. However, he then finds that stochastic variation in a steady-state world seems just as reasonable an explanation for the patterns, and therefore concludes that "...both of these levels of explanation can be applied" to understanding bryozoan evolution.

Thomson is concerned about whether curves of percentage maximum diversity existing in any given geological period "might be artifacts of the volume of deposits in the geological record". Thomson finds no difference between the shape of generic and familial diversity curves, and he concludes that the "shape of the group diversity curves is not seriously affected by the 'sampling' problems of the fossil record...". He believes therefore that increases in diversity in the teleost radiation of the Cretaceous and Tertiary are owing to "entry into a whole new range of adaptive zones", because of "the possibilities of a major increase in the number of different faunal provinces available", as well as a more finely partitioned existence in old habitats. The increase in diversity is real, but it is within the conceptual framework of a steady-state number of opportunities.

(b) In contrast to the general "steady-statist" view of diversity, many authors see directionality in changes in morphology, as summarized in Table III. Most authors argue that improvement rests upon competition. If competition is the major force, then Paul and others believe that one should find that

TABLE III

Examples of groups said to exhibit directional trends (other than size) through geological time

Group	Structures with trends (often exhibiting "efficiency")	Author
Ichnofossils	Burrow morphology; efficiency with respect to foraging behaviour.	Seilacher
Bryozoans	Defensive polymorphism; levels of colonial integration.	Schopf
Echinoderms	"Initially there would have been little competition between organisms and almost any type, however bizarre and inefficient, would have been able to survive. As faunal diversity increased, competition would have eliminated the less efficient clades which would have a short fossil record early in the Phanerozoic. This idea explains many features of the echinoderm fossil record."	Paul
Graptolites	Stipe reduction; development of floats.	Rickards
Fishes	Significant morphological "advancement" from the chondrostean to holostean grade among Actinopterygii or the "cladodont" to "hybodont" grade in Chondrichthyes.	Thomson
Mammals	Progressive evolution produces a fundamental improvement (endothermy), and the improved clade radiated rapidly.	Bakker
Mammals	Molar evolution in the Cenozoic shows a series of trends towards specialization for one or a combination of the functional components, puncturing, shearing, or grinding.	Gingerich
Mammals	"We have direct evidence of [an upward spiral of efficiency] within the Cenozoic Mammalia, both for locomotory adaptions and for intelligence."	Stanley
Angiosperms	Morphological "advancement" of Cretaceous pollen and leaves reflects true progressive evolution.	Doyle

efficiency in structures improves through time.

Essentially all features in Table III exhibit increased "efficiency" with regard to some paradigm, and can be characterized as going from the r toward the K end of the ecological spectrum (see Schopf, Table I, for a listing of r and K characters).

There are taxa for which increased "efficiency" is not a commonly cited trend (e.g., brachiopods, bivalves, ammonites, graptolites). These are the very groups in which the evolutionary clock seems to be set and reset, and characters constantly reshuffled in a seemingly haphazard way. Mosaic and iterative evolution are commonly said to occur. The evolutionary mechanisms which several authors cite are neoteny and paedomorphosis, as summarized in Table IV. Recent neontological work on changes in developmental rates greatly increase the probability that neoteny and paedomorphosis are important evolutionary processes. This raises the possibility that one can recognize different modes of evolution which might be characteristic of different taxa.

For many authors, the evolutionary problem of directional morphological gradients is to see how it is that they can be accounted for in a non-deterministic "steady-state" world. Two lines of reasoning are proposed.

(i) Directionality as a sampling problem. None of the authors who comment on the topic deny that one finds *some* structures which have become more complex (and more efficient) in some lineages through geological time. Unfortunately, the question of whether directional trends have occurred, on the

TABLE IV

Authors' opinions regarding neoteny and paedomorphosis

Group	Comment	Author
Brachiopods	Examples cited from a number of groups, e.g., chonetidines, productidines, oldhaminidines, atrypidines, and terebratulides.	Williams and Hurst
Bivalves	Neotenous retention of the byssus has sometimes occurred suddenly, and in the polyphyletic reversion from epifaunal to burrowing habits neoteny seems to have been the evolutionary mechanism.	Stanley
Ammonites	Neoteny is a recurring theme, with the repeated appearance of genera of cryptic origins. Abrupt size decrease appears a common phenomenon, as a result, it would appear, of neoteny.	Kennedy
Trilobites	Neoteny important in understanding origin of some major trilobite groups. Not noted as a continually reappearing theme.	Eldredge
Graptolites	Paedomorphic and neotenous processes seem to have been widespread in graptolite lineages, but there is as yet no evidence that they gave rise to major grades except possibly in cases of stipe reduction, where more investigation is required.	Rickards
Salamanders	Most of the forms known from the Cretaceous are paedomorphic or neotenic.	Carroll
Angiosperms	Many of the basic peculiarities of angiosperms can be partially explained as juvenile characters retained by neoteny.	Doyle

average, in a *majority* of structures of a lineage is almost unknowable (beliefs aside) since we can sample only a few of the trends in the "hard parts".

A major purpose of Raup's chapter is to show that many morphological trends may owe their *pattern* of change simply to stochastic variation in a random walk through the biologically permissible limiting values. Raup utilizes a (palaeontologically) new statistical method for evaluating the probability that a random-walk type of pattern (e.g., an increase in size through time) does or does not call for a non-random biological interpretation. After assessing recently published fusulinid and scallop data, he concludes, "How many supposed instances of Cope's Rule are only illusions resulting from improper evaluation of time series data?"

(ii) Other authors see the "upward spiral in efficiency" (Stanley) as a *possible* (Gould, Stanley) or the *expected* (Schopf) outcome of a stochastic evolutionary process. Both Gould and Stanley seek the uncoupling of changes *within* species from changes *between* species and regard this as their major contribution to this book. Gould labels the process "Wright's rule" (see Wright, 1967), the précis for which "need be no longer than: 'Speciation is stochastic with respect to the direction of evolutionary trends' ". Stanley refers to this as "species selection". For both, the aim is to make macroevolution more than simply "the extension of directional selection within local populations". Gould's exuberant conclusion is that "there can scarcely be a more important task for palaeontologists than defining the way in which macroevolution depends upon processes not observed in ecologic time". For Stanley, the inferred differences in rate of speciation of different taxa (viz. clams vs. mammals) are not only *real*, but the differences themselves become the "primary determinant of major trends and patterns of evolution". Indeed nearly half of his chapter is devoted to methods for determining how different rates of speciation actually are, and the implications of this finding. The conclusion that rates of origination and extinction themselves are a most important part of macro-evolution is also strongly asserted by Bakker with regard to tetrapods.

On the other hand, Schopf borrows from ergonomic theory and seeks to demonstrate that the occurrence of greater specialization is always the expected outcome of virtually any evolutionary trends, *to the extent that the environment can be keyed upon.* In contrast to the views of Gould, Stanley and Bakker, macro-evolution is envisaged by Schopf simply as the net result of directional selection acting on individuals within local populations of particular species through time. This general view is supported by Gingerich and Doyle.

(2) Controls

The second question which Gould finds of major historical importance in palaeontological thought is: what is the *control* of organic change? Is it chiefly environmental and external to species, or alternatively, "does change arise from some independent and internal dynamic within organisms themselves?" As summarized in Table V the largest group of us emphasizes that environmental factors are crucial in controlling diversity in a world of explicitly limited

TABLE V

Viewpoints of authors who consider that environmental factors control evolutionary changes through geological time

Group	Comment	Author
Marine metazoans	Respond directly to fluctuations in trophic resources.	Valentine
Ichnofossils	Greater efficiency in foraging behaviour coincident with the conquest of terrestrial and subtidal environments by angiosperms which yielded more food.	Seilacher
Bryozoans	Respond to changing degree of endemism regulated by plate movements. Climatic control of speciation.	Schopf
Echinoderms	The pruning effect of a limited environment via competition is the controlling factor.	Paul
Ammonites	The overall pattern of rise and decline of ammonites can be linked to periods of transgression and regression.	Kennedy
Trilobites	Sea-level changes strongly influence the available number of niches at any given time.	Eldredge
Fishes	Early record due to intense competition within relatively confined adaptive zones. Mesozoic increase related to increasingly available quantitative trophic resources.	Thomson
Amphibians	Likely that adaptive zones are filled and that competition is important. Cites with approval the notion that angiosperms and insect diversities led to increased salamander diversity.	Carroll
Reptiles and mammals	Mass tetrapod extinctions coincide with major regressions of shallow seas and with a decrease in the worldwide level of orogeny.	Bakker
Mammals	Periods of high faunal turnover correlate with major periods of climatic warming.	Gingerich
Mammals, trilobites and ammonites	Controlling factor on diversity is competition in a limited environmental space.	Stanley
Angiosperms	Late Cretaceous floral provinciality and multiplication of taxa related to spread of epicontinental seas.	Doyle

TABLE VI

Viewpoints of authors who consider that non-environmental factors control evolutionary changes through geological time

Group	Comment	Author
Metazoans	Direction significant at macro-evolutionary scale in morphology.	Gould
Bivalves	Primary control is by predation: environmental restriction via competition for resources no problem.	Stanley
Angiosperms	Cretaceous radiation may have been triggered off by evolution of key biological innovations rather than large-scale environmental changes.	Doyle

resources. Competition is often the key biological element of this model.

The other, smaller, group of authors considers that some additional non-environmental factor largely controls evolution. These opinions are summarized in Table VI. One author (Stanley) believes that both types of control may operate, depending on the group considered (competition for mammals, predation for bivalves).

The constraint of environmental versus internal controls of evolution can also be seen in accounts of phylogeny. Although nearly all authors refer to specific historical geological events in discussing the evolutionary history of a group, a few chapters (especially the one on brachiopods) almost do not refer to any geological event. This way of presenting evolutionary material seems to derive from a view that the theme "Patterns of Evolution" is chiefly a phylogenetic subject which is to be handled external to geological history *per se*. For other authors, phylogenetic patterns almost seem to be a direct resultant of geological perturbations (such as changes in sea level).

(3) Methods and rates

The third question discussed by Gould in the history of palaeontology is, "what is the tempo of change in the organic world?" On the one hand, at the species level we have a gradualist view in which genetic, morphological and taxonomic change is continuous and gradual and each successive population of each successive species is only slightly different genetically from each previous species, with as much general change genetically within the lifetime of a species as occurs between species. On the other hand, we have the episodic, punctational view in which most evolutionary change is considered to occur only at "speciation events". The views of the authors are summarized in Tables VII and VIII.

At the macro-evolutionary level, Valentine explicitly considers the topic of

TABLE VII

Viewpoints of authors who consider that the rate of evolutionary change is largely punctational

Group	Comment	Author
Metazoans	Punctational at both species and macro-evolutionary levels.	Gould
Metazoans	Punctational at both species and macro-evolutionary levels.	Valentine
Brachiopods	Breaks real (for whatever reason); the emphasis on constant turnover points toward punctation, with even the Lingulacea being considered to have evolved at more nearly normal rather than very slow rates.	Williams and Hurst
Bivalves	Most evolutionary change must occur during or shortly after speciation; also punctational at macro-evolutionary level.	Stanley
Trilobites	Strongly punctational at species level, with attempt to evaluate biases.	Eldredge
Graptolites	Major structural changes related to the broad modes of life adopted as a consequence of their achieving relatively sudden planktonic status. Usually, but not always, new species appear relatively suddenly.	Rickards
Fishes	Implied punctuation at species level; highly punctational at macro-evolutionary level.	Thomson
Reptiles and mammals	Most morphological change is concentrated in short-lived speciation events in small, isolated populations; species appear in depositional basins abruptly and show only modest changes during their stratigraphic range.	Bakker
Mammals	Punctational at both species and macro-evolutionary level.	Stanley
Angiosperms	Pollen data tend to support punctuated equilibria model but morphological resolution not sufficient to rule out possibility of some gradualistic trends.	Doyle

gradualistic change versus "rather sudden evolutionary pulses or explosions". With regard to the minimum ages of metazoan body plans, he writes, "The contrast between the soft-bodied Ediacaran fauna of the Late Precambrian, depauperate in higher metazoans, and the Burgess shale fauna (Middle Cambrian), teeming with higher metazoan phyla including novel ones, certainly suggests the second model, although this may be an artifact of preservation." Valentine

TABLE VIII

Viewpoints of authors who consider that the rate of evolutionary change is largely gradational

Group	Comment	Author
Ichnofossils	Gradational and slow at species level.	Seilacher
Bryozoans	Largely based on data for modern clines extrapolated to fossil record.	Schopf
Mammals	Gradational at species level and macro-evolutionary level.	Gingerich
Angiosperms	Suspects that some of the morphological trends reflect phyletic evolution.	Doyle

continues, "Especially during the Phanerozoic, the appearances and diversification of large numbers of animal groups are clustered in time"; and he explicitly favours a model "calling on short periods of high evolutionary rates. . .".

Many authors cite the abrupt cleavage between higher taxa, and this is one of the most strikingly repeated conclusions. Historically the question has been what sort of genetic mechanism can plausibly account for these abrupt transitions (although some disagree about the necessity of searching for a special mechanism). Valentine surveys this topic in the light of recent developments in molecular biology and states ". . . it now seems likely that evolution of the regulatory apparatus of the genome may be responsible for the rise of truly novel organisms within relatively short periods of time". The general significance of regulatory mechanisms in macro-evolution is well reviewed by Wilson (1976, and references cited therein).

In many ways, it may be impossible to present any single case of gradualism which cannot be interpreted at the species level as punctational — the existence of peripheral isolates cannot be denied, and the allopatric model certainly is a sufficient condition for speciation. Indeed the best one can probably do palae-ontologically is to evaluate the biases as thoroughly as possible, and then to present the type of detailed data which Gingerich has done. No author thus far has presented a numerical estimate of the relative importance of the two processes, but I have the general impression that estimates of the proportion of morphological change in species largely owing to allopatric changes in ecological time as opposed to sympatric and gradualistic changes over evolutionary time may vary from 9 : 1 (Eldredge, Gould, Stanley, Bakker) to 1 : 9 (Gingerich, Schopf).

Some authors explicitly consider the fact that species are identified only by morphology, and that a more complicated morphology *ipso facto* provides for more species differentia and thus a higher probability of seeing "evolutionary" change. Kennedy, for example, after noting that the most rapidly evolving ammonite species change on the order of less than 1 m.y., remarks, "Records of very long ranges of 8 to 25 m.y. are usually based on taxa which have few obvious morphological features on which to subdivide lineages." Williams and Hurst tested the relation between skeletal complexity and generic discrimination in brachiopods "by estimating the number of character complexes (total of 94) used in defining the genera of the 46 superfamilies belonging to the phylum." They find that there is a "high positive correlation between the number of genera belonging to a superfamily and the range of characters used in their discrimination . . .". Thus they show that a greater range of characters by this fact alone permits a greater number of combinations, and therefore a larger number of taxa for any given morphological grade. Schopf finds that, for bryozoa, the difference in production of genera per million years is a factor of 4 in morphologically complex forms (cheilostomes) versus morphologically simple forms (cyclostomes). Previously, Schopf et al. (1975) had found that "a range in morphological complexity by a factor of 5 (as measured by richness of morphological terms) is correlated with a change in rate of evolution of about the same extent". Thus, Schopf suggests that "It may even be the case, at least for Bryozoa, that rates of speciation are approximately the same, and that the differences which are observed are (to put it boldly) purely an artifact of changing morphological complexity."

In contrast to the opinions expressed in this last paragraph, Eldredge is quite emphatic in stating that low speciation rates imply low rates of cumulative morphological change. There are certainly two ways to interpret this issue, and it may only be resolved by studies in molecular biology (see Avise, 1976; and Wilson, 1976).

Perceived rates of evolution, or real rates of evolution (depending upon one's bias) may also be a function of organism size. Kennedy supports Hallam's (1975) finding that "organisms that increase in size more rapidly should become extinct more quickly, since extreme specialization would occur earlier, and hence render them more vulnerable to extinction, whilst given food resources are likely to become scarcer, the number of individuals will be smaller, and hence larger species will once again be more liable to extinction".

Van Valen (1973) asserted that "all groups for which data exist go extinct at a rate that is constant for a given group", and this relationship was later termed Van Valen's Law (Raup, 1975). The law is considered valid by Gingerich, citing new data on rodents, and may apply for extinct families of Bryozoa. Williams and Hurst plot the data differently, but find that brachiopod longevities "approximate to a die-away curve. . .". And Thomson finds for fish that "extinction is also independent with respect to taxonomic rank, families and genera decaying at the same rate". On the other hand, Valentine is strongly opposed to any biological significance for Van Valen's Law, and writes, "The apparent correlation of extinction waves with key environmental events argues they are deterministic. Even if extinctions were so inconstant that there was only a single extinction episode, a straight survivorship curve would be created so long as old and new lineages were carried off indiscriminately."

One by-product of these chapters are new data on the mean duration of spe-

TABLE IX

Empirical data on mean duration of species and genera (most numbers based on only a few moderately well-studied taxa)

Group	Duration in millions of years		Author
	Species	Genera	
Brachiopods		shortest lived not <10 mode 16—20	Williams and Hurst
Bryozoans	10—20		Schopf
Bivalves	~7		Stanley
Ammonites	mean ~0.5—1.0 range 0.2—25		Kennedy
Graptolites	mean 1.9 range <1—8	mean ~5	Rickards
Mammals	~1	~2 (for rodents)	Gingerich
Mammals	~1		Stanley

cies, and these are summarized in Table IX. Several taxa have durations on the order of 1 m.y. for the most rapidly evolving taxa, with genera commonly persisting for 10 m.y. For species, the numbers are close to the limits of resolution of radiometric methods. Gingerich notes that the empirical record of mammals is such that there is an "overall trend toward increasing rates of both origination and extinction... since the Cretaceous ...".

Additional Comments on Macro-Evolution

In many ways, Seilacher's viewpoint personifies (1) the traditional, classical and dominant school of animal evolution, and (2) the way a viewpoint is used to interpret data. That is, changes in ichnospecies are said to be directional (increasing diversity and complexity), environmentally controlled, and gradual. The fact that Seilacher's materials are trails and burrows seems to emphasize how broadly the classical viewpoint can be applied. Indeed, the increase in both species number and in burrow complexity through time in deep-sea ichnospecies is viewed "as a final layer of community evolution, in which faunal diversity increases less by trophic niche partitioning than by coevolutionary diversification".

Stanley combines competition for limited resources and a tendency toward expansion to fill all available niches in what he calls "elastic" taxa (e.g., mammals). He compares the diversity and rates of evolution of this group with that expected of "inelastic" taxa (e.g., clams) in which limiting resources are not important (control is by predation) but the tendency toward speciation remains. The elastic taxa may be in a steady-state diversity near saturation, but inelastic taxa would not be since saturation of the habitat is not possible (or if possible is so far in the future as not yet to be recognized).

Thomson is much impressed by his finding for fishes that the "breakdown of each higher group into subgroups (roughly at the ordinal level), each of which shows only a single peak of diversification, may reflect a situation of major biological importance". In addition, "the peaks of the separate curves tend to be remarkably similar to each other"; ". . .the similarity between the curves extends between approximately the 25% points on the ascendency and descendency arms. Between these points, the two slopes are essentially linear, suggesting growth in numbers of taxa is more exponential than logistic." Also, their span is extremely short (roughly 50 m.y. from 50% increasing to 50% diversity) . . .". And "There is some indication that peaks of diversity occur in a regular manner through the Phanerozoic, rather than randomly", a point also made by Valentine. Indeed, Thomson states that "After the initial bursts of diversification, the peaks occur regularly at an average spacing of about 60 m.y.". However, "there is no overall correlation between the summary curves for fish and (those for) invertebrates".

Just as one can ask about rates of evolution which have occurred, one can also inquire, as Paul does, why it is that adaptive radiations have not proceeded at an even faster pace. Paul writes: ". . .perhaps the biggest puzzle of echinoderm evolution . . . is why did it take sea urchins so long (from the Middle Ordovician to the Early Jurassic) to exploit an infaunal mode of life, especially as the subsequent explosive radiation of 'irregular' echinoids demonstrates how successful infaunal sea urchins have become." And Stanley writes, "Why the radiation of the Bivalvia was delayed until the Early Ordovician . . . remains uncertain."

Summary

Gould believes that, historically, the most commonly held views of evolution by palaeontologists have been directional (not steady-state), environmentalist (not internalist) and gradualist (not punctational) but not necessarily all three at once. The strongest single historical linkage is between the directional—internalist viewpoint.

To judge by the authors of this book, the currently prevailing view at the species level is a steady-statist, environmentalist and punctational, and at the macro-evolutionary level is directional (in morphology) or steady-state (in diversity), environmentalist and either punctational or gradual (depending on the group). Historically it seems most surprising that the steady-statist view is now adopted explicitly or implicitly by most authors.

Many authors view their group as consisting of several distinct grades of development which are recognized in higher categorical ranks (often families to classes). It is very difficult to assign phylogenetic relationships among these higher taxa. Within some groups there commonly seems to be a reshuffling of a variety of characters in various combinations, speciation has a strong random component, and neoteny is often called upon to account for the resetting of the evolutionary clock. Within other groups, increased efficiency is strongly suggested, going from the r selection end of the scale to the K selection end, along morphological gradients. The occurrence of directional trends leads some authors to seek to uncouple "macro-evolution" from species evolution, but other authors see no need for this. Evolutionary patterns in angiosperms are no different from those described in animal groups.

Looking to the immediate future, the state of the art of understanding patterns of evolution among groups with good fossil records seems to me to be especially dependent upon advances in two areas. First, the results of work in molecular evolution should provide a strong indication of the extent to which the palaeontologists' morphological species in fact represents a good approximation of changes in the underlying genome. This has vitally important consequences for the model of speciation which one brings to the fossil record. Second, the accurate chronological description of taxonomic and morphological change has been and will remain an important contribution of palaeontologists to knowing the course of evolution; these data can be used to test many hypotheses. In particular, further testing of the relevance of stochastic models of evolution should allow a much better appreciation of the role of directional versus random changes in the history of life. We need to have well-tested null hypotheses regarding presumed deterministic evolutionary patterns.

References

Avise, J.C., 1976. Genetic differentiation during speciation. In: F.J. Ayala (Editor), Molecular Evolution. Sinauer Assoc. Inc., Sunderland, Mass., pp. 106—122.

Hallam, A., 1975. Evolutionary size increase and longevity in Jurassic bivalves and ammonites. Nature, 258: 493—496.

Raup, D.M., 1975. Taxonomic survivorship curves and Van Valen's Law. Paleobiology, 1: 82—96.

Raup, D.M. and Gould, S.J., 1974. Stochastic simulation and evolution of morphology — towards a nomothetic paleontology. Syst. Zool., 23: 305—322.

Raup, D.M., Gould, S.J., Schopf, T.J.M. and Simberloff, D.S., 1973. Stochastic models of phylogeny and the evolution of diversity. J. Geol., 81: 525—542.

Schopf, T.J.M., Raup, D.M., Gould, S.J. and Simberloff, D.S., 1975. Genomic versus morphologic rates of evolution: influence of morphologic complexity. Paleobiology, 1: 63—70.

Simberloff, D.S., 1974. Permo-Triassic extinctions: effects of area on biotic equilibrium. J. Geol., 82: 267—274.

Van Valen, L., 1973. A new evolutionary law. Evol. Theory, 1: 1—30.

Wilson, A.C., 1976. Gene regulation in evolution. In: F.J. Ayala (Editor), Molecular Evolution. Sinauer Assoc. Inc., Sunderland, Mass., pp. 225—234.

Wright, S., 1967. Comments on the preliminary working papers of Eden and Waddington. In: P.S. Moorehead and M.M. Kaplan (Editors), Mathematical Challenges to the Neo-Darwinian Interpretation of Evolution. Monograph No. 5. Wistar Inst. Press, Philadelphia, Penn., pp. 117—120.

AUTHOR INDEX

GENERAL INDEX

Acaciaephyllum, 531, 532
Acanthoceras, 270, 285, 286
Acompsoceras, 270
Adaptive radiation, 39, 41, 46, 127, 130,
　227, 229, 238, 239, 240, 242, 257,
　259, 280, 298, 312, 315, 319, 367,
　378, 380, 381, 385, 388, 391, 392,
　397, 405, 406, 407, 427, 441, 445,
　452, 456, 465, 470, 471, 473, 482,
　501, 502, 504, 509, 510, 513, 516,
　520, 521, 524, 525, 526, 528, 533,
　534, 560
Adaptive zones, 17, 41, 159, 161, 200,
　231, 233, 238, 239, 240, 242, 243,
　244, 246, 395, 396, 397, 398, 418,
　422, 473, 474, 475, 476, 497, 520,
　526, 533, 534, 536, 537, 548, 550
Aegialodon, 470
Aegiromena, 97, 99
Aegoceras, 266
Ajatipollis, 515
Akidograptus, 349, 355
Alcyonarians, 239
Alcyonidium, 179, 180
Algae, 180, 181
Aliconchidium, 103
Alisina, 90
Alleles, 307
Allometric growth, 150, 279
Amaltheus, 269
Amauroceras, 269, 286
Ambocoelia, 107
Ammonites, see ammonoids
Ammonoids, 123 187, 222, 234, 243,
　251—304, 395, 396, 399, 400, 548,
　552, 554, 558, 559
　　　Ammonitida, 251, 252, 261, 262,
　　　279
　　　Anarcestida, 252, 254
　　　Bactritida, 252
　　　Ceratitida, 252, 257
　　　Clymeniida, 252
　　　Goniatitida, 252—256
　　　Lytoceratida, 251, 252, 261, 264,
　　　279
　　　Phylloceratida, 251, 252, 257, 259
　　　Prolecanitida, 252, 256
Amoeboceras, 269
Amphibians, 130, 393, 394, 397, 398,
　400, 405—437, 548, 554
　　Aistopoda, 416, 417
　　Anthracasauria, 408, 414
　　Anura, 407, 423, 424, 429
　　Apoda, 407, 421, 422, 429
　　Ichthyostegalia, 408, 409
　　Microsauria, 416, 417
　　Nectridea, 416
　　Temnospondyli, 408, 409, 411
　　Urodela, 407, 429
Amphiblestrum, 175
Amphioxus, 127
Amphipods, 33, 179, 180, 362
Amphithyris, 111, 113
Amphiura, 155
Amygdalotheca, 134
Anabarella, 215
Anadara, 220
Anagenesis, 2, 337, 348, 491, 498
Anastrophia, 103
Ancistrorhynchia, 105
Andrias, 419
Androgynoceras, 265, 266, 282
Anetoceras, 254
Angelosaurus, 447
Angiosperms, 130, 368, 422, 441, 457,
　470, 501—546, 549, 551, 552, 554—
　557
Anisograptus, 336
Ankylosaurus, 459
Annelids, 28, 29, 30, 32, 180, 330, 372,
　373
Apes, 488
Araliopsoides, 531
Araucaria, 130
Araucariacites, 540
Arca, 220
Archegosaurus, 412
Archimedes, 181
Arcuceras, 263
Aristogenesis, 2, 15
Arrhinoceratops, 459
Arthropods, 28, 29, 30, 32, 39, 180, 243,
　308, 324, 327, 378
Artinskia, 256
Ascarina, 513
Asteriacites, 364
Asterina, 149
Astiericeras, 295